CHEMICAL REACTOR ANALYSIS AND APPLICATIONS FOR THE PRACTICING ENGINEER

CHEMICAL REACTOR ANALYSIS AND APPLICATIONS FOR THE PRACTICING ENGINEER

LOUIS THEODORE

WILEY

A JOHN WILEY & SONS, INC., PUBLICATION

Published by John Wiley & Sons, Inc., Hoboken, New Jersey
Published simultaneously in Canada

Library of Congress Cataloging-in-Publication Data:

Theodore, Louis.
Chemical reactor analysis and applications for the practicing engineer/Louis Theodore.
p. cm.
Includes bibliographical references and index.
ISBN 978-0-470-91535-6 (hardback)
1. Chemical reactors. I. Title.
TP157.T46 2012
660′.2832—dc23

2011051081

Printed in the United States of America

10 9 8 7 6 5 4 3 2 1

To

Rupert Murdoch

A remarkable individual who, in his own way, has saved our great nation from the liberal media.

CONTENTS

PREFACE

Despite the quality of the consultants and reviewers on this project, the author did not follow their advice/suggestions in every case, which explains whatever confusion and errors remain in the text.

The design and analysis of chemical reactors is one of the most important and complicated tasks facing the practicing engineer. Chemical reaction engineering is that branch of engineering that is concerned with the application of chemical reactions on a commercial scale. Topics typically addressed include:

1. what chemical changes are to take place,
2. how fast these changes/reactions occur, and
3. what is the optimum way to carry out desired changes/reactions.

This project, as is the case with other texts in this series, was a unique undertaking. Rather than preparing a textbook on chemical reaction engineering in the usual format, the author decided to write an applications-oriented text. The text would hopefully serve as a training tool for those individuals in education and industry involved directly, or indirectly, with chemical reactors. It addresses both technical and calculational problems in this field. While this text can be complemented with texts on chemical kinetics and/or reactor design, it also stands alone as a self-teaching aid.

This book is divided into four parts.

Part I: Introduction

Part II: Traditional Reactor Analysis

Part III: Reactor Applications

Part IV: Other Reactor Topics

The first part serves as an introduction to the subject title and contains chapters dealing with history, process variables, basic operations, chemical kinetic principles, and stoichometry and conversion variables. The second part of the book addresses traditional reactor analysis; chapter topics include batch, CSTRs, and tubular flow reactors, plus a comparison of these classes of reactors. Part III keys on reactor applications that include thermal effects, interpretation of kinetic data, non-ideal reactors, and reactor design. The book concludes with other reactor topics; chapter titles include catalysis, catalytic reactions, fluidized and fixed bed reactors, biochemical reactors, open-ended questions, and ABET-related topics. An Appendix is also included.

The text is supplemented with numerous (nearly 300) illustrative examples. These range in difficulty from simple substitution into equations presented in the essay portion of the chapter to detailed design analysis and open-ended problems. It should be noted that the author cannot claim sole authorship to all the illustrative examples, problems, and essay material in this text. The present book has evolved from a host of sources, including: notes, homework problems and exam problems prepared by L. Theodore for a one-semester, three-credit "Chemical Reaction Kinetics" required undergraduate chemical engineering course at Manhattan College; J. Reynolds, J. Jeris, and L. Theodore's John Wiley & Sons' text, *Handbook of Chemical and Environmental Engineering Calculations*; J. Santoleri, J. Reynolds, and L. Theodore's Wiley-Interscience text, *Introduction to Hazardous Waste Incineration*, 2nd edition; S. Fogler's Prentice-Hall text, *Elements of Chemical Reaction Engineering* (required for the course at Manhattan College); and, to a much lesser extent, Smith's McGraw-Hill text, *Chemical Engineering Kinetics*.

Although the bulk of the problems are original and/or taken from sources that the author has been directly involved with, every effort has been made to acknowledge material drawn from other sources. In particular, Fogler's book has had a significant impact on the approach employed by the author; it is highly recommended for outside reading. As indicated above, Fogler's text has been adopted at Manhattan College for the Chemical Reaction Kinetics course and will probably be supplemented in the future with this text.

It is hoped that this writing will place in the hands of government, industrial and academic personnel a text covering the principles and applications of chemical reactors in a thorough and clear manner. Upon completion of the text, the reader should have acquired not only a working knowledge of the principles of chemical reactors but also experience in their application; and, the reader should find themself approaching advanced texts, engineering literature and industrial applications (even unique ones) with more confidence.

Thanks are due to Peter Forzaglia for proofing the manuscript and to Karen Tschinkel for both proofing the manuscript and preparing the index.

Last but not least, we believe that this modest work will help the majority of individuals working in this field to obtain a reasonably complete understanding of chemical reactors. If you have come this far, and read through the Preface and Overview to follow, you have more than just a passing interest in chemical reactors. I strongly suggest you try this book; I think you will like it.

LOUIS THEODORE
August 2012

Note: You may contact the author at loutheodore1@verizon.net for over 200 additional problems and 15 hours of exams; solutions for the problems and exams are available for those who adopt the book for training and/or academic purposes.

OVERVIEW

A major goal of the practicing engineer is to increase the supply and minimize the cost of useful materials to society. Every practicing engineer should be familiar with both the processing of chemicals and the design and operation of the equipment needed to carry out the processes. This equipment can be divided into three main groups. The first group consists of large-scale equipment used to purify or separate raw materials for further processing. These separations are normally physical in nature; the design and operation of this class of equipment are studied in mass transfer operation courses(1). The second group consists of other unit operations, including heat transfer(2) and fluid flow(3) plus chemical reactors in which the processed raw materials react to create products which possess new physical and chemical properties. The reactors may be viewed as a chemical treatment step. The design and operation of chemical reactors are the main subjects of this book. The third class of equipment is similar to the first group but (generally) on a smaller scale. The sequence of steps is pictorially represented in the figure below, and it is the "chemical reactor" box that will receive treatment in the material to follow.

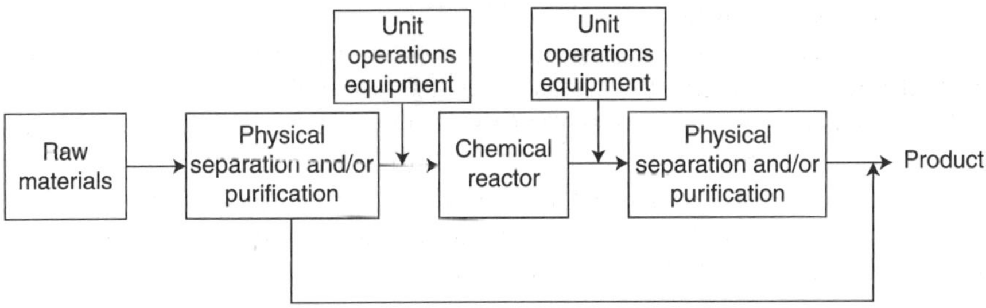

Process flow diagram.

The problem of predicting the performance and determining the size of a reactor can be divided into two sequential steps. The first is the study of the rate at which the chemical reaction occurs and the variables which affect this rate. This is the subject of chemical kinetics (see Chapter 4). The second is the problem of using reaction rate data to predict performance and/or to determine the size of the equipment to obtain the required quantity and quality of product. This is the subject of Parts II and III.

This book was prepared as both a professional book and an undergraduate text for the study of the principles and fundamentals of chemical reactors. Some of the introductory material is presented in the first part of the book. Understandably, more extensive coverage is given in the remainder of the book to applications and design. Furthermore, several additional topics were included in the last part of the book—Other Reactor Topics. Some of the these topics are now required by ABET (Accreditation Board for Engineering and Technology) to be emphasized in course offerings.

The policy of most technical societies and publications is to use SI (metric) units or to list both the common British engineering unit and its SI equivalent. However, British units are primarily used in this book for the convenience of the majority of the reading audience. Readers who are more familiar and at ease with SI units are advised to refer to the Appendix.

Ultimately the practicing engineer must consider the following eight questions relative to the reaction/reactor.

1. What reactants (new materials) are required?
2. What reaction(s) will occur?
3. What product will be formed?
4. How fast is the reaction?
5. What type and size of reactor should be employed?
6. What operating temperature, pressure, compositions, and flow rates should be selected?
7. Is the overall process safe?
8. Is the overall process economical?

REFERENCES

1. L. Theodore and F. Ricci, "*Mass Transfer Operations for the Practicing Engineer*," John Wiley & Sons, Inc., Hoboken, NJ, 2010.
2. L. Theodore, "*Heat Transfer for the Practicing Engineer*," John Wiley & Sons, Inc., Hoboken, NJ, 2012.
3. P. Abulencia and L. Theodore, "*Fluid Flow for the Practicing Engineer*," John Wiley & Sons, Inc., Hoboken, NJ, 2009.

PART I

INTRODUCTION

We must view with profound respect the infinite capacity of the human mind to resist the introduction of useful knowledge.

—Thomas Raynesford Lounsbury (1838–1915): Quoted in *The Freshman and His College* (1913), by Francis Cummins Lockwood.

THE PURPOSE of Part I can be found in its title. The book itself offers the reader the fundamentals of chemical reactor analysis with appropriate practical applications, and serves as an introduction to the specialized and more sophisticated texts in this area. The reader should realize that the contents are geared towards practitioners in this field, as well as students of science and engineering, not chemical engineers per se. Simply put, topics of interest to all practicing engineers have been included.

Part I serves as the introductory section to this book. It reviews engineering and science fundamentals that are an integral part of the field of chemical kinetics and chemical reactors. It consists of five chapters, as noted below:

1. History of Chemical Reactions
2. The Field of Chemistry
3. Process Variables
4. Kinetic Principles
5. Stoichiometry and Conversion Variables

Those individuals with a strong background in the above area(s) may choose to bypass Part I. Parts II and III are concerned with describing and designing the various classes of reactors.

Chemical Reactor Analysis and Applications for the Practicing Engineer. By Louis Theodore

CHAPTER 1

HISTORY OF CHEMICAL REACTIONS

INTRODUCTION

Until the last century, most chemicals were discovered more or less by accident. Their potential uses were based on short-term observations and their syntheses based on sketchy and simple theoretical ideas. Much of the recent progress in chemical syntheses occurred because of an increasing ability of chemists to determine the detailed molecular structure of substances and also to better understand the correlations between structure and properties. A review of how this industry arrived at its present state is presented below. Sections to follow include:

Early History

Recent History

The Chemical Industry Today

Microscopic vs Macroscopic Approach

EARLY HISTORY

As noted in the Introduction, most chemicals were discovered by accident. No one can assign with certainty a birth date to what one would classify as a "chemical reaction." However, some have claimed that the first known chemical processes were carried out by the artisans of Egypt and China. These individuals worked with metals such as gold or copper, which often occur in nature in a pure state, but they learned how to "smelt" metallic ores by heating them with carbon-bearing materials. In addition, a primitive chemical technology arose in these cultures as dyes, potting glazes and glass making were discovered. Most of these inventors also developed astronomical, mathematical, and cosmological ideas that were used to explain some of the changes that are today considered chemical.

The first to consider such ideas scientifically were the Greeks at about 600 BC. They assumed that all matter was derived from water, which could solidify to earth or evaporate to air. This theory was later expanded into the idea that the world was

Chemical Reactor Analysis and Applications for the Practicing Engineer. By Louis Theodore

composed from four elements: earth, water, air, and fire. It was Democritus who proposed that these elements combined to form atoms.

Aristotle believed that the elements formed a continuum of mass. He became the most influential of the Greek philosophers, and his ideas dominated science for nearly 1500 years. He believed that four qualities were found in nature: heat, cold, moisture, and dryness. He proposed that elements were made up of these with each element containing variable amounts of these qualities. These, in turn, combined to form materials that are visible. Because it was possible for each element to change, the elements could be combined because it was possible that material substances could be built up from the elements.

At approximately the same time a similar alchemy arose in China. The aim was to make gold, since it was believed to be a medicine that could offer long life or even immortality on anyone who consumed it. Nevertheless, the Chinese gained much practical chemical knowledge from incorrect theories.

After the decline of the Roman Empire, Greek writings were no longer studied in western Europe and the eastern Mediterranean. However, in the 7th and 8th centuries Arab conquerors spread Islam over Asia Minor, North Africa, and Spain. The Greek texts were translated into Arabic, and along with the rest of Greek learning, the ideas and practice of alchemy once again flourished.

A great intellectual reawakening began in western Europe in the 11th century. This occurred due to the cultural exchanges between Arab and Western scholars. Later, knowledge of Greek science was disseminated into Latin and ultimately reached all of Europe. Many of the manuscripts concerned alchemy.

Among the important substances discovered were alcohol and mineral acids such as hydrochloric, nitric, and sulfuric. The Chinese discovery of nitrates and the manufacture of gunpowder also came to the West through the Arabs. Gunpowder soon became a part of warfare. Thus, an effective chemical technology existed in Europe by the end of the 13th century.

During the 13th and 14th centuries the principles of Aristotle on scientific thought began to decline. The actual behavior of matter cast doubt on the relatively simple explanation Aristotle had prescribed. These doubts spread further after the invention of printing in 1450; these doubts increased into the 16th century.

It was during the first half of the 17th century that scientists began to study chemical reactions experimentally. Jan Baptista van Helmont laid the foundations of the law of conservation of mass. Van Helmont showed that in a number of reactions an "aerial" fluid was liberated which he defined as a "gas." A new class of substances with their own physical properties was shown to exist. A kinetic–molecular theory of gases began to develop. Notable in this field were the experiments of Robert Boyle whose studies, later known as Boyle's law, provided an equation describing the inverse relation between pressure and volume of gas (see the ideal gas law in Chapter 3).

During the 18th century, chemists noted that certain substances combined more easily with, or had a greater affinity for, a given chemical than did others. Tables were developed showing the relative affinities of different chemicals. The use of these tables made it possible to predict many chemical reactions before testing them in the laboratory.

It was Joseph Priestley who discovered oxygen. He realized that this gas was the component of ordinary air that was responsible for combustion and made animal respiration possible. Priestley told the chemist Antoine Laurent Lavoisier about his discovery of oxygen. He at once saw the significance of this substance and the door was opened for the chemical revolution that established modern chemistry.

RECENT HISTORY

Lavoisier showed by a series of unique experiments that combustion was due to the combination of a burning substance with oxygen and that when carbon was burned, fixed air (carbon dioxide) was produced. An earlier proposed substance phlogiston therefore did not exist, and the phlogiston theory soon disappeared to be replaced by the carbon cycle. Lavoisier used the laboratory balance to give quantitative support to his work and he used chemical equations in his papers. He further defined elements as substances that could not be decomposed by chemical means and firmly established the law of the conversation of mass. He developed a chemical nomenclature that is still used today and founded the first chemical journal.

By the beginning of the 19th century, it was shown that more than one compound could be formed between the same elements. Joseph Gay-Lussac demonstrated that the volume ratios of reacting gases were small whole numbers, implying the presence of atoms. Dalton assumed that when two elements combined, the resulting compound contained one atom of each. He arbitrarily assigned to hydrogen the atomic weight of 1 and could then calculate the relative atomic weight of oxygen. Applying this principle to other compounds, he calculated the atomic weights of other elements and actually drew up a table of the relative atomic weights of all the known elements.

In the early 19th century (1803), Dalton proposed his atomic theory. In 1811, Amedeo Avogadro made clear the distinction between atoms and molecules of elementary substances. In addition, the concepts of heat, energy, work, and temperature were developed. The first law of thermodynamics was set forth by Julius Robert von Mayer and the second law of thermodynamics was postulated by Rudolf Julius Emanuel Clausius and William Thomson (Lord Kelvin). Later in the century, Clausius, Ludwig Boltzmann, and James Clerk Maxwell related the ideal gas law in terms of a kinetic theory of matter. This led to the kinetics of reactions and the laws of chemical equilibrium.

It was Carnot who proposed the correlation between heat and work. Josiah Willard Gibbs discovered the phase rule and provided the theoretical basis of physical chemistry. And, it was Walther Hermann Nernst who proposed the third law of thermodynamics and contributed to the study of physical properties, molecular structures, and reaction rates. Jacobus Hendricus van't Hoff related thermodynamics to chemical reactions and developed a method for establishing the order of reactions. Nearing the end of this century, Syante August Arrhenius investigated the increase in the rate of chemical reactions with an increase in temperature.

The development of chemical kinetics continued into the 20th century with the contributions to the study of molecular structures, reaction rates, and chain reactions by Irving Langmuir. Another advance in chemistry in the 20th century was the foundation of biochemistry, which began with the simple analysis of body fluids; methods were then rapidly developed for determining the nature and function of the most complex cell constituents. Biochemists later unraveled the genetic code and explained the function of the gene, the basis of all life. The field has now grown so vast that its study has become a new science—molecular biology.

THE CHEMICAL INDUSTRY TODAY

The growth of chemical industries and the training of professional chemists are intertwined. In the early 19th century during the Industrial Revolution, a number of universities were established in Germany. They drew students from all over the world and other universities soon followed suit. A large group of young chemists were thus trained just at the time when the chemical industry was beginning to exploit new discoveries. This interaction between the universities and the chemical industry resulted in the rapid growth of the organic chemical industry and provided Germany with scientific predominance in the field until World War I. Following the war, the German system was introduced into all industrial nations of the world, and chemistry and chemical industries progressed rapidly.

This scientific explosion has had an enormous influence on society. Processes were developed for synthesizing completely new substances that were either better than the natural ones or could replace them more cheaply. As the complexity of synthesized compounds increased, wholly new products appeared. Plastics and new textiles were developed, energy usage increased, and new drugs conquered whole classes of disease.

The progress of chemistry in recent years has been spectacular although the benefits of this progress have included corresponding liabilities. The most obvious dangers have come from nuclear weapons and radioactive materials, with their potential for producing cancer(s) in exposed individuals and mutations in their children. In addition, some pesticides have potential damaging effects. This led to the emergence of a new industry—environmental engineering. Mitigating these negative effects is one of the challenges the science community will have to meet in the future.[1,2]

MICROSCOPIC vs MACROSCOPIC APPROACH

The history of Unit Operations is interesting. Chemical engineering courses were originally based on the study of unit processes and/or industrial technologies; however, it soon became apparent that the changes produced in equipment from different industries were similar in nature, i.e., there was a commonality in the operations in the petroleum industry as with the utility industry. These similar operations became known as *Unit Operations*. This approach to chemical engineering was promulgated in the 1922 A. D. Little report (1922) submitted to the American Institute of Chemical Engineers (AIChE), and has, with varying degrees and emphasis, dominated the profession to this day.

The Unit Operations approach was adopted by the profession soon after its inception. During the 130+ years (since 1880) that the profession has been in existence as a branch of engineering, society's needs have changed tremendously and so has chemical engineering.

The teaching of Unit Operations at the undergraduate level has remained relatively unchanged since the publication of several early- to mid-19th century texts; however, by the middle of the 20th century, there was a slow movement from the unit operation concept to a more theoretical treatment called *transport phenomena* or, more simply, engineering science. The focal point of this science is the rigorous mathematical description of all physical rate processes in terms of mass, heat, or momentum crossing phase boundaries. This approach took hold of the education/curriculum of the profession with the publication of the first edition of the Bird et al. book.[3] Some, including the author of this text, feel that this concept set the profession back several decades since graduating chemical engineers, in terms of training, were more applied physicists than traditional chemical engineers. There has fortunately been a return to the traditional approach to chemical engineering, primarily as a result of the efforts of ABET (Accreditation Board for Engineering and Technology—see also Chapter 21). Detractors to this pragmatic approach argue that this type of theoretical education experience provides answers to what and how, but not necessarily why, i.e., it provides a greater understanding of both fundamental physical and chemical processes. However, in terms of reality, nearly all chemical engineers are now presently involved with the why questions. Therefore, material normally covered earlier has been replaced, in part, with a new emphasis on solving design and open-ended problems; this approach is emphasized in this text.

The following paragraphs attempt to qualitatively describe the differences between the above two approaches. Both deal with the transfer of certain quantities (momentum, energy, and mass) from one point in a system to another. There are three basic transport mechanisms which potentially can be involved in a process. They are:

1. radiation
2. convection
3. molecular diffusion.

The first mechanism, radiative transfer, arises as a result of wave motion and is not considered, since it may be justifiably neglected in most engineering applications. The second mechanism, convective transfer, occurs simply because of bulk motion. The final mechanism, molecular diffusion, can be defined as the transport mechanism arising as a result of gradients. For example, momentum is transferred in the presence of a velocity gradient; energy in the form of heat is transferred because of a temperature gradient; and, mass is transferred in the presence of a concentration gradient. These molecular diffusion effects are described by phenomenological laws.[3]

Momentum, energy, and mass are all conserved. As such, each quantity obeys the conservation law within a system (including a chemical reactor) as provided in Equations (1.1) and (1.2):

$$\left\{\begin{matrix}\text{quantity}\\ \text{into}\\ \text{system}\end{matrix}\right\} - \left\{\begin{matrix}\text{quantity}\\ \text{out of}\\ \text{system}\end{matrix}\right\} + \left\{\begin{matrix}\text{quantity}\\ \text{generated in}\\ \text{system}\end{matrix}\right\} = \left\{\begin{matrix}\text{quantity}\\ \text{accumulated}\\ \text{in system}\end{matrix}\right\} \qquad (1.1)$$

This equation may also be written on a time rate basis

$$\left\{\begin{matrix}\text{rate}\\ \text{into}\\ \text{system}\end{matrix}\right\} - \left\{\begin{matrix}\text{rate}\\ \text{out of}\\ \text{system}\end{matrix}\right\} + \left\{\begin{matrix}\text{rate}\\ \text{generated in}\\ \text{system}\end{matrix}\right\} = \left\{\begin{matrix}\text{rate}\\ \text{accumulated}\\ \text{in system}\end{matrix}\right\} \qquad (1.2)$$

The conservation law may be applied at the macroscopic, microscopic, or molecular level.

One can best illustrate the differences in these methods with an example. Consider a system in which a fluid is flowing through a cylindrical tube reactor (see Figure 1.1) and define the system as the fluid contained within the reactor between points 1 and 2 at any time. If one is interested in determining changes occurring at the inlet and outlet of a reactor, the conservation law is applied on a "macroscopic" level to the entire system. The resultant equation (usually algebraic) describes the overall changes occurring to the system (or equipment). This approach is usually applied in the Unit Operation (or its equivalent) courses, an approach that is, as noted above, highlighted in this text and its three companion texts.[4–6]

In the *microscopic/transport phenomena approach*, detailed information concerning the behavior within a system is required; this is occasionally requested of and by the engineer. The conservation law is then applied to a differential element within the system that is large compared to an individual molecule, but small compared to the entire system. The resulting equation is differential and can then be expanded via an integration in order to describe the behavior of the entire system.

The *molecular approach* involves the application of the conservation laws to individual molecules. This leads to a study of statistical and quantum

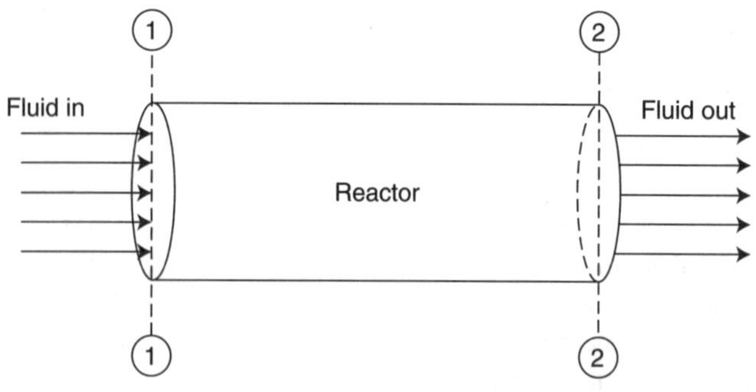

Figure 1.1 Flow reactor system.

mechanics—both of which are beyond the scope of this text. In any case, the description at the molecular level is of little value to the practicing engineer; however, the statistical averaging of molecular quantities in either a differential or finite element within a system can lead to a more meaningful description of the behavior of a system.

Both the microscopic and molecular approaches shed light on the physical reasons for the observed macroscopic phenomena. Ultimately, however, for the practicing engineer, these approaches may be justified but are akin to attempting to kill a fly with a machine gun. Developing and solving these equations (in spite of the advent of computer software packages) is typically not worth the trouble.

ILLUSTRATIVE EXAMPLE 1.1 Explain why the practicing engineer/scientist invariably employs the macroscopic approach in the solution of real world chemical reactor problems.

Solution. The macroscopic approach involves examining the relationship between changes occurring at the inlet and the outlet of a reacting system. This approach attempts to identify and solve problems found in the real world, and is more straightforward than, and preferable to, the more involved microscopic approach. The microscopic approach, which requires an understanding of all internal variations taking place within a reacting system that can lead up to an overall system result, simply may not be necessary. ■

REFERENCES

1. L. Stander and L. Theodore, "*Environmental Regulatory Calculation Handbook*," John Wiley & Sons, Hoboken, NJ, 2008.
2. M. K. Theodore and L. Theodore, "*Introduction to Environmental Management*," CRC Press/Taylor & Francis Group, Boca Raton, FL, 2010.
3. R. Bird, W. Stewart, and E. Lightfoot, "*Transport Phenomena*," John Wiley & Sons, Hoboken, NJ, 1960.
4. P. Abulencia and L. Theodore, "*Fluid Flow for the Practicing Engineer*," John Wiley & Sons, Hoboken, NJ, 2009.
5. L. Theodore and F. Ricci, "*Mass Transfer Operations for the Practicing Engineer*," John Wiley & Sons, Hoboken, NJ, 2010.
6. L. Theodore and F. Ricci, "*Thermodynamics for the Practicing Engineer*," John Wiley & Sons, Hoboken, NJ, 2009.

CHAPTER 2

THE FIELD OF CHEMISTRY

INTRODUCTION

No text concerned with chemical reactions and chemical reactors would be complete without an introduction to the field of chemistry. In a general sense chemistry involves:

1. the study of the composition, structure, and properties of material substances;
2. the interactions between materials; and,
3. the effects on the addition or removal of energy in any of its several forms.

There is general consensus that *chemistry* deals with the combination of atoms, and *physics* with the forces between atoms. Atomic combination involves atomic forces, and it is one of the objects of *physical chemistry* to see how far the chemical interactions observed between atoms and molecules can be interpreted by means of the forces existing within and between atoms. The study of atomic structure provides information of why atoms combine. For many reasons, the development of modern views concerning the structure of the atom is bound up very closely with several peripheral subjects including *organic chemistry*, *physical chemistry*, and *thermodynamics*.

In addition to the atomic or molecular approach to describing chemical reactions, which is treated superficially in this text, another important area involves the applications of *thermodynamics*, a topic which deals primarily with energy changes.[1] The treatment of the first and second laws of *thermodynamics*, including thermochemistry[1] provides an adequate basis for a consideration of the chemical change associated with gaseous, liquid, and solid states of matter.

A chemical reaction is a process by which atoms or groups of atoms are combined and/or redistributed, resulting in a change in the molecular composition and properties. The products obtained from reactants depend on the condition under which a chemical reaction occurs. The scientist and engineer have shown that although products may vary with changing conditions, some properties remain constant during any chemical reaction. These constant properties, called "conserved" properties, include the number of each kind of atom present, the electrical charge, and the total mass (conservation laws for mass).

Chemical symbols must also be understood. Almost all substances are made up of some combination of the aforementioned atoms of the elements found in nature.

Chemical Reactor Analysis and Applications for the Practicing Engineer. By Louis Theodore

Rather than full names, scientists identify elements with one- or two-letter symbols. Some common elements and their symbols are carbon, C; oxygen, O; nitrogen, N; hydrogen, H; chlorine, Cl; and, sulfur, S. These chemical symbols are derived from the letters of the name of the element. The first letter of the symbol is capitalized, and the second (if applicable) is lowercase. Symbols for some elements known from ancient times come from earlier, usually Latin, names: for example, Cu from *cuprum* (copper), Ag from *aurum* (gold), and Fe from *ferrum* (iron). This set of symbols, in referring to elements, is used universally.

Chemical formulas are occasionally represented by the symbol of the element, with the charge and mass of the atom indicated when appropriate. However, most substances are compounds in that they are composed of combinations of atoms. The formula for water, H_2O, indicates that two atoms of hydrogen are present for every atom of oxygen. The formula shows that water is electrically neutral, and it also indicates (because the atomic weights are H = 1.00, O = 16.00) that 2.00 unit weights of hydrogen will combine with 16.00 unit weights of oxygen to produce 18.00 unit weights of water. Because the relative weights remain constant, any units may be employed, e.g., pounds; in addition, all formulas that are multiples of simpler ratios can be assumed to represent molecules. The formulas O_2, N_2, H_2, H_2O_2, and C_2H_6 represent oxygen, nitrogen, hydrogen, hydrogen peroxide, and ethane, respectively.

Section topics following this introduction to the field of chemistry include:

Inorganic Chemistry

Organic Chemistry

Physical Chemistry

Other Chemistry Topics

Analysis Procedures

INORGANIC CHEMISTRY

Inorganic chemistry is that field of chemistry which is concerned with chemical reactions and properties of all the chemical elements and their compounds, with the exception of hydrocarbons (compounds composed of carbon and hydrogen) and their derivatives. The subject of carbon–hydrogen compounds is defined as organic chemistry, a topic discussed in the next section.

Inorganic chemistry is too vast a subject to form a convenient unit of study; the term would be of little importance except for the tendency in most engineering and science schools to call courses "Inorganic Chemistry" when a better title might be "Elementary Chemistry." The subject matter of such courses includes the elementary laws of chemistry and its symbols and nomenclature, and an introduction to the experimental methods that are important in experimental chemistry. The student is introduced to such fundamental chemical reaction topics as valence, ionization, reactivity, atomic theory, and the kinetic theory of gases. The properties and reactions of substances in aqueous solution also receive attention. Modern inorganic chemistry overlaps parts of many other scientific fields, including biochemistry, metallurgy, mineralogy, and solid-state physics. Finally, an increased understanding of the chemical behavior of the elements and of inorganic compounds has led to the discovery of a wide variety of new synthesizing techniques and the discovery of many new classes of inorganic substances.

ORGANIC CHEMISTRY

The branch of chemistry in which carbon compounds and their reactions are studied is defined as organic chemistry. A wide variety of classes of substances—such as drugs, vitamins, plastics, natural and synthetic fibers, as well as carbohydrates, proteins, and fats—consist of organic molecules. This subject involves:

1. the determination of the structures of organic molecules;
2. the study of their various reactions; and,
3. the development of procedures for the synthesis of organic compounds.

Organic chemistry has had a profound effect on society: it has improved natural materials and has synthesized natural and artificial materials that have, in turn, improved health, increased comfort, and added to the convenience associated with nearly every product manufactured today.

As noted earlier, the molecular formula of a compound indicates the number of each kind of atom in a molecule of that substance. Fructose, ($C_6H_{12}O_6$) consists of molecules containing six carbon atoms, 12 hydrogen atoms, and six oxygen atoms. Because at least 15 other compounds have this same molecular formula, one may distinguish one molecule from another by employing a structural formula to show the spatial arrangement of the atom. Note that an analytical analysis that indicates the percentage of carbon, hydrogen, and oxygen cannot distinguish fructose from ribose ($C_5H_{10}O_5$), another sugar in which the ratios of elements are the same, namely 1 : 2 : 1

The ability of carbon to form covalent bonds with other carbon atoms in long chains and rings distinguishes carbon from all other elements. Other elements are not known to form chains of greater than eight similar atoms. This property of carbon, and the fact that carbon nearly always forms four bonds to other atoms, accounts for the large number of known compounds. At least 80 percent of the 5 million reported chemical compounds contain carbon.

There are various classes of organic compounds, including:

1. alkanes
2. alhenes
3. alkynes.

Additional "classes" of organics include:

1. alcohols
2. aldehydes
3. amines

4. carboxylic acids
5. esters
6. ethers
7. alkyl halides
8. ketones
9. nitrites
10. sulfonic acids
11. thiols (mercaptans).

Other atoms, such as chlorine, oxygen, and nitrogen, may be substituted for hydrogen in an alkane, providing the correct number of chemical bonds is allowed—chlorine forming one bond to other atoms, oxygen forming two bonds to other atoms, and nitrogen three bonds. The chlorine atom in ethyl chloride, the $-OH$ group in ethyl alcohol, and the $-NH_2$ group in ethyl amine are called functional groups. Functional groups determine many of the chemical properties of compounds. Many of the chlorine bearing compounds are known to be carcinogenic and/or toxic.[2]

Regarding sources of organic compounds, coal tar was once the only source of aromatic and some heterocyclic compounds. Petroleum was the source of aliphatic compounds since it contains such substances as gasoline, kerosene, and lubricating oil. Natural gas provides (primarily) methane and ethane. These three categories of natural fossil compounds are still the major sources of organic compounds. When petroleum is not available, a chemical industry can be based on acetylene (if available), which in turn can be synthesized from limestone and coal. During World War II, Germany was forced into employing this process when it lost reliable petroleum and natural-gas supplies in Africa.

Covalent organic compounds are distinguished from inorganic salts by low melting points and boiling points. Hydrocarbons have low specific gravities (see also next chapter)—about 0.8 compared to water, 1.0—but functional groups may increase the densities of organic compounds. Only a few organic compounds possess specific gravities in excess of unity, e.g., carbon tetrachloride.

The practicing engineer usually designs organic reactions to be carried out at optimum conditions to produce maximum conversion or yields (terms to be defined later). One often resorts to catalysts, whether or not the reaction is reversible, and attempts to take advantage of equilibrium positions.[1] Also, catalysts are frequently essential for rapid chemical reactions.

PHYSICAL CHEMISTRY

Although chemistry deals with the combination of atoms and physics with the forces between atoms, the object of physical chemistry is to see how far the chemical interactions observed between atoms and molecules can be interpreted by means of the forces existing within and between atoms. Thus, one of the objectives of physical chemistry is to apply measurements of physical properties, such as density, surface tension, refractive index, dielectric constant, magnetism, and optical activity, to the description of chemical structure. Much of the knowledge provided in the earlier topics of this chapter can now be turned to use in this connection.

Others have defined physical chemistry as that field of science that applies the laws of physics to elucidate the properties of chemical substances and clarify the characteristics of chemical phenomena. The term physical chemistry is usually applied to the study of the physical properties of substances, such as vapor pressure, surface tension, viscosity, refractive index, density, and crystallography, as well as to the study of the so-called classical aspects of the behavior of chemical systems, such as thermal properties, equilibria, rates of reactions, mechanisms of reactions, and ionization phenomena. In its more theoretical aspects, physical chemistry attempts to explain spectral properties of substances in terms of fundamental quantum theory, the interaction of energy with matter, the nature of chemical bonding, the relationships correlating the number of energy states of electrons in atoms and molecules with the observable properties shown by these systems, and the electrical, thermal, and mechanical effects of individual electrons and protons on solids and liquids.[(3)]

Interestingly, physical chemistry can be subdivided into the study of chemical thermodynamics, chemical kinetics, the gaseous state, the liquid state, solutions, the solid state, electrochemistry, colloid chemistry, photochemistry, and statistical thermodynamics. Although providing details on each of these topics is beyond the scope of this text, a brief introductory comment on chemical kinetics is warranted at this time. This field, as the reader shall find out, is concerned with the rates of chemical reactions as a function of the concentration of the reacting species, of the products of the reaction, of any catalysts and inhibitors, of various solvent media, of temperature, and of all other variables that can affect the rate of reaction. It also seeks to relate the manner in which the reaction rate varies with time (and position) to the molecular nature of the rate-controlling intermolecular collisions involved in generating the products of reaction. Most reactions involve a series of stepwise processes, the sum of which corresponds to the overall observed reaction proportions (or stoichiometry) in which the reactants combine and the products form. Fortunately, only one of these steps is generally the rate-controlling one. By determining the nature of the rate-controlling process from a mathematical analysis of reaction kinetics data and by investigating how the reaction conditions affect the step, one can often deduce the mechanism of a reaction (see also Chapter 17).

OTHER CHEMISTRY TOPICS

There are other topics that fall under the chemistry umbrella. Some details are provided below.

1. *Electrochemistry* is concerned with the study of chemical effects produced by the flow of electric currents across interfaces (as at the boundary between an electrode and a solution) as well as the electrical effects produced by the displacement or transport of ions across boundaries or within gases, liquids, or solids.

2. *Colloid chemistry* studies the nature and effects of surfaces and interfaces on the macroscopic properties of substances. These studies involve the investigation of Brownian motion,[4,5] surface tension, interfacial tension (the tension that exists in the plane of contact between a liquid and a solid, or between two liquids), wetting and spreading of liquids on solids, adsorption of gases or of ions in solution on solid surfaces, etc.
3. *Photochemistry* is concerned with the study of the effects resulting from the absorption of electromagnetic radiation by substances, as well as the ability of substances to emit electromagnetic radiation when energized in various ways.
4. *Statistical thermodynamics and mechanics* is concerned with the calculation of the internal energy, degree of order or organization (entropy),[1] ability to do useful work (free energy), and other properties such as the equations of state of gases, the vapor pressures of liquids, the molecular shapes adopted by polymer chains, and electrical conductivities of ionic solution. These calculations are usually based on a model of the individual molecule or ion and the mathematical techniques of statistical analysis which permit the mutual interactions of large numbers of randomly arranged particles to be evaluated.

ANALYSIS PROCEDURES

Chemical analysis is concerned with the procedures and techniques used to identify and quantify the chemical composition of a sample of a substance. A chemist executing a qualitative analysis seeks to identify the substances in the sample. A quantitative analysis is an attempt to determine the quantity or concentration of a specific substance in the sample.

The measurement of chemical composition is necessary throughout the chemical industry, environmental regulatory government, and many other fields of science.

TABLE 2.1 Classes of Specialized Analysis

Clinical
Environmental
Forensic
Geochemical
Inorganic
Oceanographic
Organic
Petroleum
Pharmaceutical
Polymer
Spatial
Surface
Trace

TABLE 2.2 Separation Methods

Chromatography
Fractional distillation[5]
Precipitation[5]
Solvent extraction[5]

The practicing engineer is often required to analyze such diverse materials as stainless steel, beer, a fingernail, a rose petal, smoke, aspirin, paper, etc. The determination of the identity or quantity of a constituent in such materials is preceded by a sampling step—the selection of the amount and uniformity of material required for the analysis—and by the separation from the sample of either the desired constituent or the undesired, interfering constituents. Some typical analytical techniques are presented in Table 2.1.

Although chromatography is the most generally applicable of the separation methods available to the practicing engineer, there are a host of other procedures. These are detailed in Table 2.2.

REFERENCES

1. L. Theodore, F. Ricci, and T. VanVliet, "*Thermodynamics for the Practicing Engineer*," John Wiley & Sons, Hoboken, NJ, 2009.
2. J. Santoleri, J. Reynolds, and L. Theodore, "*Introduction to Hazardous Waste Incineration*," 2nd edition, John Wiley & Sons, Hoboken, NJ, 2000.
3. Author, title, source and date unknown.
4. L. Theodore, "*Air Pollution Control Equipment Calculations*," John Wiley & Sons, Hoboken, NJ, 2008.
5. L. Theodore and F. Ricci, "*Mass Transfer Operation for the Practicing Engineer*," John Wiley & Sons, Hoboken, NJ, 2010.

CHAPTER 3

*PROCESS VARIABLES**

INTRODUCTION

The author originally considered the title "State, Physical, and Chemical Properties" for this chapter; however, since these three properties have been used interchangeably and have come to mean different things to different people, it was decided to simply employ the title "Process Variables." The three aforementioned properties were therefore integrated into this all-purpose title eliminating the need for differentiating between the three.

This chapter provides a review of some basic concepts from physics, chemistry, and engineering in preparation for material that is covered in later chapters. All of these topics are vital to chemical kinetics and reactor applications. Because many of these topics are unrelated to each other, this chapter admittedly lacks the cohesiveness that chapters covering a single topic might have. This is usually the case when basic material from such widely differing areas of knowledge such as physics, chemistry, and engineering is surveyed. Though these topics are widely divergent and covered with varying degrees of thoroughness, all of them will find later use in this text. If additional information on these review topics is needed, the reader is directed to the literature in the reference section of this chapter.

Topics to be addressed include: temperature, pressure, moles and molecular weights, mass and volume, viscosity, heat capacity, thermal conductivity, Reynolds number, pH, vapor pressure, ideal gas law, latent enthalpy effects, and chemical reaction velocity constant. The chapter concludes with a section on property estimation.

ILLUSTRATIVE EXAMPLE 3.1 Discuss the traditional difference between chemical and physical properties.

Solution. Every compound has a unique set of *properties* that allows one to recognize and distinguish it from other compounds. These properties can be grouped into two main categories: physical and chemical. *Physical properties* are defined as those that can be measured without changing the identity and composition of the substance. Key properties include viscosity, density, surface tension, melting point,

*Adapted from "*Thermodynamics for the Practicing Engineer*," L. Theodore, F. Ricci, and T. VanVliet, John Wiley & Sons, Hoboken, NJ, 2009.

Chemical Reactor Analysis and Applications for the Practicing Engineer. By Louis Theodore

boiling point, etc. *Chemical properties* are defined as those that may be altered via chemical reaction to form other compounds or substances. Key chemical properties include upper and lower flammability limits, enthalpy of reaction, autoignition temperature, and others.

These properties may be further divided into two categories—intensive and extensive. *Intensive properties* are not a function of the quantity of the substance, while *extensive properties* depend on the quantity of the substance. ■

TEMPERATURE

Whether in the gaseous, liquid, or solid state, all molecules possess some degree of kinetic energy, i.e., they are in constant motion—vibrating, rotating, or translating. The kinetic energies of individual molecules cannot be measured, but the combined effect of these energies in a very large number of molecules can. This measurable quantity is known as *temperature*; it is a macroscopic concept only and as such does not exist on the molecular level.

Temperature can be measured in many ways; the most common method makes use of the expansion of mercury (usually encased inside a glass capillary tube) with increasing temperature. (In many thermal applications, however, thermocouples or thermistors are more commonly employed.) The two most commonly used temperature scales are the Celsius (or Centigrade) and Fahrenheit scales. The Celsius scale is based on the boiling and freezing points of water at l-atm (atmosphere) pressure; to the former, a value of 100°C is assigned, and to the latter, a value of 0°C. On the older Fahrenheit scale, these temperatures correspond to 212°F and 32°F, respectively. Equations (3.1) and (3.2) show the conversion from one scale to the other:

$$^{\circ}F = 1.8(^{\circ}C) + 32 \tag{3.1}$$

$$^{\circ}C = (^{\circ}F - 32)/1.8 \tag{3.2}$$

where $^{\circ}F$ = a temperature on the Fahrenheit scale

$^{\circ}C$ = a temperature on the Celsius scale

Experiments with gases at low-to-moderate pressures (up to a few atmospheres) have shown that, if the pressure is kept constant, the volume of a gas and its temperature are linearly related via Charles' law (see development later in chapter) and that a decrease of 0.3663% or (1/273) of the initial volume is experienced for every temperature drop of 1°C. These experiments were not extended to very low temperatures, but if the linear relationship were extrapolated, the volume of the gas would *theoretically* be zero at a temperature of approximately −273°C or −460°F. This temperature has become known as *absolute zero* and is the basis for the definition of two *absolute* temperature scales. (An *absolute* scale is one which does not allow negative quantities.) These absolute temperature scales are the Kelvin (K) and Rankine (°R) scales; the former is defined by shifting the Celsius scale by 273°C so that 0 K is

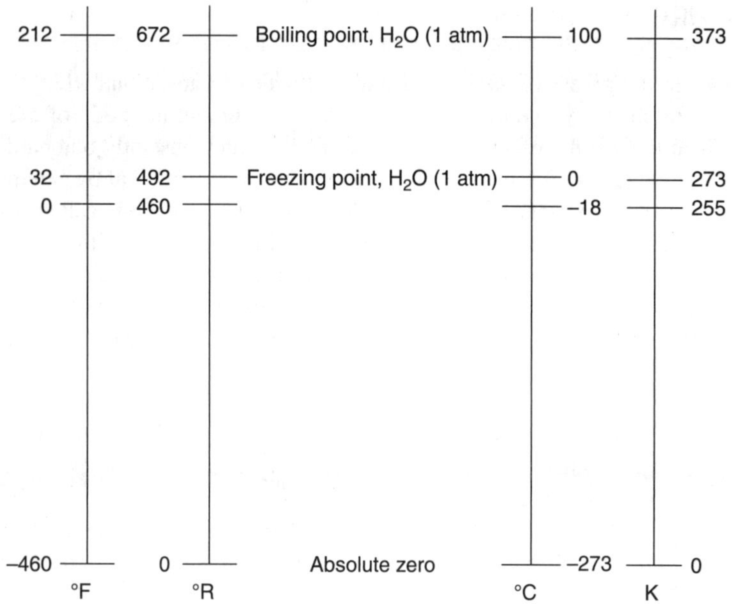

Figure 3.1 Temperature scales.

equal to −273°C; Equation (3.3) shows this relationship:

$$K = {}^{\circ}C + 273 \tag{3.3}$$

The Rankine scale is defined by shifting the Fahrenheit scale 460°, so that

$$^{\circ}R = {}^{\circ}F + 460 \tag{3.4}$$

The relationships among the various temperature scales are shown in Figure 3.1.

ILLUSTRATIVE EXAMPLE 3.2 Perform the following temperature conversions:

1. Convert 55°F to (a) Rankine, (b) Celsius, and (c) Kelvin.
2. Convert 55°C to (a) Fahrenheit, (b) Rankine, and (c) Kelvin.

Solution Employ Equations (3.3) and (3.4).

1. (a) $^{\circ}R = {}^{\circ}F + 460 = 55 + 460 = 515$
   (b) $^{\circ}C = \frac{5}{9}(^{\circ}F - 32) = \frac{5}{9}(55 - 32) = 12.8$
   (c) $K = \frac{5}{9}(^{\circ}F + 460) = \frac{5}{9}(55 + 460) = 286$

2. (a) $^{\circ}F = 1.8(^{\circ}C) + 32 = 1.8(55) + 32 = 131$
   (b) $^{\circ}R = 1.8(^{\circ}C) + 492 = 1.8(55) + 492 = 591$
   (c) $K = {}^{\circ}C + 273 = 55 + 273 = 328$ ■

PRESSURE

Molecules in the gaseous state possess a high degree of translational kinetic energy, which means they are able to move quite freely throughout the body of the gas. If the gas is in a container of some type, the molecules are constantly bombarding the walls of the container. The macroscopic effect of this bombardment by a tremendous number of molecules—enough to make the effect measurable—is called *pressure*. The natural units of pressure are force per unit area. In the example of the gas in a container, the *unit area* is a portion of the inside solid surface of the container wall and the *force*, measured perpendicularly to the unit area, is the result of the molecules hitting the unit area and giving up momentum during the sudden change of direction.

There are a number of different methods used to express a pressure measurement. Some of them are natural units, i.e., based on a force per unit area, as with pound (force) per square inch (abbreviated lb_f/in^2 or psi) or dyne per square centimeter (dyn/cm^2). Others are based on a fluid height, such as inches of water (in H_2O) or millimeters of mercury (mm Hg); units such as these are convenient when the pressure is indicated by a difference between two levels of a liquid as in a *manometer* or *barometer*. This measurement is based on the pressure at the base of a column (height) of fluid. *Barometric pressure* and *atmospheric pressure* are synonymous and measure the ambient air pressure. *Standard barometric pressure* is the average atmospheric pressure at sea level, 45° north latitude at 32°F. It is used to define another unit of pressure called the atmosphere (atm). Standard barometric pressure is 1 atm and is equivalent to 14.696 psi and 29.921 in Hg. As one might expect, barometric pressure varies with weather and altitude.

Measurements of pressure by most gauges indicate the difference in pressure either above or below that of the atmosphere surrounding the gauge. *Gauge pressure* is the pressure indicated by such a device. If the pressure in the system measured by the gauge is greater than the pressure prevailing in the atmosphere, the gauge pressure is expressed positively. If lower than atmospheric pressure, the gauge pressure is a negative quantity; the term *vacuum* designates a negative gauge pressure. Gauge pressures are often identified by the letter *g* after the pressure unit; for example, psig (pounds per square inch gauge) is a gauge pressure in psi units.

Since gauge pressure is the pressure relative to the prevailing atmospheric pressure, the sum of the two gives the *absolute pressure*, indicated by the letter *a* after the unit [e.g., psia (pounds per square inch absolute)]:

$$P = P_a + P_g \tag{3.5}$$

where P = absolute pressure (psia)
P_a = atmospheric pressure (psia)
P_g = gauge pressure (psig)

The absolute pressure scale is absolute in the same sense that the absolute temperature scale is absolute, i.e., a pressure of zero psia is the lowest possible pressure theoretically achievable—a perfect vacuum.

ILLUSTRATIVE EXAMPLE 3.3 Consider the following pressure calculations.

1. A liquid weighing 100 lb held in a cylindrical column with a base area of 3 in^2 exerts what pressure at the base in $\text{lb}_\text{f}/\text{ft}^2$?
2. If the pressure is 35 psig (pounds per square inch gauge), what is the absolute pressure?

Solution

1. By definition

$$\begin{aligned} F &= mg/g_c = 100\,\text{lb}(1\ \text{lb}_\text{f}/\text{lb}) \\ &= 100\ \text{lb}_\text{f} \end{aligned}$$

Note that g_c is a conversion factor equal to 32.2 lb · ft/$\text{lb}_\text{f} \cdot \text{s}^2$; g is the gravitational acceleration, which is equal, or close to, 32.2 ft/s^2 on Earth's surface. Therefore,

$$\begin{aligned} P &= F/\text{area} = 100\ \text{lb}_\text{f}/3\ \text{in}^2 \\ &= 33.33\,\text{lb}_\text{f}/\text{in}^2 = 33.33\,\text{psi} \\ &= 4800\,\text{lb}_\text{f}/\,\text{ft}^2 = 4800\,\text{psf} \end{aligned}$$

2. Apply the equation

$$P = P_g + P_a \tag{3.5}$$

Substituting

$$\begin{aligned} P &= 35 + 14.7 \\ &= 49.7\ \text{psia} \end{aligned}$$

■

MOLES AND MOLECULAR WEIGHTS

An atom consists of protons and neutrons in a nucleus surrounded by electrons. An electron has such a small mass relative to that of the proton and neutron that the weight of the atom (called the *atomic weight*) is approximately equal to the sum of the weights of the particles in its nucleus. Atomic weight may be expressed in *atomic mass units (amu) per atom* or *in grams per gram · atom*. One gram · atom contains 6.02×10^{23} atoms (Avogadro's number). The atomic weights of the elements are listed in Table 3.1.

The *molecular weight* (MW) of a compound is the sum of the atomic weights of the atoms that make up the molecule. Atomic mass units per molecule (amu/molecule) or grams per gram · mole (g/gmol) are used for molecular weight. One gram · mole (gmol) contains an Avogadro number of molecules. For the English system, a pound · mole (lbmol) contains $454 \times 6.023 \times 10^{23}$ molecules.

TABLE 3.1 Atomic Weights of the Elements[a,b]

Element	Symbol	Atomic weight
Actinium	Ac	227.0278
Aluminum	Al	26.9815
Americium	Am	(243)
Antimony	Sb	121.75
Argon	Ar	39.948
Arsenic	As	74.9216
Astatine	At	(210)
Barium	Ba	137.34
Berkelium	Bk	(247)
Beryllium	Be	9.0122
Bismuth	Bi	208.980
Boron	B	10.811
Bromine	Br	79.904
Cadmium	Cd	112.40
Calcium	Ca	40.08
Californium	Cf	(251)
Carbon	C	12.01115
Cerium	Ce	140.12
Cesium	Cs	132.905
Chlorine	Cl	35.453
Chromium	Cr	51.996
Cobalt	Co	58.9332
Copper	Cu	63.546
Curium	Cm	(247)
Dysprosium	Dy	162.50
Einsteinium	Es	(252)
Erbium	Er	167.26
Europium	Eu	151.96
Fermium	Fm	(257)
Fluorine	F	18.9984
Francium	Fr	(223)
Gadolinium	Gd	157.25
Gallium	Ga	69.72
Germanium	Ge	72.59
Gold	Au	196.967
Hafnium	Hf	178.49
Helium	He	4.0026
Holmium	Ho	164.930
Hydrogen	H	1.00797
Indium	In	114.82
Iodine	I	126.9044
Iridium	Ir	192.2
Iron	Fe	55.847

(Continued)

TABLE 3.1 *Continued*

Element	Symbol	Atomic weight
Krypton	Kr	83.80
Lanthanum	La	138.91
Lawrencium	Lr	(260)
Lead	Pb	207.19
Lithium	Li	6.939
Lutetium	Lu	174.97
Magnesium	Mg	24.312
Manganese	Mn	54.9380
Mendelevium	Md	(258)
Mercury	Hg	200.59
Molybdenum	Mo	95.94
Neodymium	Nd	144.24
Neon	Nc	20.183
Neptunium	Np	237.0482
Nickel	Ni	58.71
Niobium	Nb	92.906
Nitrogen	N	14.0067
Nobelium	No	(259)
Osmium	Os	190.2
Oxygen	O	15.9994
Pallasium	Pd	106.4
Phosphorus	P	30.9738
Platinum	Pt	195.09
Plutonium	Pu	(244)
Polonium	Po	(209)
Potassium	K	39.102
Praseodymium	Pr	140.907
Promcthium	Pm	(145)
Protactinium	Pa	231.0359
Radium	Ra	226.0254[c]
Radon	Rn	(222)
Rhenium	Re	186.2
Rhodium	Rh	102.905
Rubidium	Rb	84.57
Rutherium	Ru	101.07
Samarium	Sm	150.35
Scandium	Sc	44.956
Selenium	Se	78.96
Silicon	Si	28.086
Silver	Ag	107.868
Sodium	Na	22.9898
Strontium	Sr	87.62
Sulfur	S	32.064

(*Continued*)

TABLE 3.1 *Continued*

Element	Symbol	Atomic weight
Tantalum	Ta	180.948
Technetium	Tc	(98)
Tellurium	Te	127.60
Terbium	Tb	158.924
Thallium	Tl	204.37
Thorium	Th	232.038
Thulium	Tm	168.934
Tin	Sn	118.69
Titanium	Ti	47.90
Tungsten	W	183.85
Uranium	U	238.03
Vanadium	V	50.942
Xenon	Xe	131.30
Ytterbium	Yb	173.04
Yttrium	Y	88.905
Zinc	Zn	65.37
Zirconium	Zr	91.22

[a]Atomic weights apply to naturally occurring isotopic compositions and are based on an atomic mass of $^{12}C = 12$.

[b]A value given in parenthesis for radioactive elements is the atomic mass number of the isotope of longest known half-life.

[c]Geologically exceptional samples are known in which the element has an isotopic composition outside the limits for normal material.

Molal units are used extensively in chemistry calculations as they greatly simplify material balances where chemical reactions are occurring. For mixtures of substances (gases, liquids, or solids), it is also convenient to express compositions in mole fractions or mole percentages instead of mass fractions. The mole fraction is the ratio of the number of moles of one component to the total number of moles in the mixture. Equations (3.6)–(3.9) express these relationships:

$$\text{moles of A} = \frac{\text{mass A}}{\text{molecular weight of A}}$$

$$n_A = \frac{m_A}{(\text{MW})_A} \tag{3.6}$$

$$\text{mole/fraction A} = \frac{\text{moles A}}{\text{total moles}}$$

$$y_A = \frac{n_A}{n} \tag{3.7}$$

$$\text{mass fraction A} = \frac{\text{mass A}}{\text{total mass}}$$

$$w_A = \frac{m_A}{m} \tag{3.8}$$

$$\text{volume fraction A} = \frac{\text{volume A}}{\text{total volume}}$$

$$v_A = \frac{V_A}{V} \tag{3.9}$$

The reader should note that, in general, mass fraction (or percent) is *not* equal to mole fraction (or percent).

ILLUSTRATIVE EAMPLE 3.4 If a 55-gal tank contains 20.0 lb of water, (1) how many pound · moles of water does it contain? (2) how many gram · moles does it contain? and, (3) how many molecules does it contain?

Solution. The molecular weight of the water (H_2O) is

$$\begin{aligned} MW &= (2)(1.008) + (15.999) = 18.015\ \text{g/gmol} \\ &= 18.015\ \text{lb/lbmol} \end{aligned}$$

Therefore

1. $(20.0\ \text{lb})\left(\frac{\text{lbmol}}{18.015\ \text{lb}}\right) = 1.11\ \text{lbmol water}$
2. $(20.0\ \text{lb})\left(\frac{454\ \text{g}}{1\ \text{lb}}\right)\left(\frac{\text{gmol}}{18.015\ \text{g}}\right) = 503.6\ \text{gmol water}$
3. $(503.6\ \text{gmol})\left(\frac{6.023 \times 10^{23}\ \text{molecules}}{1\ \text{gmol}}\right) = 3.033 \times 10^{26}\ \text{molecules}$ ■

MASS AND VOLUME

The *density* (ρ) of a substance is the ratio of its mass to its volume and may be expressed in units of pounds per cubic foot (lb/ft^3), kilograms per cubic meter (kg/m^3), etc. For solids, density can be easily determined by placing a known mass of the substance in a liquid and determining the displaced volume. The density of a liquid can be measured by weighing a known volume of the liquid in a volumetric flask. For gases, the ideal gas law, to be discussed later, can be used to calculate the density from the pressure, temperature, and molecular weight of the gas.

Densities of pure solids and liquids are relatively independent of temperature and pressure and can be found in standard reference books.[1,2] The *specific volume* (v) of a substance is its volume per unit mass (ft^3/lb, m^3/kg, etc.) and is, therefore, the inverse of its density.

The *specific gravity* (SG) is the ratio of the density of a substance to the density of a reference substance at a specific condition:

$$SG = \rho/\rho_{\text{ref}} \tag{3.10}$$

The reference most commonly used for *solids* and *liquids* is water at its maximum density, which occurs at 4°C; this reference density is 1.000 g/cm^3, 1000 kg/m^3, or 62.43 lb/ft^3. Note that, since the specific gravity is a ratio of two densities, it is dimensionless. Therefore, any set of units may be employed for the two densities as long as they are consistent. The specific gravity of *gases* is used only rarely; when it is, air at the same conditions of temperature and pressure as the gas is usually employed as the reference substance.

Another dimensionless quantity related to density is the API (American Petroleum Institute) gravity, which is often used to indicate densities of fuel oils. The relationship between the API scale and specific gravity is

$$\text{degrees API} = \frac{141.5}{\text{SG}(60/60°\text{F})} - 131.5 \qquad (3.11)$$

where SG(60/60°F) = specific gravity of the liquid at 60°F using water at 60°F as the reference.

ILLUSTRATIVE EXAMPLE 3.5 The following information is given:

Specific gravity of liquid (methanol) = 0.92 (at 60°F)

Density of reference substance (water) = 62.4 lb/ft^3 (at 60°F)

Determine the density of methanol in lb/ft^3 at 60°F.

Solution. Calculate the density of methanol in English units by multiplying the specific gravity by the density of water—see Equation (3.10).

$$\begin{aligned}\text{Density of methanol} &= (\text{specific gravity})(\text{density of water})\\ &= (0.92)(62.4)\\ &= 57.4\ \text{lb/ft}^3\end{aligned}$$

The procedure is reversed in order to calculate specific gravity from density data. As noted above, the notation for density is usually, but not always, ρ. ■

VISCOSITY

Viscosity is a property associated with a fluid's resistance to flow; more precisely, this property accounts for energy losses which result from shear stresses that occur between different portions of the fluid that are moving at different velocities. The *absolute viscosity* (μ) has units of mass per length · time; the fundamental unit is the *poise* (P), which is defined as 1 g/cm · s. This unit is inconveniently large for many practical purposes and viscosities are frequently given in *centipoises* (0.01 poise), which is abbreviated to cP. The viscosity of pure water at 68.6°F is 1.00 cP. In English units, absolute viscosity is expressed either as pounds (mass) per foot · second (lb/ft · s) or pounds per

TABLE 3.2 Viscosity of Air at 1 Atmosphere[a]

T (°C)	Viscosity, Micropoise (μP)
0	170.8
18	182.7
40	190.4
54	195.8
74	210.2
229	263.8

[a] $1\,\text{P} = 100\,\text{cP} = 10^6\,\mu\text{P}$; $1\,\text{cP} = 6.72 \times 10^{-4}\,\text{lb/ft} \cdot \text{s}$.

foot · hour (lb/ft · h). The absolute viscosity depends primarily on temperature and to a lesser degree on pressure. The *kinematic viscosity* (ν) is the absolute viscosity divided by the density of the fluid and is useful in certain reactor problems; the units for this quantity are length squared per time, e.g., square foot per second (ft^2/s) or square meters per hour (m^2/h). A kinematic viscosity of $1\,\text{cm}^2/\text{s}$ is called a *stoke*, denoted as S. For pure water at 70°F, $\nu = 0.983$ cS (centistokes). Because fluid viscosity changes rapidly with temperature, a numerical value of viscosity has no significance unless the temperature is specified.

Liquid viscosity is usually measured by the amount of time it takes for a given volume of liquid to flow through an orifice. The *Saybolt universal viscometer* is the most widely used device in the United States for the determination of the viscosity of fuel oils and liquids. It should be stressed that Saybolt viscosities, which are expressed in *Saybolt seconds* (*SSU*), are not even approximately proportional to absolute viscosities except in the range above 200 SSU; hence, converting units from Saybolt seconds to other units requires the use of special conversion tables. As the

TABLE 3.3 Viscosity of Water

T (°C)	Viscosity, Centipoise (cP)
0	1.792
5	1.519
10	1.308
15	1.140
20	1.000
25	0.894
30	0.801
35	0.723
40	0.656
50	0.594
60	0.469
70	0.406
80	0.357
90	0.317
100	0.284

time of flow decreases, the deviation becomes more marked. In any event, viscosity is an important property because of potential flow problems that might arise with viscous liquids and/or fuel oils.

The viscosities of air at atmospheric pressure and water are presented in Tables 3.2 and 3.3, respectively, as functions of temperature. Viscosities of other substances are available in the literature.[1]

ILLUSTRATIVE EXAMPLE 3.6 What is the kinematic viscosity of a gas whose specific gravity and absolute viscosity are 0.8 and 0.02 cP, respectively?

Solution

$$\mu = \left(\frac{0.02\ \text{cP}}{1}\right)\left(\frac{6.720 \times 10^{-4}\ \text{lb/ft} \cdot \text{s}}{1\ \text{cP}}\right) = 1.344 \times 10^{-5}\ \text{lb/ft}^3 \cdot \text{s}$$

$$\rho = (\text{SG})(\rho_{\text{ref}}) = (0.8)(62.43\ \text{lb/ft}^3) = 49.94\ \text{lb/ft}^3$$

$$\nu = \mu/\rho = (1.344 \times 10^{-5}\ \text{lb/ft} \cdot \text{s})/(49.94\ \text{lb/ft}^3)$$

$$= 2.691 \times 10^{-7}\ \text{ft}^2/\text{s}$$

■

HEAT CAPACITY

The *heat capacity* of a substance is defined as the quantity of heat required to raise the temperature of that substance by one degree on a unit mass (or mole) basis. The term *specific heat* is frequently used in place of *heat capacity*. This is not strictly correct, because specific heat has been defined traditionally as the ratio of the heat capacity of a substance to the heat capacity of water. However, since the heat capacity of water is approximately 1 cal/g · °C or 1 Btu/lb · °F, the term *specific heat* has come to imply heat capacity.

For gases, the addition of heat to cause the 1° temperature rise may be accomplished either at constant pressure or at constant volume. Since the amounts of heat necessary are different for the two cases, subscripts are used to identify which heat capacity is being used—c_P for constant pressure and c_V for constant volume. For liquids and solids, this distinction does not have to be made since there is little difference in value between the two. Values of heat capacity are available in the literature.[1]

Heat capacities are often used on a *molar* basis instead of a *mass* basis, in which case the units become cal/gmol · °C or Btu/lbmol · °F. To distinguish between the two bases, uppercase letters (C_P, C_V) are used in this text to represent the molar-based heat capacities, and lowercase letters (c_P, c_V) are used for the mass-based heat capacities or specific heats.

Heat capacities are functions of both the temperature and pressure, although the effect of pressure is generally small and is neglected in almost all engineering calculations. The effect of temperature on C_P can be described by

$$C_P = \alpha + \beta T + \gamma T^2 \tag{3.12}$$

or

$$C_P = a + bT + cT^{-2} \tag{3.13}$$

Values for α, β, γ, and a, b, c, as well as average heat capacity information are provided in the literature.[(1)]

ILLUSTRATIVE EXAMPLE 3.7 The following is given:

Heat capacity of methanol = 0.61 cal/g · °C (at 60°F)

Convert the heat capacity to English units.

Solution. Note that 1.0 Btu/lb · °F is equivalent to 1.0 cal/g · °C. This also applies on a mole basis, i.e.,

$$1 \text{ Btu/lbmol} \cdot {}^\circ\text{F} = 1 \text{ cal/gmol} \cdot {}^\circ\text{C}$$

The heat capacity can be converted from units of cal/g · °C to Btu/lb · °F using appropriate conversion factors.

$$\left(\frac{0.61 \text{ cal}}{\text{g} \cdot {}^\circ\text{C}}\right)\left(\frac{454 \text{ g}}{\text{lb}}\right)\left(\frac{\text{Btu}}{252 \text{ cal}}\right)\left(\frac{{}^\circ\text{C}}{1.8{}^\circ\text{F}}\right) = 0.61 \text{ Btu/lb} \cdot {}^\circ\text{F}$$ ■

THERMAL CONDUCTIVITY

Experience has shown that when a temperature difference exists across a solid body, energy in the form of heat will transfer from the high-temperature region to the low-temperature region until thermal equilibrium (same temperature) is reached. This mode of heat transfer, where vibrating molecules pass along kinetic energy through the solid, is called *conduction*. Liquids and gases may also transport heat in this fashion. The property of *thermal conductivity* provides a measure of how fast (or how easily) heat flows through a substance. It is defined as the amount of heat that flows in unit time through a unit surface area of unit thickness as a result of a unit difference in temperature. Typical units for conductivity are Btu · ft/h · ft^2 · °F or Btu/h · ft · °F.

ILLUSTRATIVE EXAMPLE 3.8 The following data is given:

Thermal conductivity of methanol = 0.0512 cal/m · s · °C (at 60°F)

Convert the thermal conductivity to English units.

Solution. The factor for converting cal/m · s · °C to Btu/ft · h · °F can be shown to be 2.419. The thermal conductivity of methanol can be converted to Btu/ft · h · °F from cal/m · s · °C as follows:

$$k = \left(\frac{0.0512\ \text{cal}}{\text{m} \cdot \text{s} \cdot {}^\circ\text{C}}\right)\left(\frac{\text{Btu}}{252\ \text{cal}}\right)\left(\frac{0.3048\ \text{m}}{\text{ft}}\right)\left(\frac{3600\ \text{s}}{\text{h}}\right)\left(\frac{{}^\circ\text{C}}{1.8^\circ\text{F}}\right)$$
$$= (0.0512)(2.419)$$
$$= 0.124\ \text{Btu/ft} \cdot \text{h} \cdot {}^\circ\text{F}$$

Note that the usual engineering notation for thermal conductivity is k. ■

REYNOLDS NUMBER

The Reynolds number, Re, is a dimensionless number that indicates whether a moving fluid is flowing in the laminar or turbulent mode. *Laminar* flow is characteristic of fluids flowing slowly enough so that there are no eddies (whirlpools) or macroscopic mixing of different portions of the fluid. (*Note:* In any fluid, there is always *molecular* mixing due to the thermal activity of the molecules; this is distinct from *macroscopic* mixing due to the swirling motion of different portions of the fluid.) In laminar flow, a fluid can be imagined to flow like a deck of cards, with adjacent layers sliding past one another. *Turbulent* flow is characterized by eddies and macroscopic currents. In practice, moving gases are generally in the turbulent region. For flow in a pipe, a Reynolds number above 2100 is an indication of turbulent flow.

The Reynolds number is dependent on the fluid velocity, density, viscosity, and some characteristic *length* of the system or conduit; for pipes, this characteristic length is the inside diameter:

$$\text{Re} = Dv\rho/\mu = Dv/\nu \tag{3.14}$$

where Re = Reynolds number

D = inside diameter of the pipe (ft)

v = fluid velocity (ft/s)

ρ = fluid density (lb/ft^3)

μ = fluid viscosity (lb/ft · s)

ν = fluid kinematic viscosity (ft^2/s)

Any consistent set of units may be used with Equation (3.14).

ILLUSTRATIVE EXAMPLE 3.9 Calculate the Reynolds number for a fluid with a density of 50 lb/ft^3 and a viscosity of 0.65 cP flowing through a 5-inch diameter reactor at 10 fps (feet per second). Is the flow turbulent or laminar?

Solution. By definition

$$\text{Re} = Dv\rho/\mu \tag{3.14}$$

Substitution yields

$$\begin{aligned}
\text{Re} &= \left(\frac{50\,\text{lb}}{\text{ft}^3}\right)\left(\frac{10\ \text{ft}}{\text{s}}\right)\left(\frac{5\,\text{in}}{1}\right)\left(\frac{1\ \text{ft}}{12\,\text{in}}\right)\left(\frac{1}{0.65\,\text{cP}}\right)\left(\frac{1\,\text{cP}}{6.720\times10^{-4}\,\text{lb}/\text{ft}\cdot\text{s}}\right)\\
&= (50\,\text{lb}/\text{ft}^3)(10\ \text{ft/s})[(5/12)\,\text{ft}]/(0.65\times 6.72\times10^{-4}\,\text{lb}/\text{ft}\cdot\text{s})\\
&= 477{,}000
\end{aligned}$$

The Reynolds number is >2100; therefore, the flow is turbulent. ■

pH

An important chemical property of an aqueous solution is its pH. The pH measures the acidity or basicity of the solution. In a neutral solution, such as pure water, the hydrogen (H^+) and hydroxyl (OH^-) ion concentrations are equal. At ordinary temperatures, this concentration is

$$C_{H^+} = C_{OH^-} = 10^{-7}\ \text{g}\cdot\text{ion/L} \tag{3.15}$$

where C_{H^+} = hydrogen ion concentration
C_{OH^-} = hydroxyl ion concentration

The unit g · ion stands for gram · ion, which represents an Avogadro number of ions. In all aqueous solutions, whether neutral, basic, or acidic, a chemical equilibrium or balance is established between these two concentrations, so that

$$K_{eq} = C_{H^+}C_{OH^-} = 10^{-14} \tag{3.16}$$

where K_{eq} = equilibrium constant.

The numerical value for K_{eq} given in Equation (3.16) holds for room temperature and only when the concentrations are expressed in gram · ion per liter (g · ion/L). In acid solutions, C_{H^+} is $>C_{OH^-}$; in basic solutions, C_{OH^-} predominates.

The pH is a direct measure of the hydrogen ion concentration and is defined by

$$\text{pH} = -\log C_{H^+} \tag{3.17}$$

Thus, an acidic solution is characterized by a pH below 7 (the lower the pH, the higher the acidity), a basic solution by a pH above 7, and a neutral solution by a pH of 7. It should be pointed out that Equation (3.17) is not the exact definition of pH but is a close approximation to it. Strictly speaking, the *activity* of the hydrogen ion, a_{H^+}, and not the ion concentration should be employed in Equation (3.17). The reader is directed to the literature[2,3] for a discussion of chemical activities.

ILLUSTRATIVE EXAMPLE 3.10 Calculate the hydrogen ion and the hydroxyl ion concentration of an aqueous solution if the pH of the solution is 1.0.

Solution. Apply Equation (3.17). For a pH of 1.0,

$$\text{pH} = -\log(C_{\text{H}^+})$$

$$C_{\text{H}^+} = 10^{-\text{pH}} = 10^{-1} = 0.1 \text{ g} \cdot \text{ion/L}$$

$$C_{\text{H}^+} \times C_{\text{OH}^-} = 10^{-14}$$

$$C_{\text{OH}^-} = \frac{10^{-14}}{C_{\text{H}^+}} = 10^{-13} \text{ g} \cdot \text{ion/L}$$

■

ILLUSTRATIVE EXAMPLE 3.11 Process considerations require pH control in a 50,000-gal storage tank used for incoming reactor feed mixtures to a reactor. Normally, the tank is kept at neutral pH. However, the operation can tolerate pH variations from 6 to 8. The feed arrives in 5000-gal shipments.

Assume that the tank is completely mixed, contains 45,000 gal (neutral pH) when the shipment arrives, the incoming acidic feed is fully dissociated, and that there is negligible buffering capacity in the tank. What is the pH of the most acidic 5000 gal shipment that can be handled without neutralization?

Solution. The pH of the most acidic feed shipment that can be handled without neutralization is calculated as follows: 5000 gal of feed with a $[\text{H}^+] = X$ is diluted by 45,000 gal at pH = 7 (or $[\text{H}^+] = 10^{-7}$). The minimum pH of 6 that can be tolerated is equivalent to an average $[\text{H}^+] = 10^{-6}$. From an ion balance:

$$\min[\text{H}^+] = 10^{-6} = (5000/50{,}000)X + (45{,}000/50{,}000)(10^{-7}); \; [\text{H}^+] = \text{C}_{\text{H}^+}$$

$$X = \left(\frac{50{,}000}{5000}\right)\left[10^{-6} - \frac{45{,}000(10^{-7})}{50{,}000}\right] = 0.91 \times 10^{-6}$$

$$\text{pH} = 6.04$$

■

VAPOR PRESSURE

Vapor pressure is an important property of liquids, and, to a much lesser extent, of solids. If a liquid is allowed to evaporate in a confined space, the pressure in the vapor space increases as the amount of vapor increases. If there is sufficient liquid present, a point is eventually reached at which the pressure in the vapor space is exactly equal to the pressure exerted by the liquid at its own surface. At this point, a dynamic equilibrium exists in which vaporization and condensation take place at equal rates and the pressure in the vapor space remains constant.

The pressure exerted at equilibrium is defined as the vapor pressure of the liquid. The magnitude of this pressure for a given liquid depends on the temperature, but not on the amount of liquid present. Solids, like liquids, also exert a vapor pressure. Evaporation of solids (called *sublimation*) is noticeable only for those with appreciable vapor pressures.

THE IDEAL GAS LAW

The two precursors of the ideal gas law were *Boyle's* (see Chapter 2) and *Charles'* laws. Boyle found that the volume of a given mass of gas is inversely proportional to the *absolute* pressure if the temperature is kept constant:

$$P_1 V_1 = P_2 V_2 \tag{3.18}$$

where V_1 = volume of gas at absolute pressure P_1 and temperature T

V_2 = volume of gas at absolute pressure P_2 and temperature T

Charles found that the volume of a given mass of gas varies directly with the *absolute* temperature at constant pressure:

$$\frac{V_1}{T_1} = \frac{V_2}{T_2} \tag{3.19}$$

where V_1 = volume of gas at pressure P and absolute temperature T_1

V_2 = volume of gas at pressure P and absolute temperature T_2

Boyle's and Charles' laws may be combined into a single equation in which neither temperature nor pressure need be held constant.

$$\frac{P_1 V_1}{T_1} = \frac{P_2 V_2}{T_2} \tag{3.20}$$

For Equation (3.20) to apply, the mass of gas must be constant as the conditions change from (P_1, T_1) to (P_2, T_2). This equation indicates that for a given mass of a specific gas, PV/T has a constant value. Since, at the same temperature and pressure, volume and mass must be directly proportional, this statement may be extended to

$$\frac{PV}{mT} = C \tag{3.21}$$

where m = mass of a specific gas

C = constant that depends on the gas

Note that the volume terms above may be replaced by volume rate (or volumetric flow rate), q or Q.

ILLUSTRATIVE EXAMPLE 3.12 What is the final (f) volumetric flow rate of a gas that is heated at constant pressure from 100 to 300°F if its initial (i) flow is 3500 acfm.

Solution. Apply Charles' law on a volume rate basis. See Equation (3.19) and be sure to employ absolute temperature units:

$$\begin{aligned} Q_f &= Q_i(T_f/T_i) \\ &= 3500[(300+460)/(100+460)] \\ &= 4750 \text{ acfm} \end{aligned}$$

■

ILLUSTRATIVE EXAMPLE 3.13 What is the volumetric flow rate of the gas (100°F, 1 atm) in the previous example if it is compressed isothermally (constant temperature) to 3 atm?

Solution. Apply Boyle's law. See Equation (3.18).

$$\begin{aligned} Q_f &= Q_i(P_i/P_f) \\ &= 3500(1.0/3.0) \\ &= 1167 \text{ acfm} \end{aligned}$$

■

ILLUSTRATIVE EXAMPLE 3.14 What is the volumetric flow rate of the gas in the previous example if the final temperature is 300°F.

Solution. Using the combined gas law,

$$\begin{aligned} Q_f &= Q_i(P_i/P_f)(T_f/T_i) \\ &= 3500(1/3)(760/560) \\ &= 1583 \text{ acfm} \end{aligned}$$

■

Experiments with different gases showed that Equation (3.21) could be expressed in a far more generalized form. If the number of moles (n) is used in place of the mass (m), the constant amazingly is the same for all gases:

$$\frac{PV}{nT} = R \tag{3.22}$$

where R = universal gas constant.

Equation (3.22) is called the ideal gas law. Numerically, the value of R depends on the units used for P, V, T, and n (see Table 3.4). In this text, gases are generally assumed to approximate ideal gas behavior. As is usually the case in engineering practice, unless otherwise stated the ideal gas law is assumed to be valid for all illustrative and assigned problems in Parts II–IV, unless otherwise indicated. If a case is encountered in practice where the gas behaves in a nonideal fashion, e.g., a

TABLE 3.4 Values of R in Various Units

R	Temperature scale	Units of V	Units of n	Units of P	Units of PV (energy)
10.73	°R	ft^3	lbmol	psia	—
0.7302	°R	ft^3	lbmol	atm	—
21.85	°R	ft^3	lbmol	in Hg	—
555.0	°R	ft^3	lbmol	mm Hg	—
297.0	°R	ft^3	lbmol	in H_2O	—
0.7398	°R	ft^3	lbmol	bar	—
1545.0	°R	ft^3	lbmol	psfa	—
24.75	°R	ft^3	lbmol	ft H_2O	—
1.9872	°R	—	lbmol	—	Btu
0.0007805	°R	—	lbmol	—	hp · h
0.0005819	°R	—	lbmol	—	kW · h
500.7	°R	—	lbmol	—	cal
1.314	K	ft^3	lbmol	atm	—
998.9	K	ft^3	lbmol	mm Hg	—
19.32	K	ft^3	lbmol	psia	—
62.361	K	L	gmol	mm Hg	—
0.08205	K	L	gmol	atm	—
0.08314	K	L	gmol	bar	—
8314	K	L	gmol	Pa	—
8.314	K	m^3	gmol	Pa	—
82.057	K	cm^3	gmol	atm	—
1.9872	K	—	gmol	—	cal
8.314	K		gmol	—	J

high-molecular-weight gas (such as a chlorinated organic) under high pressures, one of the many *real gas* correlations should be used.[1]

Other useful forms of the ideal gas law are shown in Equations (3.23) and (3.24). Equation (3.23) applies to gas volume flow rate rather than to a gas volume confined in a container:[4]

$$PQ = \dot{n}RT \tag{3.23}$$

where $Q =$ gas volumetric flow rate (ft^3/h)

$P =$ absolute pressure (psia)

$\dot{n} =$ molar flow rate (lbmol/h)

$T =$ absolute temperature (°R)

$R = 10.73$ psia · ft^3/lbmol · °R

Equation (3.24) combines n and V from Equation (3.22) to express the law in terms of density:

$$P(MW) = \rho RT \tag{3.24}$$

where MW = molecular weight of gas (lb/lbmol)
ρ = density of gas (lb/ft^3)

ILLUSTRATIVE EXAMPLE 3.15

1. What is the density of air at 75°F and 14.7 psia? The molecular weight of air is 29.
2. Calculate the volume (in ft^3) of 1.0 lbmol of any ideal gas at 60°F and 14.7 psia.
3. Calculate the density of a gas ($MW = 29$) in g/cm^3 at 20°C and 1.2 atm using the ideal gas law.

Solution. This example is solved using the ideal gas law:

$$PV = nRT = \left(\frac{m}{MW}\right)RT \tag{3.24}$$

1. For the density,

$$\frac{m}{V} = \rho = \frac{P(MW)}{RT} = \frac{(14.7\ \text{psia})(29\ \text{lb/lbmol})}{(10.73\ \text{ft}^3 \cdot \text{psi/lbmol} \cdot {}^\circ\text{R})(75 + 460)} = 0.0743\ \text{lb/ft}^3$$

This is an important value to remember: air at or near ambient conditions has a density of approximately 0.075 lb/ft^3.

2. Solve the ideal gas law for V and calculate the volume:

$$V = \frac{nRT}{P} = \frac{(1)(10.73)(60 + 460)}{14.7}$$
$$= 379\ \text{ft}^3$$

This result is an important number to remember in many chemical reactor calculations—1 lbmol of any (ideal) gas at 60°F and 1 atm occupies 379 ft^3.

3. Calculate the density of the gas again using the ideal gas law:

$$PV = nRT = \left(\frac{m}{MW}\right)RT \tag{3.24}$$

$$\frac{m}{V} = \rho = \frac{P(MW)}{RT} = \frac{(1.2)(29)}{(82.06)(20 + 273)}$$
$$= 0.00145\ \text{g/cm}^3$$

The effects of pressure, temperature, and molecular weight on density can be obtained directly from the ideal gas law equation. Increasing the pressure and molecular weight increases the density; increasing the temperature decreases the density. ■

LATENT ENTHALPY EFFECTS

The term phase for a pure substance refers to a state of matter that is gas, liquid, or solid. Latent enthalpy (heat) effects are associated with phase changes. These phase changes involve no change in temperature but there is a transfer of energy to/from the substance. There are three possible latent effects, as detailed below:

1. vapor–liquid
2. liquid–solid
3. vapor–solid.

Vapor–liquid changes are referred to as *condensation* when the vapor is condensing and *vaporization* when liquid is vaporizing. Liquid–solid changes are referred to as *melting* when the solid melts to liquid and freezing when a liquid solidifies. Vapor–solid changes are referred to as *sublimation*. One should also note that there are enthalpy effects associated with a phase change of a solid to another solid form; however, this enthalpy effect is small compared to the other effects mentioned above. Finally, it is vapor–liquid changes that are of primary interest to the practicing engineer involved with chemical reactor analysis.

Latent enthalpy effects can be measured directly in a calorimeter for vapor–liquid changes. The notation employed is ΔH with units of energy per mole, e.g., Btu/lbmol; lower case h is used if based on mass, e.g., cal/g. However, these effects can be predicted from semi-theoretical equations provided in the literature.[3]

The Chemical Reaction Velocity Constant, *k*

Most reactions have an opposing (or reverse) reaction. The rates of the forward and reverse reactions determine the equilibrium distribution of the reactants and products. Consider one of the author's favorite elementary reversible reactions:

$$A + B \rightleftharpoons C + D \tag{3.25}$$

The *elementary* rate law describing the rate of the forward reaction for A (to be discussed in the next chapter) is

$$r_A = -k_A c_A c_B \tag{3.26}$$

where k_A is the forward reaction velocity constant for the forward reaction based on reactant A. For the reverse reaction

$$r'_A = -k'_A c_C c_D \tag{3.27}$$

where k'_A is the reaction velocity constant for the reverse reaction. At equilibrium, the rate of production of A is just equal to the rate of consumption or disappearance of A so that, at equilibrium, the rates of production and consumption are equal, i.e.,

$$r_A = r'_A \tag{3.28}$$

Combining Equations (3.26), (3.27), and (3.28) gives

$$k_A c_A c_B = k'_A c_C c_D \tag{3.29}$$

or

$$\frac{k}{k'} = \frac{c_C c_D}{c_A c_B} \tag{3.30}$$

The ratio of these rate constants is defined as the *chemical reaction equilibrium* (CRE) constant K based on concentration, i.e.,

$$K = K_c = \frac{k}{k'} = \frac{c_C c_D}{c_A c_B} \tag{3.31}$$

Both k and K will receive additional treatment in the next two chapters.

The reader is reminded that all real and naturally occurring reactions are reversible. A *reversible* reaction is one in which the products also react to form the reactants. Unlike *irreversible* reactions that proceed to the right until completion, reversible reactions achieve an equilibrium state after an infinite period of time. Reactants and products are still present in the system. The net reaction rate is zero at this equilibrium state. Thus, the rate can vary from a very large value (a fast reaction) to one that is essentially zero (at or near equilibrium).

Equations to describe the rate of reaction at the macroscopic level have been developed in terms of meaningful and measurable quantities. The reaction rate is affected not only by the concentration of species in the reacting system but also by the temperature. An increase in temperature will almost always result in an increase in the rate of reaction; in fact, the literature states that, as a general rule, a 10°C increase in reaction temperature will double the reaction velocity constant. However, this is generally no longer regarded as a truism, particularly at elevated temperatures.

The Arrhenius equation (also developed by van't Hoff) relates the reaction velocity constant with temperature. It is given by

$$k = Ae^{-E_a/RT} \tag{3.32}$$

where A = frequency factor constant and is usually assumed to be independent of temperature

R = universal gas constant

E_a = activation energy and is also usually assumed independent of temperature

This equation will be revisited several times later in the text.

ILLUSTRATIVE EXAMPLE 3.16 The reaction velocity constant data provided in Table 3.5 were obtained for the reaction between two inorganic chemicals. Using values of k at 10 and 90°C, calculate the constants of the Arrhenius equation.

Solution. The Arrhenius equation

$$k = Ae^{-E_a/RT} \tag{3.32}$$

may also be written as

$$\ln k = \ln A - E_a/RT$$

When the data are plotted on ln k vs $1/T$ (K) coordinates, an approximate straight line passing through the points ($T = 283$ K, $k = 8.5$ L/gmol · h) and ($T = 363$ K, $k = 68.3$ L/gmol · h) results.

For 10°C, substitution yields

$$\ln 8.5 = \ln A - E_a/(1.987)(283)$$

At 90°C,

$$\ln 68.3 = \ln A - E_a/(1.987)(363)$$

The two equations given above may be solved simultaneously. Subtracting the first equation from the second gives

$$\ln 8.5 - \ln 68.3 = -E_a/562.3 + E_a/721.3$$

Solving for E_a,

$$E_a = 5316 \text{ cal/gmol}$$

Solving for A yields

$$A = 108{,}400 \text{ L/gmol} \cdot \text{h}$$ ■

TABLE 3.5 Reaction Velocity Constant Data

T (°C)	k (L/gmol · h)
0	5.20
20	12.0
40	21.0
60	39.0
80	60.0
100	83.0

ILLUSTRATIVE EXAMPLE 3.17 Employ the results of the previous example and calculate the velocity constant at 75°C.

Solution. Thus, k at 75°C (348 K) is

$$\begin{aligned} k &= 108{,}400 \exp[-5316/(1.987)(348)] \\ &= 49.7 \text{ L/gmol} \cdot \text{h} \end{aligned}$$

Note that linearly interpolating between 60 and 80°C gives an approximate value of 54.8 L/gmol · h.

Accordingly, the Arrhenius equation does yield a straight line of slope $-E_a/R$ and intercept A if $\ln k$ is plotted against $1/T$. Implicit in this statement is the assumption that E_a is constant over the temperature range in question. Despite the fact that E_a generally varies significantly with temperature, the Arrhenius equation has wide applicability in industry. This method of analysis can be used to test a rate law, describe the variation of k with T, and/or evaluate E_a. The numerical value of E_a will depend on the choice and units of the reaction velocity constant. ■

ILLUSTRATIVE EXAMPLE 3.18 The activation energy for the reaction

$$C_2H_4 + H_2 \longrightarrow C_2H_6$$

is 43.2 kcal/gmol. Determine the effect on the reaction velocity constant if the reaction temperature is increased from 120°C to 500°C.

Solution. Apply Equation (3.32) to two conditions.

$$\ln\left(\frac{k_2}{k_1}\right) = \frac{E_a}{R}\left[\frac{T_2 - T_1}{T_2 T_1}\right]$$

Substituting gives

$$\ln\left(\frac{k_2}{k_1}\right) = \frac{432{,}000}{1987}\left[\frac{773 - 393}{(773)(393)}\right]$$

The value of the ratio is

$$\frac{k_2}{k_1} = 6.31 \times 10^{11}$$

How does this compare with a statement in the literature that k increases twofold for every 10°C rise in temperature? For this case,

$$\frac{k_2}{k_1} = (2)^{38} = 2.63 \times 10^{11}$$

The agreement is surprisingly reasonable considering the wide swing in temperature. ■

PROPERTY ESTIMATION

This last section provides information that will allow the practitioner to estimate key physical and chemical properties of materials. Although the scientific community has traditionally resorted to experimental methods to accurately determine the aforementioned properties, that option may very well not be available when dealing with new materials.

Predictive methods, albeit traditional ones, may be the only option available to obtain a first estimate of these properties. It should be noted that significant errors may be involved since extrapolating (or extending) satisfactory estimation procedures at the macroscale level may not always be reasonable. Notwithstanding these concerns, procedures to estimate some of the key physical and chemical properties in chemical kinetics given below are available in the literature.[3–6]

1. Vapor pressure
2. Latent enthalpy
3. Critical properties
4. Viscosity
5. Thermal conductivity
6. Heat capacity

References (3,4) are somewhat complementary, but each provides extensive information on this topic. This includes equations and procedures on several other properties not listed above. The interested reader should check these references for more details.

Is property estimation important? Absolutely. As indicated above, there are times and situations when experimental procedures cannot be implemented. For this scenario, one can turn to theoretical and semitheoretical methods and equations to obtain first estimates of important property information for some reactor studies.

One could argue that the present procedures available to estimate the properties of materials are based on questionable approaches. Nonetheless, the traditional methods available for property estimation either may be applicable or may suggest alternative theoretical approaches.

Finally, it should be noted that the Periodic Law correlates properties of *elements*. Scientists and engineers came to realize that all matter is composed of a rather limited number of basic building blocks, and the desire to discover all the fundamental units/parts became apparent. As data on the properties of elements became available, a pattern in the physical and chemical properties grew discernible, and because the pattern repeated itself in a rather well-organized fashion, it became known as the Periodic Law (see also Chapter 2). This law is one of the finest generalizations of science, and has proven extremely useful both in predicting and correlating physical and chemical properties. The law essentially states that the properties of the chemical elements are not arbitrary, but depend on the structure of the atom and vary systematically, that is, periodically, with the atomic number. The main feature is that elements and compounds exhibit structural, physical, and chemical properties that are remarkably similar, and these properties establish the periodic relations among them.

REFERENCES

1. R. Weast (Editor), "*CRC Handbook of Chemistry and Physics*," 80th edition, CRC Press, Boca Raton, FL, 1999.
2. S. Maron and C. Prutton, "*Principles of Physical Chemistry*," 4th edition, MacMillan, New York City, NY, 1970.
3. R. Perry and D. Green (Editors), "*Perry's Chemical Engineers' Handbook*," 8th edition, McGraw-Hill, New York City, NY, 2008.
4. J. Smith, H. Van Ness, and M. Abbott, "*Introduction to Chemical Engineering Thermodynamics*," 7th edition, McGraw-Hill, New York City, NY, 2005.
5. N. Chopey (Editor), "*Handbook of Chemical Engineering Calculations*," 2nd edition, McGraw-Hill, New York City, NY, 1994.
6. R. Bird, W. Stewart, and E. Lightfoot, "*Transport Phenomena*," 2nd edition, John Wiley and Sons, Hoboken, NJ, 2002.

CHAPTER 4

KINETIC PRINCIPLES

INTRODUCTION

Chemical kinetics is the subject concerned with the study of the rates at which chemical reactions occur and the variables that affect these rates. The objective of this chapter is to develop a working understanding of this subject in order to apply it to chemical reactors. Chemical reactions and rates of reaction are treated from an engineering point of view—in terms of physically measurable quantities. This information is a requisite for the study and analysis of real-world chemical reactor systems.

All chemical reactions take place at a finite rate depending on the conditions, the most important of which are the concentration of reactants and products, temperature, and presence of a catalyst, promoter, or inhibitor. Some reactions are so rapid that they appear to be instantaneous, e.g., combustion, whereas others are so slow at ordinary temperatures, e.g., the combination of two gases such as nitrogen and oxygen, that no detectable change would be observed in time. Between these two extremes are many processes involving both inorganic and organic reactions.

Specific topics covered in this chapter include:

Reaction Rates

Rate vs Equilibrium Considerations

Representation of Rate Expressions

Solutions to Rate Expressions

Reaction Rate Theories

The first section on reaction rates covers a broad spectrum of related topics. The definition of the rate of reaction, the reaction velocity constant, and the order of reaction are presented. The variables affecting the rate are also discussed. This is followed by a section entitled "Rate vs Equilibrium Considerations." The equations for many different types of reactions are developed in the third section. Particular emphasis is placed on notation, nomenclature, and units. The fourth section reviews the solution to numerous rate expressions. The chapter concludes with a section on the present day theories on reaction rates. Although many of the examples are presented in terms of species A, B, C, and so on, real systems are examined throughout the book in later chapters.

Chemical Reactor Analysis and Applications for the Practicing Engineer. By Louis Theodore

REACTION RATES

As noted earlier, chemical kinetics involves the study of reaction rates and the variables that affect these rates. It is a topic that is critical for the analysis of reacting systems. The rate of a chemical reaction can be described in any of several different ways. The most commonly used definition involves the time rate of change in the amount of one of the components participating in the reaction; this rate is also based on some arbitrary factor related to the reacting system's size or geometry, such as volume, mass, or interfacial area. However, for *homogeneous* reactions, all reactants are present in a *single* phase. Some definitions for homogeneous reactions also require that the products appear in the same phase, which may be gaseous, liquid, or solid.

A *consistent* and *correct* definition of the rate of a reaction is essential before meaningful kinetic and reactor applications can be discussed. The definitions found in most chemical kinetics and reactor kinetic texts, as well as the literature in these areas, often do not fit this description. As noted above, the rate of a chemical reaction can be described in any of several different ways. Use of the following definition is most convenient from an engineer's point of view. The rate of a chemical reaction is defined as the time rate of change in the quantity of a particular species (say A) participating in a reaction divided by a factor that characterizes the reacting system's geometry. The choice of this factor is also a matter of convenience. For homogeneous media, the factor is the volume of the reacting system. This definition is chosen for purposes of analysis in this chapter; furthermore, this volume term is often assumed to be constant. In equation form, then

$$r_A = \frac{1}{V}\frac{dN_A}{dt}; \quad \text{mole basis} \tag{4.1}$$

where r_A = rate of reaction for A

N_A = moles of A at time t

t = time

V = volume of reacting system (and *not* necessarily the volume of the reactor)

Note that attention is usually focused on rates based on moles rather than mass. If the volume term is constant, one may write Equation (4.1) as

$$\begin{aligned} r_A &= \frac{d(N_A/V)}{dt} \\ &= \frac{dC_A}{dt} \end{aligned} \tag{4.2}$$

where C_A = molar concentration of A.

Equation (4.2) states that the reaction rate is equal to rate of change of concentration with respect to time—all in consistent units. Rates expressed using concentration changes almost always require the assumption of constant volume. The units of the rate become $\text{lbmol/h} \cdot \text{ft}^3$ and $\text{gmol/s} \cdot \text{L}$ in the engineering and metric systems, respectively. For fluid–fluid surface reactions, the geometric factor is usually the interfacial area. The units of the rate of reaction become lbmol/h · ft^2. For fluid–solid reaction systems, the factor is often the mass of the solid. For example, in gas phase catalytic reactions, the factor is the mass of catalyst. The units of the rate are then lbmol/h · lbcat. A more detailed discussion of this and heterogeneous reactions can be found in Part IV.

In line with the above definition for the rate of reaction, the rate is positive if species A is being formed or produced since C_A increases with time. The rate is negative if A is reacting or disappearing due to the reaction because C_A decreases with time. The rate is zero if the system is at equilibrium.

One may now proceed to develop an equation expressing the rate in terms of measurable and/or desirable quantities. Based on experimental evidence, the rate of reaction is a function of:

1. concentration of the components existing in the reaction mixture (this includes reacting and inert species);
2. temperature;
3. pressure;
4. catalyst variables.

This may be put in equation form:

$$r_A = r_A(C_i, P, T, \text{catalyst variables})$$

or simply

$$r_A = \pm k_A f(C_i) \tag{4.3}$$

where k_A incorporates all variables other than concentration. The $\pm$ notation is included to account for the reaction or formation of A. One may think of k_A as a constant of proportionality. It was defined in the previous chapter as the specific reaction rate or more commonly the reaction velocity constant. It is a "constant" which is *independent* of concentration but *dependent* on the other variables. This approach has, in a sense, isolated one of the variables.

The reaction velocity constant, like the rate of reaction, *must* refer to one of the species in the reacting system. However, k almost always is based on the same species as the rate of reaction. This term is very definitely influenced by temperature and catalyst activity. (The subject of catalysis is considered in Part IV.) For the present, one may assume k is solely a function of temperature. Thus,

$$r_A = \pm k_A(T)\, f(C_i) \tag{4.4}$$

The functional relationship for the molar concentration, $f(C_i)$, is usually (but not always) obtained from the stoichiometric reaction equation, which assumes power law/elementary law kinetics to apply. Consider, for example, the author's favorite reaction

$$aA + bB \longrightarrow cC + dD \tag{4.5}$$

where the notation → represents an irreversible reaction; that is, if stoichiometric amounts of A and B are initially present, the reaction will proceed to the right until all A and B have reacted (disappeared) and C and D have been formed. The rate of the above elementary reaction is given by

$$r_A = -k_A C_A^a C_B^b \tag{4.6}$$

where the negative sign is introduced to account for the disappearance of A. The product concentrations do not affect the rate. The term $f(C_i)$ is then simply obtained by multiplying the molar concentrations of the reactants raised to powers of their respective stoichiometric coefficients.

The order of the above reaction with respect to a particular species is given by the exponent of that concentration term appearing in the rate expression. The above reaction is, therefore, of order a with respect to A, and of order b with respect to B. The overall order n, usually referred to as the order, is the sum of the individual orders, i.e.,

$$n = a + b \tag{4.7}$$

With regard to Equation (4.5), one may also express the rate of reaction in the following forms:

$$\begin{aligned} r_B &= -k_B C_A^a C_B^b \\ r_C &= +k_C C_A^a C_B^b \\ r_D &= +k_D C_A^a C_B^b \end{aligned} \tag{4.8}$$

where

$$\begin{aligned} r_B &= \frac{dC_B}{dt} \\ r_C &= \frac{dC_C}{dt} \\ r_D &= \frac{dC_D}{dt} \end{aligned} \tag{4.9}$$

Due to stoichiometry, one may write

$$\begin{aligned} r_A &= \left(\frac{a}{b}\right) r_B \\ r_A &= -\left(\frac{a}{c}\right) r_C \\ r_A &= -\left(\frac{a}{d}\right) r_D \end{aligned} \tag{4.10}$$

Finally, it is noted

$$\frac{k_A}{k_B} = \frac{a}{b}$$
$$\frac{k_A}{k_C} = \frac{a}{c} \tag{4.11}$$
$$\frac{k_A}{k_D} = \frac{a}{d}$$

The above analysis has assumed power law kinetics to apply. This is frequently an idealization of more complex kinetics and should therefore be looked upon as a general and first introduction to the rate laws.

All real and naturally occurring reactions are reversible. A reversible reaction is one in which products react to form reactants. Unlike irreversible reactions that proceed to the right until completion, reversible reactions achieve an equilibrium state after an infinite period of time. Reactants and products are still present in the system. At this (equilibrium) state, the reaction rate is zero.

Consider the following reversible reaction:

$$aA + bB \rightleftharpoons cC + dD \tag{4.5}$$

where the notation $\rightleftharpoons$ is a reminder that the reaction is reversible: $\rightarrow$ represents the forward reaction contribution to the total or net rate, while $\leftarrow$ represents the contribution of the reverse reaction. The notation $=$ is employed if the reaction system is at equilibrium. The rate of this reaction is given by

$$r_A = \underbrace{-k_A C_A^a C_B^b}_{\text{forward reaction}} + \underbrace{k'_A C_C^c C_D^d}_{\text{reverse reaction}} \tag{4.12}$$

The orders of the forward and reverse reactions are $(a + b)$ and $(c + d)$, respectively. The reverse reaction velocity constant is denoted by k'_A. Another convention adopted is to subscript k with the appropriate sign and order of the reaction. For the reaction

$$A \rightleftharpoons 2B$$

k_1 would represent the forward constant, while k_{-2} would represent the reverse constant. This convention suffers in that the species is not noted in k.

For equilibrium systems, the forward reaction rate is equal to the reverse rate. For Equation (4.12), note that

$$r_A = 0$$

so that

$$k_A C_A^a C_B^b = k'_A C_C^c C_D^d$$

If an equilibrium constant K (see previous chapter) is defined by

$$K = \frac{k_A}{k'_A} \tag{4.13}$$

then

$$K = \frac{C_C^c C_D^d}{C_A^a C_B^b} \tag{4.14}$$

In accordance with Equation (4.13), K is equal to the ratio of the forward to reverse reaction velocity constants. Thus, if k_A is given and K is known, then k'_A can be calculated. Note that K, expressed in terms of concentration, is *not* necessarily the true equilibrium constant.[1]

Gas phase reactions often have rates expressed in a different form. For the n-th order irreversible reaction

$$aA + bB \longrightarrow \text{products} \tag{4.15}$$

where $n = a + b$, it was previously stated that if the reaction is elementary

$$r_A = \frac{dC_A}{dt} = -k_A C_A^a C_B^b \tag{4.6}$$

This equation may also be written in terms of the partial pressures of A and B. Assuming ideal gas conditions permits the following substitution:

$$C_i = \frac{p_i}{RT} \tag{4.16}$$

where p_i = partial pressure of component i
R = universal gas constant
T = temperature

Substituting Equation (4.16) into the right-hand side (RHS) of Equation (4.6) gives

$$r_A = \frac{dC_A}{dt} = -k_{p_A} p_A^a p_B^b \tag{4.17}$$

where k_{p_A} = reaction velocity constant of A based on partial pressure. If one further substitutes Equation (4.16) into the left-hand side (LHS) of Equation (4.6), then

$$r_A^* = \frac{dp_A}{dt} = -k_{p_A}^* p_A^a p_B^b \tag{4.18}$$

where $k_{p_A}^*$ = reaction velocity constant of A based on r_A^*. One can easily show that the relationships among the above three reaction velocity constants are given by

$$\begin{aligned} k_{p_i} &= \frac{k_i}{(RT)^n} \\ k_{p_i}^* &= \frac{k_i}{(RT)^{n-1}} \\ k_{p_i} &= \frac{k_{p_i}^*}{RT} \end{aligned} \tag{4.19}$$

ILLUSTRATIVE EXAMPLE 4.1 The reaction rate constant for the first order elementary reaction $A \rightarrow$ products is 0.0037 $(s)^{-1}$. Calculate the reaction rate, in both units of gmol, L, s, and lbmol, ft^3, h if the concentration of A is equal to 0.21 gmol/L.

Solution. For a first order reaction, then

$$r_A = -k_A C_A$$

where $k_A = 0.0037\ (s)^{-1}$
$C_A = 0.21$ gmol/L at time t.

Therefore,

$$\begin{aligned} r_A &= -0.0037 \times 0.21 \\ &= -7.75 \times 10^{-4}\ \text{gmol/L} \cdot \text{s at time } t. \end{aligned}$$

The minus (−) sign indicates that A is disappearing. To convert to lbmol/h · ft^3:

$$\begin{aligned} r_A &= (-7.75 \times 10^{-4})(28.3)(3600/454) \\ &= -0.173\ \text{lbmol/ft}^3 \cdot \text{h} \end{aligned}$$

■

ILLUSTRATIVE EXAMPLE 4.2 The rate of an elementary second order combustion reaction

$$\text{organic} + \text{oxygen} \longrightarrow \text{products}$$

is 0.001 gmol/s · L for concentrations of organic and oxygen equal to 0.2 gmol/L. What is the value of k in both units of gmol, L, s, and lbmol, ft^3, h?

Solution. The rate of this elementary reaction is given by

$$r_A = -k_A C_A C_B$$

Since A is reacting

$$r_A = -0.001 \text{ gmol/L} \cdot \text{s}$$

and

$$\begin{aligned} k_A &= \frac{0.001}{(0.2)(0.2)} \\ &= 0.025 \text{ L/gmol} \cdot \text{s} \end{aligned}$$

To convert to units of lbmol, ft^3, h:

$$\begin{aligned} k_A &= 0.025 \text{ L/(gmol)(s)(3600 s/h)(1 ft}^3\text{/28.3 L)(454 gmol/lbmol)} \\ &= 1440 \text{ ft}^3\text{/lbmol} \cdot \text{h} \end{aligned}$$ ■

Remember, that if the reaction involves a number of steps in series, it is the slowest step of the series that exerts the greatest influence on the reaction rate and is said to be the controlling step. The problem is further magnified since it is often difficult to determine which variables can affect each of these steps. This information is required in order to extrapolate laboratory data in order to scale-up and predict the performance of industrial equipment.

RATE vs EQUILIBRIUM CONSIDERATIONS

With regard to chemical reactions, there are two important questions that are of concern to the engineer:

1. How *far* will the reaction go?
2. How *fast* will the reaction go?

Chemical thermodynamics provides the answer to the first question; however, it provides nothing about the second.(1)

To illustrate the difference and importance of both of the above questions in an engineering analysis of a chemical reaction, consider the following process.(2,3) Substance A, which costs 1 cent/ton, can be converted to B, which costs \$1 million/lb, by the reaction $A \rightleftharpoons B$. Chemical thermodynamics will provide information on the maximum amount of B that can be formed. If 99.99% of A can be converted to B, the reaction would then appear to be economically feasible, from a *thermodynamic* point of view. However, a *kinetic* analysis might indicate that the reaction is so slow that, for all practical purposes, its rate is vanishingly small. For example, it might take 10^6 years to obtain a $10^{-6}\%$ conversion of A. The reaction is then economically unfeasible. Thus, it can be seen that both equilibrium and kinetic effects must often be considered in an overall engineering analysis of a chemical reaction.

Equilibrium and rate are therefore both important factors to be considered in the design and prediction of the performance of equipment employed for chemical reactions. The rate at which a reaction proceeds will depend on the departure from equilibrium, with the rate at which equilibrium is established essentially dependent on a host of factors. As noted, this rate process ceases upon the attainment of equilibrium.

REPRESENTATION OF RATE EXPRESSIONS

The results of the previous section are now applied and different kinetic systems examined. Stoichiometric equations are written for irreversible, reversible, simultaneous, and consecutive reactions. Each is followed by the describing rate equation. The next section contains a detailed analysis of the different methods of solution for several of these kinetic systems. The units and notations of the various reaction velocity constants employed are carefully reviewed at the conclusion of this section.

Some of the more common and *elementary* chemical reactions are presented below for constant volume systems.

1. Zero order, irreversible reaction.

$$A \longrightarrow \text{products}$$

$$r_A = \frac{dC_A}{dt} = -k_A \tag{4.20}$$

This is an idealized reaction system. It has been shown that all real "zero" order reactions can only be represented in this form at high concentrations.

2. First order, irreversible reaction.

$$A \longrightarrow \text{products}$$

$$r_A = \frac{dC_A}{dt} = -k_A C_A \tag{4.21}$$

3. Second order, irreversible reaction.

(a) $A + B \longrightarrow$ products

$$r_A = \frac{dC_A}{dt} = -k_A C_A C_B \tag{4.22}$$

(b) $2A \longrightarrow$ products

$$r_A = \frac{dC_A}{dt} = -k_A C_A^2 \tag{4.23}$$

4. Third order, irreversible reaction.

(a) $3A \longrightarrow$ products

$$r_A = \frac{dC_A}{dt} = -k_A C_A^3 \tag{4.24}$$

(b) $2A + B \longrightarrow$ products

$$r_A = \frac{dC_A}{dt} = -k_A C_A^2 C_B \tag{4.25}$$

(c) $A + B + C \longrightarrow$ products

$$r_A = \frac{dC_A}{dt} = -k_A C_A C_B C_C \tag{4.26}$$

5. Fractional or higher order, irreversible reaction.

$$nA \longrightarrow \text{products}$$

$$r_A = \frac{dC_A}{dt} = -k_A C_A^n \tag{4.27}$$

For many complex reactions, the order is often a large integer or a fraction.

6. First order, reversible reaction.

$$A \leftrightharpoons B$$

$$\begin{aligned} r_A = \frac{dC_A}{dt} &= -k_A C_A + k'_A C_B \\ &= -k_A\left(C_A + \frac{C_B}{K_A}\right) \end{aligned} \tag{4.28}$$

where $K_A = \dfrac{k_A}{k'_A}$

7. Second order, reversible reaction.

(a) $A \leftrightharpoons 2B$

$$\begin{aligned} r_A = \frac{dC_A}{dt} &= -k_A C_A + k'_A C_B^2 \\ &= -k_A\left(C_A - \frac{C_B^2}{K_A}\right) \end{aligned} \tag{4.29}$$

(b) $A \leftrightharpoons B + C$

$$r_A = \frac{dC_A}{dt} = -k_A C_A + k'_A C_B C_C = -k_A\left(C_A - \frac{C_B C_C}{K_A}\right) \tag{4.30}$$

(c) $A + B \leftrightharpoons C$

$$r_A = \frac{dC_A}{dt} = -k_A C_A C_B + k'_A C_C = -k_A\left(C_A C_B - \frac{C_C}{K_A}\right) \tag{4.31}$$

(d) $2A \leftrightharpoons B$

$$r_A = \frac{dC_A}{dt} = -k_A C_A^2 + k_A C_B = -k_A\left(C_A^2 - \frac{C_B}{K_A}\right) \tag{4.32}$$

(e) $A + B \leftrightharpoons C + D$

$$\begin{aligned} r_A = \frac{dC_A}{dt} &= -k_A C_A C_B + k'_A C_C C_D \\ &= -k_A\left(C_A C_B - \frac{C_C C_D}{K_A}\right) \end{aligned} \tag{4.33}$$

8. Similar equations can be written for third and higher order reversible reactions.

9. Simultaneous irreversible reaction.

$$A \xrightarrow{k_{A_1}} \text{products}$$

$$A + B \xrightarrow{k_{A_2}} \text{products}$$

$$3A \xrightarrow{k_{A_3}} \text{products}$$

$$r_A = \frac{dC_A}{dt} = -k_{A_1} C_{A_1} - k_{A_2} C_A C_B - k_{A_3} C_A^3 \tag{4.34}$$

$$r_B = \frac{dC_B}{dt} = k_{A_2} C_A C_B \tag{4.35}$$

10. Consecutive irreversible reactions.

(a) Two first order reactions.

$$A \xrightarrow{k_A} B$$

$$B \xrightarrow{k_B} C$$

$$r_A = \frac{dC_A}{dt} = -k_A C_A \tag{4.36}$$

$$r_B = \frac{dC_B}{dt} = +k_A C_A - k_B C_B \tag{4.37}$$

$$r_C = \frac{dC_C}{dt} = k_B C_B \tag{4.38}$$

(b) n first order reactions.

$$A \xrightarrow{k_A} B$$

$$B \xrightarrow{k_B} C$$

$$C \xrightarrow{k_C} D$$

$$D \xrightarrow{k_D} E$$

$$r_A = \frac{dC_A}{dt} = -k_A C_A \tag{4.39}$$

$$r_B = \frac{dC_B}{dt} = k_A C_A - k_B C_B \tag{4.40}$$

$$r_C = \frac{dC_C}{dt} = k_B C_B - k_C C_C \tag{4.41}$$

$$r_D = \frac{dC_D}{dt} = k_C C_C - k_D C_D \tag{4.42}$$

$$r_i = \frac{dC_i}{dt} = k_{i-1} C_{i-1} - k_i C_i \tag{4.43}$$

11. Higher order consecutive irreversible reactions.

$$2A \xrightarrow{k_A} B$$

$$2B \xrightarrow{k_B} C$$

$$r_A = -k_A C_A^2 \tag{4.44}$$

$$r_B = k_A C_A^2 - k_B C_B^2 \tag{4.45}$$

12. Consecutive and simultaneous irreversible reaction.

$$A + B \xrightarrow{k_{A_1}} C + F$$

$$A + C \xrightarrow{k_{A_2}} D + F$$

$$r_A = -k_{A_1} C_A C_B - k_{A_2} C_A C_C \tag{4.46}$$

$$r_B = -k_{A_1} C_A C_B \tag{4.47}$$

$$r_C = k_{A_1} C_A C_B - k_{A_2} C_A C_C \tag{4.48}$$

$$r_D = r_F = k_{A_2} C_A C_C \tag{4.49}$$

A chemical (stoichiometric) equation (see next chapter) provides information on the proportions in which species eventually react with each other. However, rarely does a reaction take place in the manner implied by such an equation. The actual reaction rate equation is usually more complex, as shown in the next chapter, and in Part III.

ILLUSTRATIVE EXAMPLE 4.3 Consider the reaction:

$$3A + 2B + (1/2)C \longrightarrow \text{products}$$

The overall order of the above reaction, as written, is

A. 11.0
B. 5.5
C. 4.5
D. 10.5

Solution. Determine the order of the reaction with respect to A, n_A.

$$n_A = 3$$

Determine the order of the reaction with respect to B, n_B.

$$n_B = 2$$

Determine the order of the reaction with respect to C, n_C.

$$n_C = 0.5$$

Calculate the overall order. Apply Equation (4.7),

$$\begin{aligned} n &= n_A + n_B + n_C \\ &= 3 + 2 + 0.5 \\ &= 5.5 \end{aligned}$$

The correct answer is therefore (B). ■

ILLUSTRATIVE EXAMPLE 4.4 Assuming power law (elementary reaction rate) kinetics to apply, write the rate equation for the following simultaneous, irreversible reaction:

$$A \xrightarrow{k_{A_1}} \text{products}$$

$$A + B \xrightarrow{k_{A_2}} \text{products}$$

$$3A \xrightarrow{k_{A_3}} \text{products}$$

A. $r_A = -k_A C_A + k'_A C_B C_C$
B. $r_A = -k_{A_1} C_A - k_{A_2} C_A C_B - k_{A_3} C_A^3$
C. $r_A = +k_{A_1} C_A + k_{A_2} C_A C_B - k_{A_3} C_A^3$
D. $r_A = -k_{A_1} C_A - k_{A_2} C_A C_B + k_{A_3} C_A^3$

Solution. Write the rate equation for the first reaction.

$$r_{A_1} = -k_{A_1} C_A$$

Write the rate equation for the second reaction.

$$r_{A_2} = -k_{A_2} C_A C_B$$

Write the rate equation for the third reaction.

$$r_{A_3} = -k_{A_3} C_A^3$$

Write the rate equation for the simultaneous set of reaction.

$$r_A = r_{A_1} + r_{A_2} + r_{A_3} = -k_{A_1} C_A - k_{A_2} C_A C_B - k_{A_3} C_A^3$$

The correct answer is therefore (B).

The reader should note that in this problem,

$$r_B = -k_{A_2} C_A C_B$$

■

SOLUTIONS TO RATE EXPRESSIONS

Two of the more common conversion and conversion-related terms will now be discussed in this section. The symbol N is used to represent the number of moles of a species at time t; at initial conditions ($t = 0$), a subscript, o, can be included with N. In line with the definition of conversion the symbol α is employed to represent the change in the number of moles of a species due to chemical reaction. The term X is defined as the conversion based on the initial number of moles of a reactant, i.e.,

$$X = \frac{\text{moles of reactant that have reacted or disappeared}}{\text{initial number of moles of reactant in reacting system}} \tag{4.50}$$

There are numerous conversion variables that can be employed in developing solutions to rate equations. These as well as other conversion-related terms will be addressed in the next chapter.

Cases 2 (Equation 4.21) and 3a (Equation 4.22) in the previous section are reviewed in order to present analytical solutions to the rate equation using α and X as the conversion variables. Several other rate expressions are also examined—some of which appeared in the previous section.

The following problem-solving approach is employed. The initial moles or concentration of the reacting species is noted above the stoichiometric equation. The moles or concentration at time t is placed below the symbols.

Case 2: If the reaction

$$A \longrightarrow B$$

is to be solved in terms of concentration, one would write

$$\begin{array}{ccc} C_{A_0} & & 0 \\ A & \longrightarrow & B \\ C_A & & C_B \end{array} \tag{4.51}$$

From a material balance,

$$C_B = C_{A_0} - C_A \tag{4.52}$$

The rate equation is

$$r_A = \frac{dC_A}{dt} = -k_A C_A \tag{4.53}$$

subject to the IC (initial condition) $C_A = C_{A_0}$, at $t = 0$. The solution to Equation (4.53) is

$$C_A = C_{A_0} e^{-k_A t} \tag{4.54}$$

One may alternately write

$$\begin{array}{ccc} N_{A_0} & & 0 \\ A & \longrightarrow & B \\ N_A & & N_B \end{array}$$

where

$$N_B = N_{A_0} - N_A \tag{4.55}$$

The rate equation becomes

$$r_A = \frac{1}{V}\frac{dN_A}{dt} = \frac{k_A N_A}{V}$$

or

$$\frac{dN_A}{dt} = -k_A N_A \tag{4.56}$$

The IC is

$$N_A = N_{A_0} \quad \text{at} \quad t = 0 \tag{4.57}$$

The solution to Equation (4.56) is of the same form as Equation (4.54)

$$N_A = N_{A_0} e^{-k_A t} \tag{4.58}$$

If one substitutes

$$N_A = N_{A_0} - \alpha_A \tag{4.59}$$

$$dN_A = -d\alpha_A \tag{4.60}$$

in Equation (4.56), then

$$\frac{d\alpha_A}{dt} = k_A(N_{A_0} - \alpha_A) \tag{4.61}$$

with IC $\alpha_A = 0$ at $t = 0$.

The solution to Equation (4.61) is

$$\frac{N_{A_0}}{N_{A_0} - \alpha_A} = e^{k_A t} \tag{4.62}$$

If one uses the conversion variable X_A, Equation (4.51) is written

$$\begin{array}{ccc} N_{A_0} & & 0 \\ A & \longrightarrow & B \\ N_A & & N_{A_0} X_A \end{array}$$

where

$$N_A = N_{A_0} - N_{A_0}X_A$$
$$= N_{A_0}(1 - X_A) \tag{4.63}$$

After noting

$$dN_A = -N_{A_0}dX_A \tag{4.64}$$

Equation (4.56) becomes

$$\frac{dX_A}{dt} = k_A(1 - X_A) \tag{4.65}$$

with IC $X_A = 0$ at $t = 0$.
The solution to this equation is

$$\frac{1}{1 - X_A} = e^{k_A t} \tag{4.66}$$

Case 3a: Although more complex than Case 2, this is a more representative chemical reaction system from an analytical solution viewpoint. For this reaction, one first writes:

$$\begin{array}{lllll} C_{A_0} & & C_{B_0} & & \\ A & + & B & \longrightarrow & \text{products} \\ C_A & & C_B & & \end{array} \tag{4.67}$$

The rate equation is

$$\frac{dC_A}{dt} = -k_A C_A C_B \tag{4.68}$$

In order to integrate Equation (4.68), all concentration variables on the RHS must be expressed in terms of the dependent concentration term on the LHS. Note that from a material balance:

$$C_B = C_{B_0} - (C_{A_0} - C_A) \tag{4.69}$$

Equation (4.68) becomes

$$\frac{dC_A}{dt} = -k_A C_A[C_B - (C_{A_0} - C_A)] \tag{4.70}$$

with IC $C_A = C_{A_0}$ at $t = 0$.

The corresponding forms of Equation (4.70) in terms of α_A and X_A are

$$\frac{d\alpha_A}{dt} = \frac{k_A}{V}(N_{A_0} - \alpha_A)(N_{B_0} - \alpha_A) \tag{4.71}$$

and

$$\frac{dX_A}{dt} = \frac{k_A}{V}(1 - X_A)(N_{B_0} - N_{A_0}X_A) \tag{4.72}$$

with IC $\alpha_A, X_A = 0$ at $t = 0$.

The solution to Equation (4.71) is given by

$$t = \frac{V}{k_A(N_{A_0} - N_{B_0})} \ln\left[\frac{N_{B_0}(N_{A_0} - \alpha_A)}{N_{A_0}(N_{B_0} - \alpha_A)}\right] \tag{4.73}$$

Ideal gas phase reactions are often solved by developing the rate equation in terms of partial pressures. If the reaction is carried out at constant volume and a change in the number of moles in the reacting system results, then one can calculate or measure the rate in terms of the total pressure of the system. Keep in mind that partial pressures cannot be measured experimentally. Consider then the general gas phase reaction

$$aA + bB \longrightarrow cC \tag{4.74}$$

where $a + b \neq c$.

The rate equation for Equation (4.74) may be written in the form (see previous section)

$$\frac{dp_A}{dt} = -k_A^* p_A^a p_B^b \tag{4.75}$$

This equation may be solved analytically using the procedure set forth earlier.

ILLUSTRATIVE EXAMPLE 4.5 A first order reaction goes to $1/3$ completion after $1/2$ h. Calculate the degree of completion of this reaction after 1 h.

Solution. For this reaction,

$$A \longrightarrow \text{products}$$

The rate equation is given by

$$\frac{dC_A}{dt} = -k_A C_A$$

$$\frac{dC_A}{C_A} = -k_A dt$$

If this equation is integrated between time 0 and t, then

$$\ln\left(\frac{C_A}{C_{A_0}}\right) = -k_A t$$

Since

$$C_A = 2/3\, C_{A_0} \quad \text{at } t = 1/2\,\text{h},$$

k_A is given by

$$k_A = -\frac{1}{t}\ln\left(\frac{C_A}{C_{A_0}}\right) = -2\,\ln(2/3) = 0.81\,(\text{h})^{-1}.$$

To calculate the degree of completion after 1 h, then

$$\ln\left(\frac{C_A}{C_{A_0}}\right) = 0.81(1) = 0.81$$

$$\frac{C_A}{C_{A_0}} = 0.44$$

The degree of completion is, therefore,

$$1 - \frac{C_A}{C_{A_0}} = 0.56 = 56\%$$ ■

Solutions to several rate expressions are presented below in terms of the conversion variable α. Note, however, that the conversion variable X will be primarily employed throughout the remainder of the text. For case (2), it was shown

$$C_A = C_{A_0}e^{-k_A t} \tag{4.54}$$

since $C_{A_0} - C_A = \alpha$, one may write

$$C_{A_0} - \alpha = C_{A_0}e^{-k_A t} \tag{4.76}$$

or simply

$$\alpha = C_{A_0} - C_{A_0}e^{-k_A t} = C_{A_0}(1 - e^{-k_A t}) \tag{4.77}$$

For the reaction

$$A + B \longrightarrow p; \quad p = \text{product(s)}$$

$$t = \frac{1}{k(C_{A_0} - C_{B_0})}\ln\left[\frac{(C_{B_0})(C_A)}{(C_{A_0})(C_B)}\right] \tag{4.78}$$

For the reaction

$$A + B + C \longrightarrow p$$

$$t = \left(\frac{C_{B_0} - C_{C_0}}{k}\right)\ln\left(\frac{C_A}{C_{A_0}}\right) + (C_{C_0} - C_{A_0})\ln\left(\frac{C_B}{C_{B_0}}\right) + (C_{A_0} - C_{B_0})\ln\left(\frac{C_C}{C_{C_0}}\right) \quad (4.79)$$

For

$$2A + B \longrightarrow p$$

$$t = \frac{1}{k(C_{A_0} - 2C_{B_0})^2}\left[\frac{2\alpha(2C_{B_0} - C_{A_0})}{C_{A_0}(C_{A_0} - 2\alpha)} + \ln\frac{C_{B_0}(C_{A_0} - 2\alpha)}{C_{A_0}(C_{B_0} - \alpha)}\right] \quad (4.80)$$

where $\alpha = C_{B_0} - C_B$.

For

$$3A \longrightarrow p$$

$$t = \frac{1}{2k}\left[\frac{1}{C_A^2} - \frac{1}{C_{A_0}^2}\right] \quad (4.81)$$

For the reversible reaction

$$A \rightleftharpoons B$$

$$t = \frac{\alpha_e}{kC_{A_0}}\ln\left[\frac{\alpha_e}{\alpha_e - (C_{A_0} - C_A)}\right] \quad (4.82)$$

where $\alpha_e = C_{B_e} - C_{B_0} = C_{A_0} - C_{A_e}$ and the subscript e refers to an equilibrium value.

For the reaction

$$A \rightleftharpoons B + C$$

$$t = \frac{\alpha_e}{t(2C_{A_0} - \alpha_e)}\ln\left[\frac{C_{A_0}\alpha_e + (C_{A_0} - C_A)(C_{A_0} - \alpha_e)}{C_{A_0}(\alpha_e - C_{A_0} + C_A)}\right] \quad (4.83)$$

For

$$A + B \rightleftharpoons C$$

$$t = \frac{\alpha_e}{k(C_{A_0} - \alpha_e)}\ln\left[\frac{\alpha_e(C_{A_0}^2 - \alpha_e C_{A_0} + \alpha_e C_A)}{C_{A_0}^2(\alpha_2 - C_{A_0} + C_A)}\right] \quad (4.84)$$

For

$$A + B \rightleftharpoons C + D$$

$$t = \frac{\alpha_e}{2kC_{A_0}(C_{A_0} - \alpha_e)}\ln\left[\frac{(C_{A_0} - C_A)(C_{A_0} - 2\alpha_e) + C_{A_0}\alpha_e}{C_{A_0}(\alpha_e - C_{A_0} + C_A)}\right] \quad (4.85)$$

For the consecutive reaction

$$A \xrightarrow{k_1} B \xrightarrow{k_2} C$$

$$C_A = C_{A_0} e^{-k_1 t}$$

$$C_B = C_{A_0}\left(\frac{k_1}{k_2 - k_1}\right)(e^{-k_1 t} - e^{-k_2 t})$$

$$C_C = C_{A_0} - C_A - C_B \tag{4.86}$$

For the "side" reaction

$$A \xrightarrow{k_1} B$$

$$A \xrightarrow{k_2} C$$

$$t = \frac{1}{(k_1 + k_2)} \ln\left(\frac{C_{A_0}}{C_A}\right) \tag{4.87}$$

Regarding half-lifes (time for half the species of concern to disappear), for the reaction

$$3A \longrightarrow P$$

$$t_{1/2} = \frac{3}{2kC_{A_0}^2} \tag{4.88}$$

For

$$2A \longrightarrow P$$

$$t_{1/2} = \frac{1}{kC_{A_0}} \tag{4.89}$$

The reader should note that, in general,

$$k_A \neq k_B$$

and

$$\frac{k_A}{k_B} = \frac{a}{b}, \quad \text{and so on} \tag{4.90}$$

For the reversible reaction,

$$r_A = -k_A C_A^a C_B^b + k'_A C_C^c C_D^d \tag{4.91}$$

At equilibrium,

$$r_A = 0$$

so that

$$\frac{k_A}{k'_A} = K_A = \frac{C_C^c C_D^d}{C_A^a C_B^b} \tag{4.92}$$

where K_A is the equilibrium constant based on *concentration*.

Once the reaction rate is known, the practicing engineer is in a position to attempt to either design a chemical reactor and/or predict its performance. The analogy is similar to that of a heat exchanger once the overall heat transfer coefficient is known.(4)

Information on the reaction expression/mechanism can often be deduced from the units of the reaction velocity constant. For the elementary reaction

$$nA \xrightarrow{k} \text{products}$$

one can generate the units for k. This information (in engineering units) is provided in Table 4.1.(2)

The reader should also note that (see also Equation 4.19)

$$k_A = (RT)^n k_{p_A} = (RT)^{n-1} k^*_{p_A} \tag{4.93}$$

■

ILLUSTRATIVE EXAMPLE 4.6 Calculate the equilibrium partial pressure of the participating species given the data below:

Operating pressure = 1 atm

Operating temperature = 1950°F

Chemical reaction equilibrium constant based on partial pressures(1), K_p (1 atm, 1950°F) = 0.112 atm$^{-0.5}$

Reaction equation (elementry)

$$SO_2 + 0.5O_2 = SO_3$$

TABLE 4.1 Reaction Velocity Constant Units

Order	$\frac{dC_A}{dt} = -k_A C_A^n$ k_A	$\frac{dC_A}{dt} = -k_{p_A} p_A^n$ k_{p_A}	$\frac{dp p_A}{dt} = -k^*_{p_A} p_A^n$ $k^*_{p_A}$
0	$(\text{lbmol/ft}^3)(\text{s})^{-1}$	$(\text{lbmol/ft}^3)(\text{s})^{-1}$	$(\text{atm})(\text{s})^{-1}$
1	$(\text{s})^{-1}$	$(\text{lbmol/ft}^3)\text{s} \cdot \text{atm}$	$(\text{s})^{-1}$
2	$(\text{ft}^3/\text{lbmol})(\text{s})^{-1}$	$(\text{lbmol/ft}^3)\text{s} \cdot (\text{atm})^2$	$(\text{s} \cdot \text{atm})^{-1}$
3	$(\text{ft}^2/\text{lbmol})^2(\text{s})^{-1}$	$(\text{lbmol/ft}^3)\text{s} \cdot (\text{atm})^3$	$(\text{s})^{-1}(\text{atm})^{-2}$
⋮			
n	$(\text{ft}^3/\text{lbmol})^{n-1}(\text{s})^{-1}$	$(\text{lbmol/ft}^3)/\text{s} \cdot (\text{atm})^n$	$(\text{s})^{-1}(\text{atm})^{-n+1}$

Equilibrium partial pressure of SO_2, $O_2 = 0.0006$ atm, 0.039 atm, respectively
Ideal gas condition apply.

Solution. Write the chemical reaction equilibrium constant based on partial pressures, K_p, for the reaction,

$$SO_2 + 0.5O_2 \rightleftharpoons SO_3$$

For an ideal gas,

$$K_p = p_{SO_3}/(p_{O_2})^{0.5}(p_{SO_2})$$

where p = partial pressure, atm.

Calculate the equilibrium partial pressure of SO_3 from the above equation.

$$\begin{aligned} p_{SO_3} &= (p_{SO_2})(p_{O_2})^{0.5}K_p \\ &= (0.0006)(0.039)^{0.5}(0.112) \\ &= 1.33 \times 10^{-5}\ \text{atm} \end{aligned}$$

Strictly speaking, the SO_2 and O_2 partial pressures employed above should be corrected, since some small quantities are reacted to form SO_3. However, this iterative calculation is hard to justify based on engineering grounds with such a small quantity of SO_3 produced.[(1)]

Determine the ppm of the SO_3 in the gas assuming the system is at atmospheric pressure.

$$\begin{aligned} \text{ppm of } SO_3 &= (p_{SO_3})(10^6) \\ &= 1.33 \times 10^{-5}(10^6) \\ &= 13.3\ \text{ppm} \end{aligned}$$

■

REACTION RATE THEORIES

Equations to describe the rate of reaction at the macroscopic level have been developed in terms of meaningful and measurable quantities. Reaction rate theory *attempts* to provide some foundation from basic principles for these equations. It has, in a few isolated cases, provided information on the controlling mechanism for the rate of reaction. But keep in mind that because the engineer's concern is not with a detailed description of the reaction process at the molecular level, this approach has only rarely been used in industry. A satisfactory rigorous approach to the evaluation of reaction velocity constants from basic principles has yet to be developed. At this time, industry still relies on the procedures set forth in the last section to provide information on reactions for which data (in the form of rate equations) are not available.

The development of the Arrhenius equation provides an excellent introduction to the two theories to be discussed later. The reaction rate is affected not only by the concentration of species in the reacting system but also by the temperature. The Arrhenius equation relates the reaction velocity constant with temperature. It is given by (see Chapter 3)

$$k = Ae^{-E_a/RT} \tag{3.32}$$

where A = frequency factor constant and is usually assumed independent of temperature

R = universal gas constant

E_a = activation energy and is also usually assumed independent of temperature.

The development of this equation follows. The van't Hoff equation describes the variation of the equilibrium constant with temperature.

$$\frac{d(\ln K)}{dT} = \frac{\Delta H}{RT^2} \tag{4.94}$$

At equilibrium, however, it has been noted that

$$K = \frac{k}{k'} \tag{4.95}$$

One may, therefore, write

$$\frac{d(\ln k)}{dT} - \frac{d(\ln k')}{dT} = \frac{\Delta H}{RT^2} \tag{4.96}$$

Equation (4.96) may also be written

$$\frac{d(\ln k)}{dT} - \frac{d(\ln k')}{dT} = \frac{E_a}{RT^2} - \frac{E'_a}{RT^2} \tag{4.97}$$

where

$$E_a - E'_a = \Delta H$$

and E_a and E'_a have yet to be defined. If one separates Equation (4.97),

$$\frac{d(\ln k)}{dT} = \frac{E_a}{RT^2} + C$$

$$\frac{d(\ln k')}{dT} = \frac{E'_a}{RT^2} + C \tag{4.98}$$

Arrhenius noted, based on experimental data, that $C = 0$. He concluded

$$\frac{d(\ln k)}{dT} = \frac{E_a}{RT^2} \tag{4.99}$$

$$\frac{d(\ln k')}{dT} = \frac{E'_a}{RT^2} \tag{4.100}$$

where E_a = activation energy for the forward reaction
E'_a = activation energy for the reverse reaction.

The integrated form of Equation (4.99) becomes

$$k = Ae^{-E_a/RT} \tag{4.101}$$

or

$$\ln k = \ln A - \frac{E_a}{R}\left(\frac{1}{T}\right) \tag{4.102}$$

where A has been previously defined as the frequency factor. The term $e^{-E_a/RT}$ is simply the Boltzmann expression and can be shown to be the fraction of molecules in a system possessing energy in excess of E_a. Equation (4.101) indicates that systems with large E_a's will react slower since k is smaller. Accordingly, Equation (4.102) should yield a straight line of slope $-E_a/R$ and intercept A if $\ln k$ is plotted against $(1/T)$ (see Figure 4.1). Implicit in this statement is the assumption that E_a is constant over the temperature range in question.

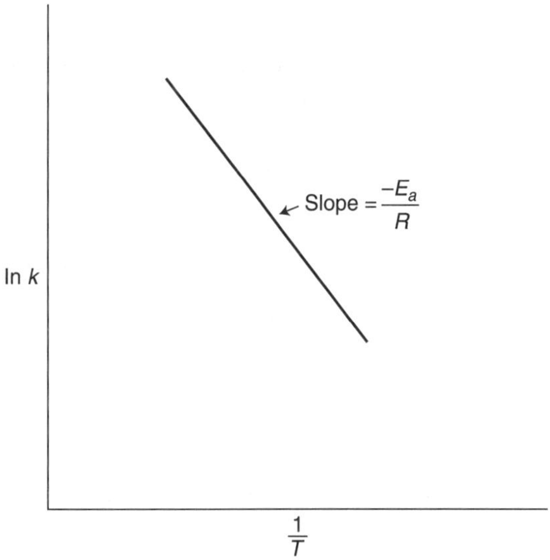

Figure 4.1 Activation energy information from reaction velocity constant-temperature data.

Despite the fact that E_a generally varies significantly with temperature, the Arrhenius equation has wide applicability in industry and in the development of the theory to follow. The above method of analysis can be used to test the law, describe the variation of k with T, and/or evaluate E_a. The numerical value of E_a will depend on the choice and units of the reaction velocity constant (k, k_p, k_p^*, and so on).

From a theoretical viewpoint, a reaction is thought of as taking place by the following two-step process:

$$\text{Reactant } (A) \longrightarrow \text{activated complex} \longrightarrow \text{products} \tag{4.103}$$

The activation energy, once again, is that energy required to activate the reactant(s) to an activated state. It may, therefore, be looked upon as a barrier to the reaction. The ΔH term in Equation (4.96) represents the difference in energy of reactants to products. However, theory assumes that a minimum energy level must be possessed by the reactants before reaction can take place. The energy level of each reactant molecule is distributed in such a manner that the fraction (of the total) of molecules possessing energy in excess of E_a is given by $e^{-E_a/RT}$.

The effect of the presence of a catalyst (see Part IV) is to reduce E_a, which in turn increases the number of molecules actively participating in the reaction. The rate of reaction is naturally increased since k is increased.

In *collision* theory, molecules are assumed to react only upon collision. The rate of reaction has been previously expressed in the form

$$R_A = k_A f(C_i) \tag{4.104}$$

Substituting Equation (4.101) into (4.104) gives

$$R_A = Ae^{-E_a/RT} f(C_i) \tag{4.105}$$

where A is viewed as a frequency factor that provides a measure of the rate of the number of molecular collisions. The exponential term represents the fraction of these collisions that results in a chemical reaction to form products. Application of kinetic theory can provide numerical values for A. For example, A is given by

$$2d_A^2\sqrt{\frac{\pi RT}{MW_A}} \tag{4.106}$$

for the reaction

$$2A \longrightarrow p$$

where d_A = distance between the centers of the two molecules (A, A) on collision

MW_A = molecular weight of A.

For the reaction

$$A + B \longrightarrow \text{products}$$

the frequency factor A is

$$A = d_{AB}^2 \left[8\pi RT \left(\frac{MW_A + MW_B}{MW_A MW_B} \right) \right]^{1/2} \tag{4.107}$$

where d_{AB} is the distance between the centers of the two molecules (A, B) on collision. Calculation shows that the numerical value of A is in the range 10^{10} to 10^{12} L/gmol · s. Despite the magnitude of this number, only a very small fraction of the molecular collisions result in reaction.

In order to account for deviations from collision theory, the rate equation is written has been written as

$$r_A = PAe^{-E_a/RT} f(C_i) \tag{4.108}$$

where P is defined as a probability or steric factor to account for the proper orientation of the molecules necessary for reaction to occur on collision. Its numerical value is in the 0.0–1.0 range and is included in the rate equation since the number of collisions is usually less than that given by kinetic theory.

A meaningful presentation of the theory of *absolute* reaction rates requires much background work in quantum and statistical mechanics. Therefore, it is preferable to discuss this theory qualitatively. Emphasis in this theory is placed on the nature and behavior of the activated complex that is hypothesized to exist during reaction. It is assumed that the complex is in thermodynamic equilibrium with the reactant(s). The concentration of the activated complex can therefore be calculated from chemical equilibrium theory. If this be so, the rate of the reaction will be dictated by the absolute energy content of the system. The free energy change for the complex is calculated in the usual manner from the enthalpy and entropy change for the reaction.[(1)] These are estimated from information on the molecular structure of the complex. In this theory, the translational, rotational, and vibrational energies of the participating species in the reaction are expressed in terms of partition functions. This permits a calculation of the equilibrium constant for the activated complex which in turn will yield information on the rate.

Other factors that can influence the rate of reaction are catalysts and radiation. Catalytic reactions are reviewed in Part IV. The reader is referred to advanced chemical kinetics texts for information on radiation-affected reactions.

The practicing engineer should realize that there is incomplete information (that is not understood) on interatomic forces for any but the most simple molecules studied as part of the activated-complex theory. Therefore, it is rarely useful for computing reaction-rate data that is sufficiently satisfactory for engineering work. While these theories possess one value, the actual evaluation of a rate expression usually requires experimental data.

To summarize, collision theory is based on the fact that molecules must collide before they can react. Although the rate of collision has been shown to be high, many reactions proceed slowly indicating that not every collision leads to reaction. Only those molecules possessing energies in excess of a critical amount, referred to earlier as the activation energy, react upon impact, and these "active" molecules may constitute only a small fraction of the whole. Scientists have concluded that the individual molecules differ from each other in their velocities and consequently in their kinetic energies. Based on kinetic theory, the *average* energy of a group of identical molecules is a function of the temperature. However, some molecules possess more than the average and others possess less. This energy distribution can be evaluated by the laws of mathematical probability, a topic beyond the scope of this text.

REFERENCES

1. L. Theodore, F. Ricci, and T. VanVliet, "*Thermodynamics for the Practicing Engineer*," John Wiley & Sons, Hoboken, NJ, 2009.
2. L. Theodore, personal notes, Manhattan College, 1968.
3. J. Reynolds, J. Jeris, and L. Theodore, "*Handbook of Chemical and Environmental Engineering Calculations*," John Wiley & Sons, Hoboken, NJ, 2004.
4. L. Theodore, "*Heat Transfer for the Practicing Engineer*," John Wiley & Sons, Hoboken, NJ, 2011.

CHAPTER 5

STOICHIOMETRY AND CONVERSION VARIABLES

INTRODUCTION

Two subjects of concern to the practicing engineer involved with chemical reactions and/or chemical reactors are stoichiometry and (a topic briefly introduced last chapter) conversion variables. The former topic is primarily concerned with the "balancing" of the chemical reaction of concern. The latter topic provides information on the extent of the reaction, i.e., the degree to which the reactant has been converted in the product formed. Interestingly, the two topics are interrelated, and to a certain degree, complement each other. It is for this reason that both are discussed in this chapter.

Intertwined with these two topics is equilibrium. If a chemical reaction is conducted in which reactants are converted to products, the products will be formed at a rate governed (in part) by the concentration of the reactants and conditions such as temperature and pressure. Eventually, as the reactants form products and the products react to form reactants, the *net* rate of reaction must equal zero. At this point, equilibrium will have been achieved. How does this relate to chemical reactor analysis? The chemical industry is usually concerned with the attainment of a product or products. [It should also be noted that the environmental industry centers on the destruction (or removal) of a reactant, often referred to as a waste—see Chapter 16.] Chemical reaction equilibrium principles allow the engineer/scientist to determine the end-products of a chemical reaction for a given set of operation conditions and initial reactant(s) if the final state is at equilibrium. However, from the standpoint of obtaining sufficient product(s) of economic value, a final state of equilibrium is almost always undesirable.

Consider the author's favorite reaction equation:

$$aA + bB \longrightarrow cC + dD \tag{5.1}$$

The upper-case letters once again represent chemical species and the lower-case letters represent stoichiometric coefficients. Taking species A as the *basis of calculation*, the reaction expression is divided through by the stoichiometric coefficient of species A in order to arrange the reaction expression in the following form:

$$A + \frac{b}{a}B \longrightarrow \frac{c}{a}C + \frac{d}{a}D \tag{5.2}$$

Chemical Reactor Analysis and Applications for the Practicing Engineer. By Louis Theodore

This essentially places every quantity on a "per mole of A" basis. The aforementioned conversion variable X_A, introduced in Chapter 4, can then be defined as the number of moles of A that have reacted per mole of A fed to the system, i.e.,

$$X_A = \frac{\text{moles of } A \text{ reacted}}{\text{moles of } A \text{ fed}} \tag{5.3}$$

It is this conversion term that receives the bulk of the treatment in this chapter and the remainder of the book.

This chapter serves to introduce the general subject of stoichiometry and conversion variables, which in a very real sense, is an extension of the material discussed in the previous chapter. To simplify the presentation to follow, some of the textual matter and illustrative examples will focus on combustion reactions because of the author's experience in this field. Topics to be reviewed include:

Stoichiometry

Conversion Variables

Volume Correction Factor

Yield and Selectively

It should again be noted that much of this material is required in the treatment of chemical reactors in the remainder of the book.

STOICHIOMETRY

The term *stoichiometry* has come to mean different things to different people. In a loose sense, stoichiometry involves the balancing of an equation for a chemical reaction that provides a quantitative relationship among the reactants and products. In the simplest stoichiometric situation, exact quantities of pure reactants are available, and these quantities react completely to give the desired product(s). In an industrial process, the reactants usually are not pure, one reactant is usually in excess of what is needed for the reaction, and the desired reaction may not go to completion because of a host of other considerations.[1,2]

The complete combustion of pure hydrocarbons yields carbon dioxide and water as the reaction products. Consider the combustion of methane in oxygen:

$$CH_4 + O_2 \longrightarrow CO_2 + H_2$$

In order to balance this reaction, *two* molecules of oxygen are needed. This requires that there be *four* oxygen atoms on the right side of the reaction. This is satisfied by introducing *two* molecules of water as product. The final balanced reaction becomes

$$CH_4 + 2O_2 \longrightarrow CO_2 + 2H_2O$$

Thus, *two* molecules (or moles) of oxygen are required to completely combust *one* molecule (or mole) of methane to yield one molecule (or mole) of carbon dioxide and two molecules (or moles) of water. Note that the numbers of carbon, oxygen, and hydrogen atoms on the right-hand side (RHS) of this reaction are equal to those on the left-hand side (LHS). The reader should verify that the total mass (obtained by multiplying the number of each molecule by its molecular weight and summing) on each side of the reaction is the same. [The term *moles* here may refer to either gram · moles (gmol) or pound · moles (lbmol); it makes no difference.]

This discussion may now be extended to reaction systems that involve the combustion of hydrocarbons and/or organics. To simplify matters, it will be assumed that both the air and feed mixture are dry. Throughout this section, air is assumed to contain 21% O_2 by volume. This is perhaps a bit high by a few hundredths of a percent. The remaining 79% is assumed to be inert and to consist of nitrogen (and a trace of the noble gases). The usual reaction products of hydrocarbons are CO_2 and H_2O. The combustion products of other organics may contain additional compounds; for example, an organic chloride will also produce chlorine and/or hydrochloric acid. Combustion products from certain fuels will yield sulfur dioxide and nitrogen. So-called complete combustion of an organic or a fuel involves conversion of all the elemental carbon to carbon dioxide, hydrogen to water, sulfur to sulfur dioxide, and nitrogen to its elemental form of N_2. Thus, the *theoretical* or *stoichiometric* oxygen described for a combustion reaction is the amount of oxygen required to burn all the carbon to carbon dioxide, all the hydrogen to water, and so on; excess oxygen (or excess air) is the oxygen furnished in excess of the theoretical oxygen required for combustion. Since combustion calculations assume dry air to contain 21% oxygen and 79% nitrogen on a mole or volume basis, 4.76 mol of air consists of 1.0 mol of oxygen and 3.76 mol of nitrogen.

In chemical reactions, atoms are neither generated nor consumed, merely rearranged with different bonding partners. The manipulation of the coefficients of a reaction equation so that the number of atoms of each element on the left is equal to that on the right is referred to as balancing the equation. The most common process of balancing an equation is called *balancing by inspection*. Balancing an equation by inspection can be accomplished by the following steps.

1. Identify all reactants and products and write their correct formulas on the left and right side of the equation, respectively.
2. Begin balancing the equation by selecting suitable coefficients that will make the number of atoms of each element the same on both sides of the equation. One can change only the coefficients but not the subscripts.
3. Start with elements that appear only once on each side of the equation and balance these elements. Next, balance elements that appear in two or more formulas on the same side of the equation.
4. Check the balanced equation to be sure that it conforms to the law of conservation of mass, i.e., the number of atoms of each element is the same on both sides of the equation.

The above may now be extended to actual reactions. When chemicals react, they do so according to a strict proportion. When oxygen and hydrogen combine to form water, the ratio of the amount of oxygen to the amount of hydrogen consumed is always 7.94 by mass and 0.500 by moles. The term *stoichiometry* refers to this phenomenon, which is sometimes called the *chemical law of combining weights*. The reaction equation for the combining the hydrogen and oxygen is

$$2H_2 + O_2 \longrightarrow 2H_2O \tag{5.4}$$

Once an equation is balanced, the small whole number molar ratio that must exist between any two components of the reaction can be determined simply by observation; these are known as *stoichiometric ratios*. There are three such ratios (not counting the reciprocals) in the above reaction. These are

$$2\,\text{mol } H_2 \text{ consumed/mol } O_2 \text{ consumed}$$
$$1\,\text{mol } H_2O \text{ generated/mol } H_2 \text{ consumed}$$
$$2\,\text{mol } H_2O \text{ generated/mol } O_2 \text{ consumed}$$

The unit *mol* represents either the *gmol* or the *lbmol*. Using molecular weights, these stoichiometric ratios (which are molar ratios) may easily be converted to mass ratios. For example, the first ratio above may be converted to a mass ratio by using the molecular weights of H_2 (2.016) and O_2 (31.999) as follows:

$$(2\,\text{gmol } H_2 \text{ consumed})(2.016\,\text{g/gmol}) = 4.032\,\text{g } H_2 \text{ consumed}$$
$$(1\,\text{gmol } O_2 \text{ consumed})(31.999\,\text{g/gmol}) = 31.999\,\text{g } O_2 \text{ consumed}$$

The mass ratio between the hydrogen and oxygen consumed is therefore

$$4.032/31.999 = 0.126\,\text{g } H_2 \text{ consumed/g } O_2 \text{ consumed}$$

Molar and mass ratios are used in material balances to find the amounts or flow rates of components involved in chemical reactions. Multiplying a balanced reaction equation through by a constant does nothing to alter its meaning. There are times, however, when care must be exercised because the solution to the problem depends in the manner the reaction equation is written. This is the case with chemical equilibrium problems and problems involving thermochemical reaction equations.

The terms used to describe a chemical reaction that do not involve stoichiometric ratios of reactants must be carefully defined in order to avoid confusion. If the reactants are not present in formula or stoichiometric ratio, one reactant is said to be *limiting*; the others are said to be *in excess*. Consider the reaction

$$CO + \frac{1}{2}O_2 \longrightarrow CO_2$$

If the starting amounts are 1 mol of CO and 3 mol of oxygen, CO is the *limiting* reactant, with O_2 present in excess. There are 2.5 mol of excess O_2, because only 0.5 mol is required to combine with the CO. Thus there is 400% excess oxygen present. The percentage of excess must be defined in relation to the amount of the reactant necessary to react completely with the limiting reactant. Thus, if for some reason only part of the CO actually reacts, this does not alter the fact that the oxygen is in excess by 400%. However, there are often several possible products. For instance, the reactions

$$C + O_2 \longrightarrow CO_2$$

and

$$C + \frac{1}{2}O_2 \longrightarrow CO$$

can occur simultaneously. In this case, if there are 3 mol of oxygen present per mole of carbon, the oxygen is again in excess. The extent of this excess, however, cannot be definitely fixed. It is customary to choose one product (e.g., the desired one) and specify the excess reactant in terms of this product. For this case, there is 200% excess oxygen for the reaction going to CO_2, and there is 400% excess oxygen for the reaction going to CO. The discussion on excess oxygen can be extended to excess air using the same approach with stoichiometric or theoretical oxygen (or air) defined as 0% excess oxygen (or air).

There are three different types of material balances that may be written when a chemical reaction is involved: the *molecular balance*, the *atomic balance*, and the "*extent of reaction*" *balance*. It is a matter of convenience which of the three types is used.

The *molecular balance* is expressed as either mass or number of moles and takes into account the occurrence of a chemical reaction. The complete balance equation (as per conservation law for mass) is:

$$I + G = O + C + A \tag{5.5}$$

where I = *Input*: the amount (mass or moles) or flow rate (mass or molar) entering the system

G = *Generation*: the amount produced or rate of production by a chemical reaction inside the system

O = *Output*: the amount or flow rate exiting the system

C = *Consumption*: the amount consumed or rate of consumption by a chemical reaction inside the system

A = *Accumulation*: the amount accumulating or rate of accumulation in the system.

Assuming a steady-state continuous reaction, the accumulation term, A, is zero and, for all components involved in the reaction, the balance equation becomes

$$I + G = O + C \tag{5.6}$$

If a total material balance is performed, the above form of the balance equation must be used if the amounts or flow rates are expressed in terms of moles, e.g., lbmol/h or gmol/h, since the total number of moles can change during a chemical reaction. If, however, the amounts or flow rates are given in terms of mass, e.g., kg or lb, the G and C terms may be dropped. Since mass cannot be lost or gained in a chemical reaction,

$$I = O \tag{5.7}$$

In general, however, when a chemical reaction is involved, it is usually more convenient to express amounts and flow rates using moles rather than mass.

A material balance which is based not on the chemicals (or molecules), but rather on the atoms that make up the molecules, is referred to as an *atomic balance*. Since atoms are neither created nor destroyed in a chemical reaction, the G and C terms equal zero and the balance again becomes

$$I = O$$

As an example, consider once again the combination of hydrogen and oxygen to form water:

$$2H_2 + O_2 \longrightarrow 2H_2O$$

As the reaction progresses, O_2 and H_2 molecules (or moles) are consumed while H_2O molecules (or moles) are generated. On the other hand, the number of oxygen atoms (or moles of oxygen atoms) and the number of hydrogen atoms (or moles of hydrogen atoms) do not change. Care must also be taken to distinguish between *molecular* oxygen and *atomic* oxygen. If, in the above reaction, one starts out with 1000 lbmol of O_2 (oxygen molecules), one is also starting out with 2000 lbmol of O (oxygen atoms).

The "*extent of reaction*" *balance* gets its name from the fact that the amounts of the chemicals involved in the reaction are described in terms of how much of a particular reactant has been consumed or how much of a particular product has been generated. As an example, take the formation of ammonia from hydrogen and nitrogen:

$$N_2 + 3H_2 \longrightarrow 2NH_3$$

If one starts with 10 moles of hydrogen and 10 moles of nitrogen, and let z represent the amount of nitrogen consumed by the end of the reaction, the output amounts of all three components are (10-z) for the nitrogen, (10-3z) for the hydrogen, and 2z for the ammonia. The following is a convenient way of representing the "extent of reaction" balance:

$$\begin{array}{lccccc} \text{Input}\rightarrow & 10 & & 10 & & 0 \\ & N_2 & + & 3H_2 & \longrightarrow & 2NH_3 \\ \text{Output}\rightarrow & 10\text{-}z & & 10\text{-}3z & & 2z \end{array} \tag{5.8}$$

Whichever of the three types of balances—molecular, atomic or "extent of reaction"—is used to solve a particular problem is mainly a matter of convenience and ease of solution. In solving some problems, the use of one type of balance yields the result more quickly and easily than using the other types. In many instances, the best procedure is to "mix" the types of balances in the same solution, being careful that the final set of balances is an *independent* set.

ILLUSTRATIVE EXAMPLE 5.1 The reaction equation (not balanced) for the combustion of butane is shown below:

$$C_4H_{10} + O_2 \longrightarrow CO_2 + H_2O$$

Determine the mole ratio of reactants to products.

Solution. A chemical equation provides a variety of qualitative and quantitative information essential for the calculation of the quantity of reactants reacted and products formed in a chemical process. As noted above, the balanced chemical equation must have the same number of atoms of each type in the reactants and products. Thus the balanced equation for butane is

$$C_4H_{10} + \left(\frac{13}{2}\right)O_2 \longrightarrow 4CO_2 + 5H_2O$$

Note that:

Number of carbons in reactants = number of carbons in products = 4

Number of oxygens in reactants = number of oxygens in products = 13

Number of hydrogens in reactants = number of hydrogens in products = 10

Number of moles of reactants is 1 mol C_4H_{10} + 6.5 mol O_2 = 7.5 mol total

Number of moles of products is 4 mol CO_2 + 5 mol H_2O = 9 mol total.

The reader should note that although the number of moles on both sides of the equation do *not* balance, the masses of reactants and products (in line with the conservation law for mass) *must* balance. ■

ILLUSTRATIVE EXAMPLE 5.2 When ethanol (C_2H_5OH) is completely combusted, the products are carbon dioxide and water.

1. Write the balanced reaction equation.
2. If 150 lbmol/h of water is produced, at what rate (molar) is the ethanol combusted?
3. If 2000 kg of the ethanol is combusted, what mass of oxygen is required?

The atomic weights of C, O, and H are 12, 16, and 1, respectively.

Solution

1. The balanced reaction equation is

$$C_2H_5OH + 3O_2 \longrightarrow 2CO_2 + 3H_2O$$

2. The stoichiometric ratio of the C_2H_5OH consumed to the water produced may now be calculated.

$$\text{Ratio} = \frac{1\ \text{lbmol } C_2H_5OH}{3\ \text{lbmol } H_2O \text{ produced}}$$

This result may be used to calculate the amount of C_2H_5OH reacted for part 2.

$$(150\ \text{lbmol/h } H_2O \text{ produced})\left(\frac{1\ \text{lbmol } C_2H_5OH}{3\ \text{lbmol } H_2O \text{ produced}}\right)$$
$$= 50\ \text{lbmol/h } C_2H_5OH \text{ reacted}$$

3. The molar amount (in kgmol) of C_2H_5OH reacted in part 3 is

$$\frac{2000\ \text{kg } C_2H_5OH}{45\ \text{kg/kgmol } C_2H_5OH} = 43.48\ \text{kgmol of } C_2H_5OH$$ ■

ILLUSTRATIVE EXAMPLE 5.3 Consider the following equation:

$$C_3H_8 + 5O_2 \longrightarrow 3CO_2 + 4H_2O$$

Determine the scf (standard cubic feet) of air required for stoichiometric combustion of 1.0 scf propane (C_3H_8). Also determine the scf of flue gas produced.

Solution. As noted earlier, stoichiometric air is the air required to assure complete combustion of a fuel, organic, and/or waste. For complete combustion:

1. no fuel or organic remains;
2. no oxygen is present in the flue gas;
3. carbon has combusted to CO_2, not CO;
4. sulfur has combusted to SO_2, not SO_3.

Excess air (fractional basis) was defined earlier and may be represented in equation form by:

$$\text{EA} = \frac{\text{Air entering} - \text{Stoichiometric air}}{\text{Stoichiometric air}} \tag{5.9}$$

Noting that, for an ideal gas, the number of moles is proportional to the volume, the scf of O_2 required for the complete combustion of 1 scf of propane is 5 scf.

The nitrogen-to-oxygen volume (or mole) ratio in air is $\frac{79}{21}$. Therefore the amount of N_2 in a quantity of air that contains 5.0 scf of O_2 is

$$\text{scf of } N_2 = \left(\frac{79}{21}\right)(5)$$
$$= 18.81\ \text{scf}$$

The stoichiometric amount of air is then

$$\begin{aligned}\text{scf of air} &= \text{scf of } N_2 + \text{scf of } O_2 \\ &= 18.81 + 5.0 \\ &= 23.81 \text{ scf}\end{aligned}$$

Therefore the amount of flue gas produced is

$$\begin{aligned}\text{scf of flue gas} &= \text{scf of } N_2 + \text{scf of } CO_2 + \text{scf of } H_2O \\ &= 18.81 + 3.0 + 4.0 \\ &= 25.81 \text{ scf}\end{aligned}$$

■

ILLUSTRATIVE EXAMPLE 5.4 A reactor produces a hydrocarbon compound. The engineering division of the company has made a decision that it would be worthwhile to combust the compound and recover the heat generated as a makeup heat source for another reactor.

If this hydrocarbon compound contains three atoms of carbon, determine its chemical formula if the flue gas composition on a dry basis formed following combustion is:

CO_2: 7.5%   CO: 1.3% O_2: 8.1% N_2: 83.1%

Solution. Assume a basis 100 mol of dry flue gas. The moles of each component in the dry product gas is then

CO_2	7.5 mol
CO	1.3 mol
O_2	8.1 mol
N_2	83.1 mol
	100.0

Since nitrogen does not react, using the ratio of oxygen to nitrogen in air will provide the amount of oxygen fed:

$$O_{2,\text{fed}} = \left(\frac{21}{79}\right)(83.1) = 22.1 \text{ mol}$$

A balanced equation for the combustion of the hydrocarbon in terms of N moles of hydrocarbon and n hydrogen atoms in the hydrocarbon yields

$$NC_3H_n + 22.1O_2 \longrightarrow 7.5CO_2 + 1.3CO + 8.1O_2 + N(n/2)H_2O$$

The moles of hydrocarbon, N, is obtained by performing an elemental carbon balance:

$$\begin{aligned}3N &= 7.5 + 1.3 \\ N &= 8.8/3 = 2.93\end{aligned}$$

Similarly, the moles of water formed is obtained by performing an elemental oxygen balance:

$$\begin{aligned} 2(22.1) &= 2(7.5) + 1.3 + 2(8.1) + N(n/2) \\ N(n/2) &= 44.2 - 15 - 1.3 - 16.2 \\ &= 11.7 \end{aligned}$$

The number of hydrogen atoms, n, in the hydrocarbon is then

$$\begin{aligned} n &= 2(11.7)/N \\ &= 23.4/2.93 \\ &= 7.99 \approx 8 \end{aligned}$$

Since $n = 8$, the hydrocarbon is C_3H_8, propane. ■

ILLUSTRATIVE EXAMPLE 5.5 A liquid stream contaminated with a pollutant has its pollutant completely removed with a reactor. If the liquid has 600 ppm (parts per million) of pollutant, and it is permissible to have 50 ppm of this pollutant in the discharge stream, what fraction of the liquid can bypass the reactor?

Solution. Using a basis of 1 lb of liquid fed to the reactor, the flow diagram in Figure 5.1 applies.

Note that:

$$\begin{aligned} B &= \text{ fraction of liquid bypassed} \\ 1 - B &= \text{ fraction of liquid treated} \end{aligned}$$

Performing a pollutant balance around point 2 in Figure 5.1 yields

$$(1 - B)(0) + 600B = (50)(1.0)$$

Solving gives

$$B = 0.0833$$

Only 8.33% can bypass the reactor. ■

ILLUSTRATIVE EXAMPLE 5.6 Ethylene oxide is produced by the catalytic oxidation of ethylene with oxygen, $C_2H_4 + \frac{1}{2}O_2 \rightarrow (C_2H_2)_2O$. The feed to the catalyst bed is a 10:1 (volume) ratio of air to ethylene, and the conversion of ethylene is 23%

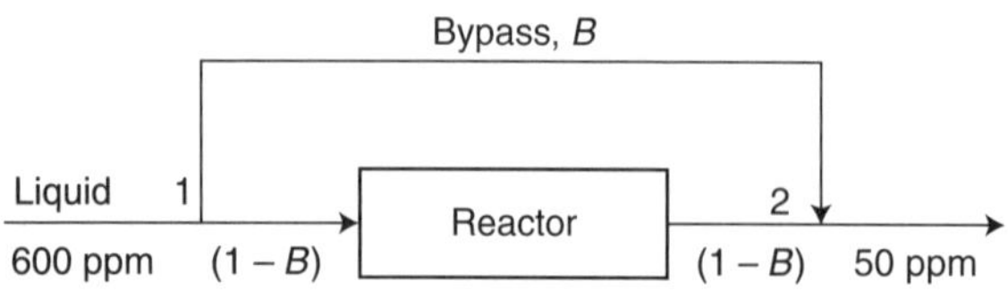

Figure 5.1 Flow diagram for Illustrative Example 5.5.

TABLE 5.1 Data/Calculations for Illustrative Example 5.6

	Into reactor					Out of reactor				
Compound	Moles	C	H	O	N	Moles	C	H	O	N
C_2H_4	1	2	4	–	–	0.77	1.54	3.08	–	–
N_2	7.9	–	–	–	15.8	7.9	–	–	–	15.8
O_2	2.1	–	–	4.2	–	1.99	–	–	3.97	–
CH_2CH_2O	0	–	–	–	–	0.23	0.46	0.92	0.23	–
Total	11.0	2	4	4.2	15.8	10.89	2.00	4.00	4.20	15.8

per pass. The ethylene oxide is selectively removed from the reaction products. The unreacted ethylene must be recycled.

What is the recycle ratio (in mol recycled/mol or total feed), and the total inlet and outlet gas compositions?

Solution. Select as a basis 1.0 lbmol of ethylene feed. Table 5.1 can then be constructed. Calculational details are left as an optional exercise for the reader.

Thus,

Ethylene recycle ratio: $0.77/1 = 0.77$

Inlet composition: 9.09% C_2H_4, 71.8% N_2, 19.1% O_2

Outlet composition: 7.08% C_2H_4, 72.5% N_2, 18.3% O_2, 2.11% CH_2CH_2O

Recycle problems can become more complicated since there can be unknowns in the overall stoichiometry of the process and "local" unknowns involving the reactor and its feed streams. These problems are solved by using two material balances: one an overall balance of the material entering and leaving the reactor complex; the other a once-through, or local, balance about the reactor itself. The latter balance includes recycle streams; the former does not. ■

ILLUSTRATIVE EXAMPLE 5.7 In one step of the Contact Process for the production of sulfuric acid, sulfur dioxide is oxidized to sulfur trioxide at high pressure in the presence of a catalyst:

$$2SO_2 + O_2 \longrightarrow 2SO_3$$

A mixture of 300 lbmol/min of sulfur dioxide and 400 lbmol/min of oxygen are fed to a reactor. The flowrate of the unreacted oxygen leaving the reactor is 300 lbmol/min. Determine the composition (in mol%) of the exiting gas stream by three methods:

1. Molecular balances
2. Atomic balances
3. "Extent of reaction" balances

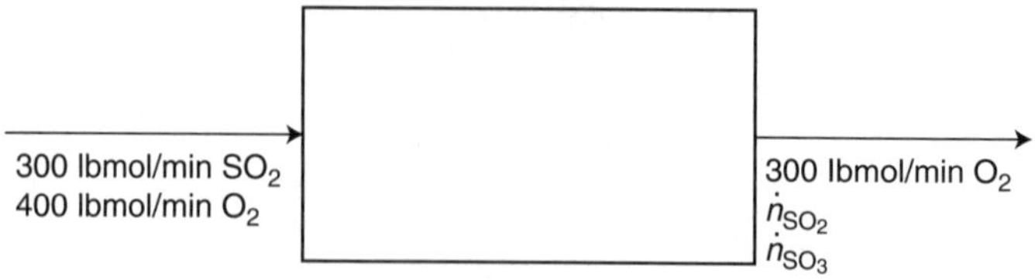

Figure 5.2 Flow diagram for Illustrative Example 5.7.

Solution. In this problem, the reader is asked to employ all three types of material balances. A flow diagram of the process is given in Figure 5.2.

1. Using the molecular balance approach, one obtains

$$\begin{aligned} O_2\text{:}\quad & I = O + C \\ & 400 = 300 + C \\ & C = 100\,\text{lbmol/min } O_2 \text{ consumed} \end{aligned}$$

$$\begin{aligned} SO_2\text{:}\quad & I = O + C \\ & 300 = \dot{n}_{SO_2} + \left(\frac{2}{1}\right)(100) \\ & \dot{n}_{SO_2} = 100\,\text{lbmol/min } SO_2 \text{ out} \end{aligned}$$

$$\begin{aligned} SO_3\text{:}\quad & G = O \\ & \left(\frac{2}{1}\right)(100) = \dot{n}_{SO_3} \\ & \dot{n}_{SO_3} = 200\,\text{lbmol/min } SO_3 \text{ out} \end{aligned}$$

Since the total flowrate of the exiting gas is 600 lbmol/min,

$$\begin{aligned} O_2\text{:}\quad & (300/600)\ (100\%) = 50.0\% \\ SO_2\text{:}\quad & (100/600)\ (100\%) = 16.7\% \\ SO_3\text{:}\quad & (200/600)\ (100\%) = 33.3\% \end{aligned}$$

2. Using the atomic balance approach, one obtains

$$\begin{aligned} S\text{:}\quad & I = O \\ & \left(\frac{1}{1}\right)(300) = \left(\frac{1}{1}\right)\dot{n}_{SO_2} + \left(\frac{1}{1}\right)\dot{n}_{SO_3} \\ & \dot{n}_{SO_2} + \dot{n}_{SO_3} = 300 \end{aligned}$$

O: $I = O$

$$\left(\frac{2}{1}\right)(300) + \left(\frac{2}{1}\right)(400) = \left(\frac{2}{1}\right)(300) + \left(\frac{2}{1}\right)\dot{n}_{SO_2} + \left(\frac{3}{1}\right)\dot{n}_{SO_3}$$

$$2\dot{n}_{SO_2} + 3\dot{n}_{SO_3} = 800$$

Solving these two equations simultaneously yields the flowrates of the SO_2 and SO_3 leaving the reactor.

As before,

$$\dot{n}_{SO_2} = 100\,\text{lbmol/min } SO_2 \text{ out}$$

$$\dot{n}_{SO_3} = 200\,\text{lbmol/min } SO_3 \text{ out}$$

Note: Since there are only two types of atoms, S and O, atomic balances can only provide two equations. This is not enough for a system with three unknowns. The oxygen output rate must be determined by one of the other methods.

3. Using the extent of reaction balances approach gives

Input→	300		400		0
	$2SO_2$	+	O_2	→	$2SO_3$
Output→	$300-z$		$400 - \left(\frac{1}{2}\right)z$		$0+z$

where z is the rate (lbmol/min) of SO_2 reacted.

Since the outlet O_2 flowrate is given as 300 lbmol/min.

$$300 = 400 - \left(\frac{1}{2}\right)z$$

$$z = 200$$

SO_2 outlet flowrate $= 300 - z = 100$ lbmol/min

SO_3 outlet flowrate $= z = 200$ lbmol/min

O_2 outlet flowrate $= 400 - \left(\frac{1}{2}\right)z = 300\,\text{lbmol/min}$ ■

ILLUSTRATIVE EXAMPLE 5.8 Ethylene oxide is produced by the oxidation of ethylene with oxygen-enriched air:

$$C_2H_4 + \frac{1}{2}O_2 \longrightarrow C_2H_4O$$

An undesired side reaction is the oxidation of ethylene to an essentially valueless carbon dioxide:

$$C_2H_4 + 3O_2 \longrightarrow 2CO_2 + 2H_2O$$

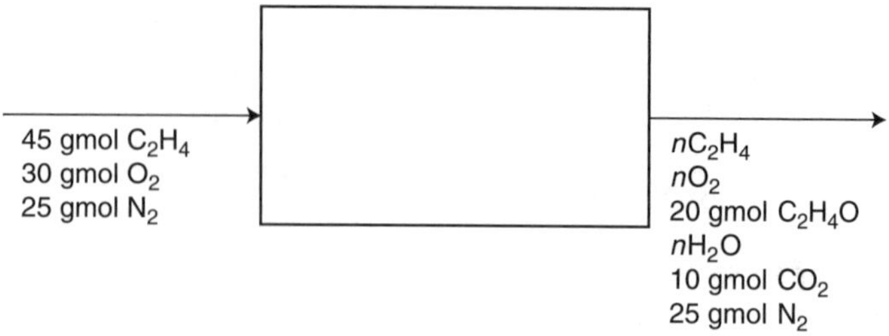

Figure 5.3 Flow diagram for Illustrative Example 5.8.

The feed stream to the ethylene oxide reactor consists of 45% (by mole) C_2H_4, 30% O_2, and 25% N_2. The amounts of ethylene oxide and carbon dioxide in the product stream are 20 gmol and 10 gmol per 100 gmol of feed stream, respectively. Determine the composite of the exiting gas stream.

Solution. Selecting a basis of 100 gmol feed steam, a flow diagram of the process may be generated as shown in Figure 5.3.

One may apply extent of reaction balances to determine expressions for each of the components leaving the reactor in terms of the amount of C_2H_4 converted to C_2H_4O and the amount converted to CO_2.

$$\begin{array}{lcccccc}
\text{Initial}\rightarrow & 45 & & 30 & & 0 & \\
 & C_2H_4 & + & \frac{1}{2}O_2 & \longrightarrow & C_2H_4O & (1) \\
\text{Final}\rightarrow & 45 - y - z & & 30 - \left(\frac{1}{2}\right)y - \left(\frac{3}{1}\right)z & & y &
\end{array}$$

$$\begin{array}{lcccccccc}
\text{Initial}\rightarrow & 45 & & 30 & & 0 & & 0 & \\
 & C_2H_4 & + & 3O_2 & \longrightarrow & 2CO_2 & + & 2H_2O & (2) \\
\text{Final}\rightarrow & 45 - y - z & & 30 - \left(\frac{1}{2}\right)y - \left(\frac{3}{1}\right)z & & 2z & & 2z &
\end{array}$$

where $y =$ amount of C_2H_4 converted to C_2H_4O in reaction (1)
$z =$ amount of C_2H_4 converted to CO_2 in reaction (2)

The amounts of all components of the product gas stream may now be calculated:

$$\text{Amount } C_2H_4O = y = 20\text{ gmol}$$

$$\text{Amount } CO_2 = 2z = 10\text{ gmol} \rightarrow z = 5$$

$$\text{Amount } H_2O = 2z = 2(5) = 10\text{ gmol}$$

$$\text{Amount } C_2H_4 = 45 - y - z = 45 - 20 - 5 = 20\text{ gmol}$$

$$\text{Amount } O_2 = 30 - \left(\frac{1}{2}\right)y - \left(\frac{3}{1}\right)z$$

$$= 30 - \left(\frac{1}{2}\right)20 - \left(\frac{3}{1}\right)5 = 5\,\text{gmol}$$

$$\text{Amount } N_2 = 25\,\text{gmol}$$

$$\text{Total amount of product gas} = 20 + 10 + 10 + 20 + 5 + 25 = 90\,\text{gmol}$$

The mole fractions are therefore:

$$\text{Mole fraction } C_2H_4O = \frac{20}{90} = 0.2222$$

$$\text{Mole fraction } CO_2 = \frac{10}{90} = 0.1111$$

$$\text{Mole fraction } H_2O = \frac{10}{90} = 0.1111$$

$$\text{Mole fraction } C_2H_4 = \frac{20}{90} = 0.2222$$

$$\text{Mole fraction } O_2 = \frac{5}{90} = 0.0556$$

$$\text{Mole fraction } N_2 = \frac{25}{90} = 0.2778$$

$$\overline{}$$
$$1.0000$$

■

ILLUSTRATIVE EXAMPLE 5.9[(3)] Four reactor tanks recently converted to mixing tanks are connected as shown in Figure 5.4. If the system is operating at a steady-state, outline how to calculate the discharge concentration at each tank. Flow data is provided below:

$$Q_{13} = 75\,\text{liters/s} \quad Q_{24} = 20\,\text{liters/s} \quad Q_{33} = 60\,\text{liters/s}$$
$$Q_{21} = 25\,\text{liters/s} \quad Q_{32} = 45\,\text{liters/s} \quad Q_{43} = 30\,\text{liters/s}$$

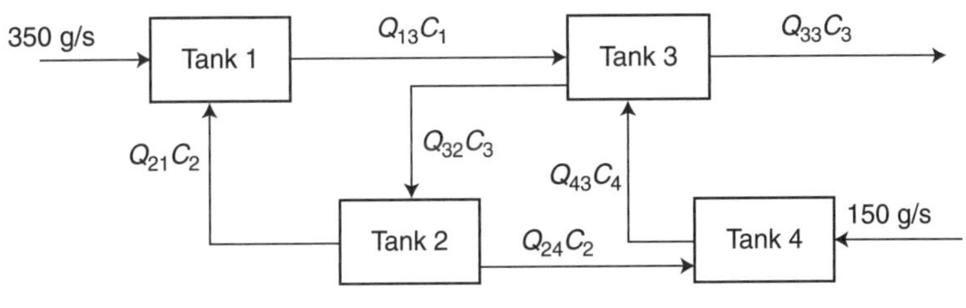

Figure 5.4 Multiple connected tanks.

Solution. Perform a material balance on each tank:

Tank 1:

$$350 + Q_{21}C_2 = Q_{13}C_1$$
$$350 + 25C_2 = 75C_1$$
$$75C_1 - 25C_2 = 350 \qquad (1)$$

Tank 2:

$$Q_{32}C_3 = Q_{21}C_2 + Q_{24}C_2$$
$$45C_3 + 25C_2 = 20C_2$$
$$45C_3 - 45C_2 = 0 \qquad (2)$$

Tank 3:

$$Q_{13}C_1 + Q_{43}C_4 = Q_{32}C_3 + Q_{33}C_3$$
$$75C_1 + 30C_4 = 45C_3 + 60C_3$$
$$75C_1 + 30C_4 - 105C_3 = 0 \qquad (3)$$

Tank 4:

$$150 + Q_{24}C_2 = Q_{43}C_4$$
$$150 + 20C_2 = 30C_4$$
$$30C_4 - 20C_2 = 150 \qquad (4)$$

Equations (1)–(4) are set up in matrix form

$$75C_1 - 25C_2 = 350$$
$$45C_3 - 45C_2 = 0$$
$$75C_1 - 105C_3 + 30C_4 = 0$$
$$-20C_2 + 30C_4 = 150$$

There are four equations and four unknowns. Any suitable method can now be employed to solve these four simultaneous linear equations (see also Chapter 21).

The solution is:[(4)]

$$C_1 = 7.44\,\text{g/L}$$
$$C_2 = 8.33\,\text{g/L}$$
$$C_3 = 8.33\,\text{g/L}$$
$$C_4 = 10.56\,\text{g/L}$$

■

CONVERSION VARIABLES

As discussed in the previous chapter, the two most common conversion variables employed in chemical reactor/kinetic studies are α and X. The term α is employed to represent the change in the number of moles of a particular species due to chemical reaction. The most commonly used conversion variable is X, and it is used to represent the change in the number of moles of a particular species (say A) relative to the number of moles of A initially present or initially introduced (to a flow reactor). Thus,

$$\begin{aligned} X = X_A &= \text{moles of } A \text{ reacted/initial moles of } A \\ &= N_A/N_{A_0} \end{aligned} \tag{5.10}$$

Other conversion (related) variables include: N_A, the number of moles of species A at some later time (or position); C_A, the concentration of A at some later time (or position); X_A^*, the moles of A reacted/*total* moles initially present. Note that all of the above conversion variables can also be based on mass, but this is rarely employed in practice. All of these variables are further defined and considered in the development that follows.

In order to define the commonly used conversion terms, one of the reactants is chosen as the basis of the calculation and then the other species involved in the reaction are related to that basis. In most instances, it is best to choose the limiting or key reactant as the basis of calculation.

Consider the author's favorite reaction equation:

$$aA + bB \longrightarrow cC + dD \tag{5.11}$$

The upper-case letters once again represent chemical species and the lower-case letters represent stoichiometric coefficients. Taking species A as the *basis of calculation*, the reaction expression is divided through by the stoichiometric coefficient of species A in order to arrange the reaction expression in the following form:

$$A + \frac{b}{a}B \longrightarrow \frac{c}{a}C + \frac{d}{a}D \tag{5.12}$$

This essentially places every quantity on a "per mole of A" basis. The conversion variable X_A can then be defined as the number of moles of A that have reacted per mole of A fed to the system, i.e.,

$$X_A = \frac{\text{moles of } A \text{ reacted}}{\text{moles of } A \text{ fed}} \tag{5.13}$$

Consider the two simultaneous and/or multiple reactions

$$aA + bB \longrightarrow cC + dD$$

that is now accompanied by a second (side or consecutive) reaction,

$$gA + fC \longrightarrow eE \tag{5.14}$$

Since species A appears as the reactant in both reactions, it is chosen as the basis of calculation, and both equations are divided by their respective stoichiometric coefficients of A:

$$\text{Reaction 1:} \quad A + \frac{b}{a}B \longrightarrow \frac{c}{a}C + \frac{d}{a}D \tag{5.15}$$

$$\text{Reaction 2:} \quad A + \frac{f}{g}C \longrightarrow \frac{e}{g}E \tag{5.16}$$

There are two conversion variables in this case:

$$X_{A1} = \frac{\text{moles of } A \text{ consumed by Reaction 1}}{\text{initial moles of } A} \tag{5.17}$$

This defines the conversion of A to form C and D in Reaction 1. Similarly, X_{A2} is defined as the conversion of A to form E in Reaction 2:

$$X_{A2} = \frac{\text{moles of } A \text{ consumed by Reaction 2}}{\text{initial moles of } A} \tag{5.18}$$

The extent of reaction can also be defined or specified in terms of the *yield*. The yield is a common term that is employed in industry since it provides information on the "yield" of a product. Although the yield term has been defined in many ways, it is usually expressed as

1. The quantity of a product formed divided by maximum quantity of the product that can be formed, or
2. The quantity of a product formed divided by the quantity of one of the reactants.

The author[(5)] has defined other conversion variables that can be employed to describe the extent of the reaction. These are listed below for reactant/species A without any explanatory details:

$$\alpha_A = \text{moles of } A \text{ reacted} \tag{5.19}$$

$$\alpha_A^* = \text{mass of } A \text{ reacted} \tag{5.20}$$

$$X_A^* = \frac{\text{mass of } A \text{ reacted}}{\text{initial mass of } A} \tag{5.21}$$

$$X_A' = \frac{\text{mass of } A \text{ reacted}}{\text{initial moles of } A} \tag{5.22}$$

$$X_A'' = \frac{\text{moles of } A \text{ reacted}}{\text{initial mass of } A} \tag{5.23}$$

For equilibrium reactions, the most convenient approach is to employ a reaction coordinate. This is discussed in the literature.[5]

Finally, it should be noted that the highest conversion that can be achieved in reversible reactions is the equilibrium conversion (which takes an infinite period of time to achieve). For endothermic (heat absorbed) reactions, the equilibrium conversion increases with increasing temperature up to a maximum of 1.0; for exothermic (heat liberated) reactions the equilibrium conversion decreases with increasing temperature. The reader is cautioned that these equilibrium concentration calculations are, for most intents and purposes, a set of "fake" or "artificial" values. They almost always represent an upper limit on the expected concentration at the temperature in question. Other chemical reactions, kinetic effects, and temperature variations in the system may render these calculations valueless. Nonetheless, these calculations serve a useful purpose since they do provide a reasonable estimate of these concentrations.

To illustrate the difference between these conversion variables, consider once again—as in the previous chapter—the elementary irreversible first order reaction

$$A \xrightarrow{k_A} p;\ k_A = k \tag{5.24}$$

The rate expression is given by

$$r_A = \frac{dC_A}{dt} = -kC_A \tag{5.25}$$

with the corresponding solution of

$$C_A = C_{A_0}e^{-kt} \tag{5.26}$$

If one choses to multiply the above by the volume, Equation (5.25) becomes

$$\frac{dN_A}{dt} - -kN_A \tag{5.27}$$

with the corresponding solution of

$$N_A = N_{A_0}e^{-kt} \tag{5.28}$$

By introducing the conversion variable X_A

$$X_A = \frac{\text{moles of } A \text{ reacted}}{\text{moles of } A \text{ initially present}} \tag{5.29}$$

one obtains

$$N_A = N_{A_0} - N_{A_0}X_A = N_{A_0}(1 - X_A) \tag{5.30}$$

Substituting into Equation (5.27) leads to

$$-\frac{dX_A}{dt} = -k(1 - X_A) \tag{5.31}$$

and

$$X_A = 1 - e^{-kt} \tag{5.32}$$

or simply substitute $N_A = N_{A_0}(1 - X_A)$ in Equation (5.28).

Employing α_A where α_A is the moles of A reacted gives

$$\alpha_A = N_{A_0} - N_A \tag{5.33}$$

Proceeding as above,

$$d\alpha_A = -dN_A \tag{5.34}$$

$$-\frac{d\alpha_A}{dt} = -k(N_{A_0} - \alpha_A) \tag{5.35}$$

and

$$\alpha_A = N_{A_0}(1 - e^{-kt}) \tag{5.36}$$

Or, once again, simply substitute Equation (5.33) into Equation (5.28). This analysis can obviously be extended to other conversion variables and systems.

Excess and Limiting Reactant

When methane is burned (or combusted) completely, the stoichiometric equation for the reaction is

$$CH_2 + 2O_2 \longrightarrow CO_2 + 2H_2O \tag{5.37}$$

The stoichiometric ratio of the oxygen to the methane is 0.5 mol methane consumed/mol oxygen consumed. If 1 mole of methane and 3 moles of oxygen are placed in a reaction vessel, only 2 moles of oxygen would be used up. Therefore, one mole of oxygen would remain unconsumed. In this case, the oxygen is called the *excess reactant* and methane is the *limiting reactant*. The *limiting reactant* is defined as the reactant that would be completely consumed if the reaction went to completion. All other reactants are *excess* reactants. The amount by which a reactant is present in excess is defined as the *percent excess* and is given by

$$\%\ \text{excess} = \left[\frac{(N - N_s)}{N_s}\right] \times 100\% \tag{5.38}$$

where N = number of moles of the excess reactant at the start of the reaction

N_s = stoichiometric number of moles of the excess reactant (i.e., the exact number of moles needed to react completely with the limiting reactant).

In the example above, the *stoichiometric* amount of the oxygen is two moles, since that is the amount that would react with the one mole of methane. The *excess* amount of oxygen is one mole, which is a *percentage excess* of 50% or a *fractional excess* of 0.50.

As another example using the same reaction equation, suppose one started with 10 moles of methane and 10 moles of oxygen in a reaction vessel. If the reaction goes to completion, all 10 moles of the oxygen would disappear, as would 5 moles of the methane. In this case, oxygen is the *limiting* reactant and methane the *excess*. The percent excess of the methane is 100%. The reader is left with the exercise of determining how 100% was arrived at.

When the amounts of reactants are present initially in *stoichiometric proportion*, there is no excess reactant because none of the reactants will remain if the reaction goes to completion. In the burning of methane, starting amounts of 10 moles of methane and 20 moles of oxygen will cause both reactants to be completely depleted.

It should be noted that in practice, few chemical reactions proceed to completion. There are two reasons for this. First, all reactions take time to occur. In some cases, an explosion for example, the reaction time is measured in fractions of a second. Many other reactions, however, are quite slow and as the reaction progresses, the reaction rate gets slower and slower. In these cases, it is not economically feasible to wait for the reaction to "finish" before removing the desired products. The second reason is that many industrial reactions are *reversible* (in fact, *all* reactions are reversible to some extent), which means that, while the reactants are combining to form products, the products are also reacting to produce the reactants. If enough time is allowed for a reversible reaction, an equilibrium point is reached where the amounts of reactants and products will no longer be changing.

Even though the reaction may never be completed, the definitions of *excess* and *limiting* reactants given earlier still apply. Note that both definitions are based on the assumption that the reaction goes to completion, even though this may not actually be the case.

How far a reaction goes to completion is measured for *each* reactant, *i*, can also be specified by the *fractional conversion*, f_i, the definition of which is

$$f_i = \text{amount } i \text{ reacted/amount } i \text{ initially present} \tag{5.39}$$

for batch (see Part II) processes, and

$$f_i = \text{rate } i \text{ reacted/rate } i \text{ input} \tag{5.40}$$

for continuous (see also Part II) processes. In many kinetic applications, the term X_i is employed, but *only* on a mole basis. The *percentage conversion* is also used and is given by $f_i \times 100\%$. Note that each reactant will usually have a different value of f_i for the same process. Suppose, in the reaction for the oxidation of sulfur dioxide to sulfur trioxide,

$$2SO_2 + O_2 \longrightarrow 2SO_3 \tag{5.41}$$

100 moles of SO_2 and 100 moles of O_2 are present initially in the reaction vessel and after a time lapse, 50 moles of the SO_2 have disappeared. Applying the above definition, the fractional conversion of the SO_2 is 50 moles reacted/100 moles initially present or 0.50. The fractional conversion of the O_2, on the other hand, is 25 moles reacted/100 moles initially present or 0.25.

A recycle stream is often used with a reactor to increase the fractional conversion of a desired product. Suppose, for example, a 100 lbmol/h stream of pure A is introduced to a reactor in which the reaction

$$A \longrightarrow B \tag{5.42}$$

is occurring. If the conversion of the A is 50%, the product stream will contain 50 lbmol/h of A and 50 lbmol/h of B. The 50% is referred to as a *single pass conversion*. If some of the A in the product stream could be separated from the B, it could (and should) be recycled and mixed with the incoming feed stream of 100 lbmol/h A. This would obviously increase the amount of desirable product, B, obtained from a given amount of the raw material, A. In other words, the recycling would result in a higher *conversion* of the A. This increased conversion resulting from recycling is referred to as the *overall conversion*. The distinction between single-pass and overall conversion is shown in the two definitions below.

$$f_{\text{single-pass}} = (A_{\text{input to reactor}} - A_{\text{output from reactor}})/A_{\text{input to reactor}} \tag{5.43}$$

$$f_{\text{overall}} = (A_{\text{input to process}} - A_{\text{output from process}})/A_{\text{input to process}} \tag{5.44}$$

The word "process" in the second definition refers to that part of the flowsheet that includes the recycle stream.

ILLUSTRATIVE EXAMPLE 5.10 Given the reaction

$$aA + bB \longrightarrow cC + dD$$

set up equations to describe X_A, N_A, N_B, N_C, and N_D in terms of N_{A_0}, N_{B_0}, N_{C_0}, and N_{D_0}. Also obtain expression for C_A, C_B, C_C, and C_D if the reaction volume is constant (as in a liquid system).

Solution. First write the reaction equation with a coefficient of unity (1.0) for species A by dividing the reaction equation by a.

$$A + \frac{b}{a}B \longrightarrow \frac{c}{a}C + \frac{d}{a}D$$

Write the equation for the conversion of A, X_A. See Introduction above, Equation (5.3).

$$X_A = (N_{A_0} - N_A)/N_{A_0}$$

Write the equation for N_A, N_B, N_C, and N_D in terms of X_A; be sure to include the "reaction" effect; it is negative if reacted and positive if it is formed.

$$
\begin{aligned}
N_A &= N_{A_0}(1 - X_A) \\
N_B &= N_{B_0} - (b/a)N_{A_0}X_A \\
N_C &= N_{C_0} + (c/a)N_{A_0}X_A \\
N_D &= N_{D_0} + (d/a)N_{A_0}X_A
\end{aligned}
$$

Write the equations in the above step in terms of θ_A, θ_B, θ_C, and θ_D where

$$
\begin{aligned}
\theta_A &= N_{A_0}/N_{A_0} = 1 \\
\theta_B &= N_{B_0}/N_{A_0} \\
\theta_C &= N_{C_0}/N_{A_0} \\
\theta_D &= N_{D_0}/N_{A_0}
\end{aligned}
$$

$$
\begin{aligned}
N_A &= N_{A_0}(1 - X_A) \\
N_B &= N_{A_0}[\theta_B - (b/a)X_A] \\
N_C &= N_{A_0}[\theta_C + (c/a)X_A] \\
N_D &= N_{A_0}[\theta_D + (d/a)X_A]
\end{aligned}
$$

Finally, rewrite the equations in terms of the concentration C_i and initial concentration, C_{i_0}.

$$
\begin{aligned}
C_A &= C_{A_0}(1 - X_A) \\
C_B &= C_{B_0} - (b/a)C_{A_0}X_A \\
C_C &= C_{C_0} + (c/a)C_{A_0}X_A \\
C_D &= C_{D_0} + (d/a)C_{A_0}X_A
\end{aligned}
$$

In addition,

$$
\begin{aligned}
C_A &= C_{A_0}(1 - X_A) \\
C_B &= C_{A_0}[\theta_B - (b/a)X_A] \\
C_C &= C_{A_0}[\theta_C + (c/a)X_A] \\
C_D &= C_{A_0}[\theta_D + (d/a)X_A]
\end{aligned}
$$

■

ILLUSTRATIVE EXAMPLE 5.11 When ammonia is burned, the products are nitric oxide and water.

$$4NH_3 + 5O_2 \longrightarrow 4NO + 6H_2O$$

Ammonia is fed to a combustion device at the rate of 150 kgmol/h and oxygen at a rate of 200 kgmol/h.

1. Which is the *limiting* reactant?
2. What is the percent excess of the *excess* reactant?

Solution

1. The *extent of reaction* balance for this reaction is

Input→	150		200		0		0
	$4NH_3$	+	$5O_2$	$\longrightarrow$	4NO	+	$6H_2O$
Output→	$150-z$		$200-(5/4)z$		z		$(6/4)z$

where z is the rate of NH_3 consumed.

To determine which is the limiting reactant, allow the output rate of NH_3 to go to zero, that is,

$$150 - z = 0$$
$$z = 150\,\text{kgmol/h}$$

The output rate of O_2 is then

$$200 - \left[\left(\frac{5}{4}\right)(150)\right] = +12.5\,\text{kgmol/h}$$

Since the output rate of O_2 is still positive when the NH_3 is all consumed, NH_3 is the limiting reactant. The stoichiometric O_2 rate is $(\frac{5}{4})(150)$ or 187.5 kgmol/h.

2. Since 12.5 kgmol/h is the "excess" rate of O_2 over and above that needed to react with all of NH_3,

$$\text{Percent excess } O_2 = \frac{12.5}{187.5}(100\%)$$
$$= 6.67\%.$$

■

ILLUSTRATIVE EXAMPLE 5.12 Refer to the previous illustrative example. If 85 kgmol/h of nitric oxide is produced, at what rate in kg/h is water produced, what is the *fractional conversion* of the ammonia, and what is the *fractional conversion* of the oxygen?

Solution. Based on the problem statement, NO is produced at the rate of 85 kgmol/h. Therefore, the reaction does not go to completion and

$$z = 85\,\text{kgmol/h}$$

$$\text{Rate } H_2O \text{ produced} = \left(\frac{6}{4}\right)z = (1.5)(85) = 127.5\,\text{kgmol/h}$$

$$\text{Mass flowrate} = (127.5\,\text{kgmol/h})(18\,\text{kg/kgmol})$$
$$= 2295\,\text{kg/h}$$

The fractional conversion of the ammonia, f_{NH_3}, becomes

$$\begin{aligned}\text{Fractional conversion of } NH_3 &= \text{rate consumed/input rate}\\ f_{NH_3} &= z/150\\ &= 85/150\\ &= 0.567\end{aligned}$$

The fractional conversion, f_{O_2}, of the oxygen is

$$\begin{aligned}f_{O_2} &= \frac{5}{4}z/200\\ &= \frac{5}{4}(85)/200\\ &= 0.531\end{aligned}$$

■

VOLUME CORRECTION FACTOR

Once again, consider the reaction

$$aA + bB \longrightarrow cC + dD \tag{5.45}$$

The volume of an ideal gas in a reactor at any time can be related to the initial conditions by the following equation:

$$V = V_0(P_0/P)(T/T_0)(N_T/N_{T_0}) \tag{5.46}$$

where P = absolute pressure
T = absolute temperature
V = volume
N_T = total number of moles
0 = subscript denotes initial conditions

If δ is defined as the increase in the total number of moles per moles of A reacted, then

$$\delta = (d/a) + (c/a) - (b/a) - 1 \tag{5.47}$$

The term ε is defined as follows:

ε = change in the *total* number of moles when the reaction is complete/the *total* number of moles initially present in (or fed to) the reactor

One may show that

$$\varepsilon = y_{A_0}\delta \tag{5.48}$$

where y_{A_0} = initial mole fraction of A.

The gas volume can now be expressed as follows:

$$V = V_0(P_0/P)(T/T_0)(1 + \varepsilon X) \tag{5.49}$$

If the pressure does not change significantly the equation becomes

$$V = V_0(T/T_0)(1 + \varepsilon X) \tag{5.50}$$

For isothermal operation, one obtains:

$$V = V_0(1 + \varepsilon X) \tag{5.51}$$

Note that both the temperature and pressure, as well as the extent or conversion of reaction (X), affect the concentration of a reacting species through the volume term.

For non-ideal gas conditions,[5] the equation for the volume is written:

$$V = V_0(P_0/P)(T/T_0)(1 + \varepsilon X)(z/z_0) \tag{5.52}$$

where z = compressibility factor.

For ideal gases $z = 1$. For most engineering applications, one may assume ideal conditions. Deviations from ideality can come into play at very high pressures.

The above analysis may be extended to multiple reactions. Consider the following two reactions

$$\begin{aligned} aA + bB &\xrightarrow{k_1} cC + dD \\ gA + fC &\xrightarrow{k_2} eE \end{aligned} \tag{5.53}$$

The procedure to follow is to rewrite the above in the form:

$$\begin{aligned} A + \frac{b}{a}B &\xrightarrow{k_1} \frac{c}{a}C + \frac{d}{a}D \\ A + \frac{f}{g}C &\xrightarrow{k_2} \frac{e}{g}E \end{aligned} \tag{5.54}$$

Define

$$X_{A_1} = X_1 = \frac{\text{mole of } A \text{ reacted via Equation (5.54)}}{\text{initial (inlet) moles of } A} \tag{5.55}$$

$$X_{A_2} = X_2 = \frac{\text{mole of } A \text{ reacted via Equation (5.54)}}{\text{initial (inlet) moles of } A} \tag{5.56}$$

Thus

$$N_A = N_{A_0}(1 - X_1 - X_2) \tag{5.57}$$

$$N_B = N_{A_0}\left(\theta_B - \frac{b}{a}X_1\right) \tag{5.58}$$

$$N_C = N_{A_0}\left(\theta_C + \frac{c}{a}X_1 - \frac{f}{g}X_2\right) \tag{5.59}$$

$$N_D = N_{A_0}\left(\theta_D + \frac{d}{a}X_1\right) \tag{5.60}$$

$$N_E = N_{A_0}\left(\theta_E + \frac{e}{g}X_2\right) \tag{5.61}$$

If the inerts are represented as N_I, then

$$N_{T_0} = N_{A_0} + N_{B_0} + N_{C_0} + N_{D_0} + N_{E_0} + N_{I_0} \tag{5.62}$$

and

$$\delta_1 = \frac{c}{a} + \frac{d}{a} - \frac{b}{a} - 1, \tag{5.63}$$

$$\delta_2 = \frac{e}{g} - \frac{f}{g} - 1 \tag{5.64}$$

Thus,

$$N_T = N_{T_0} + \delta_1 N_{A_0} X_1 + \delta_2 N_{A_0} X_2 \tag{5.65}$$

with

$$\varepsilon_1 = y_{A_0}\delta_1 \tag{5.66}$$

$$\varepsilon_2 = y_{A_0}\delta_2 \tag{5.67}$$

Finally,

$$N_T/N_{T_0} = 1 + \varepsilon_1 X_1 + \varepsilon_2 X_2 \tag{5.68}$$

$$V = V_0(1 + \varepsilon_1 X_1 + \varepsilon_2 X_2) \tag{5.69}$$

ILLUSTRATIVE EXAMPLE 5.13 Consider the following elementary, irreversible gas phase reaction.

$$A + B \longrightarrow C$$

Initial concentration of $A = 0.0013\ \text{lbmol/ft}^3$

Initial pressure = 1 atm

Final pressure = 2 atm

Initial temperature = 60°F

Final temperature = 2140°F

Initial volume = 1 ft^3

Calculate the final volume for a 50% conversion of A if the reaction mixture initially contains 50 mole percent A.

Solution. Write the equation describing the volume at any time or any conversion.

$$V = V_0(P_0/P)(T/T_0)(1 + \varepsilon X) \tag{5.49}$$

Calculate δ and ε noting that $y_{A_0} = 0.5$. For this reaction,

$$\begin{aligned}
c &= 1, b = 1, a = 1 \\
\delta &= (c/a) - (b/a) - 1 \\
&= (1/1) - (1/1) - 1 \\
&= -1 \\
\varepsilon &= y_{A_0}\delta \\
&= (0.5)(-1) \\
&= -0.5
\end{aligned}$$

Express C_A in terms of C_{A_0}. Keep in mind that the volume term appears in the denominator of the concentration term. See also Equation (5.49).

$$C_A = C_{A_0}(T_0/T)(P/P_0)(1 - X)/(1 + \varepsilon X)$$

Calculate the ratio (T/T_0) and (P_0/P)

$$\begin{aligned}
T_0/T &= (60 + 460)/(2140 + 460) \\
&= 1/5 \\
&= 0.2 \\
P/P_0 &= 2/1 \\
&= 2
\end{aligned}$$

Finally, calculate the concentration, C_A. Substitute into the equation above and note that $X = 0.5$.

$$\begin{aligned}
C_A &= (0.00132)(0.2)(2)(1 - 0.5)/[1 - (0.5)(0.5)] \\
&= 0.000528 \text{ lbmol/ft}^3
\end{aligned}$$

■

ILLUSTRATIVE EXAMPLE 5.14 Consider the irreversible, elementary, isothermal constant pressure gas phase reaction below for Questions (1–7):

$$2A + 3B \longrightarrow 3C$$

The system initially contains 40% A, 10% B, 5% I, and 45% C (by mole).

1. Calculate δ, the increase in the total number of moles per mole of A reacted.
2. Calculate ε, the change in the total number of moles the when reaction is completed per total number of moles fed to the reactor.
3. The rate expression for r_A.

4. Calculate θ_B, the initial molar ratio of A to B (note that by definition, θ_A is equal to one.)
5. Calculate the overall reaction order.
6. The units of k_A.
7. Calculate C_{B_0} in gmol/L if $P_0 = 10$ atm and $T_0 = 260$ K.

Solution

1. $$\delta = \frac{c}{a} - \frac{b}{a} - 1$$

$$\delta = \frac{3}{2} - \frac{3}{2} - 1$$

$$\delta = -1$$

2. $$\varepsilon = y_{A_0}\delta$$

$$\varepsilon = (0.4)(-1)$$

$$\varepsilon = -0.4$$

3. $$r_A = -kC_A^2 C_B^3$$

4. $$\theta_B = \frac{y_{B_0}}{y_{A_0}}$$

$$\theta_B = \frac{0.10}{0.40}$$

$$\theta_B - 0.25$$

5. The overall reaction order is

$$n = 2 + 3 = 5$$

6. The units of k_A can be calculated from the units of rate and concentrations.

$$r_A \text{ [=] moles/volume} \cdot \text{time}$$
$$C_A \text{ [=] moles/volume}$$
$$C_B \text{ [=] moles/volume}$$

Since, $r_A = -kC_A^2 C_B^3$

$$k \text{ [=] } \frac{(\text{volume})^4}{(\text{moles})^4 \cdot \text{time}}$$

7. Assuming an ideal gas,

$$C_{B_0} = \frac{P_0}{RT}$$

$$C_{B_0} = \frac{10\,\text{atm}}{(0.082\,\text{L}\cdot\text{atm/gmol}\cdot\text{K})(260\,\text{K})}$$

$$C_{B_0} = 0.47\,\frac{\text{gmol}}{\text{L}}$$

■

ILLUSTRATIVE EXAMPLE 5.15 Consider the following irreversible elementary isothermal constant pressure reaction to apply for the following gas phase reaction in Questions (1–3)

$$4A + B \longrightarrow 2D + 5E$$

The system initially contains 25% A, 25% B, 10% inerts, and 40% E (by mole).

1. Calculate δ and ε.
2. Calculate θ_B and θ_E.
3. Calculate C_E.

Solution. The solution to this example is similar to that of the previous illustrative example.

1. $\delta = [(2 + 5) - (4 + 1)]/4 = 2/4 = 0.5$

2. $\theta_B = 0.25/0.25 = 1.0$

   $\theta_E = 0.40/0.25 = 1.6$

3. $C_E = C_{A_0}\dfrac{[\theta_E + (5/4)X]}{1 + \varepsilon X}; \quad \varepsilon = y_{A_0}\delta = (0.25)(0.5) = 0.125$

   Substituting for θ_E and ε gives

   $$C_E = C_{A_0}\frac{[1.6 + 1.25X]}{1 + 0.125X}$$

■

ILLUSTRATIVE EXAMPLE 5.16 Consider the following irreversible, elementary gas phase reaction

$$\overset{1}{A} + \overset{2}{3B} \xrightarrow{k_A} \overset{3}{5C}; \quad \text{Inerts} = 2$$

Obtain a numerical rate expression for this reaction.

Solution. The rate equation is

$$r_A = -k_A C_A C_B^3$$

with

$$\begin{aligned}
\theta_B &= 2/1 = 2 \\
b/a &= 3/1 = 3 \\
\delta &= 5 - 1 - 3 = 1 \\
y_A &= 1/(1+2+3+2) = 0.125 \\
\varepsilon &= (1)(0.125) = 0.125
\end{aligned}$$

Therefore,

$$C_A = N_{A_0}\frac{(\theta_A - X)}{V_0(1+\varepsilon X)} = C_{A_0}\frac{(1-X)}{(1+0.125X)}$$

$$C_B = \frac{N_{A_0}\left[\theta_B - \left(\frac{b}{a}\right)X\right]}{V_0(1+\varepsilon X)} = C_{A_0}\left(\frac{2-3X}{1+0.125X}\right)$$

The rate expression may now be written

$$\begin{aligned}
r_A &= -k_A C_A C_B^3 \\
&= -k_A C_{A_0}^4 \frac{(1-X)(2-3X)^3}{(1+0.125X)^4}
\end{aligned}$$ ■

ILLUSTRATIVE EXAMPLE 15.17 Refer to the previous example. Obtain the rate expression if the reaction is elementary *and* reversible.

Solution. For this case,

$$r_A = -k_A C_A C_B^3 + k'_A C_C^5$$

with

$$C_C = \frac{N_{A_0}\left[\theta_C + \left(\frac{c}{a}\right)X\right]}{V_0(1+\varepsilon X)}; \quad \theta_C = 3, \left(\frac{c}{a}\right) = 5$$

Therefore

$$C_C = N_{A_0}\frac{(3+5X)}{V_0(1+\varepsilon X)} = C_{A_0}\frac{(3+5X)}{(1+0.125X)}$$ ■

The rate expression is now

$$r_A = -k_A C_{A_0}^4 \frac{(1-X)(2-3X)^3}{(1+0.125)^4} + k'_A\left[C_{A_0}\frac{(3+5X)}{(1+0.125X)}\right]^5$$ ■

ILLUSTRATIVE EXAMPLE 15.18 Provide a physical interpretation of ε, including why the term $V_0(1+\varepsilon X)$ represents the volume of a gaseous system after reacting with a conversion X.

Solution. The simplist procedure to follow is to set up the participating terms in dimensional form. Consider once again the reaction:

$$aA + bB \longrightarrow cC + dD$$

$$\delta = \left(\frac{c}{a}\right) + \left(\frac{d}{a}\right) - \left(\frac{b}{a}\right) - 1 = \frac{(c+d)-(b+a)}{a}$$

$$= \frac{\text{change in total moles}}{\text{moles of } A \text{ reacted}}$$

$$y_{A_0} = \frac{\text{moles of } A \text{ initially}}{\text{total moles initially}}$$

$$\varepsilon = y_{A_0}\delta = \left(\frac{\text{moles of } A \text{ initially}}{\text{total moles initially}}\right)\left(\frac{\text{change in total moles}}{\text{moles of } A \text{ reacted}}\right)$$

$$\varepsilon X = \frac{(\text{moles of } A \text{ initially})\ (\text{change in total moles})\ (\text{moles of } A \text{ reacted})}{(\text{total moles in initially})\ (\text{moles of } A \text{ reacted})\ (\text{moles of } A \text{ initially})};\ X = X_A$$

$$= \frac{\text{change in total moles}}{\text{total moles initially}} = \frac{\Delta n}{n_0}$$

From the ideal gas law

$$\frac{\Delta n}{n_0} = \frac{\Delta V}{V_0}$$

so that

$$\varepsilon X = \frac{\Delta V}{V_0}$$

One may write the term $V_0(1+\varepsilon X)$ based on the above expression as

$$V_0(1+\varepsilon X) = V_0\left(1+\frac{\Delta V}{V_0}\right)$$

$$= V_0 + \Delta V$$

Thus, the term $V_0(1+\varepsilon X)$ represents the change in volume ΔV relative to the original volume following a conversion X. ■

YIELD AND SELECTIVITY

When multiple reactions (either parallel or consecutive) occur, the side or undesired reactions compete with the main or desired reaction; the less predominant the main reaction is, the smaller the amount of desired product is for a given amount of reactant. Suppose the following hypothetical parallel reactions occur:

$$\begin{aligned} A &\longrightarrow B \\ A &\longrightarrow C \\ A &\longrightarrow D \end{aligned} \tag{5.70}$$

Suppose also that, of a starting amount of 10 mol of A, four form the desired product B, two form the undesired product C, one forms the undesired product D, and three remain unreacted. While this process produces 4 mol of valuable product, it could have produced 10 if everything went the way a practicing engineer would want it to, that is, if all 10 mol of A reacted to form B. The ratio of the 4 mol of B actually produced, to the 10 it potentially could have produced, is called the *yield* (0.40 or 40%). By definition, the *yield* of a reaction is a measure of how much of the desired product is produced relative to how much would have been produced if only the desired reaction occurred and if that reaction went to completion. Obviously, the fractional yield must be a number between zero and unity. As noted earlier, another definition that has been used by industry is the "ratio of product formed to initial (inlet) reactant." An additional term used in conjunction with multiple reactions is *selectivity*. Selectivity is a measure of how predominant the desired reaction is relative to one of the side reactions. The value of the selectivity is obtained by dividing the number of moles of a desired product actually generated by the number of moles of one of the undesired products produced by a side reaction. Obviously, selectivity becomes more important than conversion if "undesired" side reaction(s) take precedent over "desired" reaction(s). Other definitions of selectivity that have appeared in the literature include the popular "ratio of the amount of desired products formed to the amount of all products formed." Definitions have also included "rate of production of one product relative to another product (*point* selectivity)" and "amount of one product formed relative to another product (*overall* selectivity)." In the example above, the selectivity of B over C is 2.0 and that of B over D is 4.0. These definitions are employed in several illustrative examples to follow.

Fogler[6] provides the following excellent example. Consider the following reactions

$$A \xrightarrow{k_D} D \text{ (desirable)} \tag{5.71}$$

$$A \xrightarrow{k_U} U \text{ (undesirable)} \tag{5.72}$$

with

$$r_D = k_D C_A^{\alpha_1} \tag{5.73}$$

$$r_U = k_U C_A^{\alpha_2} \tag{5.74}$$

and

$$-r_A = r_D + r_U \tag{5.75}$$

$$= k_D C_A^{\alpha_1} + k_U C_A^{\alpha_2} \tag{5.76}$$

If one desires $r_D \ggg r_U$, then the point (or local) selectivity is

$$S = \frac{r_D}{r_U} = \frac{k_D}{k_U} C_A^{\alpha_1 - \alpha_2} \tag{5.77}$$

If $\alpha_1 - \alpha_2 = a$, then

$$S = \left(\frac{k_D}{k_U}\right)(C_A^a) \tag{5.78}$$

To maximize S, the need to maintain the product of *both* terms in Equation (5.78) is high. If a is positive, one needs to maximize C_A. This generally means no inert(s), no product(s), and for gas phase reactions operate a low temperature and high pressure. If a is negative, the reverse applies.

The effect of temperature is handled through the Arrhenius equation discussed in Chapter 3.

$$\frac{k_D}{k_U} = \frac{A_D}{A_U} e^{-\frac{1}{RT}(E_D - E_U)} \tag{5.79}$$

If $E_D > E_U$, the reaction should be run at the highest temperature economically possible. If $E_U > E_D$, it should be run at the lowest temperature possible (while maintaining a reasonable rate).

The reader should also attempt to analyze the following application:

$$A + B \xrightarrow{k_1} D \tag{5.80}$$

$$A + B \xrightarrow{k_2} U \tag{5.81}$$

where

$$r_D = k_1 C_A^{\alpha_1} C_B^{\beta_1} \tag{5.82}$$

$$r_U = k_2 C_A^{\alpha_2} C_B^{\beta_2} \tag{5.83}$$

For this case

$$S = \frac{r_D}{r_U} = \left(\frac{k_1}{k_2}\right)(C_A^{\alpha_1 - \alpha_2})(C_B^{\beta_1 - \beta_2}) \tag{5.84}$$

Hint: Consider attempting to maximize the product of C_A and C_B.

ILLUSTRATIVE EXAMPLE 15.19 Acetaldehyde can be formed by oxidizing ethane:

$$C_2H_6 + O_2 \longrightarrow CH_3CHO + H_2O$$

However, some carbon dioxide is usually an unwelcome by-product:

$$C_2H_6 + \frac{7}{2}O_2 \longrightarrow 2CO_2 + 3H_2O$$

If 65% (by mole) of the ethane feed stream forms acetaldehyde, 15% forms CO_2 and the remainder is unreacted, calculate

1. The fractional conversion of ethane
2. The fractional yield of acetaldehyde

Solution. Choose a basis of 100 lbmol/h of C_2H_6 input. Perform extent of reaction balances on the two reactions. Only the C_2H_6, CH_3CHO, and CO_2 need be considered since information on the other components is not required in the solution:

$$\begin{array}{llccccccc} \text{Input} & \rightarrow & 100 & & & & 0 & & \\ & & C_2H_6 & + & O_2 & \longrightarrow & CH_3CHO & + & H_2O \\ \text{Output} & \rightarrow & 100 - z - y & & (1/1)z & & z & & \end{array}$$

$$\begin{array}{llccccccc} \text{Input} & \rightarrow & 100 & & & & 0 & & \\ & & C_2H_6 & + & \frac{7}{2}O_2 & \longrightarrow & 2CO_2 & + & 3H_2O \\ \text{Output} & \rightarrow & 100 - z - y & & & & 2y & & \end{array}$$

where $z =$ amount of C_2H_6 that forms CH_3CHO

$y =$ amount of C_2H_6 that forms CO_2

The fractional conversion, f, of the C_2H_6 is

$$\begin{aligned} f &= \text{rate } C_2H_6 \text{ consumed/rate } C_2H_6 \text{ input} \\ &= (z + y)/100 \\ &= (65 + 15)/100 \\ &= 0.80 \end{aligned}$$

If all of the C_2H_6 had reacted to form CH_3CHO, 100 lbmol/h of the CH_3CHO would have been produced. Thus, the fractional yield Y is

$$\begin{aligned} Y &= 65/100 \\ &= 0.65 \end{aligned}$$

■

ILLUSTRATIVE EXAMPLE 5.20 Refer to the previous example. Calculate the selectivity of the acetaldehyde over carbon dioxide production.

Solution. The selectivity S of the CH_3CHO over CO_2 production is calculated as follows:

$$\begin{aligned} S &= \text{rate of } CH_3CHO \text{ produced/rate of } CO_2 \text{ produced} \\ &= z/2y \\ &= 65/[(2)(15)] \\ &= 2.17 \end{aligned}$$

The reader should note that the terms *conversion, yield*, and *selectivity* are sometimes confusing. Some people in the field use them interchangeably (which is poor practice). Note that *conversion* is measured solely on the disappearance of a particular reactant; *yield* is measured in terms of the generation of a particular product; and, *selectivity* is a comparison of the amount or flowrates of two generated products. ■

REFERENCES

1. J. Reynolds, J. Jeris, and L. Theodore, "*Handbook of Chemical and Environmental Engineering Calculations*," John Wiley & Sons, Hoboken, NJ, 2004.
2. J. Santoleri, J. Reynolds, and L. Theodore, "*Introduction of Hazardous Waste Incineration*," 2nd edition, John Wiley & Sons, Hoboken, NJ, 2000.
3. Adapted from: source, location and data unknown.
4. L. Theodore, personal notes, East Williston, NY, 1967.
5. L. Theodore, F. Ricci, and T. VanVliet, "*Thermodynamics for the Practicing Engineer*," John Wiley & Sons, Hoboken, NJ, 2009.
6. H. Fogler, "*Elements of Chemical Reaction Engineering*," 4th edition, Prentice-Hall, Upper Saddle River, NJ, 2006.

PART **II**

TRADITIONAL REACTOR ANALYSIS

Every tradition grows ever more venerable—the more remote is its origin, the more confused that origin is. The reverence due to it increases from generation to generation. The tradition finally becomes holy and inspires awe.

—FRIEDRICH NIETZSCHE, *Human, All Too Human* (1878), 1844–1900.

PART II OF this book is concerned with the analyses of traditional reactors. It contains six chapters and each serves a unique purpose in an attempt to treat nearly all the important aspects of chemical reactors. A major objective of Part II is to prepare the reader to solve real-world engineering and design problems that involve chemical reactors.

There are various classes of reactors. The three that are most encountered in practice are batch, continuous stirred tank (CSTR), and tubular. As such, they receive the bulk of the treatment. Another reactor reviewed is the semi-batch unit. Other topics reviewed include reactor classification, the conservation laws, and the comparison of reactors.

Following this introduction to Part II, the reader is introduced to the various classifications of reactors (Chapter 6) and the conservation laws (Chapter 7). The next three chapters are concerned with the application of the previous chapter to a specific reactor; Chapters 8, 9, and 10 are devoted to batch, tank flow (including semi-batch), and tubular flow reactors, respectively. The format of the presentation is the same in each of these chapters. Some analogies and similarities exist among the three reactors. For these reasons, more time is spent on the first reactor chapter. A complete and thorough treatment on batch reactors will remove the need for the repeated development of material at a later time. Part II concludes with an analysis and comparison

Chemical Reactor Analysis and Applications for the Practicing Engineer. By Louis Theodore

of the various classes of reactors that are used in practice. Numerous illustrative examples are included with each chapter.

Chapter number and title for the chapters in this Part II are listed below:

Chapter 6: Reaction and Reactor Classification
Chapter 7: The Conservation Laws
Chapter 8: Batch Reactors
Chapter 9: Continuous Stirred Tank Reactors (CSTRs)
Chapter 10: Tubular Flow Reactors
Chapter 11: Reactor Comparisons

The reader should note that the term Q (rather than q) is employed to represent the volumetric flowrate. The reader will be reminded of this throughout the presentation, particularly in Chapters 9–11.

CHAPTER 6

REACTION AND REACTOR CLASSIFICATION

INTRODUCTION

Kinetic and design calculations for reactors require information on the rate of reaction and equations to describe the concentration, temperature, and/or pressure variation within the reactor. Part I developed equations to describe the rate of a chemical reaction and qualitatively discussed energy (temperature) effects plus phase and chemical equilibrium. The contents of Part I can be combined and the results applied to the three major classifications for reactors to be addressed later.

This chapter serves to introduce the reader to the various classification of reactors. A section on reaction classification—briefly reviewed earlier—is also included; however, the chapter is highlighted with introductory material on the three major classes of reactors:

1. Batch
2. Continuous stirred-tank (CSTR)
3. Tubular flow.

As the reader might suppose, there are several approaches to classifying not only reactors but also reactions.

Topics addressed in the sections to follow include:

1. Classification of reactions
2. Classification of reactors
3. Other industrial reactors
4. Ancillary equipment.

Chemical Reactor Analysis and Applications for the Practicing Engineer. By Louis Theodore

CLASSIFICATION OF REACTIONS

Reactions may be classified in a variety of ways. On the basis of *mechanism* they may be:

1. Irreversible
2. Reversible
3. Simultaneous
4. Consecutive
5. Multiple
6. Any combinations of the above.

Another classification of reactions can be based on "order"; for example, the elementary reaction

$$r_A = -k_A C_A^n \tag{6.1}$$

is of order n—the exponent of C_A. The elementary reaction

$$r_A = -k_A C_A^n C_B^m \tag{6.2}$$

is of order n with respect to A and m with respect to B. The sum of the two exponents, i.e., $n + m$, is the (overall) order of the reaction in Equation (6.2). For the elementary reaction

$$r_A = kC_A^a C_B^b \cdots C_M^m, \qquad a + b + \cdots + m = n \tag{6.3}$$

one may conclude that the reaction is of order a with respect to A, order b with respect to B, order m with respect to M, and overall order n.

If one were to assign an integer number to the exponents above, the order could be classified as first order, second order, third order, and so on. If the exponent is zero, the reaction is classified as zero order. The order(s) need not be whole numbers. If the exponent is of fractional order, for example, 0.5, 1.25, 3.17, the reaction would then be classified as *fractional* order. When the reaction rate (and order) are based on the stoichiometric reaction equation, the reaction is termed *elementary*. If there is no direct correspondence/relation between the rate equation and the stoichiometric equation, the reaction is called *nonelementary*.

Reactions can also be classified based on operating conditions. Examples here include:

Isothermal

Isothermal at constant volume

Isothermal at constant pressure

Adiabatic

Adiabatic at constant volume

Adiabatic at constant pressure

Nonadiabatic

Nonisothermal

Nonadiabatic and nonisothermal

In *isothermal operation*, the temperature is maintained constant during the course of the reaction. This condition can be approached in practice by providing sufficient heat exchanger facilities to account for enthalpy effects arising during reaction. This mode of operation finds its major application in laboratory kinetic studies. Some energy in the form of heat is added to or removed from the reactor, but isothermal conditions are not satisfied. For *adiabatic operation*, the reactor is insulated to minimize heat transfer between the reactor contents and the surroundings. However, many industrial reactors attempt to operate in this manner. Most industrial reactors are described by *nonisothermal operation.*

Reactions may be classified according to the phases participating in the reaction. Examples include:

Homogeneous

Homogeneous, gas

Homogeneous, liquid

Homogeneous, solid

Heterogeneous

Heterogeneous, gas–liquid

Heterogeneous, gas–solid

Heterogeneous, liquid–solid

Heterogeneous, liquid–liquid

Heterogeneous, solid–solid

Heterogeneous, gas–liquid–solid

With respect to phases, the major category practicing engineers experience are gas–solid reactions. Examples include the combustion of coal and the production of hydrogen by steam reacting in the presence of a catalyst. Also note that reactions between a gas and a solid may form either a gaseous or a solid product, or both. An example of a liquid–solid reaction is water softening and ion exchange. A gas–liquid example includes the desulfurization of acid gases in power plants by scrubbing with a caustic solutions, while a liquid–liquid application could involve the desulfurization of various liquid petroleum fractions. Solid–solid applications are not extensive, but the manufacture of cement is an example.

Some reactions are spontaneous and occur extremely rapidly, e.g., explosions. Other reactions take place slowly with respect to time, e.g., rusting.

The subject of catalysis and catalytic reactions is addressed briefly later in this chapter but in more detail in Part IV, Chapters 16 and 17. Naturally, the reaction classification here simply includes:

catalyzed	catalytic
uncatalyzed	noncatalytic

The above groupings are not mutually exclusive. Thus, a reaction may be reversible and first order, operating adiabatically at constant pressure in the presence of a catalyst. This categorization may now be extended to include the type of reactor, a topic addressed in the next section.

CLASSIFICATION OF REACTORS

There have been a host of companion terms used to describe chemical reactors; among these are *reaction unit*, *reactor*, *oxidizer*, *afterburner*, *reactor device*, *burner*, and *organic (or inorganic) reactor*. Whichever term is used, the overall process of chemical reactions is best characterized by phenomena occurring in a *chemical reactor*. In effect, an incinerator is one of a number of units that fits into the class of what industry describes as a *chemical reactor*. With this in mind, one may apply chemical reactor principles to either design and/or predict the performance of this broad category. In any event, this unit is the equipment in which chemical reactions take place.

From a physical perspective, chemical reactors are usually in the form of:

1. cylindrical, in which the entire reaction mixture is contained in one part or the entire space;
2. a spherical tank, in which the entire reaction mixture is contained in one part or the entire space; and,
3. a tube, in which the reacting mixture flows from one end to the other with little or no mixing.

The differences between these three kinds of reactors are often described in terms of the degree of mixing.

As noted in the Introduction, kinetic and design calculations for reactors require information on the rate of reaction and equations to describe the concentration and temperature within the reactor. Part I developed equations to describe the rate of a chemical reaction, and briefly discussed energy (temperature) effects and chemical equilibrium. The contents of Part I can be combined and the results applied to the three major classifications of reactors: batch, tank flow, and tubular flow. These three classes of reactors are discussed in the subsections that follow.

Batch Reactors

A *batch reactor* is a vessel or container that may be open or closed. Reactants are usually added to the reactor simultaneously. The contents are then mixed (if necessary) to ensure no variations in the concentrations of the species present. The reaction then proceeds. During this period, there is no transfer of mass into or out of the reactor. Because the concentrations of reactants and products change with time, this is a *transient* or *unsteady-state* operation. The reaction is terminated when the desired chemical change has been achieved. The contents are then discharged and often sent elsewhere for further processing. Line diagrams of three typical batch reactors are given in Figure 6.1. A more detailed schematic is presented in Figure 8.1 in Chapter 8.

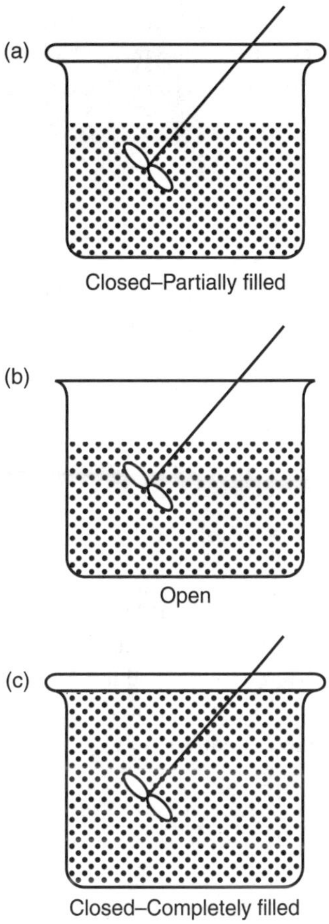

Figure 6.1 Different batch reactors.

The describing equation for this device is obtained by applying the conservation law for either mass or number of moles on a time rate basis to the contents of the reactor. Once again, it is preferable to work with moles rather than mass since the rate of reaction is most conveniently described in terms of molar concentrations.

Continuously Stirred Tank Reactor

Another reactor where mixing is important is the *tank flow* or *continuously stirred tank reactor* (CSTR). This type of reactor also consists of a tank or kettle equipped with an agitator (see Figure 6.2). It may be operated under steady-state or transient conditions. Reactants are fed continuously, and the products withdrawn continuously. The reactants and products may be in the liquid, gas, or solid state, or a combination of these. If the contents are perfectly mixed, the reactor design problem is greatly simplified for steady-state conditions because the mixing results in uniform concentration, temperature, etc., throughout the reactor. For perfect mixing, the rate of reaction is constant; therefore the describing equations (to be developed in Chapter 9) are not

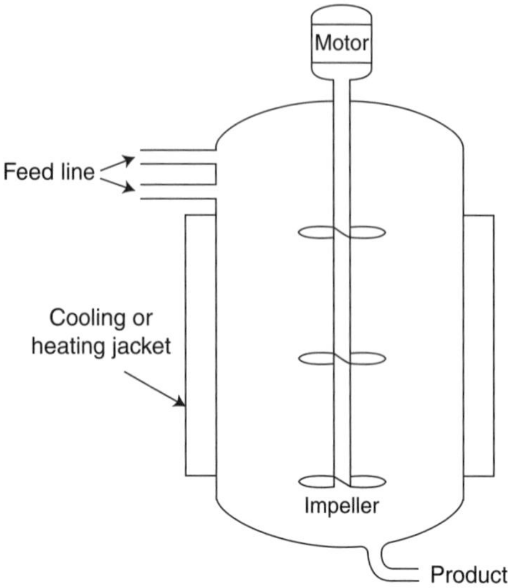

Figure 6.2 Typical tank flow reactor.

differential and do not require integration. High reactant concentrations can be maintained with low flow rates so that conversions approaching 100% may be achieved. A disadvantage of this practice, however, is that very large reactors must be required to obtain these high conversions. Note that the concentration of a species in the reactor is equal to the outlet concentration since perfectly mixed conditions are normally assumed.

Tubular Flow Reactors

The last and most important class of reactor to be examined is the *tubular flow reactor*. The most common type is the single-pass cylindrical tube (see Figure 6.3); another important type is one that consists of a number of tubes in parallel (see Figure 6.4). The reactor(s) may be vertical or horizontal. The feed is charged continuously at the inlet of the tube, and the products are continuously removed at the outlet. If heat exchange with the surroundings is required, the reactor tube is *jacketed*. If adiabatic conditions are required, the reactor is *covered* with insulation. Tubular flow reactors are usually operated under steady-state conditions so that physical and chemical properties do not vary with time. Unlike the batch and tank flow reactors, there is no mechanical mixing. Thus, the state of the reacting fluid will vary from point to

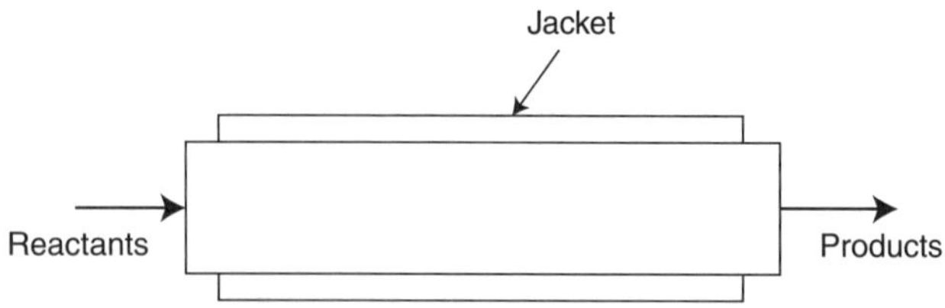

Figure 6.3 Tubular flow reactor, single pass.

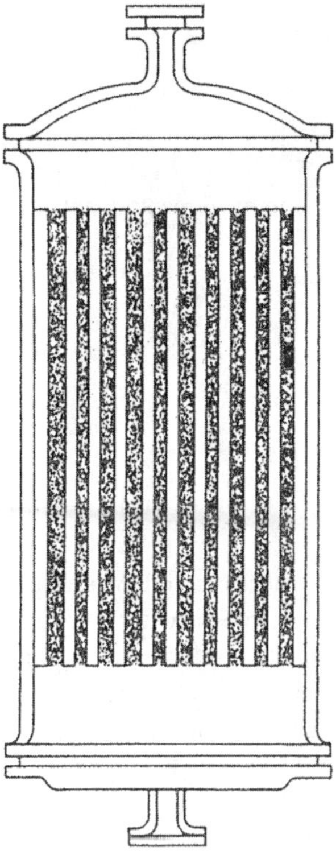

Figure 6.4 Multitube flow reactor.

point in the reactor of radius R and length L, and this variation in cylindrical coordinates may be in both the radial r and axial z directions (see Figure 6.5). The describing equations (to be developed in Chapter 10) are then differential, with position as the independent variable.

OTHER INDUSTRIAL CHEMICAL REACTORS

There are a host of other chemical reactors that are employed in industry that do not comfortably fit into the rather narrow window of three classes of reactors addressed in the previous section. Other reactors include (there are, of course, many more):

1. Thermal reactors
2. Catalytic reactors

Figure 6.5 Reactor system, cylindrical coordinates.

3. Flares
4. Incinerators
5. Kilns
6. Furnaces.

A brief description of these units is presented below. Other chemical reactors not covered include:

1. Slurry reactors
2. Gauze reactors
3. Trickle reactors
4. Transport lines.

Thermal Reactors

Chemical reactions are often conducted at elevated temperatures in order to ensure high chemical reaction rates. To achieve this temperature, it is necessary to preheat the feed stream with auxiliary energy. This type of reactor is popularly referred to by some as an *afterburner*. Along with the feed stream, air and fuel may be continuously delivered to the reactor where the fuel is combusted with air in a firing unit (burner).

Due to the high operating temperatures, the unit must be constructed of materials (usually metals) capable of withstanding this condition. These devices usually include an outer steel shell that is lined with refractory material. However, refractory material is heavy, with densities as low as 50 lb/ft^3 for lightweight insulating firebricks, and as high as 175 lb/ft^3 for castable refractories. The refractory wall thickness is often in the 3 to 9 inch range. This weight adds considerably to the cost.

Some of the advantages of the thermal reactor are:

1. Simplicity of construction
2. Small space requirements
3. Low maintenance costs.

Some of the disadvantages include:

1. High operating costs
2. Fire hazards
3. Flashback possibilities.

Catalytic Reactors

This reactor will have an entire chapter devoted to it in Part IV. Catalytic reactors are occassionally an alternative to thermal reactors. If a solid catalyst is added to the reactor, the reaction is said to be heterogeneous. For simple reactions, the effect of the presence of a catalyst is to:

1. Increase the rate of reaction.
2. Permit the reaction to occur at a lower temperature.

3. Permit the reaction to occur at a more favorable pressure.
4. Reduce the reactor volume.
5. Increase the yield of a reactant(s) relative to other species.

Metals in the platinum family are recognized for their ability to promote reactions at low temperatures. Other catalysts include various oxides of copper, chromium, vanadium, nickel, and cobalt. These catalysts are subject to poisoning, particularly from halogens, halogen and sulfur compounds, zinc, arsenic, lead, mercury, and particulates. High temperatures can at times reduce catalyst activity. It is therefore important that catalyst surfaces be clean and active to ensure optimum performance. Catalysts can usually be regenerated with superheated steam.

Catalysts may be porous pellets, usually cylindrical or spherical in shape, ranging from 1/16 to 1/2 inches in diameter. Small sizes are recommended, but the pressure drop through the reactor increases. Other shapes include honeycombs, ribbons, wire mesh, and so on. Since catalysis is a surface phenomena, a physical property of these particles is that the internal pore surface is near infinitely greater than the outside surface. Catalysts are also treated in a separate chapter in Part IV.

Some potential advantages of catalytic reactors are:

1. Low fuel requirements
2. Lower operating temperatures
3. Little or no insulation requirements
4. Reduced fire hazards
5. Reduced flashback problems.

The potential disadvantages include:

1. High initial cost
2. Catalyst poisoning
3. (Large) particles must first be removed
4. Some liquid droplets must first be removed
5. Catalyst regeneration problems.

Flares

Although not treated in any detail in this text, flares provide another unit for reacting species. Elevated, ground level, and open pit flares may be used. However, ground level flares have been outlawed in most states; open pit burning is used only under catastrophic conditions. Elevated flares are usually employed to ensure sufficient dilution and subsequent dispersion from adjacent structures and potential receptors of the energy and end products of chemical reaction(s). Flares find their primary application in the petroleum and petrochemical industries. The suggested design procedures are strictly empirical with some crude momentum and meteorological effects included in the equations. The reader is referred to the literature for additional information.[1]

Incinerators

The *incinerator* is one of the most popular forms of waste treatment in inducing a chemical reaction, especially when organics are involved. It can destroy and/or react most organic components and the resulting product is almost always less "organic" than the original feed.

The objective of the incineration process is another important factor that influences reactor selection. It may be desired, for example, to destroy the hazardous components in the waste, reduce the hazardous components and then isolate them, or merely separate the hazardous components. The end product of waste treatment must be compatible with the type of ultimate disposal to be used for the waste: sewage discharge, landfilling, deep-well injection, and so forth. For example, if the waste is to be disposed of by landfarming, the end product of the waste treatment should consist of biodegradable organic components.(2)

As mentioned earlier, the chemical reaction sequence that takes place during incineration is a complicated process. Several intermediate products appear before the original material is completely oxidized into carbon dioxide, water, nitrogen oxides, and so on. However, most of the intermediate products exist for only a very short time, and for engineering purposes complete reaction of the organics is the principal concern.

Kilns

Kilns were originally designed for lime processing. In 1885, Frederick Ransom was the first to patent a rotary kiln. Peray provides an excellent summary of the early kiln.(3) Recently, the rotary kiln has been utilized to process hazardous waste because it has the unique capability to stabilize the combustion process for many different materials that may be fed simultaneously. Kilns have demonstrated the ability to handle a wide variety of materials with minimal impact on its performance. This results in minimal preprocessing of the wastes and is one of the main reasons for the kiln's popularity.

The processing of raw materials has been a historic application of the rotary kiln. The basic building materials of a community—cement, lime, coal, iron ore, aggregates, and other bulk materials—require heating to high temperatures. These materials have been produced in the past in a rotary kiln. This high-temperature treatment and the continual mixing and blending of the raw materials provide the basis for the processing of hazardous waste. Lime, lightweight aggregate, cement, and other process kilns may operate at temperatures up to 1925°C (3500°F). Hazardous waste incinerators operate in the range of 700–1315°C (1300–2400°F) depending on the waste feed types and shapes.(4)

The kiln is a cylindrical refractory-lined shell that is mounted at a slight incline from the horizontal plane to facilitate mixing the materials and exposing the surface to the auxiliary burner flame as well as both the fuel flames and flames generated over the burning surface of solids. The kiln accepts all types of solid and liquid materials with heating values between 555 and 8333 kcal/kg (1000 and 15,000 Btu/lb), and even higher. Solid wastes and drummed wastes are usually fed by a pack-and-drum feed system, which may consist of a bucket elevator for loose solids and a conveyor system.

The kiln combines three very important aspects of waste material processing into one piece of equipment: mixing, temperature, and time. The mixing process is well known in the production of cement where lime, iron, and sand or clay are mixed while being heated. Good mixing is required to obtain a quality product. The processing equipment must allow for the wide variations in the material. The material that is placed into the kiln must have sufficient time for its components to volatilize. With proper time, the heat source within the kiln can raise the material to the desired temperature. In particular, the rotary kiln provides excellent mixing through the tumbling action to distribute the heat uniformly to all of the material being processed and to expose all of the material to the heat source. Not surprisingly, the original kilns were called "tumble burners."

The exit end of the kiln must be designed to handle the physical characteristics of the material leaving the kiln. Systems that cause slagging must keep the slag heated until it exits the kiln. If there is cooling on the exit end, the slag may freeze (solidify) and plug the end of the unit.

Vertical kilns are used in operations where no fusion or softening takes place, as in the burning of limestone or dolomite, though rotary kilns are also used for these particular operations. The vertical kiln is the antecendent of modern moving-bed equipment, such as the pebble heater and catalytic reactors of this type.

Furnaces

Hearth furnaces consist of one or more flat or concave pans, either moving or stationary, usually equipped with scraper-stirrers. Though this equipment is mostly used for ore treating and metallurgical purposes, a few inorganic chemicals are made this way, including soda ash, sodium sulfide from sodium sulfate and coal, and sodium sulfate and hydrogen chloride from salt and sulfuric acid.

Fixed-hearth incinerators, used extensively for medical waste incineration, are less commonly used for other materials. There are several units employed at commercial sites as well as plant sites. Fixed hearths can handle bulk solids and liquids; however, this unit does not have the versatility of a kiln.[5]

Fixed-hearth incinerators typically contain two furnace chambers: a primary combustion chamber (PCC) and a secondary combustion chamber (SCC). Solids and liquids may be charged into the PCC. Small units are normally batch-fed while larger units may be continuously fed with a screw feeder or moving grate or semi-continuously fed with a ram feeder. In some designs, there may be two or three step hearths on which any ash and feed are pushed with rams through the system. A controlled flow of "underfire" combustion air is introduced, usually up through the hearth on which the feed sits. In some designs, combustion air may also be provided from the front wall over the bed.

In many fixed-hearth incinerators, known as "controlled-air" or "starved-air" incinerators, 70–80% of the stoichiometric air required is provided in the PCC; thus, the PCC is operated in a reducing, substoichiometric or starved air mode. In this situation, the feed is pyrolyzed and partially combusted. In some cases, steam is injected into the PCC.

The multiple-hearth incinerator (commonly called the Herreshoff furnace) is a flexible unit that has been utilized to dispose of sewage sludges, tars, solids, gases, and liquid combustible wastes. This type of unit was initially designed to incinerate sewage plant sludges in 1934. In 1968, there were over 125 installations in operation with a total capacity of 17,000 tons/day (wet basis) for this application alone.[6]

A typical multiple-hearth furnace includes a refractory-lined steel shell, a central shaft that rotates, a series of solid flat hearths, a series of rabble arms with teeth for each hearth, a combustion air blower, fuel burners mounted on the walls, an ash removal system, and a feed system. Side ports for tar injection, liquid burners, and an SCC may also be included. Sludge and/or granulated solids are normally fed through the furnace roof by a screw feeder of belt(s) and flapgate(s). The rotating air cooled central shaft with air-cooled rabble arms and teeth distributes the feed material across the top hearth to drop holes. The feed falls to the next hearth and then the next until discharged as ash at the bottom. The feed is agitated as it moves across the hearths to make sure fresh surface is exposed to hot gases.

Units range from 6 to 25 ft in diameter and from 12 to 75 ft in height. The diameter and number of hearths are dependent on the feed, the required processing time, and the type of thermal processing employed. Generally, the uppermost hearth is used as the SCC. Normal incineration usually requires a minimum of six hearths, while pyrolysis applications require a greater number.

The rabble arms and teeth located on the central shaft all rotate in the same direction; additional agitation of the feed (back rabbling) is accomplished by reversing the angles of the rabble teeth. Retention time is controlled by the design of the rabble tooth pattern and the rotational speed of the central shaft. Liquid and/or gaseous feed may be injected into the unit through auxiliary burner nozzles; this utilization of liquids and gases represents an economic advantage because it reduces auxiliary fuel requirements, thus lowering operating costs.

ANCILLARY EQUIPMENT

In addition to the chemical reactor, certain other ancillary equipment must be included in a final design and/or for the operation of the overall process. These include:

1. Equipment for energy recovery
2. Control instrumentation
3. Prime movers
4. Ductwork
5. Valves
6. Fittings
7. Support structure(s)
8. Insulation
9. Environmental control equipment (if applicable).

Details of each of these are briefly discussed.

Equipment for Energy Recovery

Following chemical reaction, the process gas stream may be at a temperature as high as 3000°F. Heat exchangers are employed to recover the energy content of these "hot" gases before discharge elsewhere or to the atmosphere. The extent of recovery and design of the exchanger are dictated by economic considerations, i.e., equipment cost vs fuel savings. The need for process steam routinely arises.

Control Instrumentation

Operating controls and safety instrumentation must be included in the final overall design of a reaction unit. Proper controls can help lead to operation at optimum conditions (high conversion at minimum operating cost) and reduce the possibility and extent of downtime. Safety instrumentation can reduce and/or eliminate explosion possibilities.

The operation of a reactor unit requires that the controls keep the system operating within an appropriate operating envelope. The envelope is based on many of the design, process, and (possibly) regulatory constraints. Additional controls may be installed to operate additional equipment needed for energy recovery, neutralization, or product(s) concentration. The control system should also be designed to vary one or more of the process variables to maintain the appropriate conditions for the reactor. These variations are programmed into the system based on the past experience of the unit manufactured. The operational parameters that may vary include the flow rate, any system airflow rate, temperatures, and system pressure.

Prime Movers

The primary moving device to overcome frictional losses is a fan or pump, which may be induced or forced. Pressure drops can range from as low as 3 inches to as high as 25 inches water gage for most units. Generally, the pressure drop increases with an increase in throughput velocity, a decrease in the diameter of the reactor, presence of catalyst, extent of ductwork, and complexity of any heat exchanger units (number of passes and tubes). An induced (draft) unit is generally preferred since it often handles a process stream that is free of particulates and other troublesome components.

Ductwork

Duct and/or piping are necessary to deliver and discharge the process streams. The extent of the ductwork primarily depends on the location of the reactor relative to the main process and auxiliary equipment. Excessive ductwork can contribute significantly to the capital cost of the unit. In addition, it can lead to safety problems because of any potential condensation, even if it is insulated; process steam piping can also be a concern. Drainage piping often comes into play.

Valves[7]

Because of the diversity of the types of reactors, fluids, and environments in which they must operate, a vast array of valve types have been developed. Examples of the common types are the globe valve, gate valve, ball valve, plug valve, pinch valve, butterfly valve, and check valve. Each type of valve has been designed to meet specific needs. Included in this category are vents and traps (particularly steam traps). Some valves are capable of throttling flow, other valve types can only stop flow, others work well in corrosive systems, and others handle high pressure fluids. Each valve type has advantages and disadvantages.

Valves are also sometimes classified according to the resistance they offer to flow. The low resistance class of valves include the straight-through flow units, e.g., gate, ball, plug, and butterfly valves. Valves having a change in direction are usually high resistance valves; examples include globe and angle valves.

Fittings[7]

A fitting is a piece of equipment that has for its function one or more of the following:

1. The joining of two pieces of straight pipe (e.g., couplings and unions)
2. The changing of pipeline direction (e.g., elbows and Ts)
3. The changing of pipeline diameter (e.g., reducers and bushings)
4. The terminating of a pipeline (e.g., plugs and caps)
5. The joining of two streams (e.g., Ts and Ys).

Fittings may be classified as reducing, expanding, branching or deflecting. Reducing or expanding fittings are ones that change the area for flow; these include reducers, bushings, and sudden expansions and contractions. Branch fittings are Ts, crosses, or side outlet elbows. Deflecting fittings change the direction of flow, e.g., elbows and bends.

Supporting Structures

The weight of a chemical reactor and the ancillary equipment is also an important consideration. In many instances, structural support is necessary to relieve the additional load placed on the roof or floor of a building. Supports for piping (including overhead piping) is also a concern.

Insulation

Insulation prevents an excessive flow of costly heat to the surroundings from piping and the reactor in which heat is transferred from temperatures above the surrounding temperature. It also provides for protection of personnel against skin damage from high- or low-temperature surfaces. Insulation is just as useful in cryogenic operations[8] in preventing the flow of heat from the surrounding to materials which must be kept at temperatures below that of the surroundings.

The *proper* selection of insulating material is an engineering problem. Certain principles can be listed for guidance.[9]

1. Select the material to withstand working temperature ranges.
2. Choose an inexpensive low-thermal-conductivity type.
3. Choose a type and form which has sufficient durability and structural strength to withstand severe conditions; this includes resistance to moisture and (if applicable) the chemical environment.
4. Select the most economic insulation.

Environmental Control Equipment

Control equipment is occasionally used upstream from a reactor to help remove and/or recover solid and/or liquid matter in the process stream. Gaseous control problems occasionally arise. The reader is referred to the literature for a detailed review of both particulate and gaseous control equipment.[10,11]

Safety Devices and Practices

The reader is referred to Part IV, Chapter 21 in the Accident and Emergency Management Section for details.

REFERENCES

1. Control Techniques for Particulate Air Pollutants, Natural Air Pollution Control Administration Publication, Publ. no. AP-51, RTP, NC, 1969.
2. J. Santoleri, J. Reynolds, and L. Theodore, "*Introduction to Hazardous Waste Incineration*," 2nd edition, John Wiley & Sons, Hoboken, NJ, 2000.
3. K. Peray, "*The Rotary Cement Kiln*," Chemical Publishing Co., New York City, NY, 1986.
4. J. Santoleri, "Hazardous Waste Incineration Systems, Permit Procedures, Design and Operation," AWMA Continuing Education Course, WST-258 (Section 6), 84th Annual Meeting, Vancouver, BC, Canada, June, 1991.
5. U.S. EPA, Combustion Emissions Technical Resource Document, EPA 530-R-94-014, (CETRED) May, 1994.
6. C. Brunner, "*Design of Sewage Sludge Incineration Systems*," Noyes Data Corp., Park Ridge, NJ, 1980.
7. P. Abulencia and L. Theodore, "*Fluid Flow for the Practicing Engineer*," John Wiley & Sons, Hoboken, NJ, 2009.
8. L. Theodore, "*Heat Transfer for the Practicing Engineer*," John Wiley & Sons, Hoboken, NJ, 2011.
9. F. Vilbrandt and C. Dryden (adapted from), "*Chemical Engineering Plant Design*," McGraw-Hill, New York City, NY, 1959.
10. L. Theodore, "*Air Pollution Control Equipment Calculations*," John Wiley & Sons, Hoboken, NJ, 2008.
11. M. K. Theodore and L. Theodore, "*Introduction to Environmental Management*," CRC Press/Taylor & Francis, Boca Raton, FL, 2010.

CHAPTER 7

THE CONSERVATION LAWS

INTRODUCTION

The three principles underlying the describing equations of chemical reactors are: 1) the conservation law of mass, 2) the conservation law of energy, and 3) the conservation law of momentum. The first is expressed in the form of a material (mass) balance, the second in the form of an energy balance, and the third in the form of a momentum balance. These three topics are discussed in this chapter. The first law (mass) is used extensively in the rest of the chapters in Part II. The second law (energy) is discussed in the treatment of thermal effects in Part III. The third law (momentum) fundamentals assist in determining pressure drop(s) in reactors.

The remaining sections of this chapter address three key topics as they apply to chemical reactors:

Conservation of Mass

Conservation of Energy

Conservation of Momentum.

CONSERVATION OF MASS

The conservation law for mass can be applied to any process or system. The general form of this law is given by Equation (7.1):

$$\{\text{mass in}\} - \{\text{mass out}\} + \{\text{mass generated}\} = \{\text{mass accumulated}\} \tag{7.1}$$

or on a time rate basis

$$\begin{aligned}\{\text{rate of mass in}\} - \{\text{rate of mass out}\} + \{\text{rate of mass generated}\}\\ = \{\text{rate of mass accumulated}\}\end{aligned} \tag{7.2}$$

This has also come to be defined as the continuity equation; it will be applied to each of the reactors studied in the three chapters to follow.

This equation may be applied either to the total mass involved or to a particular species, on either a mole or mass basis. In many reactor processes, it is often necessary

Chemical Reactor Analysis and Applications for the Practicing Engineer. By Louis Theodore

to obtain quantitative relationships by writing mass balances on the various elements in the system.

In order to isolate a system for study, it is separated from the surroundings by a boundary or envelope. This boundary may be real (e.g., the walls of a reactor) or imaginary. Mass crossing the boundary and entering the system is part of the *mass in* term in Equation (7.2), while that crossing the boundary and leaving the system is part of the *mass out* term.

Equation (7.2) may be written for any compound, the quantity of which is not changed by chemical reaction, and for any chemical element whether or not it has participated in a chemical reaction. It may be written for one piece of equipment, around several pieces of equipment, or around an entire process. It may be used to calculate an unknown quantity directly, to check the validity of experimental data, or to express one or more of the independent relationships among the unknown quantities in a particular problem situation.

This law can be applied to steady-state or unsteady-state (transient) processes and to batch or continuous reactor systems. A steady-state process is one in which there is no change in conditions (e.g., pressure, temperature, composition) or rates of flow with time at any given point in the system. The accumulation term in Equation (7.2) is then zero. (If there is no chemical or nuclear reaction, the generation term is also zero.) All other processes are unsteady-state. In a batch reactor process, a given quantity of reactants is placed in a container, and by chemical and/or physical means, a change is made to occur. At the end of the process, the container (or adjacent containers to which material may have been transferred) holds the product or products. In a continuous process, reactants are continuously removed from one or more points. A continuous process may or may not be steady-state. A coal-fired power plant, for example, operates continuously. However, because of the wide variation in power demand between peak and slack periods, there is an equally wide variation in the rate at which the coal is fired. For this reason, power plant problems may require the use of average data over long periods of time. However, most industrial operations are assumed to be steady-state and continuous.

As indicated previously, Equation (7.2) may be applied to the total mass of each stream (referred to as an *overall* or *total* material balance) or to the individual component(s) of the streams (referred to as a *componential* or *component* material balance). The primary task in preparing a material balance in reactor engineering calculations is often to develop the quantitative relationships among the streams. The primary factors, therefore, are those that tie the streams together. An element, compound, or unreactive mass, e.g., ash, that enters or leaves in a single stream or passes through a reactor unchanged is so convenient for this purpose that it may be considered a key to the calculations. If sufficient data is given about this component, it can be used in a component balance to determine the total masses of the entering and exiting streams. Such a component is sometimes referred to as a *key component*. Since a key component does not react in any process, it must retain its identity as it passes through the reactor. Obviously, except for nuclear reactions, elements may always be used as key components because they do not change identity even though they may be involved in a chemical reaction. Thus, CO (carbon monoxide) may be used as a key component only when it does not react, but C (carbon) may be used as a key

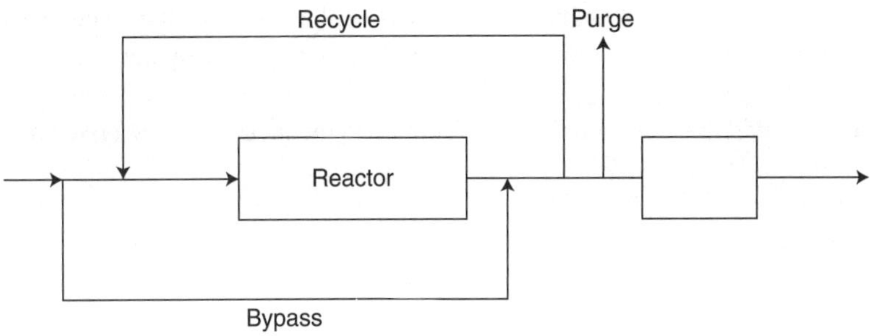

Figure 7.1 Recycle, bypass, and purge.

component. A component that enters the reactor in only one stream and leaves in only one stream is usually the most convenient choice for a key component.

Four important reactor processing concepts are *bypass*, *recycle*, *purge*, and *makeup*. With *bypass*, part of the inlet stream is diverted around the reactor to rejoin the (main) stream after the unit (see Figure 7.1), This stream effectively moves in parallel with the stream passing through the reactor. In *recycle*, part of the product stream is sent back to mix with the feed. If a small quantity of nonreactive material is present in the feed to a reactor that includes recycle, it may be necessary to remove the nonreactive material in a *purge* stream to prevent its building up above a maximum tolerable value. This can also occur in a reactor without recycle; if a nonreactive material is added in the feed and not totally removed in the products, it will accumulate until *purged*. The *purging* process is sometimes referred to as *blowdown*. *Makeup*, as its name implies, involves adding or making up part of a stream that has been removed from a reactor. *Makeup* may be thought of, in a final sense, as the opposite *of purge* and/or *blowdown*.[1,2]

ILLUSTRATIVE EXAMPLE 7.1 A continuous reactor process whose inlet, outlet, and internal system conditions do not vary with time but change as they proceed through the chemical reactors is best described by which statement provided below.

A steady, perfectly mixed
B steady, not perfectly mixed
C unsteady, perfectly mixed
D unsteady, not perfectly mixed

Solution. Consider if this system is a steady-state process. Since inlet, outlet, and internal system conditions do not vary with time, the system is at steady-state.

Consider if the system is perfectly mixed. Since conditions vary within the system, the system is imperfectly mixed.

Describe the conditions of the system. The best description of this system is steady state, imperfectly mixed. The correct answer is therefore (B). ■

ILLUSTRATIVE EXAMPLE 7.2 A gaseous material is fed into a reactor at a rate of 4000 kg/h in the presence of 8000 kg/h of air. Due to the low heating value of the feed stream, 550 kg/h of methane is added to assist the reaction. Determine the rate of product gases exiting the reactor in kg/h. Assume steady-state operation.

Solution. Apply the conservation law for mass to the incinerator on a rate basis. See Equation (7.2).

$$\{\text{rate of mass in}\} - \{\text{rate of mass out}\} + \{\text{rate of mass generated}\}$$
$$= \{\text{rate of mass accumulated}\}$$

Rewrite the equation subject to the (steady-state) conditions in the example statement.

$$\{\text{rate of mass in}\} = \{\text{rate of mass out}\}$$

or

$$\dot{m}_{\text{in}} = \dot{m}_{\text{out}}$$

Calculate $\dot{m}_{\text{in}}$

$$\dot{m}_{\text{in}} = 4000 + 8000 + 550 = 12{,}550 \text{ kg/h}$$ ■

ILLUSTRATIVE EXAMPLE 7.3 Water ($\rho = 1000 \text{ kg/m}^3$) flows in a converging circular pipe (see Figure 7.2) prior to entering a reactor. It enters at Section 1 and leaves at Section 2. At Section 1, the inside diameter is 14 cm and the velocity is 2 m/s. At Section 2, the inside diameter is 7 cm. Determine the mass and volumetric flow rates, and the mass flux of water. Assume steady-state flow.

Solution. Calculate the flow rates, q and $\dot{m}$, based on the information at Section (station) 1:

$$S_1 = \frac{\pi D_1^2}{4} = \frac{\pi (0.14)^2}{4} = 0.0154 \text{ m}^2$$
$$q_1 = S_1 v_1 = 0.0154(2) = 0.031 \text{ m}^3/\text{s}$$
$$\dot{m}_1 = \rho q_1 = 1000(0.031) = 31 \text{ kg/s}$$

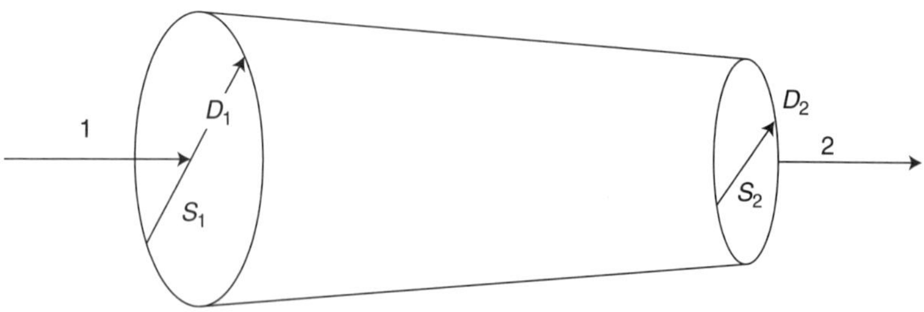

Figure 7.2 Converging pipe.

Obtain the mass flux, G, with units of $kg/m^2 \cdot s$.

$$G = \frac{m_1}{S_1} = \frac{31}{0.0154} = 2013 \text{ kg/m}^2 \cdot \text{s} \qquad \blacksquare$$

ILLUSTRATIVE EXAMPLE 7.4 Refer to the previous Illustrative Example. Calculate the velocity at Section 2.

Solution. For steady flow,

$$q_1 = q_2 = 0.031 \text{ m}^3/\text{s}$$

Since

$$v_2 S_2 = v_1 S_1$$

$$v_2 = v_1 \frac{S_1}{S_2} = v_1 \frac{D_1^2}{D_2^2} = (2)\frac{(14)^2}{(7)^2} = 8 \text{ m/s}$$

As expected, the decrease in cross-section area results in an increase in the flow velocity for steady-state flow of an incompressible fluid. ■

ILLUSTRATIVE EXAMPLE 7.5 A vertical tank 1.4 m in diameter and 1.9 m high, contains coolant water for a reactor at a height of 1.5 m. Water flows into the tank through a 9 cm pipe with a velocity of 4 m/s. Water leaves the tank through a 4 cm pipe at a velocity of 3 m/s. Is the level in the tank rising or falling?

Solution. Select the control volume (CV). Take the CV to be the instantaneous mass (or volume V) of water in the tank and apply the continuity principle to the CV. Since the generation rate = 0,

$$\left(\frac{dm}{dt}\right) = \dot{m}_{\text{in}} - \dot{m}_{\text{out}}$$

For an incompressible fluid (ρ = constant) of volume V, $q = \dot{m}/\rho$ so that

$$\frac{dV}{dt} = q_{\text{in}} - q_{\text{out}}$$

Since

$$V = Sz; \quad \text{where } z = \text{ fluid height (water)}, S = \text{ pipe area}$$

$$\frac{dV}{dt} = S\frac{dz}{dt}$$

Therefore

$$S\frac{dz}{dt} = q_{\text{in}} - q_{\text{out}}$$

Calculate the cross-section area and volumetric flow rates

$$q_{\text{in}} = \frac{\pi D^2}{4} v = \frac{\pi (0.09\ \text{m})^2}{4} 4\ \text{m/s}$$

$$q_{\text{in}} = 0.0255\ \text{m}^3/\text{s}$$

$$q_{\text{out}} = \frac{\pi D^2}{4} v = \frac{\pi (0.04\ \text{m})^2}{4} 3\ \text{m/s}$$

$$q_{\text{out}} = 0.0038\ \text{m}^3/\text{s}$$

$$S = \pi (1.4\ \text{m})^2/4$$

$$S = 1.539\ \text{m}^2$$

Substitute in the above material balance differential equation

$$1.539 \frac{dz}{dt} = 0.0255 - 0.0038$$

$$\frac{dz}{dt} = +0.0141\ \text{m/s}$$

$$\text{at} \quad t = 0, \ z = 1.5\ \text{m}$$

Because dz/dt is positive, the water level is rising in the tank from its initial height of 1.5 m. ■

ILLUSTRATIVE EXAMPLE 7.6 Given the following inlet loading and outlet loading of an air pollution particulate control unit immediately following a chemical reactor, determine the collection efficiency of the unit.

$$\text{Inlet loading} = 2.0\ \text{gr/ft}^3$$

$$\text{Outlet loading} = 0.1\ \text{gr/ft}^3$$

Solution. Collection efficiency is a measure of the degree of performance of a control device; it specifically refers to the degree of removal of a component and may be calculated through the application of the conservation law for mass. *Loading* refers to the concentration of pollutant, usually in grains (gr) of pollutant per cubic feet of contaminated gas stream.

The equation describing collection efficiency (fractional), E, in terms of inlet and outlet loading is

$$E = \frac{\text{Inlet loading} - \text{Outlet loading}}{\text{Inlet loading}} \tag{7.3}$$

Calculate the collection efficiency of the control unit in percent for the rates provided.

$$E = \frac{2.0 - 0.1}{2.0} 100 = 95\%$$

The term η is also used as the symbol for efficiency E. ■

ILLUSTRATIVE EXAMPLE 7.7 A cyclone[3,4] is used to collect catalyst particles discharged from an overhead line of a fluid bed reactor with an efficiency of 60%. A venturi scrubber[3,4] is used as a second control device. If the required overall efficiency is 99.0%, determine the minimum collection efficiency of the venturi scrubber.

Solution. Many process systems require more than one piece of equipment to accomplish a given task, e.g., recovery of a gaseous or particulate discharge from a reactor. The efficiency of each individual collector or equipment may be calculated using the procedure set forth in the previous example. The overall efficiency of multiple collectors may be calculated from the inlet stream to the first unit and the outlet stream from the last unit. It may also be calculated by proceeding sequentially through the series of collectors.

Calculate the mass of particulate leaving the cyclone using a *basis* of 100 lb of overhead particulates entering the unit. Use the efficiency equation:

$$E = (m_{\text{in}} - m_{\text{out}})/(m_{\text{in}}) \tag{7.3}$$

where E = fraction efficiency
m = loading (mass basis)

Rearranging the above gives:

$$m_{\text{out}} = (1 - E)(m_{\text{in}}) = (1 - 0.6)(100) = 40 \text{ lb}$$

Calculate the mass of particulates leaving the venturi scrubber using an overall efficiency of 99.0%:

$$m_{\text{out}} = (1 - E)(m_{\text{in}}) = (1 - 0.99)(100) = 1.0 \text{ lb}$$

Calculate the efficiency of the venturi scrubber using m_{out} from the cyclone as m_{in} for the venturi scrubber. Use the same efficiency equation above and convert to percent efficiency:

$$E = (m_{\text{in}} - m_{\text{out}})/(m_{\text{in}}) = (40 - 1.0)/(40) = 0.975 = 97.5\%$$

An extremely convenient efficiency-related term employed in recovery/control calculations is the penetration, P.[3,4] By definition:

$$\begin{aligned} P &= 100 - E \text{ (percent basis)} \\ P &= 1 - E \text{ (fractional basis)} \end{aligned} \tag{7.4}$$

Note that there is a 10-fold increase in P as E goes from 99.9 to 99%. For a multiple series of n collectors, the overall penetration is simply given by:

$$P = P_1 P_2, \ldots, P_{n-1} P_n \tag{7.5}$$

■

ILLUSTRATIVE EXAMPLE 7.8 A reactor is designed to destroy hazardous waste at 2100°F and 1 atm. Current regulations dictate that a minimum *destruction and removal efficiency* (*DRE*) of 99.99% must be achieved. The waste flowrate into the reactor is 960 lb/h while that flowing out of the reactor is measured as 0.08 lb/h. Is the reactor in compliance?

Solution. Select as a basis 1 h of operation. The mass equation employed for efficiency may also be used to calculate the minimum destruction and removal efficiency:

$$\begin{aligned} DRE &= [(m_{\text{in}} - m_{\text{out}})/m_{\text{in}}](100) \\ &= [(960 - 0.08)/960](100) \\ &= 99.992\% \end{aligned} \tag{7.3}$$

Thus, the reactor is operating in compliance with present regulations. The answer is *yes*. ■

CONSERVATION OF ENERGY

A presentation of the conservation law for energy would be incomplete without a brief review of some introductory thermodynamic principles. *Thermodynamics* is defined as that science that deals with the relationships among the various forms of energy. A system may possess energy due to its temperature (internal energy), velocity (kinetic energy), position (potential energy), molecular structure (chemical energy), surface (surface energy), and so on. Engineering thermodynamics is founded on three basic laws. As described earlier, energy, like mass and momentum, is conserved. Application of the conservation law for energy gives rise to the first law of thermodynamics. This law for batch processes, is presented below.

For batch reactor processes

$$\Delta E = Q + W \tag{7.6}$$

where potential, kinetic, and other energy effects have been neglected and Q is energy in the form of heat transferred across the system boundaries of the reactor, W is energy in the form of work transferred across system boundaries, E (often denoted as U) is the internal energy of the system and ΔE is the change in the internal energy of the system. In accordance with the recent change in convention, both Q and W are treated as *positive* terms if *added* to the system.

By definition, a flow process involves material streams entering and exiting a reactor. Work is done on the system at the stream entrance when the fluid is pushed into the system. Work is performed by the system to push the fluid out at the stream exit. The net work on the system is called flow work, W_f, and is given by

$$W_f = \sum P_{\text{out}} V_{\text{out}} - \sum P_{\text{in}} V_{\text{in}} = -\Delta(PV) \tag{7.7}$$

where P_{out} is the pressure of the outlet stream, P_{in} is the pressure of the inlet stream, V_{out} is the volume of fluid exiting the system during a given time interval, and V_{in}

is the volume of fluid entering the system during a given time interval. If the volume term is represented as the specific volume (i.e., volume/mass), the work term carries the units of energy/mass.

For practical purposes, the total work term, W, in the first law may be regarded as the sum of shaft work, W_s, and flow work, W_f

$$W = W_s + W_f \tag{7.8}$$

where W_s is work done on the fluid by some moving part within the system such as the rotating vanes of a centrifugal pump. Note that in Equation (7.8), all other forms of work such as electrical, surface tension, and so on, are neglected. The first law for steady-state *flow* processes is then:

$$\Delta H = Q + W_s \tag{7.9}$$

where H is the enthalpy of the system and ΔH is the change in the system's enthalpy.

The internal energy and enthalpy in Equations (7.6) and (7.7), as well as other equations in this section may be on a mass basis, on a mole basis, or represent the total internal energy and enthalpy of the entire system. They may also be written on a time-rate basis as long as these equations are dimensionally consistent—it makes no difference. For the sake of clarity, upper case letters (e.g., H, E) represent properties on a mole basis, while lower-case letters (e.g., h, e) represent properties on a mass basis. Properties for the entire system will rarely be used and therefore require no special symbols.

Perhaps the most important thermodynamic function the engineer works with is the above mentioned *enthalpy*. This is a term that requires additional discussion. The enthalpy is defined by the equation

$$H = E + PV \tag{7.10}$$

where P is once again the pressure of the system and V is the volume of the system. The terms E and H are state or point functions. By fixing a certain number of variables upon which the function depends, the numerical value of the function is automatically fixed, i.e., it is single-valued. For example, fixing the temperature and pressure of a one-component single-phase system immediately specifies the enthalpy and internal energy. This last statement can be verified by Gibbs's Phase Rule.[3]

The change in enthalpy as it undergoes a change in state from (T_1, P_2) to (T_2, P_2) is given by

$$\Delta H = H_2 - H_1 \tag{7.11}$$

Note that H and ΔH are independent of the path. This is a characteristic of all state or point functions, i.e., the state of the system is independent of the path by which the state is reached. The terms Q, W, and W_s in Equations (7.6) and (7.9) are path functions; their values depend on the path used between the two states.

The following can be written for a mathematical representation of this important pure component thermodynamic point function

$$H = H(T, P)$$

By the rules of partial differentiation, a differential change in H is given by

$$dH = \left(\frac{\partial H}{\partial T}\right)_P dT + \left(\frac{\partial H}{\partial P}\right)_T dP \tag{7.12}$$

The term $(\partial H/\partial P)_T$ is assumed to be negligible in most engineering applications. It is exactly zero for an ideal gas and is small for solids and liquids, and gases near ambient conditions. The term $(\partial H/\partial T)_P$ is defined as the *heat capacity at constant pressure*:

$$C_P = \left(\frac{\partial H}{\partial T}\right)_P \tag{7.13}$$

Equation (7.13) may also be written as

$$dH = C_P\, dT \tag{7.14}$$

If average molar heat capacity data are available, this equation may be integrated to yield

$$\Delta H = \overline{C_P}\Delta T \tag{7.15}$$

where $\overline{C_P}$ is the average value of C_P in the temperature range ΔT. Calculations involving enthalpy changes finds extensive application in chemical reactor analyses.

Many industrial reactor applications operate in a steady-state flow mode with no significant mechanical or shaft work added (or withdrawn) from the system. For this condition Equation (7.9) reduces to

$$Q = \Delta H \tag{7.16}$$

This equation is routinely used in many calculations. If a reactor or system is operated adiabatically, $Q = 0$ and Equation (7.16) becomes

$$\Delta H = 0 \tag{7.17}$$

Although the topics of material and energy balances have been covered separately in this and the previous section, it should be emphasized that this segregation does not exist in reality. Many processes are accompanied by heat effects, and one must work with both energy and material balances simultaneously.[5,6]

Equations (7.6) and (7.9) find application in not only many chemical reactor units but also heat exchangers and distillation columns, where shaft work plus kinetic and potential energy changes are negligible compared with heat flows and either internal energy or enthalpy changes. Energy balances on such units therefore reduce to $Q = \Delta E$ (closed system) or $Q = \Delta H$ (open system). The notation $\dot{Q}$, $\dot{E}$, and $\dot{H}$ are employed if these terms refer to a time rate basis.

ILLUSTRATIVE EXAMPLE 7.9 5.5 MW of heat is transferred from a gas as it flows through a reactor. The average heat capacity of the gas is 1090 J/(kg · °C), the gas mass flow rate, $\dot{m}$, is 9 kg/s and the gas inlet temperature, T_1, is 650°C. For this example, kinetic and potential energy effects are neglected. Furthermore, there is no shaft work. Determine the gas outlet temperature.

Solution. Since there are no kinetic, potential, or shaft work effects in this flow process, Equations (7.15) and (7.16) applies

$$\dot{Q} = \Delta\dot{H}$$

where $\Delta\dot{H} = \dot{m}\overline{C}_P\Delta T = \dot{m}\overline{C}_P(T_2 - T_1)$.

Solving for the gas outlet temperature, T_2,

$$T_2 = \frac{\dot{Q}}{\dot{m}\overline{C}_P} + T_1 = \frac{-5.5 \times 10^{+6}}{9(1090)} + 650 = 89^\circ\text{C}$$

Note that the sign of $\dot{Q}$ is negative since the heat is transferred out from the gas. ■

ILLUSTRATIVE EXAMPLE 7.10 A unique chemical reactor has three openings, as shown in Figure 7.3. The flow within the reactor is steady and the fluid has a constant density of 800 kg/m^3. The flow properties at each opening are provided in Table 7.1. What is the rate of change of the system energy in the reactor control volume? For steady-state adiabatic operation, is work being done on the system?

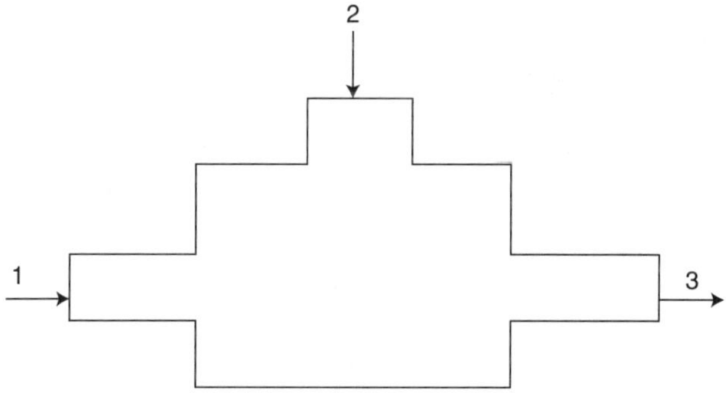

Figure 7.3 Fluid flow device for Illustrative Example 7.10.

TABLE 7.1 Flow/Energy Data for Illustrative Example 7.10

Section	Flow rate, q, m^3/s	Enthalpy, h, J/kg	Flow direction (relative to the device)
1	8	250	In
2	6	150	In
3	14	200	Out

Solution. Confirm the mass balance. For incompressible flow,

$$q_1 + q_2 = q_3$$
$$8 + 6 = 14\ \text{m}^3/\text{s}$$

Apply the total energy balance, noting that only enthalpy effects need to be considered for this flow system

$$\Delta\dot{H} = 800[(8)(250) + (6)(150) - (14)(200)] = 80{,}000\ \text{J/s}$$

For adiabatic steady operation

$$\dot{Q} = 0$$

so that

$$\Delta\dot{H} = 80\ \text{kW} = 107.2\ \text{hp}$$

Since work is positive, the surroundings must be doing work on the system through some device. ∎

ILLUSTRATIVE EXAMPLE 7.11 Determine the percentage of a river's flow available to an industry for cooling a nuclear reactor such that the river temperature does not increase more than 10°F. Fifty percent of the industrial withdrawal is lost by evaporation from the reactor and the industrial water returned to the river from the reactor is 60°F warmer than the river.

Note: This problem is a modified and edited version (with permission) of an illustrative example prepared by Ms. Marie Gillman, a graduate mechanical engineering student at Manhattan College.

Solution. Draw a flow diagram representing the process as shown in Figure 7.4. Express the volumetric flow lost by evaporation from the reactor in terms of that entering the process:

$$q_{\text{lost}} = 0.5q_{\text{in}}$$

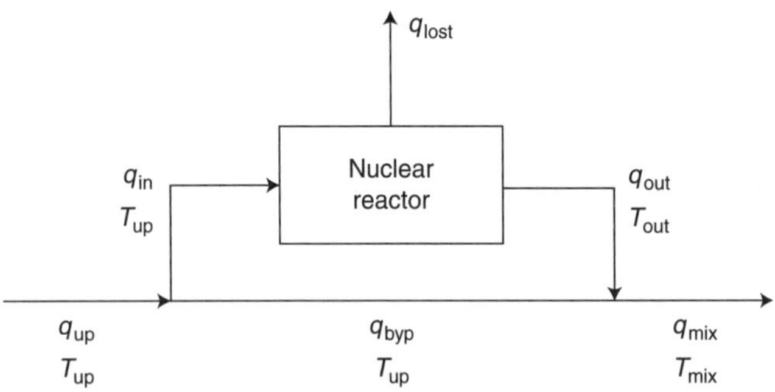

Figure 7.4 Flow diagram for Illustrative Example 7.11.

Express the process outlet temperature and the maximum river temperature in terms of the upstream temperature:

$$T_{\text{out}} = T_{\text{up}} + 60°\text{F}$$
$$T_{\text{mix}} = T_{\text{up}} + 10°\text{F}$$

Using the conservation law for mass, express the process outlet flow in terms of the process inlet flow. Also express the flow bypassing the reactor in terms of the upstream and process inlet flows:

$$q_{\text{out}} = 0.5q_{\text{in}}$$
$$q_{\text{byp}} = q_{\text{up}} - q_{\text{in}}$$
$$q_{\text{mix}} = q_{\text{up}} - 0.5q_{\text{in}}$$

The flow diagram with the expressions developed above are shown in Figure 7.5.

Noting that the enthalpy of any stream can be represented by $qc_p\rho(T - T_{\text{ref}})$, an energy balance around the downstream mixing point leads to

$$\begin{aligned}(q_{\text{up}} - q_{\text{in}})c_p\rho(T_{\text{up}} - T_{\text{ref}}) + 0.5q_{\text{in}}c_p\rho(T_{\text{up}} + 60 - T_{\text{ref}})\\ = (q_{\text{up}} - 0.5q_{\text{in}})c_p\rho(T_{\text{up}} + 10 - T_{\text{ref}})\end{aligned}$$

Note that T_{ref} is arbitrary and indirectly defines a basis for the enthalpy. Setting $T_{\text{ref}} = 0$ and assuming that density and heat capacity are constant yields

$$(q_{\text{up}} - q_{\text{in}})T_{\text{up}} + 0.5q_{\text{in}}(T_{\text{up}} + 60) = (q_{\text{up}} - 0.5q_{\text{in}})(T_{\text{up}} + 10)$$

The equation may now be solved for the inlet volumetric flow to the reactor in terms of the upstream flow:

$$\begin{aligned}q_{\text{up}}T_{\text{up}} &= q_{\text{in}}T_{\text{up}} + 0.5q_{\text{in}}T_{\text{up}} + 30q_{\text{in}}\\ &= q_{\text{up}}T_{\text{up}} + 10q_{\text{up}} - 0.5q_{\text{in}}T_{\text{up}} - 5q_{\text{in}}\end{aligned}$$

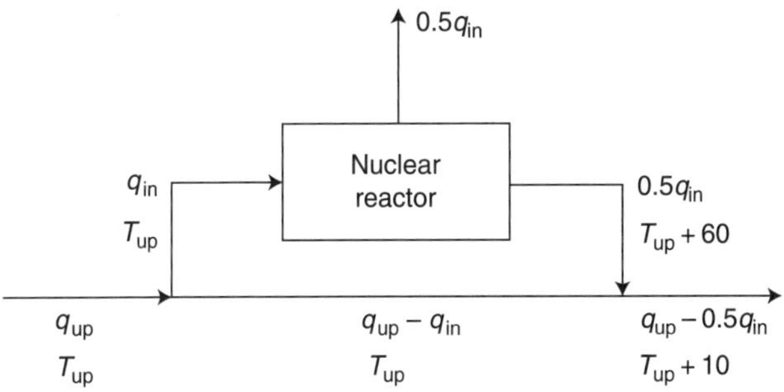

Figure 7.5 Flow diagram after applying mass balances.

Canceling terms produces

$$35q_{in} = 10q_{up}$$
$$q_{in} = 0.286q_{up}$$

Therefore, 28.6% of the original flow, q_{up}, is available for cooling the nuclear reactor.

Note that the problem can also be solved by setting $T_{ref} = T_{up}$. Since for this condition, $T_{ref} - T_{up} = 0$, the solution to the problem is greatly simplified. ■

Heat of Reaction from Heats of Formation

Of the almost infinite number of possible chemical reactions that can take place, very few have their standard heats (or enthalpies) of reaction at 298 K listed in handbooks. That would be a waste of printed space. A far more efficient method for providing this information exists. What is printed in the handbooks is *standard heat of formation* data at 298 K. A *heat of formation*, ΔH_f, for a chemical compound is the enthalpy change when 1 mole of that compound is synthesized from its elements. The heat of formation for an element such as the diatomic gases N_2 or O_2 (found naturally) would obviously be zero.

If the heats of formation for all reactants and products involved in a particular chemical reaction are known, the heat of reaction can be found by using Hess' law to combine the formation reactions. Hess's law can be stated as follows: When reactants are converted to products, the change in enthalpy is the same whether the reaction takes place in one step or in a series of steps.

Consider the following reaction:

$$C_6H_6 + (15/2)O_2 \longrightarrow 6CO_2 + 3H_2O$$

For a hypothetical path, assume that, as a first step, the reactants (one benzene and 7.5 oxygen molecules) simply break up into the component elements, that is, 6C, $3H_2$, and $(15/2)O_2$. (Actually, the oxygen does not break up.) For the second step, these elements recombine to form the products. Since the first step is the opposite of the formation reaction for one mole of benzene, its heat of reaction is the negative of benzene's heat of formation. The heat of reaction for the second step is the sum of the CO_2(g) heat of formation times 6 and the H_2O(l) heat of formation times 3. Adding the heats of reaction for each step yields the overall heat of reaction.

To generalize Hess' law, the ΔH_r for the hypothetical reaction:

$$aA + bB \longrightarrow cC + dD$$

is given by

$$\Delta H_r = c(\Delta H_f)_C + d(\Delta H_f)_D - a(\Delta H_f)_A - b(\Delta H_f)_B \tag{7.18}$$

ILLUSTRATIVE EXAMPLE 7.12 Calculate the heat of reaction (kJ) at 25°C for the reaction:

$$2Na_2O_2(s) + 2H_2O(l) \longrightarrow 4NaOH(s) + O_2(g)$$

from the following heat of formation data (also at 25°C, denoted by superscript zero):

$$NaOH(s): \quad \Delta H_f^0 = -426.8 \text{ kJ/mol}$$
$$Na_2O_2(s): \quad \Delta H_f^0 = -504.6 \text{ kJ/mol}$$
$$H_2O(l): \quad \Delta H_f^0 = -286.0 \text{ kJ/mol}$$
$$O_2(g): \quad \Delta H_f^0 = 0 \text{ kJ/mol}$$

Select the correct answer

A −1217.4
B +363.8
C −208.2
D −126.0

Solution. Write the equation for combining the heats of formation.

$$\Delta H_r^0 = 4(\Delta H_f^0)_{NaOH} + (\Delta H_f^0)_{O_2} - 2(\Delta H_f^0)_{Na_2O_2} - 2(\Delta H_f^0)_{H_2O}$$

Calculate the heat of reaction by substituting the heats of formation into the above equation, noting that the ΔH_f^0 for any element must be zero.

$$\Delta H_r^0 = (4)(-426.8) + 0 - 2(-504.6) - 2(-286.0)$$
$$= -1707.2 + 1009.2 + 572 = -126 \text{ kJ/mol}$$

The correct answer is therefore (D). ■

Standard heats of combustion, ΔH_c^0, at 25°C and 1 atm are also extensively tabulated in the literature. The main importance of these is the fact that many ΔH_f^0 s are derived by combining heat of combustion data using Hess' law. Combustion reactions are easy to carry out in the laboratory using a device called a *bomb calorimeter*.[5,6]

CONSERVATION OF MOMENTUM

Momentum transfer is introduced in this section by reviewing the units and dimensions of momentum, time rate of change of momentum, and force. A momentum balance (also termed the *impulse-momentum principle*) is important in chemical reactor problems where forces, velocities, pressure drop, and prime movers, need to be determined. This analysis is inherently more complicated than those previously presented (i.e., forces possess both magnitude and direction), because the force, *F*, and momentum, *M*, are vectors. In order to describe force and momentum vectors, both direction and magnitude must be specified; for mass and energy, only the magnitude is required.

Newton's law is applied in order to derive the linear momentum balance equation. Newton's law states that the sum of all forces equals the rate of change of linear momentum. The derivation/application of this law ultimately provides information critical to any meaningful analysis of most chemical reactors. Included in this information are such widely divergent topics as flow mechanisms, the Reynolds number, velocity profiles, two-phase flow, prime movers such as fans, pumps and compressors, pressure drop, flow measurement, valves and fittings, particle dynamics, flow through porous media and packed beds, fluidization—particularly as it applies to fluid-bed and fixed bed reactors, etc. Although much of this subject matter is beyond the scope of this text, all of these topics are treated in extensive detail by Abulencia and Theodore.(8)

ILLUSTRATIVE EXAMPLE 7.13 A liquid with a viscosity of 0.78 cP and a density of 1.50 g/cm^3 flows through a 1-in. diameter reactor tube at 20 cm/s. Calculate the Reynolds number. Is the flow laminar or turbulent?

Solution. By definition, the Reynolds number (Re) is equal to:

$$\text{Re} = \rho v D/\mu \tag{7.19}$$

where ρ = fluid density
v = fluid velocity
D = characteristic length, usually the conduit diameter of the reactor
μ = fluid viscosity

Since

$$1 \text{ cP} = 10^{-2} \text{ g/cm} \cdot \text{s}$$

$$\mu = 0.78 \times 10^{-2} \text{ g/cm} \cdot \text{s}$$

$$1 \text{ in} = 2.54 \text{ cm}$$

$$\begin{aligned} \text{Re} &= (1.50)(20)(2.54)/(0.78 \times 10^{-2}) \\ &= 9769.23 \approx 9800 \end{aligned}$$

As noted below, the value of the Reynolds number indicates the nature of fluid flow in a duct or pipe:

$\text{Re} < 2100$; flow is streamline (laminar)

$\text{Re} > 10{,}000$; flow is turbulent

$2100 \leq \text{Re} \leq 10{,}000$; transition region

For this problem, the flow is turbulent; others might classify it as not fully turbulent. ■

ILLUSTRATIVE EXAMPLE 7.14 A liquid, with specific gravity (SG) of 0.96, flows through a long reactor circular tube of radius $R = 3$ cm. The liquid has the following linear distribution of the axial velocity v (the velocity in the direction of the flow):

$$v\ (\mathrm{m/s}) = 6 - 200r$$

where r is the radial position (in meters) measured from the tube centerline.

Calculate the average velocity of the fluid in the reactor.

Solution. A real fluid in contact with a nonmoving wall will have a velocity of zero at the wall. Similarly, the fluid in contact with a wall moving at a velocity v will move at the same velocity v. This is termed "the no-slip condition" of real fluids. Thus, for a fluid flowing in a duct, the fluid velocity at the walls of a duct is zero and increases as it is displaced from the wall.

To calculate the volumetric flowrate, q, of the fluid passing through a perpendicular surface, S, one must integrate the product of the component of the velocity that is normal to the area over the whole cross-sectional area of the duct. This procedure is depicted in the equation

$$q = \int_S v\, dS \tag{7.20}$$

where q = volumetric flowrate ($\mathrm{m^3/s}$, or $\mathrm{ft^3/s}, \ldots$)

v = absolute velocity, or the normal component of the velocity to the area dS

S = surface area

The average velocity of fluid passing through the surface, S, is

$$v_{\text{av}} = q/S \tag{7.21}$$

On a differential level,

$$dq = v\, dS$$

and, for a cylindrical coordinate system[8,9]

$$dS = 2\pi r\, dr$$

Thus,

$$\begin{aligned} dq &= 2\pi r v\, dr \\ &= 2\pi r(6 - 200r)\, dr = 2\pi(6r - 200r^2)\, dr \end{aligned}$$

This equation can now be integrated between the limits of $r = 0$ and $r = R$ to give

$$q = 2\pi \int_0^R (6r - 200r^2)\, dr = 2\pi\left[3R^2 - \left(\frac{200}{3}\right) R^3\right] = 2\pi R^2\left[3 - \left(\frac{200}{3}\right) R\right]$$

The volumetric flowrate is then

$$q = 2\pi(0.03)^2(3 - 2) = 0.00565\ \text{m}^3/\text{s} = 0.200\ \text{ft}^3/\text{s}$$

Since $\rho = (\text{SG})(1000)\ \text{kg/m}^3$, the mass flowrate $\dot{m}$ is

$$\begin{aligned} \dot{m} &= (0.96)(1000)(0.00565) = 5.42\ \text{kg/s} \\ &= (5.42\ \text{kg/s})(1\ \text{lb}/0.454\ \text{kg}) = 11.95\ \text{lb/s} \end{aligned}$$

The average velocity may now be calculated:

$$S = \pi R^2 = \pi(0.03)^2$$

$$v_{\text{av}} = \frac{q}{S} = \frac{2\pi R^2(3 - 2)}{\pi R^2} = 2\ \text{m/s}$$

The reader should note that average velocity is a useful value in scale-up and flow analysis of chemical reactors. ■

ILLUSTRATIVE EXAMPLE 7.15 A 10-cm diameter horizontal line carries saturated steam at 420 m/s to a reactor. Water is entrained by the steam at the rate 0.15 kg/s. The line has a 90° bend. Calculate the force required to hold the bend in place due to the entrained water.

Solution. A line diagram of the system is provided in Figure 7.6. Select the fluid in the bend as the system and apply the conservation law for mass:

$$\dot{m}_1 = \dot{m}_2$$

Since the density and cross-sectional area are constant,

$$v_1 = v_2$$

where $\dot{m}_1, \dot{m}_2$ = mass flowrate at 1 and 2, respectively
v_1, v_2 = velocity at 1 and 2, respectively

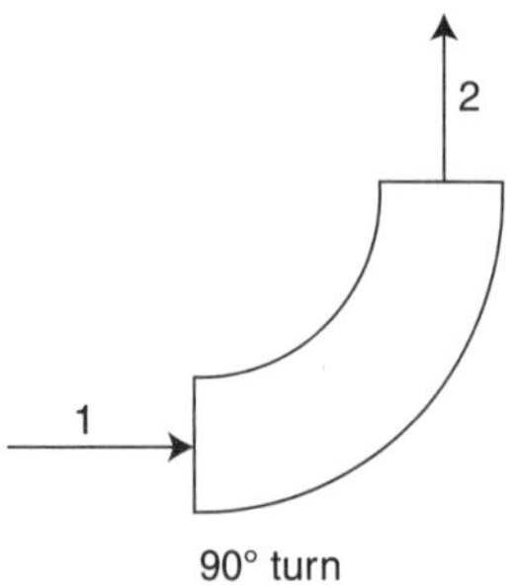

Figure 7.6 Diagram for Illustrative Example 7.15.

A linear momentum ($\dot{M}$) balance in the horizontal direction provides the force applied by the channel wall on the fluid in the x-direction, F_x:

$$F_x g_c = \dot{M}_{x,\text{out}} - \dot{M}_{x,\text{in}}$$
$$= \frac{d}{dt}(mv)_{x,\text{out}} - \frac{d}{dt}(mv)_{x,\text{in}}$$

Note that the equation assumes that the pressure drop across the bend is negligible.

Since $v_{x,\text{out}} = 0$ and $dm/dt = \dot{m}$,

$$F_x g_c = 0 - \dot{m}v_{x,\text{in}} = -\dot{m}v_2 = -(0.15)(420) = -63\ \text{N} = -14.1\ \text{lb}_\text{f}$$

The x-direction supporting force acting on the 90° elbow is 14.1 lb_f acting toward the left.

A linear momentum balance in the vertical direction results in

$$F_y g_c = \dot{M}_{y,\text{out}} - \dot{M}_{y,\text{in}}$$
$$= \dot{m}v_{y,\text{out}} - \dot{m}v_{y,\text{in}}$$
$$= \dot{m}v_2 - 0 = (0.15)(420) = 63\ \text{N} = +14.1\ \text{lb}_\text{f}$$

The y-direction supporting force on the 90° elbow is 14.1 lb_f acting upwards.

The resultant supporting force is therefore

$$F_\text{res} = \sqrt{F_x^2 + F_y^2}$$
$$= \sqrt{(-63)^2 + 63^2} = 89.1\ \text{N} = 19.1\ \text{lb}_\text{f}$$

The direction is given by

$$\tan\theta = \frac{F_y}{F_x} = \frac{63}{-63} = -1$$
$$\theta = 135^\circ$$

where θ is the angle between the positive x axis and the direction of the force. The counter-clockwise rotation of the direction from the x axis is defined as positive.

The supporting force for the feed line to the reactor is therefore 19.1 lb_f acting in the "northwest" direction. ■

REFERENCES

1. W. Badger and J. Banchero, "*Introduction to Chemical Engineering*," McGraw-Hill, New York City, NY, 1955.
2. D. Green and R. Perry (editors) "*Perry's Chemical Engineers' Handbook*," 8th edition, McGraw-Hill, New York, 2008.
3. J. Reynolds, J. Jeris, and L. Theodore, "*Handbook of Chemical and Environmental Engineering Calculations*," John Wiley & Sons, Hoboken, NJ, 2004.
4. L. Theodore, "*Air Pollution Control Equipment Calculations*," John Wiley & Sons, Hoboken, NJ, 2009.

5. R. Felder and R. Rousseau, "*Elementary Principles of Chemical Processes*," 3rd edition, John Wiley & Sons, Hoboken, NJ, 2000.
6. J. Santoleri, J. Reynolds, and L. Theodore, "*Introduction to Hazardous Waste Incineration*," 2nd edition, John Wiley & Sons, Hoboken, NJ, 2000.
7. L. Theodore, F. Ricci, and T. Van Vliet, "*Thermodynamics for the Practicing Engineer*," John Wiley & Sons, Hoboken, NJ, 2010.
8. P. Abulencia and L. Theodore, "*Fluid Flow for the Practicing Engineer*," John Wiley & Sons, Hoboken, NJ, 2009.
9. L. Theodore, "*Transport Phenomena for Engineers*," International Textbook Company, Scranton, PA, 1971.

CHAPTER 8

BATCH REACTORS

INTRODUCTION

A batch reactor is a solid vessel or container. Batch reactors were unquestionably the first type of reactor to be employed by industry. It may be open or closed. Reactants are usually added to the reactor simultaneously. The contents are then mixed (if necessary) to insure no spatial variations in the concentration of the species present. The reaction then proceeds. There is no transfer of mass into or out of the reactor during this period. The concentration of reactants and products change with time; thus, this is a *transient* or *unsteady-state* operation. The reaction is terminated when the desired chemical change has been achieved. The contents are then discharged and sent elsewhere, usually for further processing.

The batch reactor is one of the most common types of reactors used in industry. However, it is almost always the choice of the chemist working in the laboratory where processes of a chemical nature originate. It is simple, versatile, and easy to operate. Examples at the laboratory level include a test tube or a multi-neck flask. The simplest example at the industrial level is the autoclave. These units are primarily used for small-scale or complicated processes that manufacture expensive products, e.g., pharmaceuticals, dyestuff, and so on. These units almost always involve low production rates, often a major disadvantage with the conversion solely a function of reaction time. Production rates can be increased by either increasing the conversion or scaling up (increasing the size of) the reactor. Shipping limits the size of this class of reactor due to transportation problems. Larger units are therefore generally fabricated on site.

Chapter contents include:

Equipment Description and Operation

Describing Equations

Specific Reactions

Applications

It should be noted that semibatch units are not addressed in this chapter (see Chapter 9). In addition, the Applications section contains illustrative examples that require analytical solutions to the integrals appearing in some of the equations; these solutions are available in any standard table of integrals in the literature, with many of these provided in Part IV, Illustrative Example 21.18.

Chemical Reactor Analysis and Applications for the Practicing Engineer. By Louis Theodore

EQUIPMENT DESCRIPTION AND OPERATION

The batch reactor may operate:

1. Isothermally
2. Nonisothermally
3. Adiabatically
4. Any combination of the above.

A short description of these modes of operation (although previously discussed in Chapter 6), is repeated here for the convenience of the reader. In (1), the temperature is maintained constant during the course of the reaction. This condition can be approached by providing sufficient heat exchanger facilities to account for enthalpy effects arising during reaction. This mode of operation finds its major application in laboratory kinetic studies. Most industrial reactors are described by (2). Some energy in the form of heat is added to or removed from the reactor, but isothermal conditions are not satisfied. For (3), the reactor is insulated to minimize heat transfer between the reactor contents and the surroundings. Many industrial reactors attempt to operate in this manner. A detailed schematic of a batch reactor is presented in Figure 8.1.

As noted above, batch reactors are commonly used in experimental studies. Their industrial applications are somewhat limited. They are seldom used for gas

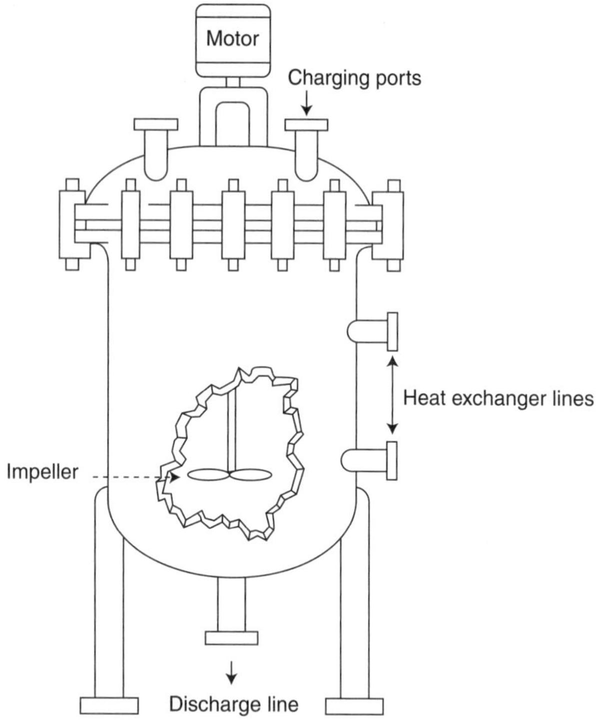

Figure 8.1 Typical batch reactor.

phase reaction, e.g., combustion, since small quantities (mass) of product are produced with even a very large-sized reactor. It is primarily used for liquid phase reactions when small quantities of reactants are to be processed and thus finds limited application in industry. As a rule, batch reactors are less expensive to purchase but more expensive to operate than either continuous stirred tank reactors (CSTRs) or tubular flow (TF) reactors.

ILLUSTRATIVE EXAMPLE 8.1 Use the following notation to answer the question below.

SS = steady state	PM = perfectly mixed
US = unsteady state	IM = imperfectly mixed

Batch reactors are most closely described by:

A SS, PM
B SS, IM
C US, PM
D US, IM

Solution

Consider answer (A). Batch reactors are *not* normally operated in the steady-state, perfectly mixed mode.

Consider answer (B). Batch reactors are *not* normally operated in the steady-state, imperfectly mixed mode.

Consider answer (C). Batch reactors are normally operated in the unsteady-state, perfectly mixed mode.

Consider answer (D). Batch reactors are *not* normally operated in the unsteady-state, imperfectly mixed mode.

The correct answer is therefore (C). ■

ILLUSTRATIVE EXAMPLE 8.2 Summarize the usage, advantages, and disadvantages of batch reactors.

Solution. Advantages include:

1. highest conversion/volume for one pass;
2. flexibility—several different products same day; and,
3. easy to clean.

Disadvantages include:

1. high operating cost; and,
2. product quality more variable than with continuous operation.

And, as noted earlier, usage includes small-scale of "intermediates" or "one shot" production. ■

DESCRIBING EQUATIONS

The describing equation for mass transfer is obtained by applying the conservation law for either mass or moles on a time rate basis to the contents of a batch reactor. As noted earlier, it is best to work with moles rather than mass in this and the following sections since the rate of reaction is most conveniently described in terms of molar concentrations.

Consider a batch reactor of arbitrary shape containing a homogeneous fluid that occupies a volume, V (see Figure 8.2). The reactor and its contents, are fixed in space. Assume:

1. The volume of the contents of the reactor remain constant during reaction.
2. The reactor contents are perfectly mixed; there are no spatial variations.

The general conservation law for moles of component A in the reaction mixture is applied to the volume V of the reactor on a time rate basis.

$$\underset{(1)}{\left\{\begin{array}{c}\text{rate of}\\ \text{moles A}\\ \text{in by}\\ \text{molecular}\\ \text{diffusion}\end{array}\right\}} - \underset{(2)}{\left\{\begin{array}{c}\text{rate of}\\ \text{moles A}\\ \text{out by}\\ \text{molecular}\\ \text{diffusion}\end{array}\right\}} + \underset{(3)}{\left\{\begin{array}{c}\text{rate of}\\ \text{moles A}\\ \text{in by}\\ \text{convection}\end{array}\right\}} - \underset{(4)}{\left\{\begin{array}{c}\text{rate of}\\ \text{moles A}\\ \text{out by}\\ \text{convection}\end{array}\right\}}$$

↓ ↓ ↓ ↓ ↓ ↓ ↓ ↓ ↓ ↓

concentration gradients bulk flow

$$+ \underset{(5)}{\left\{\begin{array}{c}\text{rate of}\\ \text{moles A}\\ \text{generated}\end{array}\right\}} = \underset{(6)}{\left\{\begin{array}{c}\text{rate of}\\ \text{moles A}\\ \text{accumulated}\end{array}\right\}}$$

↓ ↓ ↓ ↓ ↓ ↓ ↓ ↓ ↓

source inventory (8.1)

The above enumerates the rate of mass (in units of moles) of A conserved within the fluid contained in the reactor volume element V at time t. One now proceeds to evaluate the six terms in Equation (8.1).

Terms (1)–(2): This represents the net rate at which moles of A enter the system due to the presence of concentration gradients. Since the contents of the reactor are perfectly mixed, the concentration of A is uniform throughout V; there is therefore no concentration gradient at the wall. This term is zero.

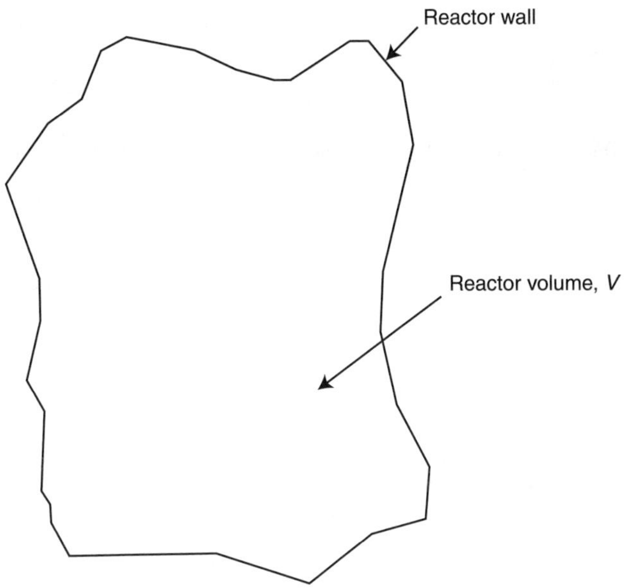

Figure 8.2 Control volume, V.

Terms (3)–(4): This is a convective term and arises due to A flowing into and/or out of the reactor. For batch reactors, no material is added to or removed from the system during the course of the reaction. This effect is also zero.

Term (5): This term represents the rate of A generated in the volume element due to chemical reaction. This source term is defined as r_A, the rate of reaction for A, and is (in engineering units) the amount of A generated per unit time per unit volume, i.e.,

$$r_A = \frac{\text{lbmol}}{\text{ft}^3 \cdot \text{s}}$$

The moles generated in V (ft^3) is given then by

$$(V)(r_A) \tag{8.2}$$

with units of lbmol/s.

Term (6): This is a measure of the rate of change of the number of moles of A within the system. If V is the volume of the reacting fluid and C_A is the molar concentration of A, lbmol A/ft^3, the term

$$C_A V$$

represents the moles of A contained in V. The rate of change of A with time at a *fixed* point in space is given by

$$\frac{\partial(C_A V)}{\partial t}$$

or

$$\frac{\partial N_A}{\partial t}$$

where N_A is the moles of A at time t. If the volume is constant, the above is

$$V\frac{\partial C_A}{\partial t}$$

Since C_A is solely a function of time and not a function of position, the above may be written as

$$V\frac{dC_A}{dt} \tag{8.3}$$

Substituting the terms of Equations (8.2) and (8.3) in Equation (8.1) give

$$\frac{dC_A}{dt} = r_A \tag{8.4}$$

or

$$\frac{1}{V}\frac{dN_A}{dt} = r_A$$

Information on r_A is usually available in kinetic and design applications.

The time necessary to obtain a given conversion may be required. Equation (8.4) is then written as

$$\int_0^t dt = \int_{C_{A_0}}^{C_A} \frac{dC_A}{r_A}$$
$$t = \int_{C_{A_0}}^{C_A} \frac{dC_A}{r_A} \tag{8.5}$$

In order to integrate the RHS (right-hand side) of Equation (8.5), r_A must be expressed in terms of C_A.

Various definitions of conversion are employed in reactor calculations. Much of this material was presented in Chapter 5 in Part I, but is repeated here for the convenience of the reader. Equations (8.4) and (8.5) are now rewritten in terms of new variables. Since

$$C_A = \frac{N_A}{V}$$

one may write

$$\frac{d\left(\frac{N_A}{V}\right)}{dt} = r_A$$

or

$$\frac{1}{V}\frac{dN_A}{dt} = r_A \tag{8.6}$$

One integrated form of Equation (8.6) is

$$t = \frac{1}{V}\int_{N_{A_0}}^{N_A} \frac{dN_A}{r_A} \tag{8.7}$$

In order to integrate the RHS of Equation (8.7), r_A must now be expressed in terms of N_A. If α_A is once again defined as the moles of A converted by chemical reaction, then

$$\alpha_A = N_{A_0} - N_A$$

and

$$-\,d\alpha_A = dN_A$$

Equation (8.6) becomes

$$-\frac{1}{V}\frac{d\alpha_A}{dt} = r_A \tag{8.8}$$

The integrated form of the above equation is

$$t = -\frac{1}{V}\int_0^{\alpha_A} \frac{d\alpha_A}{r_A} \tag{8.9}$$

And, to integrate Equation (8.9) requires expressing r_A in terms of α_A.

A conversion X_A may also be defined as the moles of A converted per mole of A initially present in the reaction mixture. For X_A, then

$$\begin{aligned} N_A &= N_{A_0} - N_{A_0}X_A \\ &- N_{A_0}(1 - X_A) \end{aligned}$$

and

$$dN_A = -N_{A_0}dX_A$$

Equation (8.6) is written

$$-\frac{N_{A_0}}{V}\frac{dX_A}{dt} = r_A$$

or

$$-C_{A_0}\frac{dX_A}{dt} = r_A \tag{8.10}$$

where C_{A_0} is the initial molar concentration of A. Equation (8.11) results if Equation (8.10) is integrated,

$$t = -C_{A_0}\int_0^{X_A} \frac{dX_A}{r_A} = C_{A_0}\int_0^{X_A} \frac{dX_A}{(-r_A)} \tag{8.11}$$

Another conversion term occasionally employed is the moles of A converted per *total* moles initially present. This conversion may be represented by X_A^* and then

$$N_A = N_{A_0} - N_0 X_A^*$$

and

$$dN_A = N_0 dX_A^*$$

where N_0 = total number of moles initially present. Equation (8.6) becomes

$$-\frac{N_0}{V}\frac{dX_A^*}{dt} = r_A$$

or

$$-C_0 \frac{dX_A^*}{dt} = r_A \tag{8.12}$$

where C_0 = initial (total) molar concentration. The integrated form of Equation (8.12) is

$$t = -C_0 \int_0^{X_A^*} \frac{dX_A^*}{(-r_A)} \tag{8.13}$$

Needless to say, r_A on the RHS of Equations (8.11) and (8.13) must be expressed in terms of X_A and X_A^*, respectively. All the above equations have their places in reactor studies. The choice is often a matter of convenience or preference, the form of the data, and the available expression for the rate of reaction. This topic is discussed in the next section.

SPECIFIC REACTIONS

The mass transfer equation developed in the previous section will now be applied to a simple kinetic system. Consider the first order *elementary irreversible* reaction:

$$A \longrightarrow \text{products}$$

For Equation (8.5), set

$$r_A = -k_A C_A$$

so that

$$\begin{aligned} t &= -\int_{C_{A_0}}^{C_A} \frac{dC_A}{k_A C_A} \\ &= \frac{1}{k_A} \ln\left(\frac{C_{A_0}}{C_A}\right) \end{aligned} \tag{8.14}$$

For Equation (8.7), then

$$r_A = -k_A\left(\frac{N_A}{V}\right)$$

and

$$t = -\int_{N_{A_0}}^{N_A} \frac{dN_A}{k_A N_A} = \frac{1}{k_A}\ln\left(\frac{N_{A_0}}{N_A}\right) \tag{8.15}$$

For Equation (8.9), then

$$r_A = -k_A\left(\frac{N_{A_0} - \alpha_A}{V}\right)$$

and

$$t = \int_0^{\alpha_A} \frac{d\alpha_A}{k_A(N_{A_0} - \alpha_A)} = \frac{1}{k_A}\ln\left(\frac{N_{A_0}}{N_{A_0} - \alpha_A}\right) \tag{8.16}$$

For Equation (8.11), set

$$-r_A = -k_A\left(\frac{N_{A_0} - N_{A_0}X_A}{V}\right) = -k_A\left(\frac{N_{A_0}}{V}\right)(1 - X_A) = k_A C_{A_0}(1 - X_A)$$

so that

$$t = \int_0^{X_A} \frac{dX_A}{k_A(1 - X_A)} = \frac{1}{k_A}\ln\left(\frac{1}{1 - X_A}\right) \tag{8.17}$$

Note that X is often used to represent X_A.

For Equation (8.13), then

$$r_A = -k_A\left(\frac{N_{A_0} - N_0 X_A^*}{V}\right)$$

so that

$$t = N_0 \int_0^{X_A^*} \frac{dX_A^*}{k_A(N_{A_0} - N_0 X_A^*)} = \frac{1}{k_A}\ln\left(\frac{N_{A_0}}{N_{A_0} - N_0 X_A^*}\right) \tag{8.18}$$

The above equations in t—Equations (8.14)–(8.18)—can easily be shown to be identical. One may also develop some of the relationships between the conversion variables.

$$X_A = \frac{\alpha_A}{N_{A_0}} \tag{8.19}$$

$$X_A^* = \frac{\alpha_A}{N_0} \tag{8.20}$$

$$X_A = \left(\frac{N_0}{N_{A_0}}\right) X_A^* \tag{8.21}$$

The solution, in terms of α_A, of the describing equations for several other kinetic systems are presented below.

1. Second order elementary irreversible reaction.

$$2A \longrightarrow \text{products}$$

For this reaction

$$r_A = -k_A C_A^2$$

and

$$t = \frac{V}{k_A N_{A_0}}\left(\frac{\alpha_A}{N_{A_0} - \alpha_A}\right) \tag{8.22}$$

2. Second order elementary irreversible reaction.

$$A + B \longrightarrow \text{products}$$

Here

$$r_A = -k_A C_A C_B$$

and

$$t = -\frac{V}{k_A(N_{B_0} - N_{A_0})} \ln\left[\frac{N_{B_0}(N_{A_0} - \alpha_A)}{N_{A_0}(N_{B_0} - \alpha_A)}\right] \tag{8.23}$$

This equation in t is valid only if

$$N_{A_0} \neq N_{B_0}$$

If

$$N_{A_0} = N_{B_0}$$

one uses the equation provided in Equation (8.22).

3. Third order elementary irreversible reaction.

$$3A \longrightarrow \text{products}$$

For this reaction

$$r_A = -k_A C_A^3$$

so that

$$t = \frac{V^2}{2k_A N_{A_0}^2}\left[\frac{2N_{A_0}\alpha_A - \alpha_A^2}{(N_{A_0} - \alpha_A)^2}\right] \tag{8.24}$$

4. First order elementary reversible reaction.

$$A \underset{k'_A}{\overset{k_A}{\rightleftharpoons}} B$$

Here

$$r_A = -k_A C_A + k'_A C_B$$

and

$$t = \frac{K}{k_A(K+1)}\ln\left[\frac{KN_{A_0} - N_{B_0}}{KN_{A_0} - (K+1)\alpha_A - N_{B_0}}\right] \tag{8.25}$$

where

$$K - \frac{k_A}{k'_A}$$

An alternative form of the above solution is

$$t = \frac{\alpha_{A_0}}{k_A N_{A_0}}\ln\left(\frac{\alpha_{A_e}}{\alpha_{A_e} - \alpha_A}\right) \tag{8.26}$$

where α_{A_e} = conversion of A at equilibrium.

The solution, in terms of concentration terms, of the describing equations for four other kinetic systems is presented below.

1. For the elementary irreversible reaction

$$A + B + C \longrightarrow \text{product}$$

one can show that

$$\begin{aligned} t = &\left[\frac{1}{k(C_{A_0} - C_{B_0})(C_{A_0} - C_{C_0})}\right]\ln\left(\frac{C_{A_0}}{C_A}\right) \\ &+ \left[\frac{1}{(C_{B_0} - C_{A_0})(C_{B_0} - C_{C_0})}\right]\ln\left(\frac{C_{B_0}}{C_B}\right) \\ &+ \left[\frac{1}{(C_{C_0} - C_{A_0})(C_{C_0} - C_{B_0})}\right]\ln\left(\frac{C_{C_0}}{C_C}\right) \end{aligned} \tag{8.27}$$

2. For the irreversible elementary reaction

$$2A + B \longrightarrow \text{product}$$

The expression for the time is

$$t = \left[\frac{1}{k(C_{B_0} - C_{A_0})^2}\right]\left[\frac{(C_{B_0} - C_{A_0})(C_{A_0} - C_A)}{C_{A_0}C_A} + \ln\left(\frac{C_{B_0}C_A}{C_{A_0}C_B}\right)\right] \quad (8.28)$$

3. For the irreversible elementary reaction

$$nA \longrightarrow \text{products}$$

the elementary equation is

$$t = \left[\frac{1}{k(n-1)C_{A_0}^{n-1}}\right]\left[\left(\frac{C_A}{C_{A_0}}\right)^{1-n} - 1\right] \quad (8.29)$$

The corresponding equation for C_A is

$$C_A = C_{A_0}\left[1 + (n-1)kC_{A_0}^{n-1}t\right]^{1/(1-n)} \quad (8.30)$$

The half life for the above reaction is given by

$$t_{\frac{1}{2}} = \left[\frac{2^{n-1} - 1}{k(n-1)}\right]C_{A_0}^{1-n} \quad (8.31)$$

4. Finally, for the irreversible elementary reaction

$$A \xrightarrow{k_1} D \xrightarrow{k_2} U \qquad C_{D_0} = C_{U_0} = 0$$

the describing equation is

$$\frac{C_A}{C_{A_0}} = e^{-k_1 t} \quad (8.32)$$

$$\frac{C_D}{C_{A_0}} = \left(\frac{k_1}{k_2 - k_1}\right)(e^{-k_1 t} - e^{-k_2 t}) \quad (8.33)$$

$$C_U = C_{A_0} - C_A - C_D \quad (8.34)$$

This reaction will receive additional treatment in a later illustrative example. The development of the corresponding form of the above equation in terms of C_A, N_A, X_A, and X_A^* is left as an exercise for the reader.

Finally, the reader should keep in mind that reactions which are *not* elementary and irreversible often have rate expressions that are quite complex. In effect, the rate expression cannot be expressed in terms of a simple order of reaction in numerous applications. The practicing engineer then generally employs an empirical equation.

APPLICATIONS

This last section consists of a total of 22 illustrative examples. They vary from solutions that require simple substitutes to others that require more detatiled mathematics.

ILLUSTRATIVE EXAMPLE 8.3 The author annually hosts a party (his family refer to it as the "animal" party) that primarily consists of former basketball players he coached during the 1950s and 1960s and old friends from Astoria, Queens (the area where most of the attendees grew up). It is usually held on the first Sunday in January during which there is an NFL playoff double header. Several of the players have been known to drink excessively. On one occasion, an attendee's blood alcohol concentration was 0.21%. The player knew he should not drive home until his blood alcohol concentration dropped to 0.02%. If his blood alcohol drops to 0.17% in 45 minutes and he stops drinking at 8:00 PM, at what time can he leave the party and drive home? The reader may assume that the elimination of alcohol in one's body is a first order reaction process that is elementary.

Solution. For the alcohol A,

$$\frac{dC_A}{dt} = -k_A C_A \tag{8.4}$$

For this equation

$$k = \frac{\ln(C_A/C_{A_0})}{t} \tag{8.14}$$

solving,

$$k = \frac{\ln(0.17/0.21)}{0.75}$$
$$= 0.282\ (\text{h})^{-1}$$

To calculate the time, rewrite the above equation and solve for t.

$$t = \frac{\ln(0.02/0.21)}{-0.282}$$
$$= 8.34\ \text{h}$$

The player should plan on leaving the Theodore household at approximately 4:30 AM the following morning. ■

ILLUSTRATIVE EXAMPLE 8.4 A first order irreversible elementary reaction

$$A \xrightarrow{k_A} B; \qquad MW_A = 72, \qquad N_{A_0} = 3.0\ \text{lbmol}$$

with a reaction velocity constant of 0.376 $(\text{min})^{-1}$ takes place in a constant volume batch reactor. Calculate the time required to achieve a conversion of 80%. The reactor contents consists of an equimolar ratio of A and I (inerts)

Solution. Apply the batch equation

$$t = N_{A_0} \int \frac{dX}{(-r_A V)} = C_{A_0} \int \frac{dX}{(-r_A)}; \quad X = X_A \tag{8.11}$$

with

$$r_A = -kC_A = -kC_{A_0}(1 - X); \quad k = k_A$$

Substituting and integrating gives

$$t = \frac{1}{k} \int_0^X \frac{dX}{(1 - X)} = \ln\left(\frac{1}{1 - X}\right) \tag{8.17}$$

$$t = \left(\frac{1}{0.376}\right) \ln\left(\frac{1}{1 - 0.80}\right)$$

$$t = 4.28 \text{ min}$$

■

ILLUSTRATIVE EXAMPLE 8.5 Resolve the previous example if the reactor contents consist of pure A.

Solution. Since t is independent of C_{A_0}, the reaction time remains the same. ■

ILLUSTRATIVE EXAMPLE 8.6 Refer to the two previous illustrative examples. If the density of the reactor contents mixture is 55 lb/ft^3, Calculate the volume of the reaction mixture.

Solution. The volume V is given by

$$V = \frac{m_A}{\rho_A} = \frac{(N_{A_0})(MW_A)}{\rho_A}$$

substituting gives

$$V = \frac{(3)(72)}{55} = 3.93 \text{ ft}^3$$

Note that this represents the volume of the mixture in the reactor, *not* the actual reactor volume. ■

ILLUSTRATIVE EXAMPLE 8.7 Consider the following elementary irreversible *gas* phase reaction:

$$A + B \longrightarrow C$$

It is desired that the reaction be carried out in a batch reactor until a conversion of 90% is achieved. If the reaction vessel is initially filled with equal volumes of components A and B, determine the time for the reaction to take place. The following information is known:

System initial and final pressure = 1.0 atm
System temperature = 60°F (isothermal)
Reaction velocity constant, $k_A = 0.1\ \text{ft}^3/\text{lbmol}\cdot\text{s}$ @ 60°F

Solution. Note that this is a variable volume reactor since it operates at a constant temperature and pressure. The rate expression for the reaction, r_A, is

$$-r_A = kC_AC_B$$

The component concentrations in terms of the conversion variable X_A are (see Chapter 5, Part I, for volume correction effects).

$$C_A = C_{A_0}(1 - X)/(1 + \varepsilon X)$$
$$C_B = C_{A_0}(\theta_B - X)/(1 + \varepsilon X)$$

For this reaction system,

$$\theta_B = 1$$

and

$$\delta = 1 - 1 - 1$$
$$= -1$$

In addition, since equal volumes indicates equal moles,

$$y_{A_0} = 0.5$$

so that

$$\varepsilon = y_{A_0}\delta = -0.5$$

Combine the above and rewrite the rate expression.

$$-r_A = kC_{A_0}^2(1 - X)^2/(1 - 0.5X)^2$$

The batch reactor design equation is now

$$t = N_{A_0}\int_0^X [1/(-r_A)V]dX; \quad V = V_0(1 + \varepsilon X)$$

$$= (1/k_AC_{A_0})\int_0^X [(1 - 0.5X)/(1 - X)^2]dX; \quad N_{A_0}/V = C_{A_0}$$

Calculate initial concentration of A, C_{A_0}, using the ideal gas law

$$C_{A_0} = y_{A_0}P/RT; \qquad y_{A_0} = 0.5$$
$$= 0.00132 \text{ lbmol/ ft}^3$$

Integrate the above equation and solve for the required reaction time at a conversion of 90%. Refer to Illustrative Example 21.18 for integration details

$$\int_0^X [(1+\varepsilon X)/(1-X)^2]dX = (1+\varepsilon)X/(1-X) - \varepsilon \ln[1/(1-X)]$$
$$= (1-0.5)(0.9)/(1-0.9) + 0.5\ln[1/(1-0.9)]$$
$$= (0.5)(0.9)/(0.1) + 0.5\ln(10)$$
$$= 5.65$$

Therefore,

$$t = 5.65/kC_{A_0}$$
$$= (5.65)/(0.1)(0.00132)$$
$$= 42{,}813 \text{ s}$$
$$= 714 \text{ min}$$
$$= 11.9 \text{ h}$$

The reader is left the exercise of resolving this problem without taking volume effects into account. ■

ILLUSTRATIVE EXAMPLE 8.8 The irreversible elementary reaction

$$A + B \longrightarrow C$$

takes place in a constant pressure batch reactor which contains 50 gmols of an ideal gas mixture. The reactor operates at 100°C and 3.0 atm. The feed composition is 25% A, 25% B, and 50% inerts. The reaction is first order with respect to both A and B, and has a reaction velocity constant of $k = 62.7 \times 10^{-2}$ L/gmol · min at 100°C. Calculate the time required to achieve 75% conversion.

Solution. The reaction may be written as

$$[I] + A + B \longrightarrow C + [I]$$

Since the reaction is first order with respect to both A and B, the batch design equation is

$$t = N_{A_0}\int_0^X \frac{dX}{(k_A C_A C_B V)}$$

If the volume of a gas is a variable,

$$V = V_0(1 + \varepsilon X)\left(\frac{T}{T_0}\right)\left(\frac{P_0}{P}\right)$$

For this system

$$\theta_A = 1, 0$$
$$\theta_B = 1$$
$$\theta_C = 0$$
$$\varepsilon = \delta y_{A_0} = (-1)(0.25) = -0.25$$
$$N_{A_0} = (0.25)(50) = 12.5$$

Apply the ideal gas law to solve for V_0

$$V_0 = \frac{N_0RT_0}{P_0} = \frac{(50 \text{ gmol})(0.082 \text{ atm L/gmol K})(373 \text{ K})}{3.0 \text{ atm}} = 509.8 \text{ L}$$

After simplifying (and noting that the pressure and temperature are constant) the design equation becomes

$$t = (V_0/k_AN_{A_0})\int_0^X [(1 + \varepsilon X)/(1 - X)^2]dX; \qquad \frac{kN_{A_0}}{V_0} = \frac{(0.627)(12.5)}{509.8} = 0.01537$$

The integral (see Part IV, Chapter 21) of the design equation is

$$t = \frac{1}{0.01537}\{(1 + \varepsilon)X/(1 - X) - \varepsilon \ln[1/(1 - X)]\}$$

Finally, by substituting $X = 0.75$ into the above expression, one can solve for the time necessary to achieve the desired conversion,

$$t = 168.8 \text{ min} = 10{,}130 \text{ s} = 2.81 \text{ h}$$

■

ILLUSTRATIVE EXAMPLE 8.9 Consider the elementary liquid reaction:

$$A + B \longrightarrow \text{products}; \qquad k_A = 0.5 \text{ ft}^3/\text{lbmol} \cdot \text{h}$$

with

$$MW_B = 30; \qquad \rho_B = 60 \text{ lb/ft}^3$$
$$MW_A = 20; \qquad \rho_A = 30 \text{ lb/ft}^3$$

A 10 ft^3 batch reactor is to be operated full when the reaction starts. The reaction is to proceed for one hour. Is more product produced if the unit is operated by filling the reactor one third (by volume) full with B and then the rest with A, or by filling equally (by volume) with A and B?

Solution. The reaction in terms of the conversion variable X is

$$\begin{aligned} -r_A &= k_A C_A C_B \\ &= k_A C_{A_0}(1 - X) C_{A_0}(\theta_B - X) \\ &= k_A C_{A_0}^2 (1 - X)(\theta_B - X) \end{aligned}$$

with the characteristic batch reactor design equation given by

$$t = N_{A_0} \int_0^X [-1/(r_A V)] dX \tag{8.11}$$

For a liquid phase reaction, V is a constant and may be taken out from under the integral sign.

$$\begin{aligned} t &= (N_{A_0}/k_A C_{A_0}^2) \int_0^X [1/(1 - X)(\theta_B - X)] dX; \quad C_{A_0} = N_{A_0}/V \\ &= (V/k_A N_{A_0}) \int_0^X [1/(1 - X)(\theta_B - X)] dX \\ &= (1/k_A C_{A_0}) \int_0^X [1/(1 - X)(\theta_B - X)] dX \end{aligned}$$

Integrating gives (see Illustrative Example 12.18)

$$t = \left[V/k_A N_{A_0}(1 - \theta_B)\right] \ln\left\{[\theta_B(1 - X)]/[\theta_B - X]\right\}; \qquad \theta_B = N_{B_0}/N_{A_0}$$

Rewrite this equation in terms of α_A. Since $\alpha_A = (N_{A_0})(X)$, the above equation becomes

$$t = \left[V/k_A(1 - \theta_B)\right] \ln\left\{[N_{B_0}(N_{A_0} - \alpha_A)]/[N_{A_0}(N_{B_0} - \alpha_A)]\right\}$$

Consider first the case where the reactor contains one third B and two thirds A. Calculate the initial number of moles of A, N_{A_0}, and the initial number of moles of B, N_{B_0}.

$$\begin{aligned} N_{A_0} &= (6.67)(30)/20 \\ &= 10.0 \text{ lbmol } A \text{ initially} \\ N_{B_0} &= (3.33)(60)/30 \\ &= 6.67 \text{ lbmol } B \text{ initially} \\ \theta_B &= 0.667 \end{aligned}$$

Calculate α_A for the first case using the above equation for t.

$$1.0 = [10/0.5(10 - 6.67)] \ln \{[6.67(10 - \alpha_A)]/[10(6.67 - \alpha_A)]\}$$

By trial and error (or any suitable numerical method),

$$\alpha_A = 2.27 \text{ lbmol } A \text{ reacted}$$

Now calculate α_A for the case when the reactor is half filled with A and with B.

$$\begin{aligned} N_{A_0} &= (5)(30)/20 \\ &= 7.5 \text{ lbmol} \\ N_{B_0} &= (5)(60)/30 \\ &= 10 \text{ lbmol} \\ \theta_B &= 1.33 \\ 1.0 &= [10/0.5(7.5 - 10)] \ln\{[10(7.5 - \alpha_A)]/[7.5(10 - \alpha_A)]\} \end{aligned}$$

Solving

$$\alpha_A = 2.65 \text{ lbmol } A \text{ reacted}$$

Obviously, more A is reacted in the second case.

This problem was solved using the conversion variable α_A to once again introduce the reader to another method of solving the batch reactor design equation. The most convenient conversion variable to work with in chemical reactor applications is X_A; the bulk of the problems in this text are solved using this variable. ■

ILLUSTRATIVE EXAMPLE 8.10 A certain hydrolysis reaction is known to proceed via the reversible mechanism

$$A + B \rightleftharpoons R$$

where B is water, and the value of the equilibrium constant (based on concentration) K_C at 298°K is 10.0 (L/g mol). However, when 1 gmole of A is added to 2 liters of B (water) and later titrated for A, essentially no A is found. The detection limit of the titration is approximately 1×10^{-3} gmole A per liter. Offer a reasonable explanation for this result, and support your answer with appropriate (simple) calculations. The volume of the A/B mixture is 2 liters.

Solution. For this reaction,

$$A + B \rightleftharpoons R$$

Calculate the equilibrium concentration based on concentration:

$$K_C = \frac{C_{R,eq}}{C_{A,eq}C_{B,eq}}$$

Note that $C_{A_0} = 0.5$, $C_{B_0} = 55.5$ (2 liters of B is equivalent to 2000 g H_2O = 111 gmol), $C_{R_0} = 0$, $\theta_B = 111$, $\theta_R = 0$.

At equilibrium,

$$C_A = C_{A_0}(1 - X); \quad X = X_e$$
$$C_B = C_{A_0}(\theta_B - X)$$
$$C_R = C_{A_0} = (\theta_R + X)$$

Therefore

$$\frac{C_{A_0}X}{C_{A_0}(1 - X)C_{A_0}(\theta_B - X)} = 10.$$

Solve for X.

$$X = X_e = 0.001 \text{ gmol/L}$$

Since the above is at or below the detection limit of the analysis, no A would be found! ■

ILLUSTRATIVE EXAMPLE 8.11 Compound B is prepared from pure compound A according to the following liquid-phase elementary reversible reaction:

$$A \underset{k_2}{\overset{k_1}{\leftrightarrows}} B$$

Determine the amount of time necessary to achieve a conversion of 40% if a 10-liter batch reactor is used to perform the reaction.

Additional information:

$$C_{A_0} = 3.0 \text{ lbmol/L}$$
$$k_1(\text{forward}) = 6.0 \text{ (h)}^{-1}$$
$$k_2(\text{reverse}) = 0.53 \text{ (h)}^{-1}$$

Solution. The general rate expression for the reaction, r_A, in terms of the concentrations of the participating species is

$$-r_A = k_1 C_A - k_2 C_B$$

The expressions for the concentrations of A and B in terms of the conversion variable X_A can also be written:

$$C_A = C_{A_0}(1 - X_A)$$
$$C_B = C_{A_0}(\theta_B + X_A)$$
$$= C_{A_0}X_A$$

where θ_B is once again initial moles of B/initial moles of A.

Note that $\theta_B = 0$ since the initial amount of B present is zero. The reaction rate equation may be rewritten in terms of the conversion:

$$\begin{aligned} -r_A &= 6.0C_{A_0} - 6.0C_{A_0}X_A - 0.53C_{A_0}X_A \qquad C_{A_0} = 3.0 \\ &= 18(1 - 1.0883X_A) \end{aligned}$$

This expression may be inserted into the characteristic batch reactor design equation:

$$\begin{aligned} t &= C_{A_0}\int_0^{X_A} \frac{1}{(18)(1 - 1.0883X_A)}\,dX_A \\ &= \frac{C_{A_0}}{18}\int_0^{X_A} \frac{1}{(1 - 1.0883X_A)}\,dX_A \end{aligned}$$

The integrated form of this equation is:

$$\begin{aligned} t &= (C_{A_0}/18)[1/(-1.0883)]\ln[1 - 1.0883X_A] \\ &= (3.0/18)[1/(-1.0883)]\ln[1 - (1.0883)(0.4)] \\ &= 0.0875\,\text{h} \\ &= 5.25\,\text{min} \end{aligned}$$

■

ILLUSTRATIVE EXAMPLE 8.12 When batch reactors are employed in industry the times required to charge and empty the reactor can become significant relative to the reaction time. The time required to complete one sequence of events is called the cycle time t_c and can consist of several terms when applied to chemical reactors:

$$\begin{aligned} t_c &= t_{\text{fill}} + t_{\text{empty}} + t_{\text{react}} + t_{\text{standby}} + t_{\text{heat,cool}} + \cdots \\ &= t_{\text{react}} + t_{\text{downtime}} \\ &= t + t_d \end{aligned} \tag{8.35}$$

Expand the above development in terms of maximizing the productive rate in a batch reactor.[1]

For the reactor

$$aA \longrightarrow pP$$

the "production" rate, PR, for reactant A is given by

$$\text{PR}(A) = N_{A_0}X_A/(t + t_d) \tag{8.36}$$

and is a function of both X and t. Normally the production rate refers to one of the products formed. This production rate, using the same conversion variable, is then (for product P)

$$\text{PR}(P) = \text{PR} = (p/a)N_{A_0}X_A/(t + t_d) \tag{8.37}$$

where p and a are the stoichiometric coefficients in the reaction equation for species P and A, respectively.

To maximize production, one needs to express PR in terms of X (replace t by the design equation that contains X) and solve the equation

$$d(\text{PR})/dX_A = 0 \tag{8.38}$$

or alternately

$$d(\text{PR})/dt = 0 \tag{8.39}$$

For example, if

$$A \xrightarrow{k_A} \text{products}$$

then

$$X_A = 1 - e^{-k_A t}$$

or

$$t = -\ln(1 - X_A)/k_A$$

To maximize P, one can either set

$$\frac{d(\text{PR})}{dt}; \quad X_4 = 1 - e^{-k_A t}$$

or

$$\frac{d(\text{PR})}{dX_A}; \quad t = -\ln(1 - X_A)/k_A$$

equal to zero and then solve for either X_A or t.

One should check to see that the second order derivative is negative, indicating that the solution is a maximum and not a minimum. ■

ILLUSTRATIVE EXAMPLE 8.13 Consider the elementary reaction of

$$A \rightarrow D, k_A = 0.04\,(\text{min})^{-1}$$

It is desired to produce 5000 g mol of D per day, and 98% of A entering the reactor is to be converted. To charge the reactor and to heat it to reaction temperature requires 0.30 h. To discharge the reactor and to prepare it for the next run takes 1.0 h. Calculate the required volume of the reactor. Pure A with a molar density of 6.0 g mol/L is charged to the reactor.

Solution. The reaction time for this first out reactor is calculated as follows:

$$t_{\text{react}} = \left(\frac{1}{0.04}\right)\ln(1-0.98) = \left(\frac{1}{0.04}\right)\ln(0.02) = \frac{3.912}{0.04} = 97.8 \text{ min}$$

or

$$t = 1.63\,\text{h}$$

Thus the total cycle time is

$$t_c = 0.30 + 1.0 + 1.63 = 2.93\,\text{h}$$

Assume two cycles per 8-hour work shift can be completed. Each cycle then requires that 2500 gmol of D must be produced. Therefore

$$N_{A_0} = \frac{N_D}{0.98} = \frac{2500}{0.98}$$
$$= 2550 \text{ gmol}$$

The reactor volume V is given by

$$V = \frac{2550 \text{ gmol}}{6.0 \text{ gmol/L}} = 425 \text{ L}$$
$$= 112 \text{ gal} = 15\,\text{ft}^3$$

■

ILLUSTRATIVE EXAMPLE 8.14 Product B is being produced from reactant A in a batch reactor in a first order irreversible reaction. The materials are allowed to react for a period of 30 minutes at which time the batch (products) are removed. This shutdown period lasts approximately 15 minutes. Calculate the size of the reactor necessary to yield B at a rate of 250 lb/h. Assume the temperature and pressure of the reaction to be constant. The following data is given:

$$\text{Molecular weight of } A = 100$$
$$\text{Molecular weight of } B = 100$$
$$\text{Density of } A = 1.0 \text{ g/cm}^3$$
$$\text{Density of } B = 1.0 \text{ g/cm}^3$$
$$\text{Reaction velocity constant} = 0.05\,(\text{min})^{-1}$$

Solution. The production rate of B is

$$\text{PR}(B) = \frac{250}{100} = 2.50\,\text{lb moles/h}$$

The time for one cycle $= t_c = 30 + 15 = 45 \text{ min} = 0.75 \text{ h}$.

The mole of B produced is

$$N_B = \text{PR}(B)(t_c) = (2.50)(0.75)$$
$$= 1.875 \text{ lb moles in one cycle}$$

For this first order reaction,

$$t = \frac{1}{k_A}\int_0^{X_A} \frac{dX_A}{(1 - X_A)} = -\frac{1}{k_A}\ln(1 - X_A) = t_{\text{react}} = t_r$$

$$30 \text{ min} = -\frac{1}{0.05}\ln(1 - X_A)$$

$$X_A = 0.777$$

By definition,

$$N_B = N_{A_0}X_A$$

$$N_{A_0} = \frac{N_B}{X_A} = \frac{1.875}{0.777} = 2.41 \text{ lb moles}$$

The volume of the reacting mixture is then

$$V = \frac{(N_{A_0})(MW)}{\rho} = \frac{(2.41)(100)}{62.4} = 3.86 \text{ ft}^3$$ ■

ILLUSTRATIVE EXAMPLE 8.15 Provide rough estimate of industrial "down" time for batch reactors.

Solution. As noted, the reaction time, t_r, may be only a small fraction of the total cycle, t_c. Typical "down" times are shown below in Table 8.1. ■

ILLUSTRATIVE EXAMPLE 8.16 It is desired to produce 12,000 kg/day of ethyl acetate in an isothermal batch reactor by the following elementary reversible liquid reaction:

$$CH_3COOH + C_2H_5OH \underset{k_2}{\overset{k_1}{\leftrightarrows}} CH_3COOC_2H_5 + H_2O$$

(A) (B) (C) (D)

TABLE 8.1 Approximate Batch Down Times

Operation	Time, min
Fill	5–30
Empty	5–30
Clean	5–120
Standby	5–15
Heat/cool	15–90

where $k_1 = 8.0 \times 10^{-6}$ and $k_2 = 2.7 \times 10^{-6}$, both with units of $m^3/kmol \cdot s$. The initial charge to the reactor contains an *aqueous* solution of ethanol at concentration 550 kg/m^3, and of acetic acid at concentration 300 kg/m^3, and a small quantity of HCl to act as a catalyst. The time between batches for discharging, cleaning, and recharging is 45 min. Assuming that the density of the reaction mixture is constant at 1145 kg/m^3, determine the volume of the reactor required for 40% conversion of the acetic acid.

Solution. Assume as a basis, 1.0 m^{3}of reactor volume. First calculate the initial molar concentrations of A, B, C, and D.

$$C_{A_0} = 300/60 = 5.0\ \text{kgmol/m}^3; \qquad MW_A = 60$$

$$C_{B_0} = 550/46 = 11.96\ \text{kgmol/m}^3; \quad MW_B = 46$$

$$C_{C_0} = 0.0$$

$$C_{D_0} = (1145 - 300 - 550)/18; \qquad MW_D = 18$$

$$= 16.39\ \text{kgmol/m}^3$$

Calculate θ_A, θ_B, θ_C, and θ_D.

$$\theta_A = 1.0$$

$$\theta_B = 11.96/5.0 = 2.393$$

$$\theta_C = 0.0$$

$$\theta_D = 16.39/5.0 - 3.278$$

Since the reaction is elementary,

$$-r_A = +k_1 C_A C_B - k_2 C_C C_D$$

$$C_A - C_{A_0}(1 - X_A)$$

$$C_B = C_{A_0}(\theta_B - X_A)$$

$$C_C = C_{A_0}(\theta_C + X_A)$$

$$C_D = C_{A_0}(\theta_D + X_A)$$

so that

$$-r_A = +8.0 \times 10^{-6} C_{A_0}^2 (1 - X_A)(2.392 - X_A) - 2.7 \times 10^{-6} C_{A_0}^2 (X_A)(3.278 + X_A)$$

Substitute $-r_A$ into the batch design equation. Expanding yields

$$t = \left[1/(C_{A_0})(10^{-6})\right] \int_0^{0.4} \left[1/(a + bX + cX^2)\right] dX; \quad X = X_A$$

where $a = 19.136$

$b = -27.987$

$c = 5.3 \times 10^2$

The analytical solution for the reaction time, t, required to achieve a conversion of $X = 0.4$ is

$$t = t_r = 5979 \text{ s}$$

The total cycle time, t_c is now calculated.

$$\begin{aligned} t_d &= 45 \text{ min} \\ &= 2700 \text{ s} \end{aligned}$$

Therefore,

$$\begin{aligned} t_c &= t_r + t_d \\ &= 5979 + 2700 \\ &= 8679 \text{ s} \end{aligned}$$

The kgmoles of ethyl acetate (C) produced is

$$\begin{aligned} \text{acetic acid } (A) \text{ converted} &= (5.0)(0.4) \\ &= 2.0 \text{ kgmol} \end{aligned}$$

From stoichiometry,

$$\text{ethyl alcohol } (B) \text{ converted} = 2.0 \text{ kgmol}$$

Calculate the production rate of ethyl acetate (C) in kg/day.

$$\begin{aligned} \text{PR}(C) &= (2.0)(88)(3600)(24)/(8679) \\ &= 1.752 \text{ kg/day} \end{aligned}$$

Since the result of the above is based on 1.0 m^3,

$$\begin{aligned} \text{Volume of reactor } V &= (12{,}000 \text{ kg/day})/(1752.1 \text{ kg/day} \cdot \text{m}^3) \\ &= 6.85 \text{ m}^3 \end{aligned}$$

ILLUSTRATIVE EXAMPLE 8.17 The following data applies to an isothermal, constant volume batch reactor.

$$A + B \longrightarrow 2P \qquad \text{(elementary)}$$

$$\begin{aligned} N_{A_0} &= 100 \text{ lb moles} \qquad & k_A &= 0.04 \text{ ft}^3/\text{lb mol} \cdot \text{min} \\ N_{B_0} &= 100 \text{ lb moles} \qquad & \text{down-time} &= 10 \text{ min} \end{aligned}$$

If the reactor can operate with a reaction time of either 10, 20, 30, 40, or 50 minutes at a time, what is the minimum reactor volume to achieve a production rate of 150 lb moles/h of P? Also calculate the maximum volume that will provide the desired results.

Solution. For a batch reactor with a second order reactors and $\theta_B = 1.0$

$$t = \frac{V}{N_{A_0} k_A}\left(\frac{X_A}{1 - X_A}\right)$$

$$PR(A) = \frac{N_{A_0} X_A}{\text{total cycle time}} = \frac{N_{A_0} X_A}{t_c} = 150/2 \tag{1}$$

$$PR(P) = \frac{N_{A_0}(2X_A)}{t_c}; \quad tct = \frac{t_r + 10}{60}, \text{h}; N_{A_0} = 100$$

where $t_c = t + 10$.

Substituting,

$$PR(A) = \frac{N_{A_0} X_A}{(t + 10)/60} = 75$$

Rearranging

$$X_A = \frac{(t_r + 10)}{60}\left(\frac{150}{2}\right)\Big/100$$

$$= (t_r + 10)/80 \tag{2}$$

Combining Equations (1) and (2) gives

$$V = N_{A_0} k_A t \left[\frac{1 - \dfrac{t_r + 10}{80}}{\dfrac{t_r + 10}{80}}\right] = N_{A_0} k_A \left[80\left(\frac{t_r}{t_r + 10}\right) - t_r\right]$$

The results are provided below in Table 8.2 for $N_{A_0} = 100$ and $k_A = 0.04$. The minimum volume over the 10–50 min reaction time space is 66.67 ft^3. The maximum appears to be around 133 ft^3.

To determine min/max, set

$$\frac{dV}{dt} = 0 = N_{A_0} k_A (80)\left[\frac{10}{(t_r + 10)^2}\right] - N_{A_0} k_A$$

TABLE 8.2 Time–Volume Calculations for Illustrative Example 8.17

t, min	V, ft^3
10	120
20	133.33
30	120
40	96
50	66.67

Dividing by $N_{A_0}k_A$ gives

$$\frac{800}{(t_r + 10)^2} = 1.0$$

Thus quadratic may now be solved for t_r

$$t_r^2 + 20t_r - 700 = 0$$

$$\begin{aligned} t_r &= \frac{-b \pm \sqrt{b^2 - 4ac}}{2a} \\ &= \frac{-20 \pm \sqrt{400 + 2800}}{2} \\ &= \frac{-20 + 56.6}{2} \\ &= 18.3 \text{ min} \end{aligned}$$

Thus,

$$X_A = \frac{18.3 + 10}{80} = 0.354$$

and

$$\begin{aligned} V &= (100)(0.04)(18.3)\left[\frac{1 - 0.354}{0.354}\right] \\ &= 133.5 \text{ ft}^3 \end{aligned}$$

■

ILLUSTRATIVE EXAMPLE 8.18 It is necessary to product 100 lb/h of acetaldoxime from acetaldehyde and hydroxylamine in a constant volume batch reactor. The reaction mechanism is as follows:

$$\underset{\substack{\text{acetaldehyde} \\ \text{(A)}}}{CH_3COH} + \underset{\substack{\text{hydroxylamine} \\ \text{(B)}}}{H_2NOH} \longrightarrow \underset{\substack{\text{acetaldoxime} \\ \text{(C)}}}{CH_3CHNOH} + \underset{\substack{\text{water} \\ \text{(D)}}}{H_2O}; \quad k_A = 1.74 \text{ ft}^3/\text{lbmol} \cdot \text{h}$$

1. Show that the volume of the contents of the reactor V_c is 4.11 ft^3.
2. What conversion of A is required for a reaction time and down-time of 40 minutes and 10 minutes, respectively?
3. Obtain the equation describing the reaction time as a function of conversion.

Data: $N_{A_0} = N_{B_0} = 3.0$ lb moles; ρ (constant) $= 56.2$ lb/ft^3.

Solution

1.

$$V_c = \frac{m}{\rho}$$
$$= \frac{(3 \text{ lb moles } A)(44) + (3 \text{ lb moles } B)(33)}{56.2 \text{ lb/ft}^3}$$
$$= 4.11 \text{ ft}^3$$

2.

$$PR(A) = PR(C) = \frac{N_{A_0} X_A}{t_c}; \qquad MW(C) = 59$$

with

$$t_c = 10 + 40 = 50 \text{ min}$$

Substituting

$$100\frac{\text{lb}}{\text{h}} = 1.695\frac{\text{lb moles}}{\text{h}} = \frac{3 \text{ lb moles } A(X_A)}{(50/60)}$$
$$X_A = 0.471$$

3. The rate expression is

$$-r_A = kC_A C_B = kC_{A_0}^2(1 - X)(\theta_B - X); \quad k = k_A, X = X_A$$
$$= kC_{A_0}^2(1 - X)^2$$

The design equation is

$$t_r = \frac{N_{A_0}}{V}\frac{1}{kC_{A_0}^2}\int_0^X \frac{dX}{(1 - X)^2}; \qquad C_{A_0} = \frac{3.0}{4.11} = 0.73\frac{\text{lb moles}}{\text{ft}^3}$$

Integrating and substituting gives

$$t = \frac{3.0}{(4.11)(1.74)(0.73)^2}\left[\frac{X}{1 - X}\right]$$
$$t = t_r = 0.787\left[\frac{X}{1 - X}\right]$$

■

ILLUSTRATIVE EXAMPLE 8.19 Refer to the previous illustrative example. What conversion will maximize the production rate of C?

Solution. Once again,

$$PR(C) = \frac{N_{A_0}X_A}{t_{\text{cycle}}} = \frac{N_{A_0}X_A}{0.787\left(\dfrac{X_A}{1-X_A}\right)+\dfrac{10}{60}}$$

$$= \frac{(3.0)X_A}{(0.787)\left(\dfrac{X_A}{1-X_A}\right)+0.167}$$

To obtain $PR(\max)$,

$$\frac{d(PR)}{dX} = 0; \qquad X = X_A$$

Set

$$f(X) = 3X$$

$$g(X) = 0.787\left(\frac{X}{1-X}\right)+0.167$$

Note that

$$\frac{d[f(X)/g(X)]}{dX} = \frac{g(X)df(X) - f(X)dg(X)}{[g(X)]^2}$$

Substituting

$$\frac{d(PR)}{dX} = \frac{(0.787)\left(\dfrac{X}{1-X}\right)(3) - (3X)(0.787)\left[\dfrac{1}{(1-X)^2}\right]}{0.62\dfrac{X^2}{(1-X)^2} + 0.262\left(\dfrac{X}{1-X}\right) + 0.027} = 0$$

Solving for X by either trial and error or Newton–Raphson gives

$$X = X_A \approx 0.212$$

■

ILLUSTRATIVE EXAMPLE 8.20 As part of a hazardous waste incinerator project,[3] a test substance, C_{A_0} (pure) = 10.33 gmol/L, is to be decomposed in a batch reactor. Calculate the conversion (to four significant figures) for the following conditions if the reaction time is 0.25 s.

1. Pure A, zero order, $k = 10.328\ (\text{gmol/L})(\text{s})^{-1}$
2. 50% A/50% inerts, zero order, $k = 10.328\ (\text{gmol/L})(\text{s})^{-1}$
3. 10% A/90% inerts, zero order, $k = 10.328\ (\text{gmol/L})(\text{s})^{-1}$

Solution. For a zeroth order

$$X_A = X = \frac{kt}{C_{A_0}}$$

1. For Pure A:

$$X = \frac{(10.328)(0.25)}{10.33} = 0.2499$$

2. For 50% A, $C_A = 0.5, C_{A_0} = 5.165\,\text{lgmol/L}$:

$$X = \frac{kt}{C_A} = \frac{(10.328)(0.25)}{5.165} = 0.4999$$

3. For 10% A, $C_A = (0.1), C_{A_0} = 1.033\ \text{gmol/L}$:

$$X = \frac{kt}{C_A} = \frac{(10.328)(0.25)}{1.033} = 2.499$$

This is impossible. The reaction must occur in less than 0.25 s.

Note that the test substance is carbon tetra chloride.[3] ■

ILLUSTRATIVE EXAMPLE 8.21 Refer to the previous example. Calculate the conversion (to four significant figures) for the following conditions if the reaction time is 0.25 s.

1. Pure A, $\frac{1}{2}$ order, $k = 6.364\ (\text{L/gmol})^{1/2}\ (\text{s})^{-1}$
2. 50% A, 50% inerts, $\frac{1}{2}$ order, $k = 6.364\ (\text{L/gmol})^{1/2}\ (\text{s})^{-1}$

Solution. For a half order reaction

$$X = 1 - \left[\pm\left(kt/2C_{A_0}^{1/2}\right) + 1\right]^2$$

1. For pure A, i.e., $C_{A_0} = 10.33\ \text{gmol/L}$

$$X = 1 - \left\{\pm\left[\frac{6.364(0.25)}{(2)(10.33)^{1/2}}\right] + 1\right\}^2$$

$$= 0.4338$$

2. For 50% A, $C_{A_0} = 0.5(C_{A_0}) = 5.165$ gmol/L

$$X = 1 - \left\{ \pm \left[\frac{(6.364)(0.25)}{(2)(5.165)^{1/2}} + 1 \right] \right\}^2$$

$$X = 0.5775$$

■

ILLUSTRATIVE EXAMPLE 8.22 Refer to Illustrative Examples (8.20) and (8.21). Calculate the conversion for 0.25 s.

1. Pure A, fourth order, $k = 3.024 \times 10^8$ $(\text{L/gmol})^3$ $(\text{s})^{-1}$
2. 0.1% A, 99.9% inerts, fourth order, $k = 3.024 \times 10^8$ $(\text{L/gmol})^3$ $(\text{s})^{-1}$
3. 0.01% A, 99.99% inerts, fourth order, $k = 3.024 \times 10^8$ $(\text{L/gmol})^3$ $(\text{s})^{-1}$

Solution. For a fourth order reaction

$$X = 1 - \left[\frac{1}{3ktC_{A_0}^3 + 1} \right]^{1/3}$$

1. For pure A, i.e., $C_A = 10.33$

$$X = 1 - \left[\frac{1}{3(3.024 \times 10^8)(0.25)(10.33)^3 + 1} \right]^{1/3}$$

$$X = 0.9998$$

2. For 0.1% A, $C_A = (0.001)(10.33) = 0.01033$ gmol/L

$$X = 1 - \left[\frac{1}{3(3.024 \times 10^8)(0.25)(0.01033)^3 + 1} \right]^{1/3}$$

$$X = 0.8415$$

3. For 0.01% A, $C_A = (0.0001)(10.33) = 0.001033$ gmol/L

$$X = 1 - \left[\frac{1}{3(3.024 \times 10^8)(0.25)(1.033 \times 10^{-3})^3 + 1} \right]^{1/3}$$

$$X = 0.0717$$

■

ILLUSTRATIVE EXAMPLE 8.23 Consider the series reaction

$$A \xrightarrow{k_1} D \xrightarrow{k_2} U$$

where A, D, and U represent the reactant, the desirable product, and the undesirable product, respectively. The rate equation terms for the reactions are $C_{A_0} = 2.0$ gmol/L and $C_{D_0} = C_{U_0} = 0.0$ with $k_1 = 0.2\ (\text{h})^{-1}$ and $k_2 = 0.4\ (\text{h})^{-1}$.

Calculate the maximum concentrate of D that one can achieve in a batch reactor.

Solution. The rate equations are:

$$\frac{dC_A}{dt} = -k_1 C_A$$

$$\frac{dC_D}{dt} = k_1 C_A - k_2 C_D$$

$$\frac{dC_U}{dt} = k_2 C_D$$

As presented in an earlier section

$$C_A = C_{A_0} e^{-k_1 t} \tag{8.32}$$

$$C_D = C_{A_0}\left(\frac{k_1}{k_2 - k_1}\right)(e^{-k_1 t} - e^{-k_2 t}) \tag{8.33}$$

$$C_U = C_{A_0} - C_A - C_D \tag{8.34}$$

The maximum concentration of D is obtained by setting $dC_D/dt = 0$ and noting that $de^u/dt = e^u du/dt$. The resulting solution is[3]

$$C_{D\max} = C_{A_0}\left(\frac{k_1}{k_2}\right)^{k_2/(k_2-k_1)}$$

Substituting

$$C_{D\max} = 2.0(0.2/0.4)^{0.5}$$
$$= 1.4 \text{ gmol/L}$$

■

ILLUSTRATIVE EXAMPLE 8.24 Refer to the previous example. Calculate the reaction time in a batch reactor to achieve the maximum concentration of D.

Solution. The answer is obtained by solving the following equation for t

$$1.4 = 2.0\left(\frac{k_1}{k_2 - k_1}\right)(e^{-k_1 t} - e^{-k_2 t})$$

with $k_1 = 0.2$ and $k_2 = 0.4$. The result is

$$t(\max) = 0.34\,\text{h} = 20.8 \text{ min}$$

The reader should verify that the solution for t reduces to[3]

$$t(\max) = \frac{\ln(k_2/k_1)}{k_2 - k_1}$$ ■

REFERENCES

1. L. Theodore, personal notes, Manhattan College, 1967.
2. J. Santoleri, J. Reynolds, and L. Theodore, "*Introduction to Hazardous Waste Incineration*," 2nd edition, John Wiley & Sons, Hoboken, NJ, 2000.
3. L. Theodore, personal notes, 2010.

CHAPTER 9

CONTINUOUS STIRRED TANK REACTORS

A reactor where mixing is important is the tank flow or continuously stirred tank reactor (CSTR); it is also referred to as a "back-mix" reactor. This type of reactor, like the batch reactor, essentially consists of a tank or kettle equipped with an agitator.

As with the batch reactor, perfect mixing is normally assumed so that both the concentration and temperature within the reactor is the same, i.e., there is no spatial variation so that both terms are not a function of position within the reactor. This perfect mixing assumption also requires that both the concentration and temperature at the discharge point (the outlet pipe) is identical to that in the reactor. Fogler[1] (p. 87) notes that for perfect mixing: "The idea that the composition is identical everywhere in the reactor and in the exit pipe requires some thought. It might seem that, since the concentration changes instantly at the entrance where mixing occurs, reaction occurs there and nothing else happens in the reactor because nothing is changing. However, reaction occurs throughout the reactor, but mixing is so rapid that nothing appears to change with time or position." One can extend this reasoning to also conclude that the rate of reaction within the reactor is uniform and the extent of conversion of reactant is the same within the reactor and the discharge port.

This class of reactor finds application in industry when:

1. the production rate (or quantity) is high; and,
2. the reaction time is relatively low.

The bottom line is that, in addition to producing greater quantities with smaller equipment, most industrial reactors are operated in a continuous rather than batch mode. They also require less labor and maintenance; however, flow processes are more difficult to start and stop than their batch counterparts.

In addition to the traditional CSTR, this chapter will address several other topics that many classify under the CSTR umbrella. This point was noted in both the Introduction to Part II and the introductory section to Chapter 8—Batch Reactors. These topics include:

1. Semi-batch reactors.
2. Unsteady state operations.
3. Cascades of CSTRs.

Chemical Reactor Analysis and Applications for the Practicing Engineer. By Louis Theodore

The following sections are discussed in this chapter:

Equipment Description and Operation

Describing Equations

Applications

EQUIPMENT DESCRIPTION AND OPERATION

As noted previously, the CSTR essentially consists of a tank or kettle equipped with an agitator (see Figure 9.1). It may be operated under steady or transient conditions. Reactants are fed continuously, and the product(s) are withdrawn continuously. The reactant and product may be liquid, gas, or solid, or a combination of them all. If the contents are perfectly mixed, the reactor design problem is greatly simplified for steady conditions because the mixing results in the aforementioned uniform concentration, temperature, and so on, throughout the reactor. This means that the rate of reaction is also constant and the describing equations are not differential and, therefore, do not require integration. In general, CSTRs are used for liquid phase reactions. High reactant concentrations can be maintained with low flow rates so that conversions approaching 100% can be achieved. However, the overall economy of the system may be reduced because of low throughput rates.

CSTRs (as well as tubular flow reactors to be addressed in the next chapter) are often connected in series in such a manner that the exit stream of one reactor is the feed stream for another reactor. This series or cascade of reactors is pictured in Figure 9.2 and discussed later in this chapter. Under these conditions it is convenient to define the conversion at any point downstream in the battery of CSTR reactors in terms of *inlet* conditions, rather than with respect to any one of the reactors in the series. The

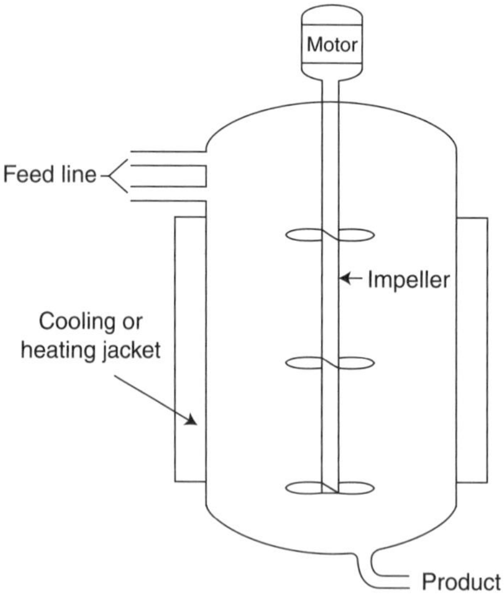

Figure 9.1 Typical tank flow reactor.

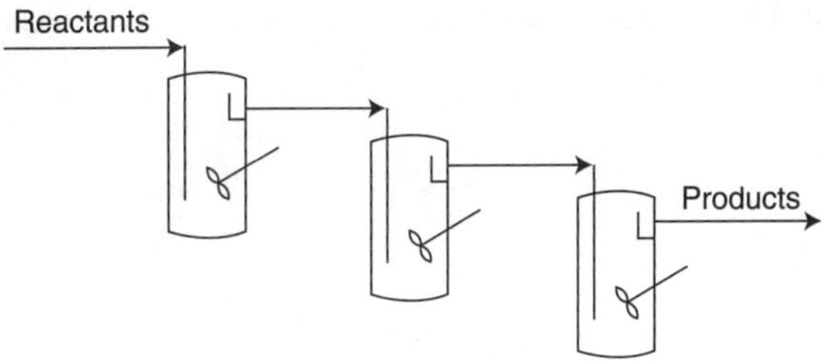

Figure 9.2 Three CSTRs in series.

conversion X is then the total moles of A that have reacted up to that point per mole of A fed to the first reactor. However, this definition should only be employed if there are no side stream withdrawals and the only feed stream enters the first reactor in the series. The conversion from reactors 1, 2, 3, ... in the series are usually defined as X_1, X_2, X_3, ... , respectively, and effectively represent the overall conversion for that reactor relative to the feed stream to the first reactor.

ILLUSTRATIVE EXAMPLE 9.1 Use the following notation to answer the question below.

SS = steady state	PM = perfectly mixed
US = unsteady state	IM = imperfectly mixed

CSTRs are most closely described by

A SS, PM
B SS, IM
C US, PM
D US, IM

Solution

Consider answer (A). CSTRs *are* normally operated in the steady-state, perfectly mixed mode.

Consider answer (B). CSTRs are *not* normally operated in the steady-state, imperfectly mixed mode.

Consider answer (C). CSTRs are *not* normally operated in the unsteady-state, perfectly mixed mode.

Consider answer (D). CSTRs are *not* normally operated in the unsteady-state, imperfectly mixed mode.

The correct answer is therefore (A). ■

ILLUSTRATIVE EXAMPLE 9.2 List some of the advantages and disadvantages of CSTRs.

Solution. Advantages include:

1. Continuous operation.
2. Good temperature control.
3. Can handle two phase reactions.
4. Good operational control.
5. Simplicity of construction.
6. Low operating (labor) costs.
7. Easy to clean.

Disadvantages include:

1. Low conversion per unit volume of reactor.
2. Bypassing.
3. Channeling.
4. Hot spots are possible with poor agitation.

CSTR usage include liquid phase, liquid–gas phase, and some solid–liquid reactions. ■

DESCRIBING EQUATIONS

Consider the tank flow reactor pictured in Figure 9.3. The equation describing the concentration changes in this system are obtained by applying the conservation law for mass to the reactor at any interval of time. The following is assumed in this development.

1. The volume of the contents of the reactor are unchanged during the process.
2. The reactor contents are perfectly mixed.
3. The (total) mass density of the system is constant.

The six terms in Equation (7.1) or Equation (7.2) are again examined.

Term 1: Although a concentration gradient exists at the inlet to the reactor (point 0), no mass transfer into the system by this mechanism is usually assumed.

Term 2: This term is zero since there is no gradient at the outlet (point 1).

Term 3: If Q_0 is the volumetric flow rate into the reactor and C_{A_0} is the molar concentration of A, then

$$Q_0 C_{A_0} \tag{9.1}$$

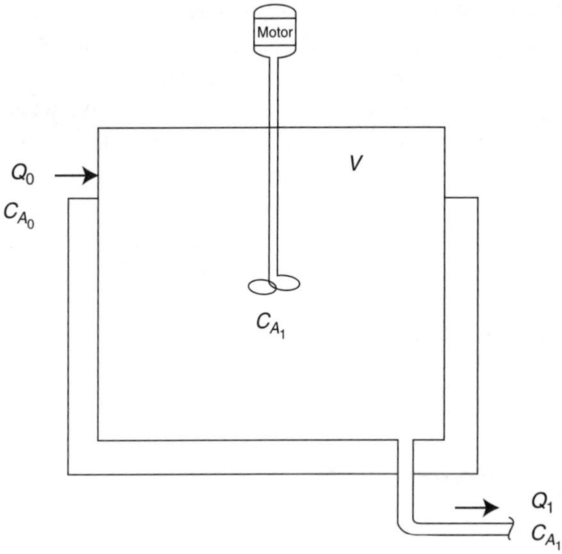

Figure 9.3 Line diagram; tank flow reactor.

is the moles of A flowing into the reactor per unit time. Note that the volumetric flowrate is now represented by Q (*not* q) for this chapter since it is the notation adopted in many texts as well as industry.

Term 4: Similarly,

$$Q_1 C_{A_1} \tag{9.2}$$

represents the molar flow rate of A leaving the reactor.

Term 5: This term is again given by

$$r_A V \tag{9.3}$$

Term 6: This term is zero for steady-state conditions.

Substituting terms in Equations (9.1)–(9.3) in Equation (7.2) gives

$$Q_0 C_{A_0} - Q_1 C_{A_1} + r_A V = 0 \tag{9.4}$$

For a constant density system

$$Q_0 = Q_1 = Q$$

Equation (9.4) may then be rewritten as

$$\tau = \frac{V}{Q} = \frac{C_{A_1} - C_{A_0}}{r_A} \tag{9.5}$$

The term on the left-hand side (LHS) of the above equation has the units of time and represents the average holdup (residence) time in the reactor. It is denoted by the symbol τ. The reciprocal of τ is defined as the space velocity (SV). (An average holdup time is usually assumed for the species participating in the reaction. However, there is a distribution around this average, and it is often important that this distribution effect be included in the analysis of certain type systems. For example, a high or variable residence time for certain organic reactions in the pharmaceutical industry can lead to undesirable side reactions. For polymeric reactions, this distribution effect can lead to a product of variable chain length.)

There are other forms of Equation (9.5). An equivalent equation may also be derived on a mass basis containing any of several conversion variables. These equations can be used, provided they are dimensionally consistent. The choice is generally one of convenience.

Whenever a tank flow reactor is started up, shut down, or has its operating conditions changed, the state and properties of the system change with time. The mass transfer equation becomes more complicated because Term 6 in Equation (7.2) is no longer zero. Rather, it is given by

$$\frac{d(VC_{A_1})}{dt}$$

The describing equation then becomes

$$Q_0 C_{A_0} - Q_1 C_{A_1} + r_A V = \frac{d(VC_{A_1})}{dt} \tag{9.6}$$

where Q_0, Q_1, C_{A_0}, and V *can* vary with time, and r_A and C_{A_1} *will* then vary with time. However, if the density and volume of the system are constant, Equation (9.6) becomes

$$Q(C_{A_0} - C_{A_1}) + r_A V = V\frac{dC_{A_1}}{dt}$$

or

$$\frac{C_{A_0} - C_{A_t}}{\tau} + r_A = \frac{dC_{A_1}}{dt} \tag{9.7}$$

Note that if Q is equal to zero, a batch reactor results with the describing equation reducing to (as expected)

$$r_A = \frac{dC_A}{dt}$$

Alternately, for steady conditions, Equation (9.7) reduces to

$$\frac{C_{A_0} - C_{A_1}}{\tau} + r_A = 0 \tag{9.8}$$

If the enthalpy of reaction is not negligible, or if it is required to supply or remove thermal energy in the form of heat from the reactor, an energy balance around the system must be obtained in order to determine the temperature within the reactor. This is addressed later in the text (Part III, Chapter 12).

The classical material on cascade reactors in the literature addresses first order irreversible reactions. For example, for

$$A \longrightarrow P; \qquad -r_A = k_A C_A = kC_A$$

one may now substitute into Equation (9.4)

$$\frac{V}{Q} = \frac{C_{A_0} - C_{A_1}}{-r_A} = \tau$$

or

$$\tau = \frac{C_{A_0} - C_{A_1}}{kC_{A_1}} \tag{9.5}$$

$$C_{A_1} = \frac{C_{A_0}}{1 + k\tau} \tag{9.9}$$

For a cascade of CSTRs of equal volume,

$$C_{A_2} = \frac{C_{A_1}}{1 + k\tau} \tag{9.10}$$

$$\vdots \qquad \vdots$$

$$C_{A_n} = \frac{C_{A_{n-1}}}{1 + k\tau} \tag{9.11}$$

$$= \frac{C_{A_0}}{(1 + k\tau)^n} \tag{9.12}$$

If the rate expression is not first order, the equations must be solved by trial-and-error.

These equations can usually be solved graphically as shown in Figure 9.4. Figure 9.4 provides a method of solution for reaction order $n > 1$. Starting at the point a ($C_A = C_{A_0}$, $-r_A = 0$), a line of slope $(-1/\tau)$ is extended until it reaches

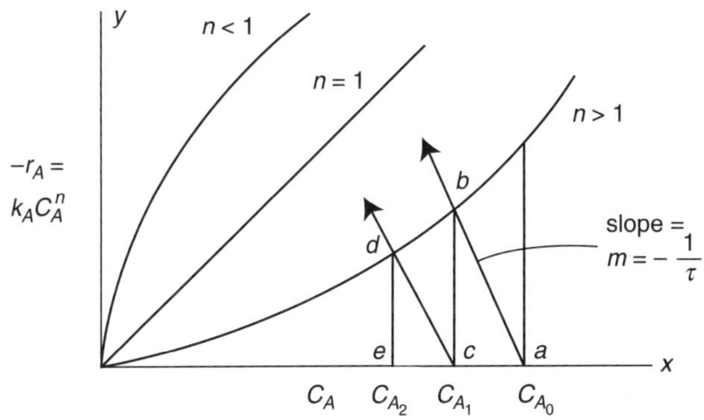

Figure 9.4 Graphical approach to describing a cascade of CSTRs.

the rate curve for $n > 1$ (point b). The concentration leaving reaction 1 is then C_{A_1} at points $b-c$. This procedure is repeated to obtain C_{A_2}, i.e., points $d-e$, etc., are drawn. In effect, a "staircase" or "staged" approach, similar to that employed in some distillation calculations[2] is used. For constant volume CSTRs, the lines ab, cd, etc., are parallel. If the volumes are not equal, the slope of the line will vary for each stage. The validity of this approach is noted by simply examining the following equation for the calculation of C_{A_1}

$$m = -\frac{1}{\tau} = \frac{\Delta y}{\Delta x} = \frac{-r_A - 0}{C_{A_1} - C_A} \tag{9.13}$$

The reader should note that calculations involving a cascade (series) of CSTRs are normally solved algebraically, not graphically.

CSTRs in series are usually designed so that the volumes of the individual reactors are equal. For almost all reactions, the total volume requirement for achieving a given conversion decreases as the number of reactors in series increases. This can significantly impact the economics, particularly the capital cost (see Part IV, Chapter 21). However, the total volume requirement to achieve a particular conversion can be further reduced, particularly for non-elementary reactions, if the constraint of equal reactor volumes is removed, i.e., the volumes of each reactor need not be the same. Although these systems can be designed to lower volume requirements, the impact on the overall economics can be negative. Several illustrative examples in the Applications section address this issue.

Another type of the CSTR class of reactors is a *semibatch* unit. A semibatch reactor is defined as a reactor that is initially charged with reactants and then the same or other reactants are added continuously to the vessel. Products, may also be withdrawn (see Figure 9.5). Since the concentration of the reaction mixture in the reactor changes with time, the operation can be viewed as batch; because another stream is

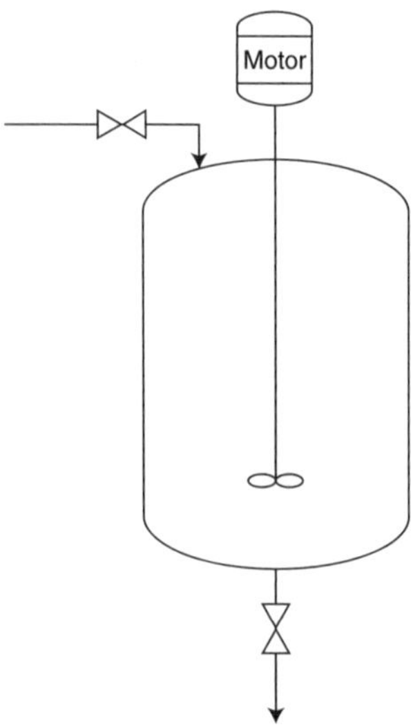

Figure 9.5 Semibatch reactor.

added to or removed continuously from the vessel, the operation can be labelled as continuous. The procedure is therefore referred to as a *semibatch operation*.

The equipment used in a semibatch operation is similar to that employed in a batch operation except that the added and/or removed stream is pumped to/from the batch reactor at a controlled rate. This reactor is almost always equipped with a stirrer, a possible heating or cooling jacket, and control valves at the inlet and outlet in order for the streams to be varied, if necessary.

This unit is usually employed as a single stirred tank. Some of the reactants are fed into the reactor as a single charge, and the remaining can be fed in gradually. This method of operation is especially favored when:

1. large heat effects occur since exothermic reactions may be slowed down via the addition of a high heat capacity inert substance;
2. endothermic or exothermic rates are constrained by limiting the concentration of one of the reactants;
3. the reaction may thus be kept within controllable limits with heat transfer;
4. when high concentrations result in the formation of undesirable side products; and,
5. when one of the reactants is a gas of limited solubility so that it may be fed only at particular rate.

As noted, some reactions may yield a product in a different phase from the reaction mixture, e.g., the liberation of a gas from a liquid-phase reaction or the formation of a precipitate in a liquid reaction. This unit may also assist in moving the reaction to completion in order to continuously separate the raw product phase. The describing equation is provided in Equation (9.7)

$$\frac{C_{A_0} - C_{A_1}}{\tau} + r_A = \frac{dC_{A_1}}{dt} \tag{9.7}$$

This equation has been solved for a number of cases.[1]

Unsteady-state conditions arise when a reactor is either started up or shut down, or operating conditions are changed. If it is necessary to determine how long it will take for a reactor to reach steady state or concentrations suitable for content removal, Equation (7.2) is applied with the transient term included, i.e., Equation (9.7) again applies.

The describing equations for a CSTR can also be expressed in terms of the feed rate F on a mass or mole basis for this case,

$$F_0 = Q_0 C_0$$

and

$$F_{A_0} = Q_0 C_{A_0} \tag{9.14}$$

ILLUSTRATIVE EXAMPLE 9.3 Provide two examples of why a semibatch reactor is occasionally employed in industry.

Solution.

1. Semibatch operation can be used to prevent the formation of by-products.
2. Semibatch reactors can be employed to increase the yield of a reversible reaction by removing one of the products. ■

APPLICATIONS

The final section of this chapter is concerned with applications. A total of 18 illustrative examples follow.

ILLUSTRATIVE EXAMPLE 9.4 A company has two CSTRs of unequal size which it would like to use in the production of a specified product formed by a irreversible first-order reaction. How should these reactors be hooked up in series to achieve maximum conversion? Justify your answer.

Solution. The two CSTRs are pictured in Figure 9.6.

For (a) in Figure 9.6,

$$\tau_S = \frac{C_{A_0} - C_{A_1}}{kC_{A_1}}$$

$$C_{A_1} = \frac{C_{A_0}}{1 + k\tau_S}$$

$$\tau_L = \frac{C_{A_2} - C_{A_1}}{kC_{A_2}}$$

$$C_{A_2} = \frac{C_{A_1}}{1 + k\tau_L}$$

and

$$C_{A_2} = \frac{C_{A_0}}{(1 + k\tau_S)(1 + k\tau_L)}$$

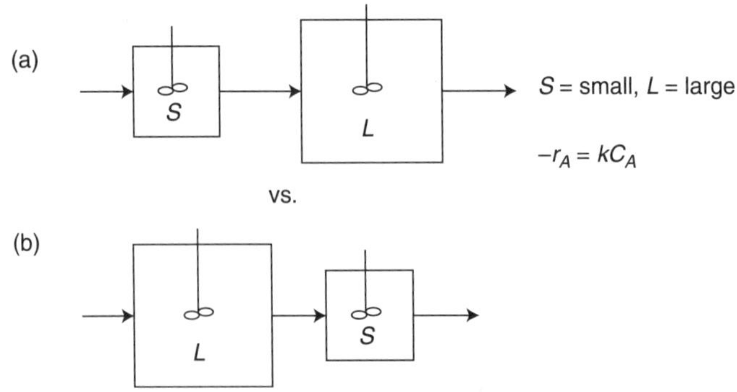

Figure 9.6 Two CSTRs of unequal volume.

For (b) in Figure 9.6,

$$\tau_L = \frac{C_{A_0} - C_{A_1}}{kC_{A_1}}$$

$$C_{A_1} = \frac{C_{A_0}}{1 + k\tau_L}$$

$$\tau_S = \frac{C_{A_2} - C_{A_1}}{kC_{A_2}}$$

$$C_{A_2} = \frac{C_{A_1}}{1 + k\tau_S}$$

and

$$C_{A_2} = \frac{C_{A_0}}{(1 + k\tau_L)(1 + k\tau_S)}$$

It makes no difference! This can also be demonstrated graphically. ■

ILLUSTRATIVE EXAMPLE 9.5 A company has two CSTRs of equal size which it would like to use in the production of a specified product formed by a irreversible first-order reaction. Should these reactors be hooked up as provided in Figure 9.7 or as in Figure 9.8 to achieve maximum conversion?

Solution. Case (a) is considered in Figure 9.7. Note that $F_A = QC_A$ (constant units). For a first order reaction,

$$\frac{kVC_{A_0}}{FA_0} = \frac{X}{1 - X}$$

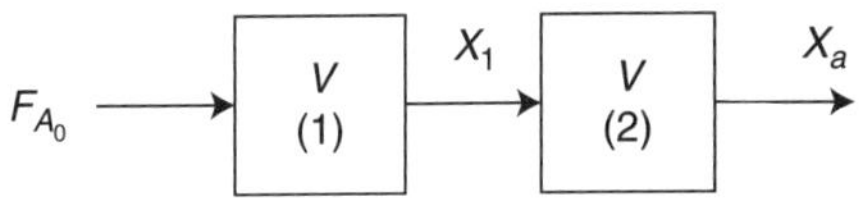

Figure 9.7 Two equal volume CSTRs in series.

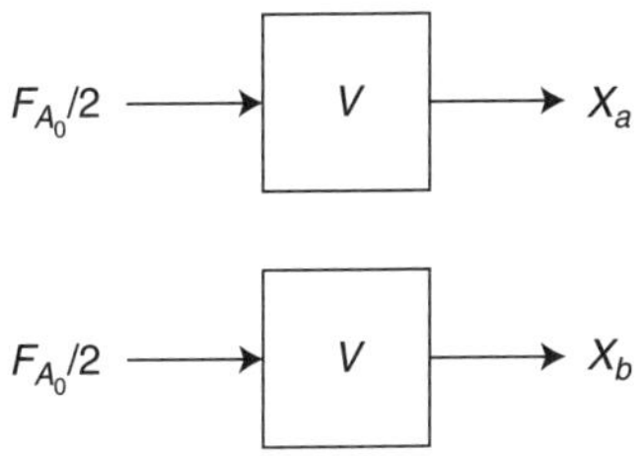

Figure 9.8 Two equal volume reactors receiving half the feed.

let

$$T = \frac{X}{1 - X}$$

For reactor (1)

$$X_1 = \frac{T}{1 + T}$$

For reactor (2)

$$T = \frac{X_a - X_1}{1 - X_a}; \qquad X_1 = \frac{T}{1 + T}$$

Solution for X_a leads to

$$X_a = \frac{2T + T^2}{1 + 2T + T^2} = \frac{T^2 + 2T}{(T + 1)^2} = \frac{T^2 + 2T}{T^2 + 2T + 1}$$

Case (b) is pictured in Figure 9.8.

For either reactor, one notes that

$$2T = \frac{X_b}{1 - X_b}$$

Therefore

$$X_b = \frac{2T}{1 + 2T}$$

Compare conversion X_a to conversion X_b

$$X_a = \frac{T^2 + 2T}{T^2 + 2T + 1}$$

$$X_b = \frac{2T}{2T + 1}$$

Since $2T/(2T + 1)$ is <1, i.e., a fractional number, any addition to both numerator and denominator will increase its numerical value. Thus,

$$X_a > X_b$$

A higher conversion is obtained for case (a). The reader is left the exercise of proving the above graphically. ■

ILLUSTRATIVE EXAMPLE 9.6 The following third order elementary reaction is to be carried out in a continuous stirred tank reactor.

$$A + 2B \xrightarrow{k_A} C; \quad k_A = 1.5\ \text{L}^2/(\text{gmol})^2 \cdot \text{h}$$

The concentrations in the feed to the reactor are as follows:

$$C_{A_0} = 1.0\ \text{gmol/L}$$
$$C_{B_0} = 2.5\ \text{gmol/L}$$
$$C_{C_0} = 0.0\ \text{gmol/L}$$

It is known that the reaction takes place at constant density and temperature.

Calculate the volume of the reactor if the feed is 3 gal/min and the conversion of A is 70%.

Solution. The rate equation is

$$r_A = -k_A C_A C_B^2$$

The describing equation for the system is

$$\frac{V}{Q} = \frac{C_{A_1} - C_{A_0}}{-k_A C_{A_1} C_{B_1}^2}$$

where

$$C_{A_1} = C_{A_0}(1 - X_A) = (1.0)(1 - 0.7) = 0.3.$$

If 0.7 g/mol of A react, 1.4 gmol of B must react from stoichiometric considerations. Therefore,

$$C_{B_1} = C_{B_0} - 2(C_{A_0} - C_{A_1}) = 2.5 - 2(1.0 - 0.3) = 1.1\ \text{gmol/L}$$

Substituting these two values into the above equation and solving for V (exercising care with units) gives

$$V = 877\,\text{L}$$

■

ILLUSTRATIVE EXAMPLE 9.7 Refer to the previous example. If the desired conversion is once again 70% and the reaction is carried out in a 400-L reactor, what is the necessary feed rate, Q?

Solution. Rearranging (1) in the previous example gives

$$Q = V\frac{(-k_A C_{A_1} C_{B_1}^2)}{(C_{A_1} - C_{A_0})} = 1.37\ \text{gal/min}$$

One could just as easily note that τ is a constant in both examples. Thus

$$\tau = \frac{877}{3.0} = \frac{400}{Q_2}$$

Solving gives

$$Q_2 = Q = 1.37 \text{ gal/min}$$ ■

ILLUSTRATIVE EXAMPLE 9.8 The elementary reversible liquid phase reaction

$$A + B \underset{k'}{\overset{k}{\rightleftharpoons}} C$$

is conducted in a CSTR with an equimolar feed mixture of A and B.

Data:

$k = 1.0 \text{ ft}^3/\text{gmol} \cdot \text{s}$

$K = 100 \text{ ft}^3/\text{gmol}$ (equilibrium constant based on concentration)

$C_{A_0} = 2.0 \text{ gmol/ft}^3$

Calculate the reactor volume requirement if $X_A = 0.9$ and $F_{A_0} = 0.171$ gmol/s.

Solution. Write the general rate expression for the reaction, $-r_A$, in terms of the reactant and product concentration(s).

$$-r_A = kC_AC_B - k'C_C$$

At equilibrium,

$$-r_A = 0$$

and

$$k/k' = C_C/C_AC_B$$

Note that

$$K = k/k'$$

so that

$$-r_A = kC_AC_B - k(C_C/K)$$

The expressions for the component concentrations in terms of the conversion variable X_A (use X) are:

$$\begin{aligned} C_A &= C_{A_0}(\theta_A - X); \qquad \theta_A = 1 \\ &= C_{A_0}(1 - X) \\ C_B &= C_{A_0}(\theta_B - X); \qquad \theta_B = 1 \\ &= C_{A_0}(1 - X) \\ C_C &= C_{A_0}(\theta_C + X); \qquad \theta_C = 0 \\ &= C_{A_0}X \end{aligned}$$

The reaction rate in terms of the conversion variable, X, is therefore

$$-r_A = k\left[C_{A_0}^2(1 - X)^2 - (C_{A_0}X/K)\right]$$

Substitute this rate expression into the characteristic CSTR design equation.

$$\begin{aligned} V &= F_{A_0}X/(-r_A) \\ &= F_{A_0}X\Big/k\left[C_{A_0}^2(1 - X)^2 - (C_{A_0}X/K)\right] \end{aligned}$$

Substitute known values into the design equation and solve for the volume.

$$\begin{aligned} V &= (0.171)(0.9)/(1)[(2.0)^2(1 - 0.9)^2 - (2)(0.9)/(100)] \\ &= 7.0 \text{ ft}^3 \end{aligned}$$

■

ILLUSTRATIVE EXAMPLE 9.9 The following liquid phase reaction takes place in a CSTR:

$$A + 2B \longrightarrow C + D$$

It is irreversible and elementary. What is the conversion of A if the volume of the reactor is 3.0 ft^3? Data is given below.

$$\begin{aligned} k_A &= 6\,\text{ft}^6/(\text{lbmol})^2(\text{min}) \\ Q = Q_0 &= 20\text{ ft}^3/\text{min};\ \ 50\% \text{ by volume is } A \\ (\text{MW}_A) = (\text{MW}_B) &= 35\text{ lb/lbmol} \\ \rho(\text{mixture}) &= 50\text{ lb/ ft}^3 = \rho_A = \rho_B \end{aligned}$$

Solution. Based on the example statement,

$$\begin{aligned} Q &= Q_0 = 20\ \text{ft}^3/\text{min} \\ Q_{A_0} &= (0.5)(20) = 10\ \text{ft}^3/\text{min} \\ F_{A_0} &= (10)(50)/35 = 14.3\ \text{lbmol/min} = F_{B_0} \\ \theta_B &= 1.0, \quad b/a = 2.0 \end{aligned}$$

Since

$$\begin{aligned} r_A &= -k_A C_A C_B^2 \\ -r_A &= k_A C_{A_0}(1 - X)C_{A_0}^2(1 - 2X)^2; \quad X = X_A \\ &= k_A C_{A_0}^3(1 - X)(1 - 2X)^2 \end{aligned}$$

In addition,

$$C_{A_0} = \left(14.3 \frac{\text{lbmol}}{\text{min}}\right) \Big/ \left(\frac{\text{min}}{20\ \text{ft}^3}\right) = 0.715\ \text{lbmol/ft}^3$$

The describing equation is

$$V = F_{A_0}X/-r_A; \ V = 3.0\ \text{ft}^3$$

$$3.0 = \frac{(14.3)(X)}{(6)(0.715)^3(1 - X)(1 - 2X)^2}$$

Solving

$$\frac{X}{(1 - X)(1 - 2X)^2} = 0.460$$

by trial and error (or any other suitable method)

$$X = X_A = 0.168 = 16.8\%$$

■

ILLUSTRATIVE EXAMPLE 9.10 The reaction

$$A + B \underset{k'}{\overset{k}{\rightleftharpoons}} C + D$$

is being conducted in a continuous stirred tank reactor. What initial concentration of A is necessary if the maximum allowable retention time is 5 minutes and a conversion of 50% of A is required? The inlet concentrations of B, C, and D are 0.5, 0.0, and 0.0 gmol/L, respectively.

Data:

$k = 0.75$ L/gmol · min;

$k' = 0.35$ L/gmol · min;

Solution. Assume an elementary reaction since not stated.

$$-r_A = kC_AC_B - k'C_CC_D$$

with

$$\begin{aligned} C_A &= C_{A_0}(1 - X) \\ C_B &= C_{A_0}(\theta_B - X) \\ C_C &= C_{A_0}(\theta_C + X); \quad \theta_C = 0 \\ C_D &= C_{A_0}(\theta_D + X); \quad \theta_D = 0 \end{aligned}$$

and

$$\theta_B = \frac{0.5}{C_{A_0}}$$

Thus, for 50% conversion of A

$$\begin{aligned} C_A &= 0.5C_{A_0} \\ C_B &= C_{A_0}\left(\frac{0.5}{C_{A_0}} - 0.5\right) = 0.5 - 0.5C_{A_0} \\ C_C &= 0.5C_{A_0} \\ C_D &= 0.5C_{A_0} \end{aligned}$$

Substituting into the rate expression gives

$$\begin{aligned} r_A &= -0.75\left[0.5C_{A_0}(0.5 - 0.5C_{A_0})\right] + 0.35(0.5C_{A_0})^2 \\ &= C_{A_0}(0.275C_{A_0} - 0.1875) \end{aligned}$$

The CSTR design equation is

$$\tau = V/Q = \frac{C_{A_0} - C_A}{-r_A}$$

Substituting

$$\tau = 5.0 = \frac{0.5}{0.1875 - 0.275C_{A_0}}$$

Solving,

$$C_{A_0} = 0.318 \text{ gmol/L}$$ ■

ILLUSTRATIVE EXAMPLE 9.11 Based on the fact that a color indicator shows when the concentration of A falls below 0.1 gmol/L, the following scheme is devised to explore the kinetics of the decomposition of A. A feed of concentration 0.6 gmol A/L is introduced into the first of two CSTRs in series, each having a volume of 400 cc. The color change first occurs in the first reactor for a steady-state feed rate of 10 cc/min and in the second reactor for a steady-state feed rate of 50 cc/min. Outline how to obtain the rate equation for the decomposition of A based on this information.

Solution. Assume a reaction rate equation of the form

$$A \xrightarrow{k_A}; \qquad -r_A = k_A C_A^n$$

The system is pictured in Figure 9.9. The change in color occurs at (A) and (B) for flowrate of 10 cc/mm and 50 cc/mm, respectively.

For (A)

$$V = \frac{Q_0(C_{A_0} - C_{A_1})}{-r_A} = \frac{Q_0(C_{A_0} - C_{A_1})}{k_A C_{A_1}^n}$$

Substituting

$$40Q = 10(0.6 - 0.1)/k_A(0.1)^n \tag{1}$$

For (B), $Q_0 = 50$ cc/min. For reactor (1),

$$400 = \frac{50(C_{A_1} - 0.1)}{k_A C_{A_1}^n} \tag{2}$$

Apply to both reactors

$$V = V_1 = V_2 = \frac{Q_0(C_{A_0} - C_{A_1})}{-k_A C_{A_1}^n} = \frac{Q_0(C_{A_1} - C_{A_2})}{-k_A C_{A_2}^n}$$

$$400 = \frac{0.6 - C_{A_1}}{C_{A_1}^n} = \frac{C_{A_1} - 0.1}{(0.1)^n} \tag{3}$$

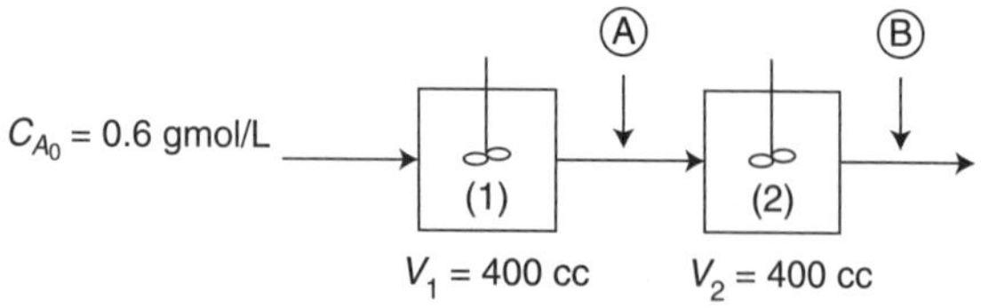

Figure 9.9 CSTR data for Illustrative Example 9.11.

Three equations result with three unknowns—k_A, C_{A_1}, and n! These may be solved simultaneously. The values of k_A and n are employed to generate the rate equation, i.e.,

$$r_A = -k_A C_A^n$$

The reader is left the exercise of numerically solving the problem. ■

ILLUSTRATIVE EXAMPLE 9.12 The *gas* phase pyrolysis reaction

$$A \longrightarrow 4R$$

is known to be *second order* with respect to A. What size CSTR operating at constant temperature and pressure is required to effect 90% conversion of A with a feed consisting of 50% A (by mole) and 50% inerts?

Data:

$$F_{A_0} = 100 \text{ gmol/ min}$$
$$C_{A_0} = 1.0 \text{ gmol/L}$$
$$k_A = 5.0 \text{ L/gmol} \cdot \text{min at the operating temperature}$$

Solution Write the design equation for a CSTR.

$$V = F_{A_0}\Delta X/-r_A = F_{A_0}X/-r_A$$

Note once again that ΔX represents the difference between the outlet conversion and the inlet conversion.

Determine and/or calculate y_{A_0}, δ, and ε for this gas phase reaction at the conditions specified and express C_A in terms of X and ε.

$$y_{A_0} = 0.5$$
$$\delta = 4 - 1 = 3$$
$$\varepsilon = (0.5)(3) = 1.5$$
$$C_A = C_{A_0}(\theta_A - X)/(1 + \varepsilon X); \text{ by definition, } \theta_A = 1.$$
$$= C_{A_0}(1 - X)/(1 + \varepsilon X)$$

Since this is a gas phase reaction, the volume correction factor must be included in the rate expression

$$-r_A = k_A C_A^2$$
$$= kC_{A_0}^2\left[(1 - X)/(1 + \varepsilon X)\right]^2$$

Calculate the rate of reaction in the reactor for $X = 0.9$.

$$-r_A = (5)(1)^2\{(1-0.9)/[1+(1.5)(0.9)]\}^2$$
$$= 9.05 \times 10^{-3} \text{ gmol/L} \cdot \text{min}$$

The volume requirement for this reactor is therefore

$$V = (0.9)(100)/(9.05 \times 10^{-3})$$
$$= 9.94 \times 10^3 \text{ L}$$

■

ILLUSTRATIVE EXAMPLE 9.13 Refer to the previous example. Is the outlet concentration, C_{A_1}, from the reactor 0.1 gmol/L since there is 90% conversion?

Solution. It is *not* 0.1 gmol/L since this is a gas phase reaction and the volume of the reacting mixture has changed. The concentration of A is

$$C_A = C_{A_0}[(1-X_A)/(1+\varepsilon X_A)]$$
$$= 1(1-0.9)/[1+(1.5)(0.9)]$$
$$= 0.043 \text{ gmol/L}$$

■

ILLUSTRATIVE EXAMPLE 9.14 Consider the elementary irreversible reaction between benzoquinone (A) and cyclopentadiene (B)

$$A + B \longrightarrow \text{products}$$

If one employs a feed containing equimolar concentrations of reactants, the reaction rate expression can be written as

$$-r_B = kC_AC_B = kC_B^2; \; k = k_B$$

Calculate the reactor size requirements for one CSTR. Also calculate the volume requirements for a cascade composed of two identical CSTRs. Assume isothermal operation at 25°C where the reaction rate constant is equal to 9.92 $m^3/(\text{kgmol} \cdot \text{ks})$. Reactant concentrations in the feed are each equal to 0.08 kgmol/m^3, and the liquid feed rate is equal to 0.278 m^3/ks. The desired degree of conversion is 87.5%.

Solution. The rate equation, $-r_B$, in terms of the conversion variable X is

$$-r_B = -r = kC_B^2$$
$$C_B = C_{B_0}(1-X)$$
$$-r = kC_{B_0}^2(1-X)^2$$

If only one reactor is employed,

$$V = \frac{F_{B_0}X}{-r}$$
$$= \frac{F_{B_0}X}{kC_{B_0}^2(1-X)^2}$$

Since

$$F_{B_0} = C_{B_0}Q$$
$$= (0.08)(0.278)$$
$$= 0.02224 \text{ kgmol/ks}$$
$$= 0.02224 \text{ gmol/s}$$

the volume may be calculated:

$$V = (0.02224)(0.875)/\left[(9.92)(0.08)^2(1-0.875)^2\right]$$
$$= 19.6 \text{ m}^3$$

The design equation above may also be applied to two reactors in series noting that $X_1 - 0 = X_1$.

$$V_1 = \frac{F_{D_0}}{kC_{B_0}^2}\frac{X_1}{(1-X_1)^2}$$

$$V_2 = \frac{F_{B_0}}{kC_{B_0}^2}\frac{\Delta X}{(1-X_2)^2}; \quad \Delta X = X_2 - X_1$$

For reactor 1,

$$\frac{kC_{B_0}^2V_1}{F_{B_0}} = \frac{X_1}{(1-X_1)^2}$$

For reactor 2,

$$\frac{kC_{B_0}^2V_2}{F_{B_0}} = \frac{X_2 - X_1}{(1-X_2)^2}$$

Since the LHSs of both of the above equations are equal, ($V_1 = V_2$),

$$\frac{X_1}{(1-X_1)^2} = \frac{X_2 - X_1}{(1-X_2)^2}$$

Solving with $X_2 = 0.875$ yields

$$X_1 = 0.7251$$

Thus,

$$
\begin{aligned}
V_1 &= \frac{(0.278)(0.08)(0.7251)}{(9.92)(0.08)^2(1-0.7251)^2} \\
&= 3.36\ \text{m}^3 \\
V_T &= V_1 + V_2 \\
&= (2)(3.36) \\
&= 6.72\ \text{m}^3
\end{aligned}
$$

■

ILLUSTRATIVE EXAMPLE 9.15 Refer to the previous example. Compare the results—in terms of the volume requirements—for the two cases.

Solution. Using one reactor requires a volume of

$$V_1 = 19.6\ \text{m}^3$$

Use of two equal volume reactors in series produced a volume requirement of

$$V_2 = 3.36 + 3.36 = 6.72\ \text{m}^3$$

As expected, the latter case provided the smaller volume requirement. ■

ILLUSTRATIVE EXAMPLE 9.16 Refer to Illustrative Example 9.14. Determine the effect of using a cascade of two CSTRs, that *differ* in size, on the volume requirements for the reactor network. If the same feed composition and flow rate above are employed and if the reactors are operated isothermally at 25°C, determine the *minimum* total volume required and the manner in which the volume should be distributed between the two reactors. A fractional overall conversion of 0.875 is to be achieved again.

Solution. Set up an equation to determine the total volume requirement if the two reactor volumes are not equal.

$$V_T = V_1 + V_2\,; \quad V_1 \neq V_2$$

$$V_T = \left(\frac{F_{B_0}}{kC_{B_0}^2}\right)\left(\frac{X_1}{(1-X_1)^2} + \frac{(X_2 - X_1)}{(1-X_2)^2}\right); \quad X_2 = 0.875$$

To solve the above equation for the intermediate conversion X_1 under minimum volume conditions, one needs to minimize V_T, by setting

$$\frac{dV_T}{dX_1} = 0\,; \quad V_T = V_T(X_1)$$

and then solve for X_1. Analytically differentiating yields

$$\frac{dV_T}{dX_1} = \frac{1}{(1 - X_1)^2} + \frac{2X_1}{(1 - X_1)^2} - 64 = 0$$

Solving by trial and error gives

$$X_1 = 0.702$$

The volumes of the two individual reactors and the total minimum volume requirement can now be calculated.

$$V_1 = \frac{(0.02224 \text{ gmol/s})(0.702)}{(9.92 \text{ L/gmol} \cdot \text{s})(0.08 \text{ gmol/L})^2(1 - 0.702)^2}$$

$$V_1 = 2.77 \text{ L}$$

$$V_2 = \frac{(0.02224 \text{ gmol/s})(0.875 - 0.702)}{(9.92 \text{ L/gmol} \cdot \text{s})(0.08 \text{ gmol/L})^2(1 - 0.875)^2}$$

$$V_2 = 3.88 \text{ L}$$

$$V_T = V_1 + V_2$$

$$V_T = 2.77 \text{ L} + 3.88 \text{ L}$$

$$V_T = 6.65 \text{ L}$$

The above result is slightly lower than that calculated earlier ($V_T = 6.72$ L). ■

ILLUSTRATIVE EXAMPLE 9.17 Refer once again to Illustrative Example 9.14. Determine the effect of using a cascade of three CSTRs that differ in size on the volume requirements for the reactor network. If the same feed composition and flow are employed and if the reactors are operated isothermally at 25°C, determine the minimum total volume required and the manner in which the volume should be distributed between the three reactors. An overall conversion of 0.875 is to be achieved again. For this problem, first *outline the method of solution*. Then solve the problem.

Solution. Write the equation for the volume of each reactor and total volume requirement for three reactor volumes that are not equal.

For each reactor:

$$V_1 = \left(\frac{F_{B_0}}{kC_{B_0}^2}\right)\left(\frac{X_1}{(1 - X_1)^2}\right)$$

$$V_2 = \left(\frac{F_{B_0}}{kC_{B_0}^2}\right)\left(\frac{X_2 - X_1}{(1 - X_2)^2}\right)$$

$$V_3 = \left(\frac{F_{B_0}}{kC_{B_0}^2}\right)\left(\frac{X_3 - X_2}{(1 - X_3)^2}\right)$$

The total volume requirement is

$$V_T = V_1 + V_2 + V_3$$

$$V_T = \left(\frac{F_{B0}}{kC_{B0}^2}\right)\left(\frac{X_1}{(1-X_1)^2}\right) + \left(\frac{F_{B0}}{kC_{B0}^2}\right)\left(\frac{X_2-X_1}{(1-X_2)^2}\right) + \left(\frac{F_{B0}}{kC_{B0}^2}\right)\left(\frac{X_3-X_2}{(1-X_3)^2}\right)$$

Take the partial derivative of the equation obtained above with respect to X_1. Note that the partial derivative must now be employed since the total volume, V_T, depends on *two* variables, X_1 and X_2, i.e., $V_T = V_T(X_1, X_2)$.

$$\frac{\partial V_T}{\partial X_1} = \left(\frac{F_{B0}}{kC_{B0}^2}\right)\left(\frac{1}{(1-X_1)^2} + \frac{2X_1}{(1-X_1)^3} - \frac{1}{(1-X_2)^2}\right)$$

Set this derivative equal to zero.

$$\frac{\partial V_T}{\partial X_1} = \left(\frac{F_{B0}}{kC_{B0}^2}\right)\left(\frac{1}{(1-X_1)^2} + \frac{2X_1}{(1-X_1)^3} - \frac{1}{(1-X_2)^2}\right) = 0$$

Take the partial derivative of the equation for V_T with respect to X_2.

$$\frac{\partial V_T}{\partial X_2} = \left(\frac{F_{B0}}{kC_{B0}^2}\right)\left(\frac{1}{(1-X_2)^2} - \frac{2(X_2-X_1)}{(1-X_2)^3} - \frac{1}{(1-X_3)^2}\right)$$

Also, set this derivative equal to zero.

$$\frac{\partial V_T}{\partial X_2} = \left(\frac{F_{B0}}{kC_{B0}^2}\right)\left(\frac{1}{(1-X_2)^2} - \frac{2(X_2-X_1)}{(1-X_2)^3} - \frac{1}{(1-X_3)^2}\right) = 0$$

Solve the resulting two partial derivative equations—that are nonalgebraic, simultaneously. For $X_3 = 0.875$, X_2 is obtained as follows by trial and error (see Table 9.1). Therefore, $X_1 = 0.585$, $X_2 = 0.7879$.

TABLE 9.1 Trial and Error Calculation for Illustrative Example 9.17

X_1	X_2	Residual
0.6	0.8	11
0.59	0.7918	4
0.58	0.7835	−2.5
0.585	0.7879	0.8

Finally, calculate the volume of the three individual reactors and the total minimum volume requirement.

$$V_1 = \left(\frac{F_{B0}}{kC_{B0}^2}\right)\left(\frac{X_1}{(1-X_1)^2}\right)$$

$$V_1 = \left(\frac{0.02224 \text{ gmol/s}}{(9.92 \text{ L/gmol} \cdot \text{s})(0.08 \text{ gmol/L})^2}\right)\left(\frac{0.585}{(1-0.585)^2}\right)$$

$$V_1 = 1.19 \text{ L}$$

$$V_1 = \left(\frac{F_{B0}}{kC_{B0}^2}\right)\left(\frac{X_2 - X_1}{(1-X_2)^2}\right)$$

$$V_2 = \left(\frac{0.02224 \text{ gmol/s}}{(9.92 \text{ L/gmol} \cdot \text{s})(0.08 \text{ gmol/L})^2}\right)\left(\frac{0.7879 - 0.585}{(1-0.7879)^2}\right)$$

$$V_2 = 1.58 \text{ L}$$

$$V_3 = \left(\frac{F_{B0}}{kC_{B0}^2}\right)\left(\frac{X_3 - X_2}{(1-X_3)^2}\right)$$

$$V_3 = \left(\frac{0.02224 \text{ gmol/s}}{(9.92 \text{ L/gmol} \cdot \text{s})(0.08 \text{ gmol/L})^2}\right)\left(\frac{0.875 - 0.7879}{(1-0.875)^2}\right)$$

$$V_3 = 1.95 \text{ L}$$

$$V_T = 1.19 \text{ L} + 1.58 \text{ L} + 1.95 \text{ L}$$

$$V_T = 4.72 \text{ L}$$

Note the reduction in the volume requirement when three CSTRs in series are employed. The reader is left the exercise of showing that the total volume requirement for three CSTRs of *equal* volume is

$$V_1 = V_2 = V_3 = 2.17 \text{ m}^3$$

$$V_T = V_1 + V_2 + V_3 = 6.51 \text{ m}^3$$

with

$$X_1 = 0.6710$$

$$X_2 = 0.7782$$

■

ILLUSTRATIVE EXAMPLE 9.18 Refer to Illustrative Example 9.14. Determine the reactor size requirements for a cascade composed of two CSTRs, with the first inlet reactor THRICE the size of the second.

Solution. The results for this case are

$$V_1 = 1.425 \text{ m}^3$$

$$V_2 = (3)(1.425) = 4.28 \text{ m}^3$$

$$V_T = V_1 + V_2 = 5.71 \text{ m}^3$$

with

$$X_1 = 0.64$$

The details of the solution are left as an exercise for the reader. ■

ILLUSTRATIVE EXAMPLE 9.19 Refer to Illustrative Example 9.14. Determine the reactor size requirements for a cascade composed of two CSTRs, with the first inlet reactor ONE THIRD the size of the second.

Solution. The solution to this problem is similar to that employed earlier. The results are

$$V_1 = 4.93 \text{ m}^3$$

$$V_2 = \frac{4.93}{3} = 1.64 \text{ m}^3$$

$$V_T = V_1 + V_2 = 6.57 \text{ m}^3$$

The calculation of X_1 is left as an exercise for the reader. The reader may also choose to compare this result with that of the previous example. ■

ILLUSTRATIVE EXAMPLE 9.20 Refer to Illustrative Example 9.14. Estimate the volume required if a very large number (approaching infinity) of CSTRs are employed.

Solution. As will be demonstrated in the next chapter, an infinite (∞) number of finite volume CSTRs connected in series approaches the behavior of a tubular flow (TF) reactor. Furthermore, an infinite number of CSTRs of differential (approaching zero) volume also approaches the behavior of a TF reactor. The volume of a TF reactor for this application as shown in Illustrative Example 10.16 is:

$$V(TF) = 2.45 \text{ L}$$

This volume is significantly below those volumes calculated in the previous illustrative examples. ■

ILLUSTRATIVE EXAMPLE 9.21 Once again, consider the system in Illustrative Example 9.14. If the reaction is reversible, rather than irreversible, calculate the volume requirements (for two equal sized CSTRs). The rate expression is given by

$$-r = kC_{B_0}^2(1 - X)^2 - 0.04kC_{B_0}(2X)^2$$

The overall conversion for this last example is to be set at 0.70, not 0.875.

Solution. For reactor (1),

$$V_1 = F_{B_0}X_1/kC_{B_0}^2\left[(1 - X_1)^2 - 0.16X_1^2\right]$$

or

$$V_1kC_{B_0}^2/F_{B_0} = X_1/\left[(1 - X_1)^2 - 0.16X_1^2\right]$$

For reactor (2),

$$V_2kC_{B_0}^2/F_{B_0} = (X_2 - X_1)/\left[(1 - X_2)^2 - 0.16X_2^2\right]; \quad X_2 = 0.70$$

Equate the above two equations since $V_1 = V_2$. Employ a trial and error procedure (see Table 9.2).

TABLE 9.2 Trial and Error Calculation for Illustrative Example 9.18

X_1	$kC_{B_0}^2V_1/F_{B_0}$	$kC_{B_0}^2V_2/F_{B_0}$
0.6	5.86	8.62
0.625	8.00	6.47
0.62	7.48	6.90
0.6175	7.239	7.112

Using $X_1 = 0.6175$, yields

$$V_1 = 2.536\ \text{m}^3, \quad V_2 = 2.491\ \text{m}^3$$

Therefore,

$$V_1 = V_2 = 2.5\ \text{m}^3$$

and

$$V_T = 5.0\ \text{m}^3$$

■

REFERENCES

1. S. Fogler, "*Elements of Chemical Reaction Engineering*," 4th edition, Prentice-Hall, Upper Saddle River, NJ, 2006.
2. L. Theodore and F. Ricci, "*Mass Transfer Operations for the Practicing Engineer*," John Wiley & Sons, Hoboken, NJ, 2010.

CHAPTER **10**

TUBULAR FLOW REACTORS

INTRODUCTION

The last reactor to be examined in Part II is the tubular flow (TF) reactor. The most common type is the single-pass cylindrical tube; another type is one which consists of a number of tubes in parallel. Both will be reviewed in more detail in the next section.

The tubular flow reactor shown in Figure 10.1 can be conceptualized by two different views. The first is to consider a CSTR (see previous chapter) of given volume V, and feed rate Q (or F) and divide the volume into a number of much smaller CSTRs in series. In the limit, an infinite number of these CSTRs with a total volume V reduces to a TF reactor (see also Illustrative Example 9.21). In other words, the time of reaction in a flow system may be visualized as the time required for an infinitesimal parcel of mass to pass through the reactor in an imaginary compartment (a CSTR of differential volume) that moves with the flowing stream and reacts producing changes to the temperature, pressure, and composition of the stream. The second approach is to envision a pipe (or tube) of uniform diameter with reactants and products flowing axially along the pipe. This simpler perspective is also a tubular flow reactor.

In terms of kinetic behavior, the TF reactor may be viewed as "identical" to a batch reactor. The time t for reaction which was defined for a batch reactor is now expressed in terms of τ (the residence time) and represents the length of time necessary for any given increment of feed to travel the entire length of the reactor.

Tubular flow reactors are characterized by a continuous and decreasing concentration of reactants in the direction of flow. This is in contrast to the discontinuous characteristic of the CSTR reactor. Most of these units consist of one or several pipes or tubes in parallel. Either horizontal or vertical orientation is common. The reactants are charged continuously at one end, and the products are removed continuously at the other end. The unit almost always operates in a steady-state mode. This greatly simplifies design and predictive calculations. It is a unit that is amendable to automatic control and to experimental work. When heat transfer is required, a jacketed tube or a construction similar to that of a shell-and-tube heat exchanger is employed.[1] In the latter case the reactants may be on either the tube or shell side.

The key assumption with a CSTR is that it is completely *mixed*. The key assumption with a TF unit is that plug flow is present, i.e., the fluid is completely *unmixed* along the reactor length and flows down the tube as a plug since it is perfectly

---

Chemical Reactor Analysis and Applications for the Practicing Engineer. By Louis Theodore

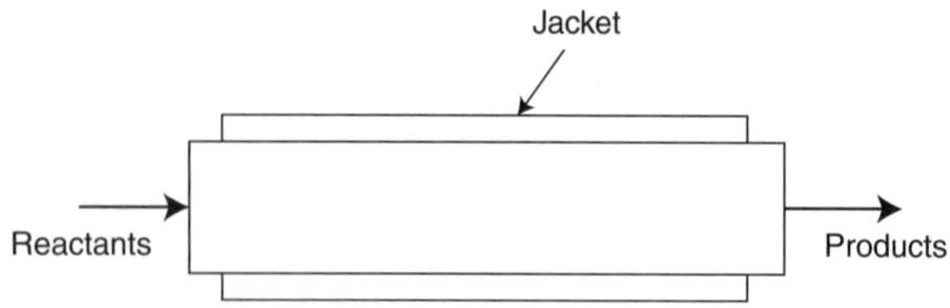

Figure 10.1 Tubular flow reactor, single pass.

mixed radially. Even though the plug-flow approximation may not apply rigorously in some applications, the simplicity of solutions in the limit of perfect plug-flow makes it a very useful model from both an academic and industrial perspective. The more complex flow profiles are discussed in Part III, Chapter 14.

Three additional topics are discussed in this chapter:

Equipment Description and Operation

Describing Equations

Applications

EQUIPMENT DESCRIPTION AND OPERATION

As noted in the Introduction to this chapter, the most common type of TF reactor is the single-pass cylindrical tube, as shown in Figure 10.1. Another type is one which consists of a number of tubes in parallel (see Figure 10.2). The reactor(s) may be vertical or horizontal. The feed is charged continuously at the inlet of the tube, and the products are continuously removed at the outlet. If heat exchange with surroundings is required, the reactor setup includes a jacketed tube and effectively acts as either a double pipe or a single pass heat exchanger.(1) If the reactor is "empty", a homogeneous reaction—one phase present—usually occurs. If the reactor contains catalyst particles, the reaction is said to be heterogenous. This type will be considered in Part IV.

TF reactors are usually operated under steady conditions so that physical and chemical properties do not vary with time. Unlike the batch and tank flow reactors, there is no mechanical mixing. Thus, and as noted in the previous section, the state of the reacting fluid will vary from point to point in the system, and this variation may be in both the radial r and axial z direction (see Figure 10.3). The describing equations are then differential, with position as the independent variable.

In the describing equations section to follow, the reacting system is assumed to move through the reactor in plug flow (see Figure 10.4(a)). It is further assumed that there is no mixing in the axial direction and complete mixing in the radial direction so that the concentration, temperature, etc., do not vary through the cross section of the tube. Thus, the reacting fluid flows through the reactor in an undisturbed plug of mass. Note that the time for this hypothetical plug to flow through this type of reactor is the same as the contact or reaction time in a batch reactor. Under these conditions, the form of the describing equations for batch reactors will also apply to TF reactors. From a qualitative point of view, as the length of the reactor approaches infinity, the

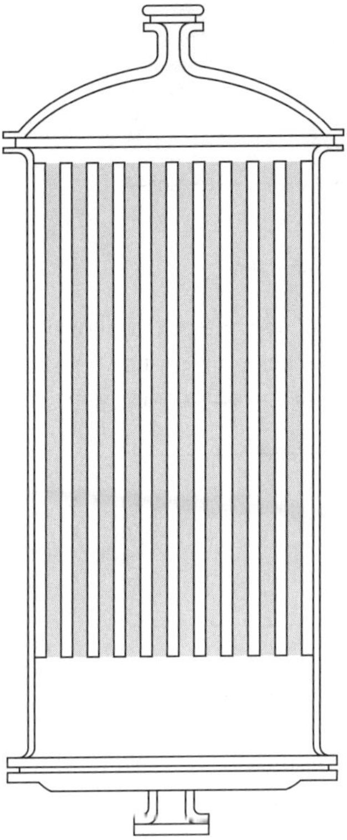

Figure 10.2 Multitube flow reactor.

concentration of a (single) reactant approaches zero for irreversible reactions (except zero order) and the equilibrium concentration for reversible reactions.

In a plug flow reactor the entire feed stream moves with the same radially uniform axial velocity along parallel streamlines. The entire feed stream therefore has the same residence time since there is no mixing in the axial direction but (effectively) complete mixing radially.

In actual practice, TF reactors deviate from the plug flow model because of velocity variations in the radial direction (see Figure 10.4(b)–(d)). For any of these conditions, the residence time for annular elements of fluid within the reactor will vary from some minimal value at a point where the velocity is a maximum to a maximum value near the wall where the velocity approaches zero. The concentration and temperature profiles, as well as the velocity profile are therefore also not constant across the reactor. The describing equations based on the plug flow assumption are then not applicable.

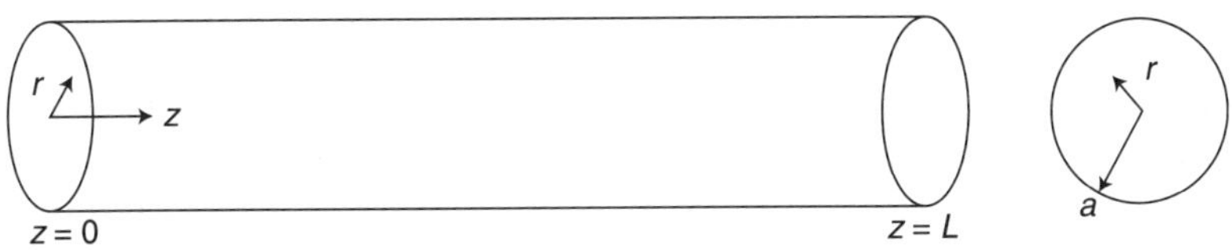

Figure 10.3 Reactor system, cylindrical coordinates.

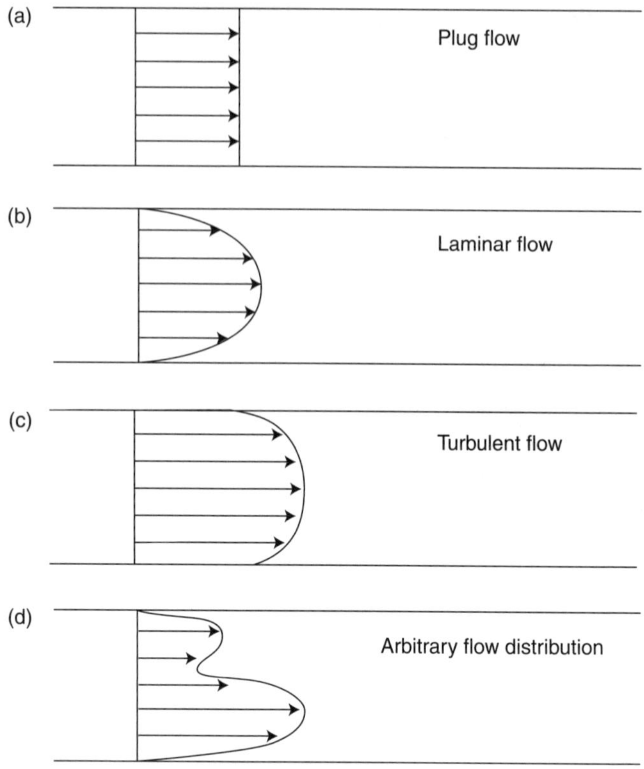

Figure 10.4 Velocity profiles in flow reactors.

The design equation for tubular reactors operating in a steady-state mode will shown to be differential in form, unlike the algebraic design equation for CSTRs. The describing equation is of an ordinary differential form for variations along the axial (longitudinal) length of the reactor. A partial differential equation arises if there are variations along both the axial and radial directions.

DESCRIBING EQUATIONS

The describing equation for mass transfer is obtained by applying the conservation law for mass to a differential volume element in a flow reactor (see Figure 10.5). Assume

1. Plug flow.
2. Steady conditions.
3. No radial or axial diffusion effects.

The conservation of mass law for *A* presented in the terms in Equation (7.2) are now evaluated.

Terms 1–2: This is zero since diffusion effects are neglected.
Term 3: This term is given by

$$F_{A_0}X_A \tag{10.1}$$

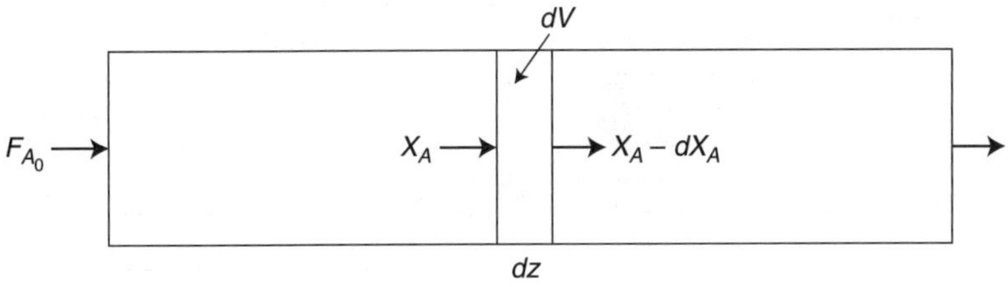

Figure 10.5 Mass balance on a differential reactor element.

where

$$F_{A_0} = \text{Inlet molar (or mass) flow rate of } A$$

$$X_A = \frac{\text{Moles of } A \text{ converted}}{\text{Mole (or mass) of } A \text{ entering the reactor}}$$

Term 4: The moles leaving the differential volume element is

$$F_{A_0}(X_A - dX_A) \tag{10.2}$$

Term 5: This term is again given by (see Equation 8.2)

$$r_A dV \tag{10.3}$$

Term 6: This is zero for steady conditions.

Substituting the above terms in Equation (7.2) yields

$$F_{A_0} dX_A + r_A dV = 0 \tag{10.4}$$

The usual procedure is to set this equation up in the following form:

$$dV = -F_{A_0} \frac{dX_A}{r_A} \tag{10.5}$$

$$V = -F_{A_0} \int \frac{dX_A}{r_A} = F_{A_0} \int \frac{dX_A}{(-r_A)} \tag{10.6}$$

In order to integrate the RHS, the rate of reaction r_A must be expressed in terms of the conversion variable X_A. For example, consider the reaction

$$A \xrightarrow{k_A} \text{product}$$

with

$$r_A = -k_A C_A$$

Replace C_A by

$$C_A = \frac{N_A}{V_c} \tag{10.7}$$

$$= \frac{N_{A_0}(1 - X_A)}{V_c} \tag{10.8}$$

where

N_{A_0} = moles of A at inlet conditions

V_c = volume term upon which the concentration is based at inlet conditions

For complex reactions, it becomes difficult to integrate Equation (10.6). It is then more convenient to write this equation in the form

$$\frac{dX_A}{dV} = -\frac{r_A}{F_{A_0}} \tag{10.9}$$

which is more suitable for analytical or numerical solutions.

Various other forms of the describing equations are also available for analysis and/or design purposes. For the notation in Figure 10.6, one obtains

$$V = F_0 \int \frac{dX_A^*}{(-r_A)} \tag{10.10}$$

For Figure 10.7, one may write (*with Q once again representing the volumetric flowrate*)

$$\frac{V}{Q_0} = -C_{A_0} \int \frac{dX_A}{r_A} = C_{A_0} \int \frac{dX_A}{(-r_A)} \tag{10.11}$$

or

$$\frac{V}{Q_0} = -C_0 \int \frac{dX_A^*}{r_A} = C_0 \int \frac{dX_A^*}{(-r_A)} \tag{10.12}$$

If the volume rate of flow is constant

$$Q_0 = Q$$

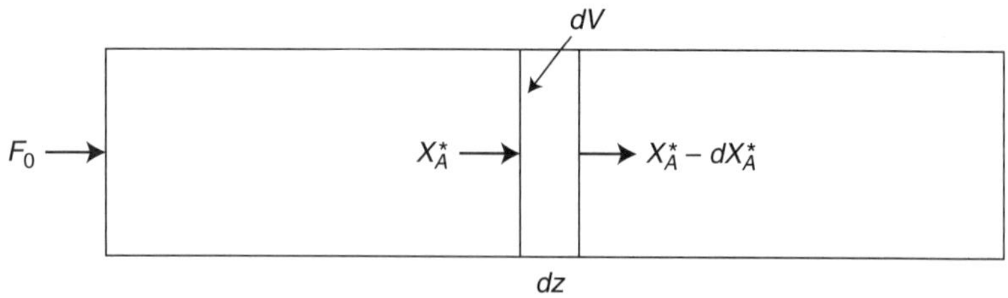

Figure 10.6 Mass balance on differential element; X_A^*, F_0.

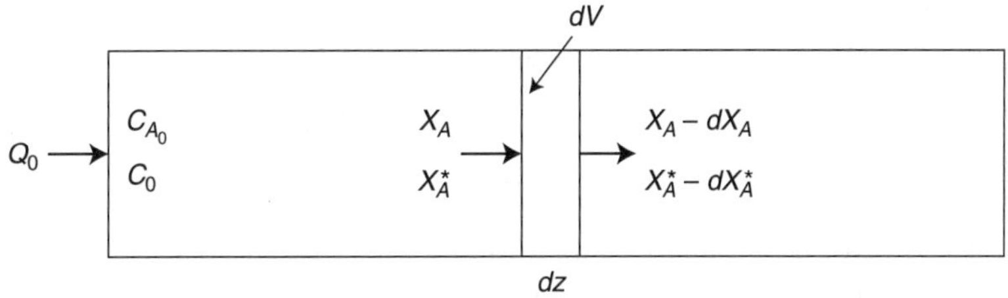

Figure 10.7 Mass balance on differential element; C_{A_0}, Q_0.

Equations (10.11) and (10.12) become

$$\frac{V}{Q} = \int \frac{dC_A}{r_A} \tag{10.13}$$

Note that the LHS of this equation is simply the average holdup (residence) time, τ, that is,

$$\frac{V}{Q} = \tau = \int \frac{dC_A}{r_A} = C_{A_0} \int \frac{dX_A}{(-r_A)} \tag{10.14}$$

For the system pictured in Figure 10.8, the describing equation becomes

$$v_z \frac{dC_A}{dz} = r_A \tag{10.15}$$

This is occasionally a more convenient form of the equation to work with since the concentration (or equivalent) is obtained as a function of the cylindrical coordinate position variable z rather than the volume.

The average residence time τ, introduced in Equation (10.14) deserves further comment. Another convenient term relating the feed rate and reactor volume in any type of flow process (including the tubular flow unit) is the *space velocity* (SV). This term is defined as *the volumetric feed rate per unit volume of reactor*. The SV has the units of reciprocal time and is therefore equal to $1/\tau$. In a variable-density

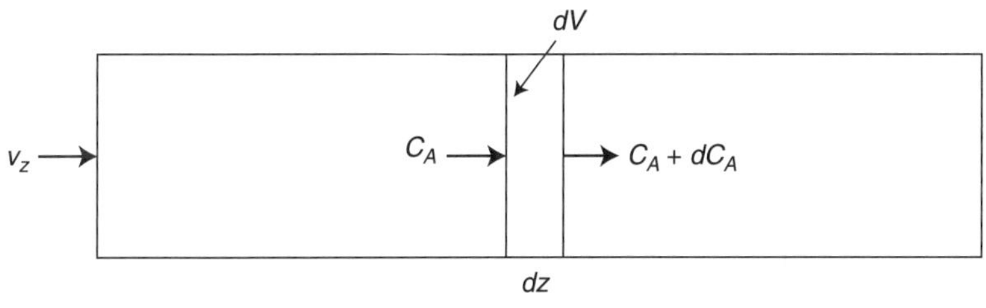

Figure 10.8 Mass balance on differential element; C_A, v_z.

TF reactor the residence time τ and space velocity SV vary as the fluid passes through the reactor. Therefore, these terms are often based on inlet conditions. Both terms are therefore constant while reactor parameters such as X, T, and P vary.

Pressure Drop

The calculation of pressure drop and pressure variation in reactors can be a difficult problem, particularly if a change of phase occurs or more than one phase is present. This effect is often small and can be safely neglected for most flow reactors. However, this effect may be estimated using Fanning's equation[2]

$$\Delta P = \frac{4fLv^2}{2g_cD} \tag{10.16}$$

where (in consistent units):

ΔP = pressure drop
f = Fanning friction factor
L = reactor length
v = average flow velocity
g_c = gravity conversion constant
D = inside reactor diameter

This equation can be derived from a mechanical energy balance equation which is an extension of the conservation law for energy.

The pressure drop for tubular reactors in which gases are flowing at relatively high Reynolds numbers can be estimated from the following equation:

$$\Delta P = \frac{0.0235\,\mathrm{L}}{D^{4.8}}\left(\frac{\dot{m}}{1000}\right)^{1.8}\frac{(\mu)^{0.2}}{\rho} \tag{10.17}$$

where

$\dot{m}$ = mass rate of flow
μ = viscosity
ρ = density

For some gas phase reactions, the pressure drop can be significant because of changes in the temperature and the number of moles. The reader is referred to fluid flow texts in the literature for additional details on this subject.[2]

ILLUSTRATIVE EXAMPLE 10.1 Which of the following is a design equation for a plug-flow tubular reactor ($X = X_A$)?

A $V = \int \frac{F_{A_0}\,dX}{(-r_A)}$

B $V = \int \frac{r_A}{F_{A_0}X}$

C $V = \frac{F_{A_0}X}{r_A|_X}$

D $V = \int F_{A_0} r_A dX$

E None of the above

Solution. The differential design equation for a tubular flow reactor based on $X = X_A$ is as follows:

$$\frac{dF_A}{dV} = r_A$$

Substitute for $F_A = F_{A_0}(1 - X)$,

$$\frac{d\left[F_{A_0}(1 - X)\right]}{dV} = r_A$$

Separation and integration yields the following:

$$V = \int \frac{F_{A_0} dX}{(-r_A)}$$

Alternatively, the reader is referred to Equation (10.6). The correct answer is (A). ■

ILLUSTRATIVE EXAMPLE 10.2 Discuss pressure drop relative to the various types of reactors employed in practice.

Solution. The reader should note that the pressure drop associated with both batch reactors and CSTRs may be safely assumed to be negligible. However, the pressure drop for catalytic reactors—whether fixed bed or fluidized bed—is naturally higher than that for the (empty) TF unit. ■

ILLUSTRATIVE EXAMPLE 10.3 Is it more advantageous to connect two plug flow reactors in series than in parallel with each reactor receiving half the feed? How does reaction order and degree of conversion affect the answer. Explain.

Solution. The series(s) system is pictured in Figure 10.9; note that (a) is equivalent to (b).

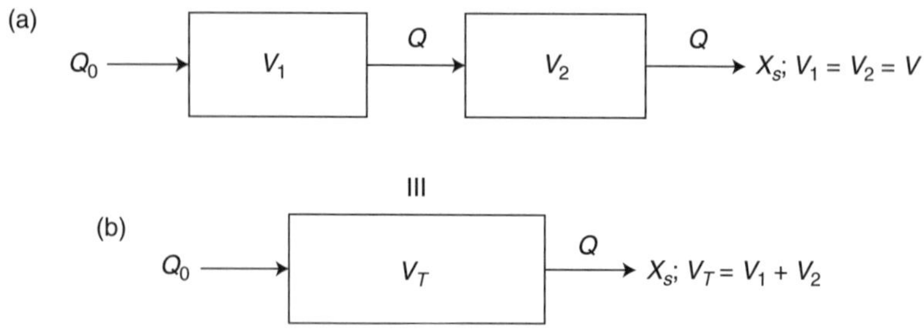

Figure 10.9 TFs in series.

The design equation for a tubular flow reactor is

$$\tau_s = \frac{V}{Q} = C_{A_0}\int \frac{dX}{(-r_A)} \tag{10.14}$$

$$\tau_s = \frac{V_T}{Q}$$

For the parallel (p) system see Figure 10.10. For this case,

$$\tau_p = \frac{V_1}{Q_0/2} = \frac{V}{Q/2} = \frac{V_T/2}{Q/2} = \frac{V_T}{Q} = C_{A_0}\int \frac{dX}{(-r_A)}$$

Since the two residence times are the same, one concludes that the degree of conversion is the same. In addition, since r_A appears in both equations, the conversion in either case is unaffected by the reaction order (or reaction mechanism). ■

ILLUSTRATIVE EXAMPLE 10.4 You have been asked to maximize the conversion of a product formed from a first order irreversible reaction. Which of the two systems in Figure 10.11 will do that job? Justify your answer.

Solution. Refer to Figure 10.12 for case (1). In order to solve this problem, the three reactors in (1) and (2) must be viewed as occurring in 2 reactors. Note that in Figure 10.12, (a) and (b) are identical.

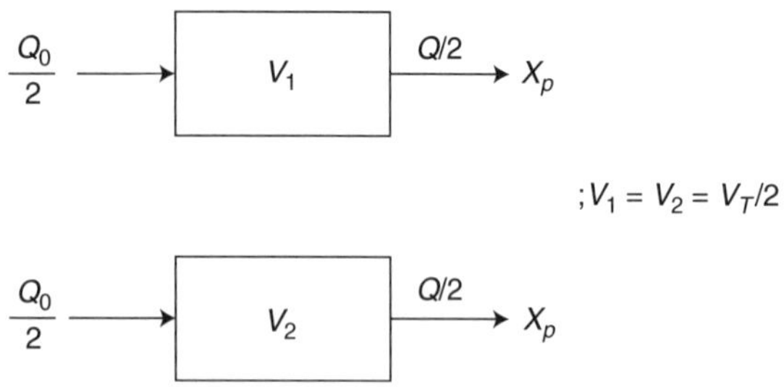

Figure 10.10 TFs in parallel.

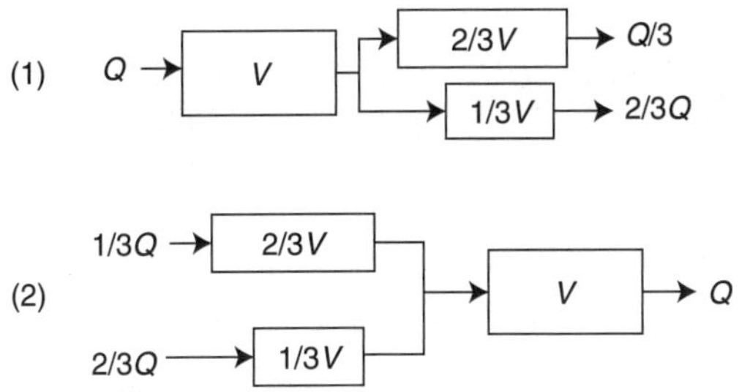

Figure 10.11 Tubular reactions in series.

For the higher conversion X_{H1}, the residence time is

$$\tau_{H1} = \frac{V + (2/3)V}{Q/3} = \frac{(5/3)V}{Q/3} = 5\left(\frac{V}{Q}\right)$$

The lower conversion is

$$\tau_{L1} = \frac{V + (1/3)V}{2/3Q} = \frac{4/3\,V}{2/3Q} = 2\left(\frac{V}{Q}\right)$$

Refer to Figure 10.13 for case (2). Note once again that (a) and (b) are equivalent. For the higher conversion

$$\tau_{H2} = \frac{(5/3)V}{Q/3} = 5\left(\frac{V}{Q}\right)$$

$$\tau_{L2} = \frac{(4/3)V}{(2/3)Q} = 2\left(\frac{V}{Q}\right)$$

Conclusion: Both systems are identical! Is the above analysis limited to first order reactions? The reader can ponder the question related to the effect of the reaction order and/or mechanism. ■

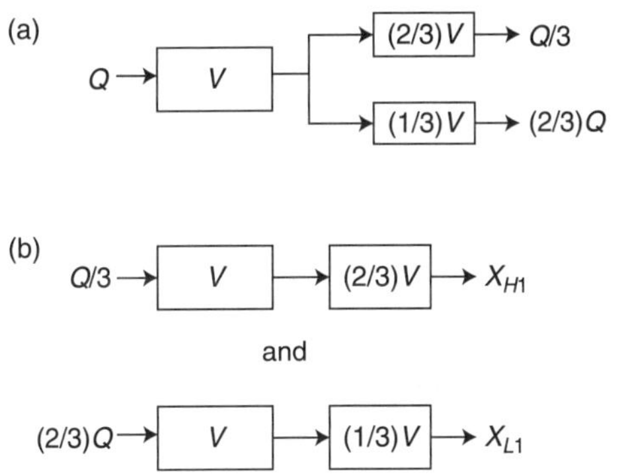

Figure 10.12 TFs in series with split at back end.

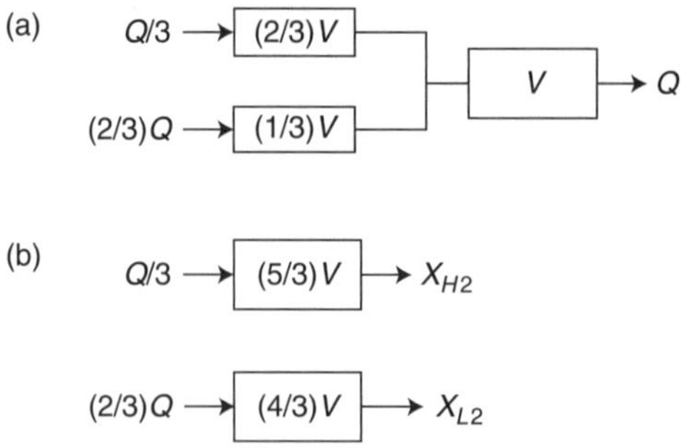

Figure 10.13 TFs in series with split at front end.

APPLICATIONS

This last section is concerned with applications. A total of 19 illustrative examples are provided.

ILLUSTRATIVE EXAMPLE 10.5 It is proposed to decompose pure diethyl peroxide (A) at 225°C in a bench-scale tubular reactor. The organic will be entering the reactor at a flow rate of 12.1 L/s. It is desired to decompose 99.995% of the diethyl peroxide. The following data are available:

$$r_A = -k_A C_A \text{ gmol/L} \cdot \text{s} \qquad k_A = 38.3 \text{ (s)}^{-1} \text{ at } 225^\circ\text{C}$$

The inside diameter of the reactor is 8.0 cm. Calculate the reactor volume.

Solution. Assume a plug flow reactor and use Equation 10.14. Since the reactor is first order and irreversible, the integrated result is

$$\tau = -\frac{1}{k_A}\ln\left(\frac{C_A}{C_{A_0}}\right)$$

For 99.995% destruction,

$$C_A = 5 \times 10^{-5} C_{A_0}$$

so that

$$\tau = -\left(\frac{1}{38.3}\right)\ln(5 \times 10^{-5})$$
$$= 0.259 \text{ s}$$

Calculate the reactor volume:

$$V = (\tau)(Q_0) = (0.259)(12.1)$$
$$= 3.13 \text{ L} = 3130 \text{ cm}^3$$

■

ILLUSTRATIVE EXAMPLE 10.6 Refer to the previous Illustrative Example. Calculate the length of the reactor.

Solution. The length of the reactor is given by

$$L = \frac{V}{\pi D^2/4}$$

Substituting

$$L = \frac{3130}{(\pi)(8^2)/(4)}$$
$$= 62.3\,\text{cm}$$

■

ILLUSTRATIVE EXAMPLE 10.7 The elementary, irreversible reaction

$$2A + B \xrightarrow{k_A} C$$

is conducted in a tubular flow reactor with a volume of 650 cm^3. A 75% conversion is achieved for an initial concentration of 1.0 gmol/L of A and 0.5 gmol/L of B, with a feed flow rate of 8.5 gmol A/min. If the desired conversion is reduced to 50%, what should be the new flow rate?

Solution. Note that k_A is *not* provided. For this third order reaction,

$$C_A^2 = C_{A_0}^2(1 - X_A)^2 = (1 - X_A)^2$$
$$C_B = C_{B_0} - (b/a)C_{A_0}X_A = C_{B_0} - 0.5C_{A_0}X_A = 0.5(1 - X_A)$$

The describing equation is

$$\frac{V}{F_A} = \int_0^{X_A} \frac{dX_A}{(-r_A)}$$

Integrating and substituting,

$$\frac{0.650}{8.50} = \frac{1}{k_A}\left[\frac{2}{0.5(1 - X_A)^2}\right]; \qquad X_A = 0.75$$
$$= \frac{0.30}{0.5k_A}$$

Solving for k_A gives

$$k_A = 837\ \text{L}^2/(\text{gmol})^2 \cdot \text{min}$$

For the desired conversion, then

$$\frac{V}{F_A} = \frac{1}{0.5k_A}\left[\frac{2}{(1-X_A)^2}\right]; \qquad X_A = 0.5$$

Solving gives

$$F_A = 34.0\,\text{gmol/min}$$

■

ILLUSTRATIVE EXAMPLE 10.8 Calculate the volume of a reactor required for 80% conversion of propane to propylene and hydrogen. The reactor is operated at 1752°R and 1 atm with an inlet propane flow rate of 2 lbmol/h. No air is present. The reaction velocity constant for propane for this reaction is given by

$$k_A = 459\,(\text{h})^{-1}$$

Solution. For this first order reaction the describing equation is

$$\frac{V}{2.0} = \int_0^{0.8} \frac{dX_A}{(459)(7.83\times 10^{-4})(1-X_A)}$$

Solving for V gives

$$V = 9.0\ \text{ft}^3$$

■

The reader is left the exercise of justifying the term 7.83×10^{-4}. [Hint: Apply the ideal gas law.]

ILLUSTRATIVE EXAMPLE 10.9 The reversible elementary liquid-phase reaction

$$A + B \underset{k'}{\overset{k}{\rightleftharpoons}} P$$

takes place in a plug flow reactor. The feed rate is 100 gmol/h, while the feed molar composition is 50% A, 35% B, and 15% inerts. The initial concentration of A is 4.0 gmol/L. The equilibrium constant based on concentration, K, is 20 L/gmol, and the forward specific reaction rate constant, k, is 0.5 L/gmol · h. Calculate the volume needed to achieve 90% of the equilibrium conversion.

Solution. First calculate θ_A, θ_B and θ_P.

$$\begin{aligned}\theta_A &= 1.0\,(\text{by definition})\\ \theta_B &= 0.35/0.50\\ &= 0.7\\ \theta_P &= 0.0\end{aligned}$$

Develop the rate equation, $-r_A$, in terms of X_A, noting that this is a liquid-phase reaction. Therefore, volume effects can be neglected.

$$\begin{aligned} C_A &= C_{A_0}(1 - X) \\ C_B &= C_{A_0}(\theta_B - X) \\ C_P &= C_{A_0}(\theta_P + X) \\ -r_A &= kC_AC_B - k'C_p \\ &= k\left[C_AC_B - (C_p/K)\right] \\ &= k\left[C_{A_0}^2(1 - X)(\theta_B - X) - (C_{A_0}X/K)\right] \end{aligned}$$

Calculate the equilibrium conversion, X_e. At equilibrium, $-r_A = 0$, so that

$$K = C_{A_0}X/\left[C_{A_0}^2(1 - X)(\theta_B - X)\right]; \qquad X = X_e$$

Substituting gives

$$20 = (4)(X)/\left[(4)^2(1 - X)(0.7 - X)\right]$$

Solving the quadratic or by trial and error yields

$$X = X_e = 0.674$$

The desired conversion is then

$$\begin{aligned} X &= (0.9)(X_e) \\ &= (0.9)(0.674) \\ &= 0.6066 \end{aligned}$$

Substitute into the tubular flow design equation.

$$V = F_{A_0}\int_0^X (-1/r_A)dX \qquad (10.6)$$

$$\begin{aligned} &= (F_{A_0}/k)\int_0^X \left\{1/\left[C_{A_0}^2(1 - X)(\theta_B - X) - (C_{A_0}X/K)\right]\right\}dX \\ &= \left[(100)(0.5)/(0.5)\right]\int_0^{0.6066} \left\{1/\left[(4)^2(1 - X)(0.7 - X) - (4X/20)\right]\right\}dX \\ &= 100\int_0^{0.6066} \left[1/(16X^2 - 27.4X + 11.2)\right]dX \end{aligned}$$

Integrate this equation analytically using information provided in Chapter 21.

$$V = (100)\{1/[a(p-q)]\} \ln\{(q/p)[(X-p)/(X-q)]\}$$
$$a = 16$$
$$b = -27.4$$
$$c = 11.2$$

$$p, q = \{-b \pm (b^2 - 4ac)^{1/2}\}/2a$$
$$= \{27.4 \pm (33.96)^{1/2}\}/32$$
$$p = 1.038$$
$$q = 0.675$$

The volume requirement in L is therefore

$$V = (100)\{1/[16(1.038 - 0.675)]\}$$
$$\ln\{(0.675/1.038)[(0.6066 - 1.038)/(0.6066 - 0.675)]\}$$
$$= 24.5\,\text{L}$$

■

ILLUSTRATIVE EXAMPLE 10.10 The following elementary gaseous reaction

$$A \longrightarrow 2B \qquad k_A = 0.7\,\text{s}^{-1}$$

is conducted in a TF reactor. The volumetric flowrate and inlet concentration of A are 2.0 L/s and 0.1 gmol/L, respectively. Determine the volume of reactor required to achieve a conversion of 90%. Perform the calculations *without* the volume effects included in the analysis. Assume constant temperature and pressure conditions.

Solution. For this gas phase reaction,

$$C_A = C_{A_0}(1 - X)/(1 + \varepsilon X)$$

and

$$-r_A = k_A C_{A_0}(1 - X)/(1 + \varepsilon X)$$

Without volume effect changes,

$$C_A = C_{A_0}(1 - X)$$

and

$$-r_A = k_A C_{A_0}(1 - X)$$

The tubular flow reactor design equation is

$$V = F_{A_0} \int_0^X (-1/r_A) dX$$

Since $F_{A_0} = C_{A_0} Q_0$,

$$V = C_{A_0} Q_0 \int_0^X (-1/r_A) dX$$

The volume of the reactor, neglecting any volume effect changes, is:

$$\begin{aligned} V &= (Q_0/k_A) \int_0^{0.9} [1/(1-X)] dX \\ &= -(Q_0/k_A) \ln(1-X) \\ &= -(2.0/0.7) \ln(0.1) \\ &= 6.57\,\text{L} \end{aligned}$$

■

ILLUSTRATIVE EXAMPLE 10.11 Perform the calculations in the previous example *with* the volume effects included in the analysis.

Solution. In order to include volume effect changes, first determine y_{A_0}, δ, and ε.

$$\begin{aligned} y_{A_0} &= 1 \\ \delta &= 2 - 1 \\ &= 1 \\ \varepsilon &= (1)(1) \\ &= 1.0 \end{aligned}$$

Calculate the volume of the reactor, but this time include volume effect changes.

$$\begin{aligned} V &= (Q_0/k_A) \int [(1 + \varepsilon X)/(1 - X)] dX; \qquad \varepsilon = 1.0 \\ &= (Q_0/k_A) \int_0^{0.9} [(1 + X)/(1 - X)] dX \end{aligned}$$

Once again, repeat Part IV, Chapter 21.

$$\begin{aligned} V &= (Q_0/k_A) \Big[-X - (2) \ln(1 - X) \Big]_0^{0.9} \\ &= (2.0/0.7)[-0.9 - (2)\ln(0.1) + 0 - (2)\ln(1)] \\ &= 10.59\,\text{L} \end{aligned}$$

■

ILLUSTRATIVE EXAMPLE 10.12 Comment on the results of the two previous illustrative examples.

Solution. The reader should note the difference in the two calculations for the volume. Some researchers have incorrectly neglected this effect. This effect becomes more pronounced for reactions involving large changes in the moles of the reacting system (e.g., petrochemical reactions) and where there are significant changes in the temperature and/or pressure. ■

ILLUSTRATIVE EXAMPLE 10.13 The gas phase pyrolysis

$$A \longrightarrow 5R$$

is known to be *second* order with respect to A. What size isothermal plug flow reactor operated at constant pressure is required to effect 90% conversion of a feed consisting of 50% A and 50% inerts?

Data:

$$F_{A_0} = 100\,\text{gmol/min}$$
$$C_{A_0} = 1.0\,\text{gmol/L}$$
$$k_A = 50\,\text{L/gmol}\cdot\text{min}$$

Solution. The design equation is

$$V = F_{A_0}\int_0^{X_A} \frac{dX_A}{(-r_A)}$$

The rate equation (accounting for volume changes) is

$$-r_A = kC_A^2 = k\frac{C_{A_0}^2(1-X_A)^2}{(1+\varepsilon X_A)^2}$$

For this gas phase reaction,

$$y_{A_0} = 0.50$$
$$\delta = 5 - 1 = 4$$
$$\varepsilon = y_{A_0}\delta = 2$$

Substituting the rate equation into the design equation gives

$$V = \frac{C_{A_0}Q_0}{kC_{A_0}^2}\int_0^X \frac{(1+\varepsilon X)^2}{(1-X)^2}dX_A;\ F_{A_0} = C_{A_0}Q_0,\ X_A = 0.9,\ \ Q_0 = 100\,\text{L/min}$$

After integration, the volume is determined to be

$$V = 114\,\text{L}$$

■

ILLUSTRATIVE EXAMPLE 10.14 In order to manufacture a commercial pesticide, a toxic inorganic compound is reacted with an inert component to create the final product. The rate of the reaction can be expressed in terms of conversion as

$$-r_A = kC_{A_0}e^{7X}/(6X^2+5); \qquad k = 0.075\,(\text{s})^{-1}$$

If the reaction is to be carried out in a 2.0 ft ID (inside diameter) tubular reactor with an inlet volume flow rate of 2 ft^3/s, what reactor volume is required to attain an 80% conversion? Also calculate the length of the reactor.

Solution. Write the tubular flow reactor design equation and substitute the rate expression.

$$\begin{aligned} V &= F_{A_0}\int_0^X \frac{dX}{(-r_A)}\,dX \\ &= Q_0 C_{A_0}\int_0^X \left\{1\Big/\left[kC_{A_0}e^{7X}/(6X^2+5)\right]\right\}dX \\ &= (Q_0/k)\int_0^{0.8}\left[(6X^2+5)/e^{7X}\right]dX \end{aligned}$$

Evaluate the integral using Simpson's three point rule (see Chapter 21).

$$f(X) = (6X^2+5)/e^{7x}$$

$$\int_0^X f(X)dX = (h/3)\left[f(0) + 4f(1/2X) + f(X)\right]; \; h = 0.4$$

$$f(X=0.0) = 5.0$$
$$f(X=0.4) = 0.362$$
$$f(X=0.8) = 0.032$$

$$\begin{aligned} \int_0^{0.8}\left[(6X^2+5)/e^{7X}\right]dX &= (0.4/3)\left[(5.0) + 4(0.362) + (0.032)\right] \\ &= 0.864 \end{aligned}$$

Calculate the volume in ft^3.

$$\begin{aligned} V &= (2.0)(0.864)/0.075 \\ &= 23.0\,\text{ft}^3 \end{aligned}$$

Calculate the length of the reactor in ft.

$$\begin{aligned} L &= 4V/\pi D^2 = 4(23.0)/\pi(2)^2 \\ &= 7.3\,\text{ft} \end{aligned}$$

■

ILLUSTRATIVE EXAMPLE 10.15 Comment on the integration applied in the previous illustrative example.

Solution. The integral was evaluated using Simpson's three point rule. The integral could be evaluated using any other suitable numerical integration method, including: the general form of Simpson's rule, the Trapezoidal rule, and so on. In addition, the resulting integral equation could have been solved by rewriting the equation in differential equation form, i.e.,

$$(dV/dX) = (Q_0/k)\Big[(6X^2 + 5)/e^{7X}\Big]$$

or

$$(dX/dV) = (k/Q_0)\Big[e^{7X}/(6X^2 + 5)\Big]$$

Equations in this form can sometimes be solved analytically. If this is not possible, a numerical solution may be accomplished. One very common numerical technique that is employed in practice to solve non-linear differential equations is the Runge–Kutta method. See Part IV, Chapter 21 for additional details. ■

ILLUSTRATIVE EXAMPLE 10.16 Calculate the volume required if a TF reactor is employed in the solution to Illustrative Example 9.14.

Solution. For this case, employ the design equation for a TF reactor

$$V_T = F_{B_0}\int_0^{0.875}\frac{dX}{(-r_B)}; \qquad -r_B = k_B C_{B_0}^2(1-X)^2$$

Integrating and substituting,

$$V_T = \left(\frac{F_{B_0}}{kC_{B_0}^2}\right)\left(\frac{X}{(1-X)}\right); \quad X = 0.875$$

$$V_T = \left(\frac{0.02224\,\text{gmol/s}}{(9.92\,\text{L/gmol}-\text{s})(0.08\,\text{gmol/L})^2}\right)\left(\frac{0.875}{1-0.875}\right)$$

$$V_T = 2.45\,\text{L}$$

As noted in Chapter 9, note the smaller volume required to accomplish the required conversion when a TF reactor is employed. This topic will be revisited in the next chapter. ■

ILLUSTRATIVE EXAMPLE 10.17 Refer to Illustrative Example (8.23). Calculate the maximum concentration of A that can be achieved if a TF (rather than a batch) reactor is employed.

Solution. The reaction of interest is

$$A \xrightarrow{k_1} D \xrightarrow{k_2} U$$

with

$$C_{A_0} = 2.0\,\text{gmol/L}$$
$$C_{D_0} = C_{U_0} = 0.0$$
$$k_1 = 0.4\,(\text{min})^{-1}$$
$$k_2 = 0.2\,(\text{min})^{-1}$$

For a tubular reactor,

$$\tau = C_{A_0}\int_0^X \frac{dX}{(-r)} \tag{10.14}$$

The companion design equation for a batch reactor is

$$t = C_{A_0}\int_0^X \frac{dX}{(-r)} \tag{8.11}$$

Since the integrals are the same, the answer is also the same, i.e.,

$$C_D(\,\max\,) = 1.4\,\text{gmol/L}$$

■

ILLUSTRATIVE EXAMPLE 10.18 Refer to Illustrative Example 8.24. Calculate the residence time required to achieve the maximum concentration of A in a TF reactor.

Solution. Once again, the describing equations for both the batch and tubular are identical. Therefore, the residence time for the tubular reactor is the same as the reaction time for a batch reactor, i.e.,

$$\tau(\max) = 0.34\text{ h} = 20.8\text{ min}$$

■

ILLUSTRATIVE EXAMPLE 10.19 The liquid reaction

$$A \longrightarrow \text{products}$$

follows the rate law

$$-r_A = \frac{k_1 C_A^{1/2}}{1 + k_2 C_A^2}$$

where $k_1 = 5.0\,(\text{gmol/L})^{1/2}/\text{h}$ and $k_2 = 10(\text{L/gmol})^2$. The initial concentration of A is 0.5 gmol/L, and the feed rate is 200 gmol/h of A.

Find the volume necessary to achieve 60% conversion in a tubular flow reactor.

Solution. The rate of reaction, $-r_A$, in terms of the conversion variable X_A is

$$C_A = C_{A_0}(1 - X_A)$$

$$-r_A = \frac{k_1 C_{A_0}^{1/2}(1 - X_A)^{1/2}}{1 + k_2 C_{A_0}^2 (1 - X_A)^2}$$

The design equation for a TF reactor is once again

$$V = F_{A_0} \int_0^X \frac{dX}{(-r_A)} \tag{10.6}$$

with

$$V = \frac{F_{A_0}}{k_1 C_{A_0}^{1/2}} \int_0^X \frac{1 + k_2 C_{A_0}^2 (1 - X)^2}{(1 - X)^{1/2}} dX$$

The integral I may be defined as

$$I = \int_0^{0.6} \frac{1 + k_2 C_{A_0}^2 (1 - X)^2}{(1 - X)^{1/2}} dX$$

$$= \int_0^{0.6} f(X) dX$$

Simpon's three point rule is again used to evaluate the integral:

$$I = (h/3)[f(0.0) + (4)f(0.3) + f(0.6)]; \quad h = 0.3$$

$$= (0.3/3)[(3.5) + (4)(2.659) + 2.214]$$

$$= 1.635$$

The volume is then

$$V = \frac{F_{A_0} I}{k_1 C_{A_0}^{1/2}} = \frac{(200)(1.635)}{(5.0)(0.5)^{0.5}}$$

$$= 92.5\,\text{L}$$

■

ILLUSTRATIVE EXAMPLE 10.20 A test substance, carbon tetrachloride (A), has a pure molar concentration of $C_A = 10.33$ gmol/L. It is to be decomposed in a tubular flow reactor. Calculate the conversion (to four significant figures) for the following conditions if the average residence time in the reactor is 1.0 s.

Assume isothermal, constant density conditions. Since the actual mechanism for this unique decomposition reaction is not known, three proposed reaction mechanisms are provided below.

1. Pure A, first order, $k = 9.21\ (\mathrm{s})^{-1}$.
2. 50% A, 50% inerts, first order, $k = 9.21\ (\mathrm{s})^{-1}$.
3. 10% A, 90% inerts, first order, $k = 9.21\ (\mathrm{s})^{-1}$.

Solution.

1. Calculate the conversion for case (1)

$$-r_A = kC_{A_0}(1 - X)$$

$$\tau = (C_{A_0}/kC_{A_0})\int_0^X 1/(1 - X)dX; \quad X = X_A$$

$$= (1/k)\int_0^X 1/(1 - X)dX$$

Integrating yields

$$X = 1 - \exp(-k\tau); \quad k = 9.21$$

Solving for X with $\tau = 1.0$ s gives

$$X = 0.9999$$

2. Since C_{A_0} does not appear in the integrated form of the equation,

$$X = 0.9999$$

3. Again, since C_{A_0} does not appear in the integrated form of the equation,

$$X = 0.9999$$

■

ILLUSTRATIVE EXAMPLE 10.21 Refer to the previous example. Two other proposed reaction mechanisms are provided below.

1. Pure A, second order, $k = 968\ (\mathrm{L/gmol})(\mathrm{s})^{-1}$
2. 1% A, 99% inerts, second order, $k = 968\ (\mathrm{L/gmol})(\mathrm{s})^{-1}$

Calculate the conversion for both cases.

Solution

1. For a second order reaction, the integrated form of the tubular flow design equation becomes

$$X = 1 - 1.0/(k\tau C_{A_0} + 1)$$

For $k = 968$ and $C_{A_0} = 10.33$,

$$X = 0.9999$$

2. Calculate the conversion for the second case

For $C_{A_0} = 0.1033$,

$$X = 0.9900$$

■

ILLUSTRATIVE EXAMPLE 10.22 Refer to Illustrative Example 10.20. Another two mechanisms are proposed.

1. Pure A, third order, $k = 4.686 \times 10^5$ (L/gmol)^2 (s)$^{-1}$.

2. 0.1% A, 99.9%, inerts, third order, $k = 4.686 \times 10^5$ (L/gmol)^2 (s)$^{-1}$.

Calculate the conversion for both cases.

Solution

1. Calculate the conversion for case (1). For a third order reaction, the integrated form of the tubular flow design equation is:

$$X = 1 - \left[1.0/(2k\tau C_{A_0}^2 + 1)\right]^{1/2}$$

For $k = 4.686 \times 10^5$ and $C_{A_0} = 10.33$,

$$X = 0.9999$$

2. Calculate the conversion for case (2). For $C_{A_0} = 0.01033$,

$$X = 0.9005$$

■

ILLUSTRATIVE EXAMPLE 10.23 Once again, refer to Illustrative Example 10.20. One final proposed mechanism is provided:

Pure A, fourth order, $k = 3.024 \times 10^8$ (L/gmol)^3 s^{-1}

Calculate the conversion.

Solution. For a fourth order reaction,

$$X = 1 - \left[1.0/(3k\tau C_{A_0}^3 + 1)\right]^{1/3}$$

For $k = 3.024 \times 10^8$ and $C_{A_0} = 10.33$;

$$X = 0.9999$$

■

REFERENCES

1. L. Theodore, "*Heat Transfer for the Practicing Engineer*," John Wiley & Sons, Hoboken, NJ, 2011.
2. P. Abulencia and L. Theodore, "*Fluid Flow for the Practicing Engineer*," John Wiley & Sons, Hoboken, NJ, 2009.

CHAPTER 11

REACTOR COMPARISONS

INTRODUCTION

There are certain inherent advantages and disadvantages to using each of the three reactors discussed in the three previous chapters—batch reactors (B), CSTRs (C), and tubular flow (TF) reactors (T). The advantages and disadvantages of these three classes of reactor is provided in Table 11.1.

These reactors are sometimes used in combination for any one of a variety of reasons. The sequence in which the reactors are placed can impact on the design or the final conversion. The three possible combinations of two different reactors is provided in Figures 11.1(a)–(c).

This final chapter in Part II contains three additional sections that are primarily concerned with reactor comparisions. Section titles include:

Specific Comparisons—Batch, CSTR, and TF

Graphical Analysis

Applications

SPECIFIC COMPARISONS: BATCH, CSTR, AND TF

The three combinations of two reactor sets are briefly reviewed in the three subsections titled: Batch Reactor vs TF Reactor, Batch Reactor vs CSTR, and CSTR vs TF Reactor.

TABLE 11.1 Comparison of Advantages (A) and Disadvantages (D) of Batch (B), CSTR (C), and TF (T) Reactors

	B	C	T
Reactor size for similar conversion	A	D	A
Simplicity and cost	A	A	D
Continuous operation	D	A	A
Ability to handle large throughputs	D	A	A
Instrumentation	A	D	D
Flexibility	A	D	D

Chemical Reactor Analysis and Applications for the Practicing Engineer. By Louis Theodore

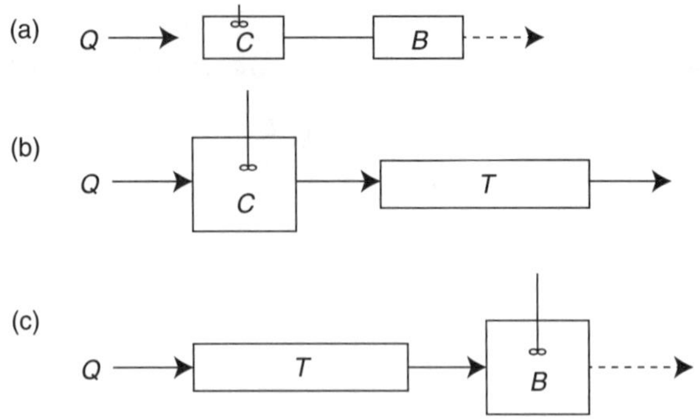

Figure 11.1 Possible combinations of two reactor.

Batch vs TF Reactor

In order to compare the batch and tubular reactor, one need simply examine the describing operations for both. For the batch reactor,

$$t = C_{A_0} \int \frac{dX}{(-r_A)} \tag{11.1}$$

For the TF reactor (with Q once again employed for the volumetric flow rate)

$$\tau = \frac{V}{Q} = C_{A_0} \int \frac{dX}{(-r_A)} \tag{11.2}$$

This comparison produces the following conclusions: the behavior of these two reactors may be viewed as identical since the reaction time, t, for a batch reactor is equal to the residence time, τ, for a TF reactor. In effect, and since the tubular reactor is assumed to operate in plug-flow mode, the residence time of the reacting species is simply the reaction time for the reacting species in a batch reactor. Since there is no difference one may conclude that the performances in terms of the conversion, X, are identical. These can be verified by substituting various r_A terms in Equations (11.1) and (11.2).

Batch Reactor vs CSTR

The performance of these two classes of reactors is significant. This was discussed in both Chapters 8 and 9, and also noted in Table 11.1. The describing equations for these two units may also be examined. For the batch reactor (once again),

$$t = C_{A_0} \int \frac{dX}{(-r_A)} \tag{11.1}$$

and, for the CSTR

$$\tau = \frac{C_{A_0} \Delta X}{(-r_A)} \tag{11.3}$$

An analytical analysis is not quite as obvious but the reader is encouraged to compare reaction/residence times for both reactors for various normally associated reaction rates, r_A. For example, the ratio of CSTR residence time to batch reactor time for various conversions for a first order reaction is provided in Table 11.2.

Obviously, the CSTR becomes extremely large with respect to volume and/or residence time when compared to a batch reactor for high conversions for first order kinetics. This analysis can be extended to other reaction orders. Alternatively, one may deduce that a batch reactor provides a higher conversion for similar reaction/residence times. The reader should note that the above analysis/conclusion may not apply if the reaction kinetics are complex.

CSTRs vs TF Reactors

The CSTR has certain advantages because of the near uniform temperature, concentration, etc., that results because of mixing. However, for perfect mixing, the reaction occurs at a rate determined by the concentration of the discharge (or exit) stream from the reactor. Since the rate usually decreases with the extent of reaction (conversion), the CSTR operates in the range between the high reaction rate corresponding to the concentration in the feed stream and the (normally) lower rate corresponding to the concentration leaving the reactor. On the other hand, the tubular flow reactor takes maximum advantage of the high reaction rate corresponding to the high concentration(s) near the entrance to the reactor. The CSTR therefore requires a larger volume than a tubular reactor to accomplish a given degree of conversion. This statement applies to all positive reaction orders. The reaction/residence time ratio for a zero order reaction is unity, i.e., both perform identically. However, the above analysis occasionally does not apply, particularly if the reaction mechanism is complex and/or nonelementary.

The analysis and conclusion presented in the previous subsection, i.e., batch reactors vs tubular flow reactors applies since the residence time of a tubular flow reactor is identical to the reaction time for a batch reactor. Table 11.3 is therefore a slight modification of Table 11.2 with t_B replaced by τ_T.

TABLE 11.2 Ratio of Residence Times to Reaction Times in Batch Reactors vs CSTRs vs Conversion for a First-Order Irreversible Reaction

X	$\tau_{\mathrm{CSTR}}/t_{\mathrm{B}}$
0.0	1.0
0.5	1.44
0.9	3.91
0.95	6.34
0.99	21.5
0.0	∞

TABLE 11.3 Ratio of Residence Times to Reaction Time in a TF Reactor vs CSTR Conversion for a First-Order Irreversible Reaction

X	τ_{CSTR}/τ_T
0.0	1.0
0.5	1.44
0.9	3.91
0.95	6.34
0.99	21.5
0.0	∞

General Comments

Ultimately, the performance of the three different classes of reactor is based on a host of considerations. They include:

1. Conversion
2. Production rate
3. Volume requirement(s)
4. Selectivity (and the effect of unwanted by-products).

Energy considerations, space, power requirements, etc., also play a role but superimposed on all of the above is the economics associated with the choice. The reader is referred to Part IV, Chapter 21 for the section concerned with Economics and Finance.

Walas,[1] in an earlier article, provides some general "rules of thumb" regarding reactor selection. Some of his (unedited) suggestions are noted below.

1. The rate of reaction in every instance must be established in the laboratory, and the residence time or space velocity and product distribution be found in a pilot plant.
2. The optimum proportions of stirred-tank reactors are liquid level equal to tank diameter; at high pressures, slimmer proportions are more economical.
3. Power input to a homogeneous-reaction stirred tank is 0.5–1.5 hp/1000 gal, but input is three times this amount when heat is to be transferred.
4. Ideal continuous-stirred-tank-reactor (CSTR) behavior is approached when the mean residence time is 5–10 times the time needed to achieve homogeneity, which is accomplished by 500–2000 revolutions of a properly designed stirrer.
5. Batch reactions are conducted in stirred tanks for small, daily production rates, or when the reaction times are long, or when some condition, such as feedrate or temperature, must be programmed in some way.
6. Relatively slow reactions of liquids and slurries are conducted in continuous stirred tanks. A battery of four or five units in series is most economical.

7. Tubular-flow reactors are suited to high production rates at short residence times (seconds or minutes) and when substantial heat transfer is needed. Jacketed tubes or shell-and-tube construction are used.
8. For conversions under about 95% of equilibrium, performance of a five-stage CSTR unit approaches plug flow.

Walas also offers the following regarding catalytic reactors (see Part III).

1. Catalyst particles are 0.1 mm dia. in fluidized beds, 1 mm in slurry beds, and 2–5 mm in fixed beds.
2. In granular-catalyst packed reactors, the residence-time distribution often is no better than that of a five-stage CSTR battery.

GRAPHICAL ANALYSIS

The design and performance capabilities of batch, CSTR, and TF reactors for any arbitrary kinetics can also be compared using a graphical procedure defined by Fogler[2] as the $(-1/r_A)$ plot. This is illustrated in the figures that follow.

The describing equation for the above three classes of reactor are first noted.

Batch:

$$\frac{t}{C_{A_0}} = \int_{X_0}^{X_1} \frac{dX}{(-r_A)} \tag{11.4}$$

CSTR:

$$\frac{\tau}{C_{A_0}} = \left(\frac{\Delta X}{-r_A}\right); \qquad \Delta X = X_1 - X_0 \tag{11.5}$$

TF:

$$\frac{\tau}{C_{A_0}} = \int \frac{dX}{-r_A} \tag{11.6}$$

The LHS of these equations may be obtained by simply plotting $(-1/r_A)$ vs X. The analysis to follow is to be based on the kinetics, i.e., $(-1/r_A)$ described in Figure 11.2 for a typical reactor. The same rate expression will be employed in the development to follow.

A comparison of either a batch or TF vs a CSTR is presented in Figure 11.3. For a specified final conversion of X, one notes that the rectangular ▨ area represents the τ/C_{A_0} for a CSTR. This may also be deduced from Equation (11.5) for a conversion from 0 to X. Alternately, the area ▩ represents the t/C_{A_0} or τ/C_{A_0} for either a batch or TF reactor, respectively. One can clearly see that the time for reaction or holdup (residence) time is greatest for a CSTR, i.e., a larger reactor is required for a given conversion if a CSTR is employed. This is typically the case for most reactions.

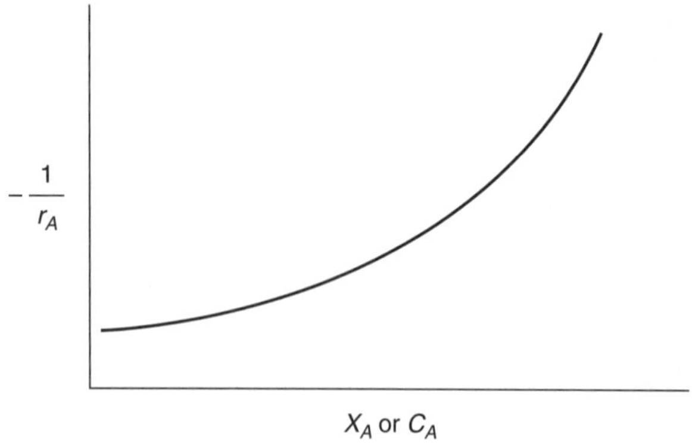

Figure 11.2 Rate^{-1} vs conversion for a typical reactor.

This analysis may be applied to a CSTR and TF reactor connected in series and achieving the same conversion X. Figure 11.4 results, if the CSTR is placed first, while Figure 11.5 results if the CSTR follows the TF reactor.

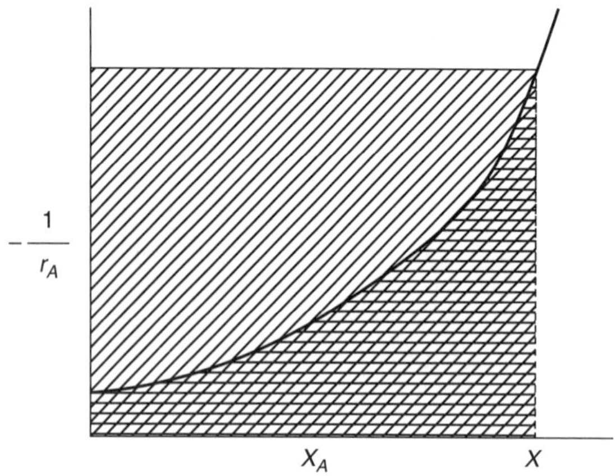

Figure 11.3 CSTR vs TF reactor.

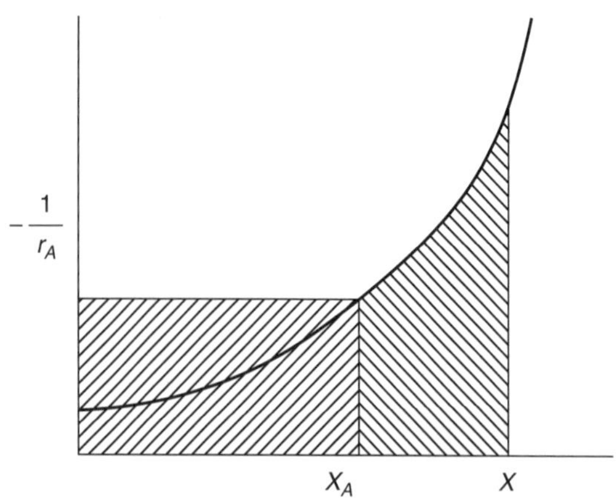

Figure 11.4 CSTR followed by a TF reactor.

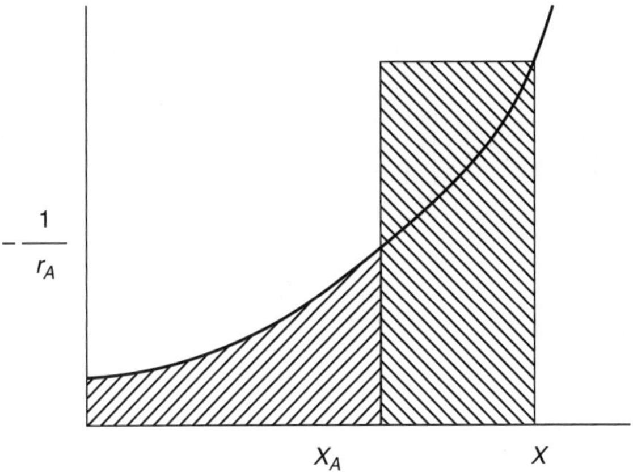

Figure 11.5 TF followed by a CSTR.

This may also be extended to a number of CSTRs in series achieving the same final conversion X. This is demonstrated in Figure 11.6(a)–(d). As noted in Chapter 9, an infinite number of CSTRs in series is equivalent to a TF reactor. One need only compare Figure 11.6(d) with Figure 11.3.

The reader may now return to the CSTR illustrative examples in Chapter 9 dealing with CSTRs in series and attempting to minimize the volume requirements for these cases. Figure 11.7 provides a graphical representation of the solutions to Illustrative Examples 9.14, 9.15, and 9.16. Figure 11.7(a) provides the solution for the volume of one reactor  and for two reactors 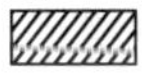 of equal volumes.

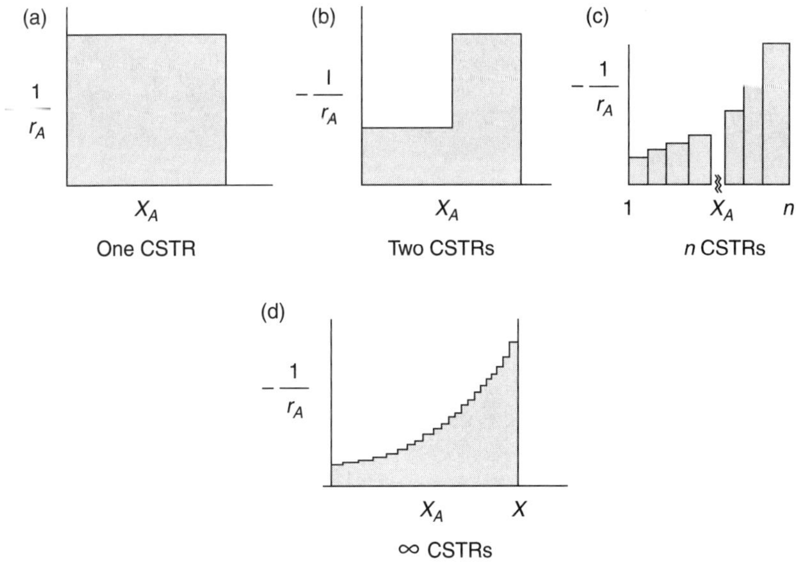

Figure 11.6 CSTRs in series.

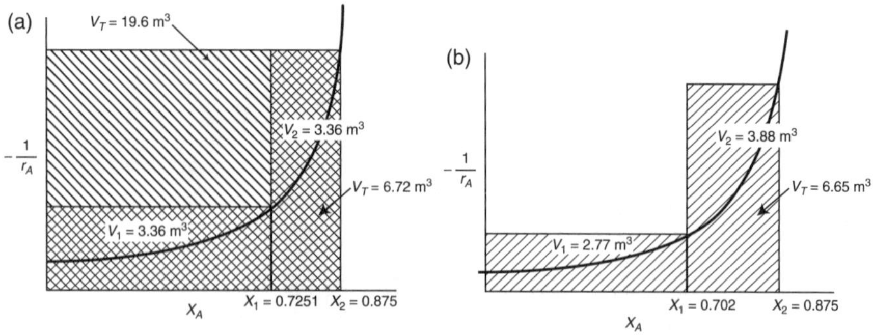

Figure 11.7 Two CSTRs in series.

Figure 11.7(b) provides the solution for the minimum of two reactors in series accomplishing the same final conversion of 0.875.

The reader can also return to Illustrative Example 9.17 for the three CSTRs problem. Figure 11.8 provides the graphical solution for the case of equal volumes (a) and the case where the total volume of three reactors has been minimized (b).

The material presented above was based on a typical rate expression. Occasionally an "oddball" rate expression, as presented in Figure 11.9, may occur. As noted, one observes that the volume requirement for the TF is greater than a CSTR—both operating at the conversion.

Another rate expression producing surprising results is provided in Figure 11.10. The minimum volume reactors to achieve a final conversion of X_f is a TF operating to a conversion X_i followed by a larger CSTR operating in the conversion range of X_i to X_f. Would this system be more attractive than one CSTR operating over the range 0 to X_f? Probably not. The one CSTR requires only a slightly larger reactor (see shaded area); this would probably make the single unit more attractive from an economic perspective. The reader should also note the large volume requirement if a single TF unit is employed.

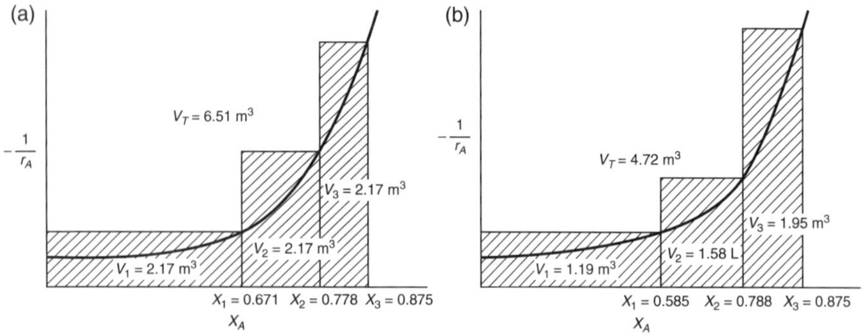

Figure 11.8 Three CSTRs in series.

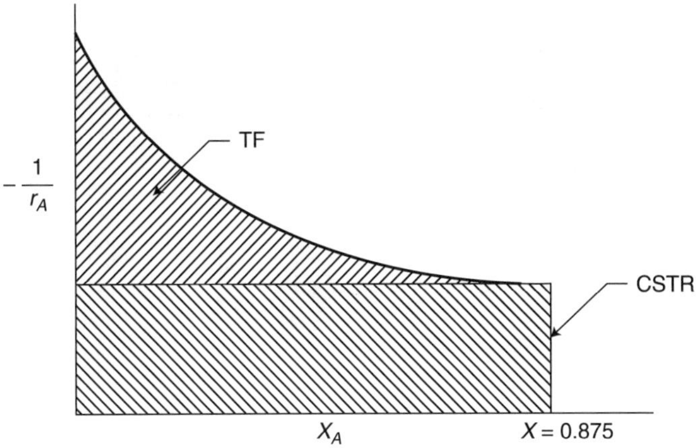

Figure 11.9 Oddball rate expression.

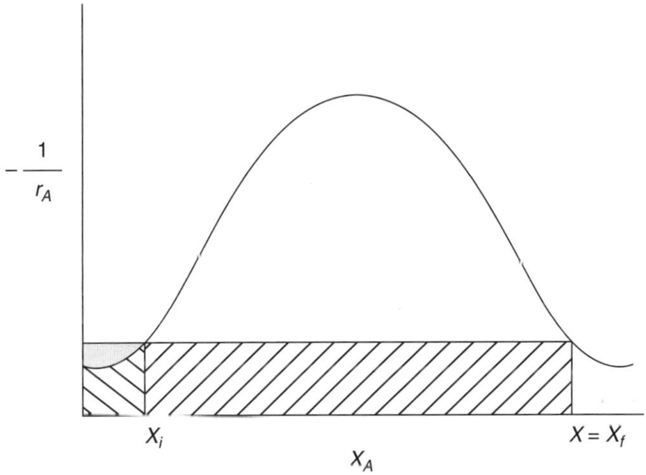

Figure 11.10 TF followed by CSTR.

ILLUSTRATIVE EXAMPLE 11.1 Your company has two reactors of *equal* volume which it would like to use in the production of a specified product formed by a first order irreversible liquid reaction. One reactor is a CSTR and the other is a TF reactor. How should these reactors be hooked up to achieve maximum conversion (see Figure 11.11)? Justify your answer using concentration (not conversion) terms in the design equations.

Solution. Write the design equation for a CSTR in terms of the concentration (not the conversion) and the residence time τ.

$$\begin{aligned} V &= F_{A_0}(X - 0)/(-r_A) \\ &= QC_{A_0}(X - 0)/kC_{A_1} \end{aligned}$$

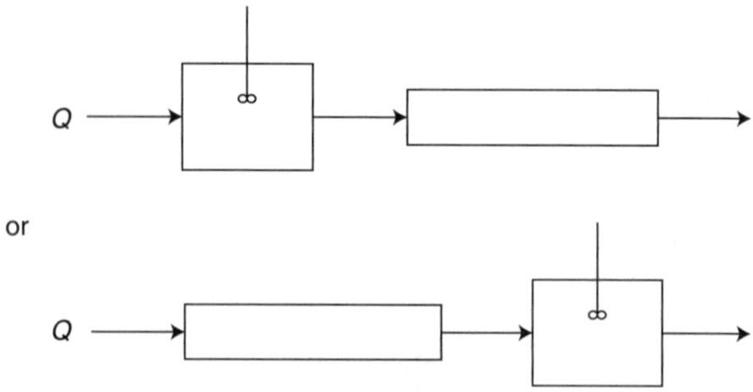

Figure 11.11 Order of CSTR and TF.

Since $C_{A_0}X = C_{A_0} - C_{A_1}$ and $V/Q = \tau$,

$$\tau = (C_{A_0} - C_{A_1})/kC_{A_1}$$
$$C_{A_1} = C_{A_0}/(1 + k\tau)$$

Write the design equation for a TF reactor in terms of the concentration (not the conversion) and the residence time τ.

$$V = F_{A_0}\int_0^X (-1/r_A)\,dX$$
$$V/Q = C_{A_0}\int_0^X (-1/r_A)\,dX$$

Since $C_{A_0}\,dX = -dC_A$

$$\tau = -(1/k)\int_{C_{A_0}}^{C_{A_1}} (1/C_A)\,dC_A$$
$$-k\tau = \ln(C_{A_1}/C_{A_0})$$
$$C_{A_1}/C_{A_0} = \exp(-k\tau)$$
$$C_{A_1} = C_{A_0}\exp(-k\tau)$$

Consider the first case (a)—a CSTR followed by a TF reactor—and determine the outlet concentration C_{A_1} from the CSTR in terms of τ_1.

$$C_{A_1} = C_{A_0}/(1 + k\tau_1)$$

Determine the outlet concentration, C_{A_2}, for the TF reactor. Set $\tau = \tau_2$ for the TF reactor.

$$C_{A_2} = C_{A_1}\exp(-k\tau_2)$$
$$= C_{A_0}\exp(-k\tau_2)/(1 + k\tau_1)$$

Since $\tau_1 = \tau_2 = \tau$,

$$C_{A_2} = C_{A_0} \exp(-k\tau)/(1 + k\tau)$$

Consider the second case (b)—a tubular flow reactor followed by a CSTR—and determine C_{A_1} and C_{A_2}.

$$\begin{aligned} C_{A_1} &= C_{A_0} \exp(-k\tau_1) \\ C_{A_2} &= C_{A_1}/(1 + k\tau_2) \\ &= C_{A_0} \exp(-k\tau_1)/(1 + k\tau_2) \end{aligned}$$

Since (once again) $\tau_1 = \tau_2 = \tau$,

$$C_{A_2} = C_{A_0} \exp(-k\tau)/(1 + k\tau)$$

The results above are identical, i.e., both cases provide the same outlet conversion. ■

ILLUSTRATIVE EXAMPLE 11.2 How would the analysis in the previous example be affected if the volumes of the two reactors are not equal?

Solution. Repeat the previous calculation if the two volumes are not equal. From the first case,

$$C_{A_2} = C_{A_0} \exp(-k\tau_I)/(1 + k\tau_C)$$

From the second case,

$$C_{A_2} = C_{A_0} \exp(-k\tau_T)/(1 + k\tau_C)$$

The results are once again identical. ■

ILLUSTRATIVE EXAMPLE 11.3 How would the results of the previous two examples be affected if the reaction is not first order?

Solution. The reader should not interpret the previous two results too loosely. For any order reaction other than first order, the sequence in which the reactors are placed may affect the performance. The reader is left the exercise of verifying this statement; however, for a zero order or ∞ order reaction, i.e., for n equal to either zero or ∞, the "performance" or degree of conversion would still be identical. ■

ILLUSTRATIVE EXAMPLE 11.4 Consider the rate-conversion data provided in Figure 11.12. Calculate the volume requirement for a reactor with a flow rate of 20 lbmol/s over the 0.4–0.6 conversion range. Perform the calculation for both a CSTR and TF.

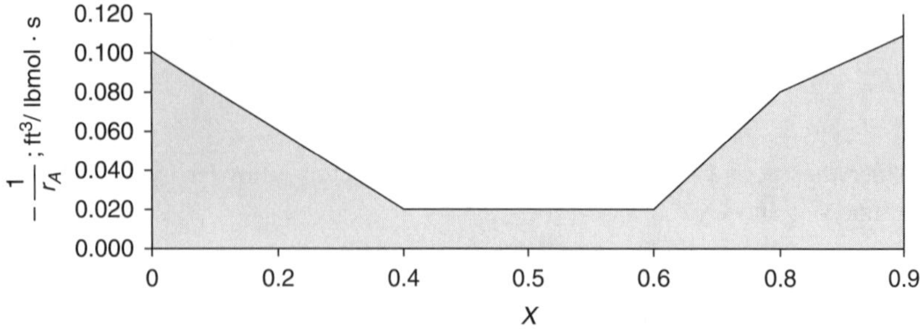

Figure 11.12 Rate-conversion plot for Illustrative Example 11.4.

Solution. From a graphical perspective, the CSTR volume and the TF volume are equal. The details of the calculations follow:
For a CSTR, apply Equation (11.3) and solve.

$$V = F_{A_0}(\Delta X)/(-r_A) = 20(0.6 - 0.4)/(0.02)$$
$$= 0.08 \text{ ft}^3$$

For the TF reactor, apply Equation (11.2) and solve.

$$V = F_{A_0}\int_{X_1}^{X_2} \frac{dX}{(-r_A)} = \frac{F_{A_0}\Delta X}{(-r_A)} = 0.08 \text{ ft}^3$$

As expected, the volumes are equal. ■

ILLUSTRATIVE EXAMPLE 11.5 Employing the rate data provided in the previous example, calculate the volume of a TF reactor to achieve a conversion of 0.75. The rate data was regressed to equation form as shown below.

$$-\frac{1}{r_A} = -2.1346X^5 + 3.8767X^4 - 1.8104X^3 + 0.2994X^2 - 0.2174X + 0.1$$

Solution. An analysed solution is now possible by using the rate expression provided. Since

$$F(X) = -\frac{1}{r_A} = \Big[\left(-2.1346X^5\right) + \left(3.8767X^4\right) - \left(1.8104X^3\right)$$
$$+ \left(0.2994X^2\right) - (0.2174X) + 0.1\Big]$$

$$F_{A_0} = 20 \text{ lbmol/s}$$

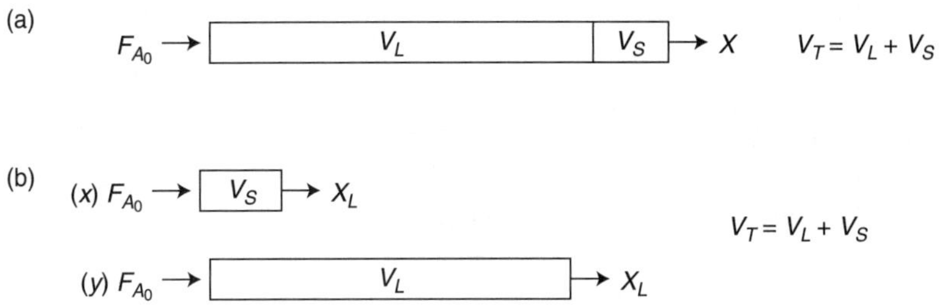

Figure 11.13 Two plug flow reactor in series.

Substituting into the TF and integrating (by any suitable method—see also Chapter 21) gives

$$\begin{aligned} V &= F_{A_0} \int_{0.0}^{0.75} f(x)\,dx \\ &= (20)(0.0335) \\ &= 0.67\ \text{ft}^3 \end{aligned}$$

■

ILLUSTRATIVE EXAMPLE 11.6 Is it more advantageous to connect two plug flow reactors of different volumes in series than in parallel? How does degree of conversion and reaction order affect your answer. Explain.

Solution. Consider the situation pictured in Figures 11.13(a) and (b). The terms x and y are fractional quantities that satisfy the requirement

$$x + y = 1.0; \qquad S = \text{Small}, \quad L = \text{Large}$$

Consider the limiting case in (b) of $x = 1.0$ and $y = 0.0$. Because of the low τ, the conversion would be low. If $x = 0.0$ and $y = 1.0$, the conversion would be higher due to the higher τ. One can therefore conclude that the two systems are different because of the ratio of the feed outlet to the reactors. However, if the fraction of the feed to each reactor is equal to the fraction of the reactor's volume to the total voume, i.e.,

$$x = \frac{V_S}{V_T}, \qquad y = \frac{V_L}{V_T}; \qquad V_T = V_L + V_S$$

the performances of both would be identical. ■

APPLICATIONS

This section presents applications that are concerned with reactor comparisions. Thirteen Illustrative Examples (11.7–11.19) follow.

ILLUSTRATIVE EXAMPLE 11.7 The elementary, irreversible, liquid phase reaction

$$A + B \xrightarrow{k} C + D$$

is carried out at 70°F. The reactants are fed to the reactor at a feed rate of 50 ft^3/h with no inerts present. The initial concentrations of A and B are 2.0 lbmol/ft^3 and 4.0 lbmol/ft^3, respectively. The reaction velocity constant k is 0.010 $\text{ft}^3/\text{lbmol}\cdot\text{h}$ at 100°F. The activation energy is 3000 Btu/lbmol A.

1. Calculate the volume of a TF plug flow reactor required to achieve a conversion of 75%.
2. Calculate the volume of a CSTR to achieve the same conversion.

Solution. The reaction is assumed to follow the Arrhenius temperature dependence.

$$k(T) = k_1 \exp(-E/R)\left(\frac{T_1 - T}{T_1 T}\right)$$

Substituting

$$k(70^\circ\text{F}) = \left(0.010\frac{\text{ft}^3}{\text{lbmol}\cdot\text{h}}\right)\exp\left(\frac{-3000}{1.987}\right)\left(\frac{560^\circ\text{R} - 530^\circ\text{R}}{(560)(530)}\right)$$

$$k = (0.010)(0.858)$$

$$k = k_A = 2.0086\frac{\text{ft}^3}{\text{lbmol}\cdot\text{h}} \quad \text{at } 70^\circ\text{F}$$

1. The TF reactor design equation is

$$V = F_{A_0}\int\frac{dX}{(-r_A)} = C_{A_0}Q_0\int\frac{dX}{(-r_A)} \tag{11.2}$$

Since the reaction is elementary

$$-r_A = kC_AC_B$$

Since the reaction involves liquids, there are no volume effect changes, so that

$$C_A = C_{A_0}(1 - X)$$

$$C_B = C_{A_0}(\theta_B - X); \qquad \theta_B = \frac{C_{B_0}}{C_{A_0}} = \frac{4.0}{2.0} = 2.0$$

Substituting the rate expression into the design equation gives

$$V = C_{A_0}Q_0 \int_0^X \frac{dX}{kC_A C_B} = C_{A_0}Q_0 \int_0^X \frac{dX}{kC_{A_0}^2(1-X)(\theta_B - X)}$$

$$V = \frac{Q_0}{kC_{A_0}} \int_0^X \frac{dX}{(1-X)(\theta_B - X)}$$

See Illustrative Example 21.18 for integration details.

$$V = \frac{Q_0}{kC_{A_0}} \left[\frac{1}{\theta_B - 1} \ln\left(\frac{\theta_B - X}{\theta_B(1-X)} \right) \right] \Bigg|_{X=0.0}^{X=0.75}$$

Substituting

$$V = \frac{(50\,\text{ft}^3/\text{h})}{(0.0086\,\text{ft}^3/\text{lbmol}\cdot\text{h})(2.4\,\text{lbmol/ft}^3)} \left[\left(\frac{1}{2-1} \right) \ln\left(\frac{2-0.75}{2(1-0.75)} \right) \right]$$

$$V = 2664\,\text{ft}^3$$

2. The CSTR design equation is

$$V = \frac{F_{A_0}X}{-r_A} = \frac{C_{A_0}Q_0X}{-r_A} \tag{11.3}$$

Substituting

$$-r_A = kC_{A_0}^2(1-X)(\theta_B - X)$$

into the above equation leads to

$$V = \frac{C_{A_0}Q_0X}{kC_{A_0}^2(1-X)(\theta_B - X)}$$

Substituting

$$V = \frac{(50\,\text{ft}^3/\text{h})(0.75)}{(0.0086\,\text{ft}^3/\text{lbmol}\cdot\text{h})(2\,\text{lbmol/ft}^3)(1-0.75)(2-0.75)}$$

$$V = 6977\,\text{ft}^3$$

■

ILLUSTRATIVE EXAMPLE 11.8 Refer to the previous example. Which volume is larger? Give a qualitative answer as to why, by sketching a graph of $(-1/r_A)$ vs X.

Solution. As expected, the volume of the CSTR is larger. A plot of $(1/-r_A)$ vs X would appear as shown in Figure 11.14 for the rate expression.

$$-\frac{1}{r_A} = \frac{1}{kC_{A_0}^2(1-X)(2-X)}$$

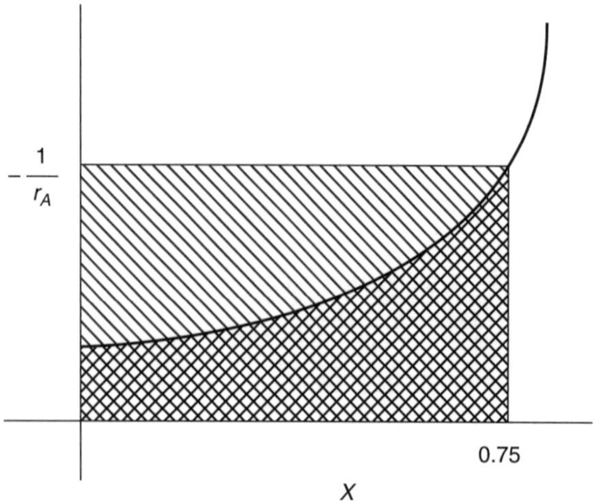

Figure 11.14 Rate-conversion plot for Illustrative Example 11.8.

For CSTR, the equation

$$V = \frac{F_{A_0}X}{(-r_A)} \tag{11.3}$$

represents the area of the rectangle that is shaded ▨ times F_{A_0}. For the TF reactor, the equation

$$V = F_{A_0}\int \frac{dX}{(-r_A)}$$

represents the area under the curve that is shaded ▩ times F_{A_0}.

Thus, the volume of the CSTR is larger since that area of the rectangle is larger than the area under the curve. See also Figure 11.3 for a similar plot. ■

ILLUSTRATIVE EXAMPLE 11.9 It is desired to carry out the following gas phase reaction:

$$A \xrightarrow{k} B + C; \quad k = 3.2\ (\text{h})^{-1}$$

by passing a volumetric flow of 6.25 gal/min, containing 70% A and 30% inerts into a CSTR followed by a tubular flow reactor. The volume of the CSTR is 40 ft^3, and the final conversion desired at the end of the TF plug flow reactor is set at 85%.

1. Calculate the intermediate conversion and the volume of the TF reactor.
2. Calculate the conversion associated solely with the TF reactor.

Solution. Preliminary calculations include:

$$Q_0 = \left(6.25\frac{\text{gal}}{\text{min}}\right)\left(60\frac{\text{min}}{\text{h}}\right)\left(\frac{1\ \text{ft}^3}{7.482\ \text{gal}}\right) = 50.1\ \text{ft}^3/\text{h}$$

$$y_{A_0} = 0.70, \quad \delta = 2 - 1 = 1; \quad \varepsilon = (0.7)(1) = 0.7$$

The rate equation is

$$-r_A = kC_A = kC_{A_0}\frac{(1-X)}{(1+\varepsilon X)}; \qquad X = X_A$$

1. Substituting into the CSTR design equation gives

$$V = \frac{F_{A_0}X}{(-r_A)} = \frac{C_{A_0}Q_0X}{kC_{A_0}}\frac{(1+\varepsilon X)}{(1-X)} = \frac{Q_0X(1+\varepsilon X)}{k(1-X)}$$

Solving this nonlinear equation leads to

$$Vk(1-X) = Q_0X(1+\varepsilon X)$$
$$Vk - VkX = Q_0X + Q_0\varepsilon X^2$$
$$V_0\varepsilon X^2 + (V_0 + Vk)X - Vk = 0$$

Substitute and solve the quadratic equation.

$$50.1(0.70)X^2 + [(50.1 + 40(3.2))]X - 40(3.2) = 0$$
$$35.07X^2 + 178.1X - 128 = 0$$
$$X = 0.64$$

For the TF reactor:

$$V = F_{A_0}\int_{X_1}^{X_2}\frac{dX}{(-r_A)} = F_{A_0}\int_{X_1}^{X_2}\frac{(1+\varepsilon X)}{kC_{A_0}(1-X)}dX; \; X_1 = 0.64$$

Substituting and integrating (see Chapter 21, Illustrative Example 21.18)

$$V = \frac{Q_0}{k}\left[\left((1+\varepsilon)\ln\left(\frac{1}{1-X_2}\right) - EX_2\right) - \left((1+\varepsilon)\ln\left(\frac{1}{1-X_1}\right) - \varepsilon X_1\right)\right]_{X_1=0.64}^{X_2=0.85}$$

$$V = \frac{50.1}{3.2}\left[\left(1.7\ln\left(\frac{1}{0.15}\right) - 0.595\right) - \left(1.7\ln\left(\frac{1}{0.36}\right) - 0.448\right)\right]$$

$$V = 15.66(2.63 - 1.29) = 21\ \text{ft}^3$$

2. By definition,

$$X_{\text{overall}} = 1 - (1 - X_1)(1 - X_2)$$

where X_2 is the conversion of the TF reactor based on inlet conditions to the TF reactor. Substituting,

$$0.85 = 1 - (1 - 0.64)(1 - X_2)$$
$$0.85 = 1 - 0.36(1 - X_2)$$
$$0.85 = 0.64 - 0.36X_2$$
$$X_2 = \frac{0.85 - 0.64}{0.36} = 0.58$$

■

ILLUSTRATIVE EXAMPLE 11.10 The reaction $A + B \rightarrow R$ is first order with respect to each of the reactants. At present, the reaction is being effected in the liquid phase in a 5 ft^3 plug flow reactor using a mole ratio $M = C_{B_0}/C_{A_0} = 2$ with 90% conversion of A. A new engineer is considering replacing the present system with a 25 ft^3 CSTR. What mole ratio M would allow her to produce the same amount of product using the same initial flowrate of A and fractional conversion of A?

Solution. For a TF reactor,

$$V = F_{A_0}\int \frac{dX}{(-r_A)}$$
$$= F_{A_0}\int \frac{dX}{kC_{A_0}^2(1 - X)(M - X)}; \qquad M = 2.0$$
$$V = \frac{F_{A_0}}{kC_{A_0}^2}\int_0^{0.9} \frac{dX}{(1 - X)(2 - X)}$$

For the CSTR,

$$5V = \frac{F_{A_0}X}{(-r_A)}$$
$$V = \frac{F_{A_0}X}{5kC_{A_0}^2(1 - X)(M - X)}; \qquad X = 0.9$$

Equating the above two equations and solving for M ultimately gives

$$M = 1.96$$

■

ILLUSTRATIVE EXAMPLE 11.11 Consider the second order, liquid phase reaction

$$2A \longrightarrow \text{products}$$

which has a rate constant $k = 0.5\ \text{L/gmol} \cdot \text{min}$ at 300°K. For isothermal operation at 300°K and the following process conditions

$$F_{A_0} = 100\ \text{gmol/min}$$
$$C_{A_0} = 1.0\ \text{gmol/L}$$
$$X_A = 95\%$$

1. What size plug flow reactor would be required?
2. What size CSTR would be required?
3. What would be the total reactor volume needed if two equal sized CSTRs were placed in series?

Solution.

1. For a TF reactor,

$$V = F_{A_0}\int \frac{dX}{-r_A} = \int_0^{0.95} \frac{dX}{kC_{A_0}^2}(1 - X)^2; \qquad X = X_A$$

$$= \frac{F_{A_0}}{kC_{A_0}^2}\int_0^{0.95} \frac{dX}{(1 - X^2)} = \left(\frac{100}{0.5}\right)\left[\frac{1}{1 - X} - 1\right]; \qquad X = 0.95$$

$$= 3800\ \text{L}$$

2. For the CSTR,

$$V = \frac{F_{A_0}X}{kC_{A_0}^2(1 - X)^2}; \qquad X = 0.95$$

$$= \left(\frac{100}{0.5}\right)(0.95)\left(\frac{1}{0.05}\right)^2 = (3800)(20)$$

$$= 76{,}000\ \text{L}$$

3. For two equal volume CSTRs (as with Illustrative Example 9.14)

$$V_1 = \frac{F_{A_0}X_1}{kC_{A_0}^2(1 - X_1)^2}$$

$$V_2 = \frac{F_{A_0}(X_2 - X_1)}{kC_{A_0}^2(1 - X_2)^2}$$

Set $V_1 = V_2 = V$ and $X_2 = 0.95$.

Solving yields,

$$X_1 = 0.85$$
$$V_1 = 7500 \text{ L}$$
$$V_{\text{TOTAL}} = 15{,}000 \text{ L}$$ ■

ILLUSTRATIVE EXAMPLE 11.12 Refer to the previous example. Briefly explain why the answers to (1), (2), and (3) compare as they do.

Solution. Note the tremendous increase in volume requirement in going from (1) to (3), to (2). Generally, (1) is lowest since the rate is based on the concentration variation though the TF reactor. Reactor (2) is highest since the rate is based on the outlet (lower) concentration. With staging, the CSTRs begin to approach TF behavior and volume requirements decrease with increasing staging. ■

ILLUSTRATIVE EXAMPLE 11.13 The elementary irreversible liquid-phase reaction

$$A + B \xrightarrow{k_A} C; \qquad MW_A = MW_B = 40$$

takes place at constant temperature in a 100 ft^3 reactor. Feed (A and B) of density 20 lb/ft^3 enters at a linear velocity of 0.053 ft/s through a vessel that may be viewed as a pipe with an inside diameter of 1 ft at a rate of 75 lbmol/h. If 80% of the entering A is converted, in what mode (CSTR or plug flow) is the reactor operating? Data are provided below:

$$y_{A_0} = 0.2; \ y_{B_0} = 0.8$$
$$k = k_A = 6.93 \text{ ft}^3/\text{lbmol} \cdot \text{h}$$

Solution. Preliminary calculations are first provided:

$$Q = Q_0 = \left(0.053 \frac{\text{ft}}{\text{s}}\right)\left[\frac{\pi(1)^2}{4} \text{ft}^2\right] = 0.04163 \frac{\text{ft}^3}{\text{s}} = 150 \frac{\text{ft}^3}{\text{h}}$$

$$\dot{m} = \left(150 \frac{\text{ft}^3}{\text{h}}\right)\left(\frac{20 \text{ lb}}{\text{ft}^3}\right) = 3000 \ \frac{\text{lb}}{\text{h}}$$

$$F_{A_0} = \left(3000 \frac{\text{lb}}{\text{h}}\right)\left(\frac{\text{lbmol}}{40 \text{ lb}}\right)(0.2) = 15 \text{ lbmol/h}$$

$$F_{B_0} = \left(\frac{3000}{40}\right)(0.8) = 60 \text{ lbmol/h}$$

$$\theta_B = \frac{60}{15} = 4.0$$

$$C_{A_0} = \frac{F_{A_0}}{Q_0} = \frac{15}{150} = 0.1 \frac{\text{lbmol}}{\text{ft}^3}$$

The rate expression is

$$-r_A = kC_AC_B$$

with

$$\begin{aligned} C_A &= C_{A_0}(1 - X);\ X_A = X \\ C_B &= C_{A_0}(\theta_B - X) \\ &= C_{A_0}(4 - X) \end{aligned}$$

$$-r_A = 6.93(0.1)^2(1 - X)(4 - X) = 0.0693(1 - X)(4 - X)$$

For the CSTR,

$$V = \frac{F_{A_0}X}{-r_A} \tag{11.3}$$

$$= \frac{15(0.8)}{0.0693(1 - 0.8)(4 - 0.8)}$$

$$= 270.6\ \text{ft}^3$$

For the TF reactor,

$$V = F_{A_0}\int \frac{dX}{(-r_A)} \tag{11.2}$$

$$= \frac{15}{0.0693}\int \frac{dX}{(1 - X)(4 - X)}$$

Integrating and substituting,

$$V = \frac{15}{0.0693}\frac{1}{(4 - 1)}\left(\ln\frac{4 - 0.8}{4(1 - 0.8)}\right)$$

$$= 100.0\,\text{ft}^3$$

It appears that the reactor is a TF reactor. ∎

ILLUSTRATIVE EXAMPLE 11.14 A chemical engineering firm contacts a consultant, hoping to obtain her insights on improving the design of one of its reactors. The firm provides the following information: the liquid-phase reaction

$$A + B \longrightarrow P$$

takes place in a reactor of volume 10 L. The reaction follows the rate law

$$-r_A = kC_A{C_B}^{0.5}$$

where $k_A = 7.0\,(\text{L/gmol})^{0.5}\,(\text{min})^{-1}$ with the initial concentrations of A and B equal to 2.0 and 3.0 gmol/L, respectively. A conversion of 0.75 is being achieved with a feed rate of *about* 70 L/min.

With a second meeting to follow, the consultant has not yet been asked for any specific analyses; however, she would like to prepare herself for certain possibilities ahead of time. She realizes, though, that the firm has failed to indicate which type of reactor it is operating. Based on what she does know, what type of reactor does the consultant suspect?[3]

Solution. This problem involves determining the input flow rate F_{A_0} to a CSTR and comparing it with the input flow rate to a TF. The rate expression is

$$-r_A = kC_A C_B^{0.5}; \qquad \theta_B = 1.5$$

where

$$C_A(X) = C_{A_0}(1 - X); \qquad X_A = X$$
$$C_B(X) = C_{A_0}(\theta_B - X)$$

Thus,

$$-r_A = k_A C_{A_0}^{1.5}(1 - X)(\theta_B - X)^{0.5}$$

The design equations for a CSTR and a TF can now be solved for the initial molar flow rate of A, F_{A_0}.

For the CSTR,

$$V = F_{A_0}\frac{\Delta X}{(-r_A)} \tag{11.3}$$

or

$$F_{A_0} = -r_A\left(\frac{V}{\Delta X}\right)$$

Substituting ultimately gives

$$F_{A_0} = 57.155 \text{ gmol A/min}$$

The volumetric flow rate of the feed is therefore

$$\begin{aligned} Q_0(\text{CSTR}) &= \frac{F_{A_0}}{C_{A_0}} \\ &= \frac{57.155}{2.0} \\ &= 28.577 \text{ L/min} \end{aligned}$$

For the TF reactor

$$V = F_{A_0}\int_0^X \frac{1}{(-r_A)}dX \tag{11.2}$$

or

$$F_{A_0} = \frac{V}{\displaystyle\int_0^X \frac{1}{(-r_A)}\,dX}$$

The integral may be solved numerically by Simpson's rule (see Chapter 21), either by hand or by computer. MathCad's version of Simpson's rule gives

$$F_{A_0} = 143.52 \text{ gmol A/min}$$

The feed rate for TF reactor is therefore

$$Q_0(\text{TF}) = \frac{F_{A_0}}{C_{A_0}} = 71.76 \text{ L/min}$$

The consultant concludes that the reactor in question is a TF (plug-flow) reactor. ■

ILLUSTRATIVE EXAMPLE 11.15 A chemical engineering student decided to model the human stomach as a reactor vessel. After studying anatomical and physiological data, the student was unsure whether to model the stomach as a CSTR or a batch reactor. To reach a definitive conclusion, he took representative data (see parameters given below) and decided to calculate the residence time for both a batch reactor and a CSTR, and compare these data to the biological information he obtained.

Using the data below, calculate the residence time as if the human stomach were a batch reactor, and also if it were a CSTR. Compare these values to the known physiological and anatomical values, and discuss any discrepancies as they relate to the best suited model for the stomach. Assume digestion follows first order kinetics with 50% digestion occurring.[4]

$$k = k_A = 0.17\ (\text{h})^{\;1}$$
$$Q_0 = 100 \text{ ml/h}$$

Stomach volume range: 100–600 mL

Stomach residence time range: 3–5 h

Solution. For a batch reactor following first order kinetics

$$t = C_{A_0}\int_0^X \frac{dX}{(-r_A)};\ X = X_A$$

$$t = C_{A_0}\int_0^X \frac{dX}{k_A C_{A_0}(1-X)}$$

$$t = \frac{C_{A_0}}{k_A C_{A_0}}\int_0^X \frac{dX}{(1-X)} = \left(\frac{1}{k}\right)\ln\left(\frac{1}{(1-X)}\right)$$

Substituting

$$t = \left(\frac{1}{0.17}\right) \ln \frac{1}{(1 - 0.5)}$$
$$= 4.07 \text{ h}$$

For a CSTR,

$$V = \frac{F_{A_0}X}{-r_A} = \frac{Q_0 C_{A_0} X}{k_A C_{A_0}(1 - X)} = \frac{Q_0 X}{k_A(1 - X)}; \; k = k_A$$

Substituting,

$$V = \frac{(100)(0.5)}{(0.17)(0.5)} = 588 \text{ ml}$$

As both answers fall within "acceptable bounds" of the known data, one may initially assume the stomach operates in a manner involving a positive combination of both reactor modes. ■

ILLUSTRATIVE EXAMPLE 11.16 Refer to Illustrative Example 11.1. How should the reactors of equal volume in the figure be arranged for maximum conversion if the reaction is second order? A numerical example is an acceptable solution.

Solution. First, apply the second order equation to a TF reactor

$$V = \frac{F_{A_0}}{k_A C_{A_0}^2} \int_0^{X_1} \frac{dX_A}{(1 - X_A)^2}; \; X = X_A, \; k = k_A$$

$$V = \frac{F_{A_0}}{k C_{A_0}^2} \left(\frac{X_1}{1 - X_1}\right)$$

Arbitrarily set $F_{A_0} = k = C_{A_0} = V = 1.0$ so that

$$\frac{VkC_{A_0}^2}{F_{A_0}} = 1.0 = \frac{X_1}{1 - X_1}$$

$$X_1 = 0.5$$

This is followed by a CSTR so that

$$\frac{VkC_{A_0}^2}{F_{A_0}} = 1.0 = \frac{X_2 - X_1}{(1 - X_2)^2}; \; X_1 = 0.5$$

Solving for X_2 gives

$$X_2 = 0.634$$

If the CSTR is the first reactor

$$\frac{VkC_{A_0}^2}{F_{A_0}} = 1 = \frac{X_1}{(1 - X_1)^2}$$

Solution for X_1 gives

$$X_1 = 0.382$$

For the TF reactor

$$\frac{VkC_{A_0}^2}{F_{A_0}} = 1 = \int_{X_1}^{X_2} \frac{dX}{(1 - X)^2}$$

$$1.0 = \frac{1}{1 - X_2} - \frac{1}{1 - X_1}; \; X_1 = 0.382$$

Solving for X_2 gives

$$X_2 = 0.618$$

Placing the TF reactor first produces a higher conversion; therefore, the TF unit should be placed first for maximum conversion (for this system). ■

ILLUSTRATIVE EXAMPLE 11.17 Refer to the previous example and Illustrative Example 11.1. How should the reactors of equal volume in the figure be arranged for maximum conversion if the reaction is half order? A numerical example is an acceptable solution.

Solution. The rate expression is

$$-r_A = k_A C_{A_0}^{1/2}(1 - X)^{1/2}$$

Once again, set $F_{A_0} = k_A = C_{A_0} = V = 1.0$. For the TF unit followed by the CSTR,

$$V = F_{A_0} \int_0^{X_1} \frac{dX_A}{k_A C_{A_0}^{1/2}(1 - X_A)^{1/2}}$$

$$= \frac{F_{A_0}}{k_A C_{A_0}^{1/2}} \left[-2(1 - X_1)^{1/2} - 1\right]$$

Setting $F_{A_0} = k_A = C_{A_0} = V = 1$ results in

$$-2(1 - X_1)^{1/2} + 2 = 1$$

Solving

$$X_1 = 0.299$$

For the CSTR,

$$V = \frac{F_{A_0}(X_2 - X_1)}{k_A C_{A_0}^2 (1 - X_2)^{1/2}}; X_1 = 0.299$$

This results in the equation

$$\frac{X_2 - 0.299}{(1 - X_2)^{1/2}} = 1.0$$

Solving for X_2 gives

$$X_2 = 0.7742$$

For the CSTR–TF combination

$$V = \frac{F_{A_0} X_1}{k C_{A_0}^{1/2} (1 - X_1)}$$

Solving for X_1

$$X_1 = 0.3196$$

For the TF reactor,

$$V = F_{A_0} \int_{X_1}^{X_2} \frac{dX}{k C_{A_0}^{1/2} (1 - X)^{1/2}}$$

$$= -\frac{2F_{A_0}}{k C_{A_0}^{1/2}} \left[(1 - X_2)^{1/2} - (1 - X_1)^{1/2}\right]$$

Solving for X_2 with $X_1 = 0.3196$,

$$X_2 = 0.8217$$

This latter case provides the higher conversion, i.e., a comparison of two results indicates that the latter (second) case of a CSTR followed by a TF reactor is more efficient from a conversion perspective. ■

ILLUSTRATIVE EXAMPLE 11.18 The liquid reaction

$$A \longrightarrow \text{products}$$

follows the rate law

$$-r_A = \frac{k_1 C_A^{1/2}}{1 + k_2 C_A^2}$$

where $k_1 = 5(\text{gmol/L})^{1/2}/\text{h}$ and $k_2 = 10(\text{L/gmol})^2$. The initial concentration of A is 0.5 gmol/L, and the feed rate is 200 gmol/h of A.

1. Find the volume necessary to achieve 60% conversion in a CSTR.

2. Would a larger or smaller tubular flow reaction be required to achieve the same degree of conversion?[5]

Solution.

1. Part (1) of this problem was solved earlier in Chapter 10, Illustrative Example 10.19. The result is presented below:

$$V(\text{CSTR}) = 75.1 \text{ L}.$$

2. The design equation if the reaction is conducted in a tubular flow reactor is once again

$$V = F_{A_0} \int_0^X -\frac{dX}{r_A} \tag{11.2}$$

$$= \frac{F_{A_0}}{k_1 C_{A_0}^{1/2}} \int_0^X \frac{1 + k_2 C_{A_0}^2 (1 - X)^2}{(1 - X)^{1/2}} \, dX$$

The integral I may be defined as

$$I = \int_0^{0.6} \frac{1 + k_2 C_{A_0}^2 (1 - X)^2}{(1 - X)^{1/2}} \, dX$$

$$= \int_0^{0.6} f(X) \, dX$$

Simpson's three-point rule (see also Chapter 21) is used to evaluate the integral:

$$\begin{aligned} I &= (h/3)[f(0.0) + (4)f(0.3) + f(0.6)]; \qquad h = 0.3 \\ &= (0.3/3)[(3.5) + (4)(2.659) + (2.214)] \\ &= 1.635 \end{aligned}$$

The volume is then

$$\begin{aligned} V &= \frac{F_{A_0} I}{k_1 C_{A_0}^{1/2}} \\ &= 92.5 \text{ L}. \end{aligned}$$

■

ILLUSTRATIVE EXAMPLE 11.19 Comment on the results of the previous example.

Solution. Surprisingly, the CSTR requires a smaller volume. This result arises because of the unique nature of the rate expression. Also note that the integral could have been evaluated by any one of several methods.

The reader should confirm that a rate–conversion plot results in an inverted parabola. ■

REFERENCES

1. S. Walas, "Rules of Thumb: Selecting and Designing Equipment," *Chem. Eng.*, New York City, NY, 1987.
2. S. Fogler, "*Elements of Chemical Reaction Engineering*," 4th edition, Prentice-Hall, Upper Saddle River, NJ, 2006.
3. W. Blackburn, assignment submitted to L. Theodore, Manhattan College, NY, 1991.
4. P. Antonecchia, assignment submitted to L. Theodore, Manhattan College, NY, 1991.
5. J. Reynolds, J. Jeris, and L. Theodore, "*Handbook of Chemical and Environmental Engineering Calculations*," John Wiley & Sons, Hoboken, NJ, 2002.

PART III

REACTOR APPLICATIONS

The bearings of this observation lays in the application on it.

—Charles Dickens, 1812–70
Dombey and Son

THE DESIGN and analysis of chemical reactors is one of the most important and complicated tasks facing the practicing engineer. Part III of the text is concerned with reactor applications and focusses on subject matter that include:

1. Thermal Effects (Chapter 12)
2. Interpretation of Kinetic Data (Chapter 13)
3. Nonideal Reactors (Chapter 14)
4. Reactor Design (Chapter 15)

Each chapter is supplemented with numerous illustrative examples.

The reader should note that Q and $\dot{Q}$ are employed to represent heat and heat rate, respectively. Unlike Part II q once again represents the voulmetric flow rate.

Chemical Reactor Analysis and Applications for the Practicing Engineer. By Louis Theodore

CHAPTER 12

THERMAL EFFECTS

INTRODUCTION

Why does the practicing engineer require information on thermal effects as they apply to chemical reactors? The answer to this question can best be understood by examining the series of actions that accompany chemical reactions:

1. reaction occurs;
2. energy is liberated (usually);
3. temperature 'T' increases (usually);
4. the reactor velocity constant(s) 'k' increases;
5. the rate of reaction 'r' increases;
6. more energy is liberated.

Keep in mind that it was previously shown that

$$r = r(k) \quad \text{and} \quad k = k(\mathrm{T})$$

Chemical reactions are almost always accompanied by thermal effects, and because of this, the temperature tends to either increase (usually) or decrease as the reaction proceeds. The reader should note that isothermal (constant temperature) operation has been assumed at this point in the text. In experimental work, it is desirable to maintain isothermal conditions so that kinetic studies are not complicated by the variation of the reaction rate with temperature. This latter effect can be determined by experiments conducted at several different but constant temperatures. It is relatively easy to maintain isothermal conditions in the laboratory since the area to volume ratio is high, thus allowing temperature control. However, in large industrial equipment, maintaining isothermal conditions is often difficult.

The three thermal modes of operating are:

1. isothermal,
2. adiabatic,
3. non-isothermal.

Details on modes two and three follow.

Chemical Reactor Analysis and Applications for the Practicing Engineer. By Louis Theodore

Adiabatic reaction is practical when the operating temperatures stay in a reasonable range, i.e., when they decrease (endothermic reactions) so that the rate becomes too small, or increase (exothermic reactions) when the rate becomes increasingly high. The following conditions are favorable to adiabatic operations.

1. When the heat of reaction (to be discussed in the next section) is small.
2. When it is possible to adjust the initial temperature so that the changes will not take the system out of an acceptable temperature range.
3. When the heat capacity of the equipment or of the solvent or of any inert materials that may be present provides sufficient thermal buffering to moderate any temperature effects.
4. When an inert material is present or can be introduced without harm and can moderate temperature changes due to latent enthalpy changes.

The following data are needed for nonisothermal operation.

1. The reaction rate and/or the reaction velocity constant(s) as a function of temperature.
2. The heat capacities and sensible and latent enthalpies of all reactant and product species plus inert materials present, are a function of temperature.
3. The enthalpy of reaction at some base or reference temperature.
4. Heat fluxes (if applicable).
5. Heat-transfer coefficients (if applicable).

The practicing engineer should realize that industrial reactors usually approach adiabatic behavior where no heat exchange occurs between the reactor and its surroundings. Although some heat transfer to the environment occurs, it is not unusual to assume that the temperature profiles in the reactors approach truly adiabatic operation. If the reactor approximates adiabatic operation, the describing energy balance equation leads to simple calculations, unlike those associated with nonisothermal operation. Nonetheless, both modes are considered in sections addressing individual reactors. It should be noted that it is usual to operate nonisothermally during exothermic reactions to reap the benefit of the energy released due to reaction that produces temperatures that, motion leads, in turn, to higher rates and smaller reactor volumes.

The equation describing the amount of heat removed from or added to a reactor across the heat exchanger walls is similar to that employed in the design of conventional heat exchangers.

$$\dot{Q} = U(T - T_c)A \tag{12.1}$$

where U = the overall heat transfer coefficient,

A = the area across which heat exchange occurs,

T = the reaction temperature,

T_c = the "coolant" temperature and

$\dot{Q}$ = the rate of heat removal.

On a differential basis

$$d\dot{Q} = U(T - T_c)\,dA \tag{12.2}$$

In any event, the best or optimum operating temperature in a reactor is dictated by economics, a topic that receives treatment in Part IV. The cost-effective analysis during an economic study is affected by:

1. the process flow rate,
2. the materials of construction,
3. safety considerations (pressure, explosion, etc.), other side effects,
4. potential impurities, and
5. energy factors, etc.

In addition to this introductory section, the chapter reviews the following topics:

Thermal fundamentals and principles

Batch reactors

CSTRs

Tubular flow reactors

THERMAL FUNDAMENTALS AND PRINCIPLES

It was shown in Chapter 4 that the rate of reaction r_A is a function of temperature and concentration. The application of the subsequent equations developed were simplest for isothermal conditions since r_A is then generally solely a function of concentration. If nonisothermal conditions exist, another equation must be developed to describe any temperature variations with position and time in a reactor. For example, in adiabatic operation, the enthalpy (heat) effect accompanying the reaction can be completely absorbed by the system and result in temperature changes in the reactor. As noted earlier, in an exothermic reaction, the temperature increases, which in turn increases the rate of reaction, which in turn increases the conversion for a given interval of time. The conversion, therefore, would be higher than that obtained under isothermal conditions. When the reaction is endothermic, the decrease in temperature of the system results in a lower conversion than that associated with the isothermal case. If the endothermic enthalpy of reaction is large, the reaction may essentially stop due to the sharp decrease in temperature.

Two key topics need to be addressed before developing the equations describing (individually) batch, CSTR, and tubular flow reactors.

1. Enthalpy (heat) of reaction.
2. Conservation law for energy.

The latter topic received treatment in Part II, Chapter 7. However, this analysis will be applied to, and focus on, chemical reactors.

Standard Enthalpy of Reaction

Standard enthalpy of formation data is employed to calculate the energy liberated (usually) or absorbed during a chemical reaction. If the reactants and products are at the same temperature and in a standard (or reference) state, the enthalpy (heat) of reaction is termed the standard enthalpy (heat) of reaction. For engineering purposes, the standard state of a chemical may be taken as the *pure* chemical at 1 atm pressure. A superscript zero is often employed to identify a standard heat of reaction, e.g., ΔH^0. A T subscript (ΔH^0_T) is sometimes used to indicate the temperature; standard heat of reaction data are meaningless unless the temperatures are specified. ΔH^0_{298} data (i.e., for 298K or 25°C) for many reactions are available in the literature.[1,2]

As described in Chapter 7, the first law of thermodynamics provides that, in a steady-flow process with no mechanical work,

$$Q = \Delta H \tag{12.3}$$

Since enthalpy is a point function, it is independent of the path for any process. If heat of reaction determinations are made in a flow reactor, the energy (in the form of heat) transferred across the reactor boundary or surface, with inlet and outlet temperatures equal, is exactly equal to the heat of reaction. This is not the case for batch or nonflow systems. For this reason, *heat of reaction* is a misleading term. More recently, it had been referred to as the *enthalpy of reaction* although a more accurate term would be *enthalpy change* of reaction. The terms heat and enthalpy of reaction are used interchangeably in this subsection.

The *heat of formation* (ΔH_f) is defined as the enthalpy change occurring during a chemical reaction where 1 mol of a product is formed from its elements. The *standard heat of formation* (ΔH^0_f) is applied to formation reactions that occur at constant temperature with each element and the product in its standard state.

Consider the formation reaction for CO_2 at standard conditions at 25°C:

$$\begin{array}{lccc} \text{Enthalpy:} & 0 \quad 0 & & \\ & C + O_2 & \xrightarrow{\Delta H^0_{f298}} & CO_2 \\ \text{Enthalpy:} & & & H_{CO_2} \end{array} \tag{12.4}$$

Once again, this reaction reads: "1 mol of carbon and 1 mol of oxygen react to form 1 mol of carbon dioxide." The enthalpies of reactants and products are printed above and below, respectively, the symbols in the reaction, and the enthalpy change for the reaction, ΔH^0_f, placed above the arrow. Note that the enthalpies of elements in their standard states (pure, 1 atm) at 25°C have arbitrarily been set to zero. The enthalpy change accompanying this reaction is the *standard heat of formation* and is given by

$$\Delta H^0_{f298} = H^0_{CO_2} - (H^0_C + H^0_{O_2}) = H^0_{CO_2} \tag{12.5}$$

Note that this formation reaction for CO_2 is a chemical reaction; the heat of formation, in this case, is therefore equal to the heat of combustion. Since a combustion reaction is one type of chemical reaction, the development to follow will concentrate on chemical reactions in general.

Chemical (stoichiometric) equations may be combined by addition or subtraction. The standard heat (enthalpy) of reaction (ΔH^0) associated with each equation may likewise be combined to give the standard heat of reaction associated with the resulting chemical equation. This is possible, once again, because enthalpy is a point function, and these changes are independent of path. In particular, formation equations and standard heats of formation may always be combined by addition and subtraction to produce any desired equation and its accompanying standard heat of reaction. This desired equation, however, cannot itself be a formation equation. Thus, the enthalpy change for a chemical reaction is the same whether it takes place in one or several steps. This is referred to as *the law of constant enthalpy summation* and is a direct consequence of the first law of thermodynamics.

Consider the author's favorite general reaction

$$aA + bB + \cdots \longrightarrow cC + dD + \cdots \tag{12.6}$$

where $A, B =$ formulas for the reactants (r)

$C, D =$ formulas for the products (p)

$a, b, c, d =$ stoichiometric coefficients of the balanced reaction

To simplify the presentation that follows, Equation (12.6) is shortened to

$$aA + bB \quad \longrightarrow \quad cC + dD \tag{12.7}$$

Although this presentation plus those in later chapters of this book will deal with the hypothetical species, A, B, C, etc., application to real systems can be found in the Illustrative Examples in this and later chapters. This reaction reads: "a moles of A react with b moles of B to form c moles of C and d moles of D." The standard heat of reaction for this chemical change is given by

$$\Delta H^0 = cH_C^0 + dH_D^0 - aH_A^0 - bH_B^0 \tag{12.7}$$

where $H_C^0 =$ enthalpy of C in its standard state, etc.

$C, D, A, B =$ subscripts indicating chemical species C, D, A, and B, respectively

If the temperature is 25°C, the enthalpies of the elements at standard state are, by convention, equal to zero. Therefore, as developed in Equation (12.5),

$$(\Delta H_f^0)_{i298} = H_{i298}^0 \tag{12.8}$$

Substituting Equation (12.8) for each component into Equation (12.7) yields

$$\Delta H_{298}^0 = c(\Delta H_f^0)_C + d(\Delta H_f^0)_D - a(\Delta H_f^0)_A - b(\Delta H_f^0)_B \tag{12.9}$$

or

$$\Delta H_{298}^0 = \sum_p n_p(\Delta H_f^0)_p - \sum_r n_r(\Delta H_f^0)_r \tag{12.10}$$

where p = products

r = reactants

n_p, n_r = coefficients from the chemical equation

The standard heat of a reaction is therefore obtained by taking the difference between the standard heat of formation of the products and that of the reactants. If the standard heat of reaction or formation is negative (*exothermic*), as is the case with most reactions, then energy is liberated as a result of the reaction. Energy is absorbed if ΔH is positive (*endothermic*). Standard heat of formation and standard heat (enthalpy) of combustion data at 25°C are available in the literature. Both of these heat (or enthalpy) effects find extensive application in thermodynamic calculations.[1]

ILLUSTRATIVE EXAMPLE 12.1 Calculate the standard enthalpy of reaction (25°C, 1 atm) for the combustion of methane

$$CH_4 + O_2 \longrightarrow CO_2 + H_2O(g)$$

Standard (25°C, 1 atm) enthalpy of formation data are given below.

$$(\Delta H_f^0)_{CH_4} = -17{,}889 \text{ cal/gmol}$$

$$(\Delta H_f^0)_{CO_2} = -94{,}052 \text{ cal/gmol}$$

$$(\Delta H_f^0)_{H_2O(g)} = -57{,}798 \text{ cal/gmol}$$

Solution. Balance the above reaction equation.

$$CH_4 + 2O_2 \longrightarrow CO_2 + 2H_2O(g)$$

Write the equation for the standard enthalpy of reaction, exercising care to multiply by the appropriate stoichiometric coefficients.

$$\Delta H^0 = (\Delta H_f^0)_{CO_2} + 2(\Delta H_f^0)_{H_2O} - 2(\Delta H_f^0)_{O_2} - (\Delta H_f^0)_{CH_4}$$

Calculate the standard enthalpy of reaction. Note once again that the standard enthalpy of formation for all pure elements in their standard state is, by definition, zero. Substitute the data into the above equation. Note that the water is assumed to be in the gaseous (vapor) phase.

$$\begin{aligned}\Delta H^0 &= -94{,}052 + (2)(-57{,}798) - (2)(0) - (-17{,}889)\\ &= -191{,}759 \text{ cal/gmol}\end{aligned}$$

The effect of temperature on the standard enthalpy of reaction is also available in the literature.[1] ■

Conservation Law for Energy

The equation describing temperature variations in reactors is obtained by applying the conservation law for energy on a time-rate basis to the reactor contents, as provided in Equation (12.11).

$$\begin{Bmatrix} \text{rate of energy} \\ \text{in by} \\ \text{conduction} \\ (1) \end{Bmatrix} - \begin{Bmatrix} \text{rate of energy} \\ \text{out by} \\ \text{conduction} \\ (2) \end{Bmatrix} + \begin{Bmatrix} \text{rate of energy} \\ \text{in by} \\ \text{convection} \\ (3) \end{Bmatrix} - \begin{Bmatrix} \text{rate of energy} \\ \text{out by} \\ \text{convection} \\ (4) \end{Bmatrix}$$

$$\underbrace{\downarrow \quad \downarrow \quad \downarrow \quad \downarrow \quad \downarrow \quad \downarrow}_{\text{temperature gradients}} \qquad\qquad \underbrace{\downarrow \quad \downarrow \quad \downarrow \quad \downarrow \quad \downarrow \quad \downarrow}_{\text{bulk flow}}$$

$$\begin{Bmatrix} \text{rate of energy} \\ \text{generated due} \\ \text{to chemical} \\ \text{reaction} \\ (5) \end{Bmatrix} = \begin{Bmatrix} \text{rate of energy} \\ \text{accumulated} \\ \text{in system} \\ (6) \end{Bmatrix} \tag{12.11}$$

$$\underbrace{\downarrow \quad \downarrow \quad \downarrow \quad \downarrow}_{\text{source}} \qquad\qquad \underbrace{\downarrow \quad \downarrow \quad \downarrow \quad \downarrow}_{\text{inventory}}$$

One may now proceed to evaluate the terms in the above equation.

Terms 1–2: This is zero for an adiabatic reactor. However, energy (heat) transfer across the walls of the reactor can occur if a temperature difference exists between the reactor contents and the surroundings. Terms 1–2 are then given by

$$UA(T - T_a) \tag{12.12}$$

where T_a = temperature of surroundings, °F

T = temperature of reactor contents, °F

A = effective area available for heat transfer, ft^2

U = overall heat transfer coefficient, $\text{Btu}/\text{ft}^2 \cdot \text{s} \cdot {}^\circ\text{F}$

Terms 3–4: This is a convective term and arises due to the energy a fluid possesses when flowing into or out of a reactor. For a batch reactor, no material is added to or removed from the system during the course of the reaction; this term is then zero. For a flow system Terms 3–4 are given by

$$FC_p(T_0 - T_i) \tag{12.13}$$

where F = flow-rate, assumed constant

C_p = heat capacity

T_o = outlet (discharge) temperature

T_i = inlet temperature

Term 5: This represents the rate of energy generated in V due to chemical reaction. Electrical and radiation effects can be included in this term. The source term A (not the reactant) is defined as the amount of energy generated per unit time per unit volume, i.e.,

$$A = \frac{\text{Btu}}{\text{ft}^3 \cdot \text{s}}$$

The enthalpy of reaction (ΔH) was previously described as the energy change due to chemical reaction per mole of A reacted. In accordance with the convention adopted by the engineering profession, ΔH is *negative* if energy is liberated. The moles of A reacted per unit time per unit volume is given by r_A. The term A is therefore

$$(-\Delta H_A)|r_A| \tag{12.14}$$

The rate of energy generated in V is then

$$V(-\Delta H_A)|r_A| \tag{12.15}$$

Term 6: This is a measure of the rate of change of energy within the reactor. The mass within the reactor may possess energy due to its

1. temperature,
2. velocity,
3. position.

The energies corresponding to these states are

1. internal,
2. kinetic,
3. potential.

Since batch reactors are stationary (fixed in space), kinetic and potential effects can be neglected. The relationship between internal energy and temperature for single-phase one-component liquids may be approximated by

$$C_V(T - T_b) \tag{12.16}$$

where T_b is an arbitrary base or reference temperature. (For liquids C_V may be replaced by C_p.) The energy per unit volume is

$$\rho C_p(T - T_b) \tag{12.17}$$

while

$$\rho C_p(T - T_b)V \tag{12.18}$$

represents the energy contained in V. The rate of change of this energy content with time is given by

$$\frac{d}{dt}[\rho C_p(T - T_b)V] \tag{12.19}$$

Note that it was assumed internal energy changes *solely* produce sensible enthalpy (temperature) changes within the reactor. The reference temperature T_b is a constant and does not contribute to the above term. If ρ and C_p are constant, the above becomes

$$\rho V C_p \frac{dT}{dt}$$

or

$$mC_p \frac{dT}{dt} \tag{12.20}$$

where m is the total mass in the reactor. Substituting terms 12.12, 12.13, and 12.20 in Equation (12.11) gives

$$mC_p \frac{dT}{dt} = -UA(T - T_a) + V(-\Delta H_A)|r_A| \tag{12.21}$$

This equation describes the temperature variation in reactors due to energy transfer, subject to the assumptions in its development.

For nonisothermal reactors, one (or more) of the mass transfer equations, the energy transfer equation, Equation (12.10), and an expression for the rate in terms of concentration and temperature, must be solved simultaneously to give the conversion as a function of time. Note that the equations are interdependent; each contains terms that depend on the other (second) equation. These equations, except for simple systems, are usually too complex for analytical treatment.

Regarding the energy equation, note that for an adiabatic system Equation (12.21) reduces to

$$mC_p \frac{dT}{dt} = V(-\Delta H_A)|r_A| \tag{12.22}$$

If the rate is expressed in terms of the conversion, one obtains

$$mC_p\, dT = (-\Delta H_A) N_{A_0}\, dX_A \tag{12.23}$$

This equation may be integrated directly to give the temperature as a function of conversion.

$$mC_p(T - T_0) = (-\Delta H_A) N_{A_0}(X_A - X_{A_0}) \tag{12.24}$$

The subscript zero refers to initial condition. The need for the simultaneous solution of the mass and energy equations is, therefore, removed for this case. Note that all the terms in the above equation must be dimensionally consistent.

Although there is internal motion (due to mixing) of the reacting fluid, no bulk flow into or out of the reactor occurs. Momentum considerations are therefore not important when examining batch reactors.

BATCH REACTORS

The equation describing temperature variations in batch reactors is obtained by applying the conservation law for energy on a time-rate basis to the reactor contents (see previous section for a qualitative discussion of temperature effects). The equation describing the temperature variation in reactors due to energy transfer, subject to the assumptions in its development, with units of energy/time is

$$mC_p(dT/dt) = -UA_e(T - T_a) + V(-\Delta H_A)|-r_A| \tag{12.25}$$

where m = mass of the reactor contents

C_p = heat capacity of the reactor contents = C_V (for liquids)

V = reactor volume

$-\Delta H_A$ = enthalpy of reaction of species A

$|-r_A|$ = absolute value of the rate of reaction of A

T = reactor temperature

and

$$\dot{Q} = UA(T - T_a) \tag{12.26}$$

where $\dot{Q}$ = heat transfer rate across the walls of the reactor

U = overall heat transfer coefficient

A = area available for heat transfer

T_a = temperature surrounding the reactor walls

In practice, the area term should be sufficient to prevent any unacceptable temperature excursions.

As developed in the previous section, note that for an adiabatic system the above equation reduces to

$$mC_p(dT/dt) = V(-\Delta H_A)|-r_A| \tag{12.27}$$

Since, by definition,

$$\begin{aligned} -r_A &= (1/V)(dN_A/dt); \quad N_A = N_{A_0}(1 - X) \\ &= (1/V)d[N_{A_0}(1 - X)]/dt \end{aligned} \tag{12.28}$$

Thus,

$$V|-r_A| = N_{A_0}(dX/dt) \tag{12.29}$$

and the above equation becomes

$$mC_p\,dT = (-\Delta H_A)N_{A_0}dX \tag{12.30}$$

This equation may be integrated directly to give the temperature as a function of conversion,

$$mC_p(T - T_o) = (-\Delta H_A)N_{A_0}(X - X_o) \tag{12.31}$$

where the subscript zero refers to initial conditions. This equation directly relates the conversion with the temperature. The need for the simultaneous solution of the mass and energy equations is, therefore, removed for this case. Note that all the terms in the above equation must be dimensionally consistent. The reader should also note that both the heat capacity and enthalpy of reaction have been assumed to be constant.

Fogler, in his Prentice-Hall text *Elements of Chemical Reaction Engineering*,[2] has derived a somewhat similar equation relating the conversion to temperature.

$$T = T_o - \left[\Delta H^0(T_o)\right]X/\left[\Sigma\theta_i C_{pi} + X\Delta C_p\right] \tag{12.32}$$

where $\Delta H^0(T_o)$ = standard enthalpy of reaction at inlet temperature T_o

ΔC_p = heat capacity difference between products and reactants

$\theta_i C_{pi}$ = applies only to the initial reaction feed mixture (including inerts)

An alternate form of this equation is given by:

$$T = T_0 - X[\Delta H^0(T_R) + \Delta C_p(T - T_R)]/\Sigma\theta_i C_{pi};$$

$$T_R = \text{a reference temperature}$$

For the reaction: $aA + bB \rightarrow cC + dD$

$$\Delta C_p = cC_{pC} + dC_{pD} - aC_{pA} - bC_{pB} \tag{12.33}$$

Note that this term accounts for enthalpy of reaction variation with temperature.

Less time is required to achieve a specified conversion for exothermic reaction under adiabatic conditions than for isothermal operation. For this case, the exothermic reaction produces a rise in temperature in the adiabatic reactor as the reaction proceeds and, as noted earlier, this rise in temperature produces an increase in the rate of reaction and consequently a decrease in time to achieve a specified conversion than would be the case in isothermal conditions. For endothermic adiabatic operation, a longer time would be required to achieve a specified conversion and, as Fogler has indicated, "we see why it is essential to consider the energy balance very carefully in designing chemical reactors. The isothermal reactor assumption, while a good starting point for estimating reactor performance, is seldom adequate for real reactors, and neglect of heat release and possible temperature increase can have very dangerous consequences."

Although there is internal motion (due to mixing) of the reacting fluid, no bulk flow into or out of the reactor occurs. As discussed earlier, momentum considerations and pressure drop concerns are therefore not important when examining batch reactors.

ILLUSTRATIVE EXAMPLE 12.2 The elementary liquid-phase reaction

$$A + B \longrightarrow P$$

is carried out to a conversion of A of 0.5 in a constant volume, insulated batch reactor. The reactor is charged with equimolar concentrations of A and B at 27°C.

Additional information:

Rate constant $k_0 = 1.0 \times 10^{-4}\ \mathrm{L/gmol \cdot min}$ at 27°C

Activation energy $E_a = 3000\ \mathrm{cal/gmol}$

Heat capacities $C_{PA} = C_{PB} = 10\ \mathrm{cal/(gmol \cdot K)}$

$C_{PP} = 20\ \mathrm{cal/(gmol \cdot K)}$

Heat of reaction $\Delta H_R = -5000\ \mathrm{cal/gmol}$ of A at 27°C

Initial concentrations $C_{A_0} = C_{B_0} = 5\ \mathrm{gmol/L}$

Calculate the reaction temperature at $X_A = 0.5$.

Solution. In line with the development above, the equation describing temperature variations in batch reactors is obtained by applying the conservation law for energy on a time–rate basis to the reactor contents. The equation describing the temperature variation in reactors due to energy transfer, subject to the assumptions in its development, is given by Equation (12.25).

$$nC_P(dT/dt) = \dot{Q} + V(-\Delta H_A)|-r_A|$$

where
n = number of moles of the reactor contents at time t

C_P = heat capacity of the reactor contents, mole basis

$\dot{Q}$ = heat transfer rate across the walls of the reactor

V = reactor volume

$-\Delta H_A$ = enthalpy of reaction of species A

$|-r_A|$ = absolute value of the rate of reaction of A

T = reactor temperature

In addition,

$$\dot{Q} = UA(T - T_a) \tag{12.26}$$

where
U = overall heat transfer coefficient

A = area available for heat transfer

T_a = temperature surrounding reactor walls

is employed to calculate $\dot{Q}$. First calculate ΔC_P for this reaction:

$$\begin{aligned}\Delta C_P &= C_{PP} - C_{PA} - C_{PB}\\ &= 20 - 10 - 10\\ &= 0\end{aligned}$$

Since the enthalpy of reaction is not a function of temperature,

$$\begin{aligned} \Delta H(T) &= \Delta H^0(T_R) + \Delta C_P(T - T_R) \\ &= \Delta H^0(T_R) \\ &= -5000 \text{ cal/gmol A} \end{aligned}$$

Thus, $\Delta H(T)$ is a constant.

By definition,

$$\theta_A = 1.0$$

and for an equimolar feed,

$$\theta_B = 1.0$$

The temperature in the reactor may be obtained as a function of the conversion:

$$T = T_0 - \frac{X\Delta H(T)}{\Sigma\theta_i C_{Pi}} \tag{12.32}$$

$$\begin{aligned} \Sigma\theta_i C_{Pi} &= (1)(10) + (1)(10) \\ &= 20 \\ T &= 300 - [(X)(-5000)/(20)] \\ &= 300 + 250X \end{aligned}$$

The temperature when the conversion is 0.5 becomes

$$\begin{aligned} T &= 300 + (250)(0.5) \\ &= 300 + 125 \\ &= 425\text{K} \end{aligned}$$

■

ILLUSTRATIVE EXAMPLE 12.3 Refer to the previous Illustrative Example. Calculate the time required to achieve a conversion of 0.5.

Solution. The rate of reaction, $-r_A$, is

$$-r_A = k_A C_A C_B$$

Since

$$\begin{aligned} C_A &= C_{A_0}(1 - X) \\ C_B &= C_{A_0}(\theta_B - X) \\ &= C_{A_0}(1 - X) \\ -r_A &= kC_{A_0}^2(1 - X)^2 \end{aligned}$$

The design equation for this batch reactor becomes

$$\frac{dC_A}{dT} = r_A; \quad C_A = C_{A_0}(1 - X)$$

$$C_{A_0}\frac{dX}{dt} = -r_A$$

$$\frac{dX}{dt} = kC_{A_0}(1 - X)^2$$

The reaction velocity constant, k, may be expressed as a function of the conversion X through the Arrhenius equation:

$$k = k_0 \exp\left[-\frac{E}{R}\left(\frac{1}{T} - \frac{1}{T_0}\right)\right]; \qquad T_0 = 300\,\text{K}$$

$$T = 300 + 250X$$

Therefore,

$$= k_0 \exp\left[-\frac{E}{R}\left(\frac{1}{300 + 250X} - \frac{1}{T_0}\right)\right]$$

$$= (10^{-4})\exp\left[-\frac{3000}{1.987}\left(\frac{1}{300 + 250X} - \frac{1}{300}\right)\right]$$

The term k in the design equation can now be replaced by the expression obtained above. Thus,

$$\frac{dX}{dt} = (5)(10^{-4})(1 - X)^2 \exp\left(-\frac{1510}{300 + 250X} + 5.03\right) = I$$

Therefore,

$$dt = (1.0/I)\,dX = Y\,dX; Y = 1.0/I$$

This resulting integral may be evaluated to obtain the time required to achieve an $X = 0.5$.One procedure would be to employ the trapezoidal rule. However, the Runge–Kutta method would produce more accurate results. (See Chapter 21 for additional details.)

Table 12.1 is generated using the trapezoidal rule with $\Delta X = 0.05$ to obtain the time required to achieve an $\Delta X = 0.5$.

Thus,

$$\begin{aligned} t &= (\Delta X)(Y) = (\Delta X)(\Sigma Y) \\ &= (0.05)(\Sigma Y) \\ &= (0.05)(16{,}380) \\ &= 819 \text{ min} = 13.35\,\text{h} \end{aligned}$$

■

TABLE 12.1 Trapezoidal Rule Calculations; Illustrative Example 12.3

X	Y
0.0	2000
0.05	1814
0.10	1681
0.15	1588
0.20	1529
0.25	1501
0.30	1501
0.35	1530
0.40	1592
0.45	1691
0.50	1838
	$Y = \Sigma Y =$ 16,380

ILLUSTRATIVE EXAMPLE 12.4 A gas phase reactor which is operating adiabatically changes temperature from 1000 to 900°K as the limiting reaction conversion becomes 75% (X is zero initially). The enthalpy of the reaction is given by the formula

$$\Delta H = 18{,}900 + 5.8\text{T};\ \text{cal/gmol};\ \text{T} \equiv {}^\circ\text{K}$$

and the heat capacity at constant volume of the mixture can be assumed constant at 0.62 cal/g · °C. If the molecular weight of the limiting gaseous reactant is 54, determine the weight *fraction* of this reactant initially present. Determine this value by assuming that ΔH is constant at 1000°K.

Solution. The enthalpy of reaction at 1000°K is first calculated.

$$\Delta H_{1000^\circ\text{K}} = 18{,}900 + 5.8(1000) = 24{,}700\ \text{cal/gmol}$$

A modified form of Equation (12.31)—noting that C_V should be employed since it is a gas phase reaction—is

$$mC_V(T - T_0) = -(\Delta H)\frac{m_{A_0}}{MW_A}(X_A - X_{A_0})$$

where m_{A_0} = initial mass of A

MW_A = molecular weight of A

If one takes a basis of $m = 1$ g, then

$$m_{A_0} = \frac{mC_V(T - T_0)(MW_A)}{(-\Delta H)(X_A - \text{X}_{A_0})} = \frac{(1)(0.62)(100)(54)}{(24{,}700)(0.75)} = 0.1805$$

■

ILLUSTRATIVE EXAMPLE 12.5 Refer to the previous illustrative example. Determine the weight fraction by integrating the ΔH equation from 1000 to 900°K.

Solution. If the enthalpy of reaction is a function of temperature, the differential form of the equation in the previous example is employed:

$$mC_V\,dT = (-\Delta H)\left(\frac{m_{A_0}}{MW_A}\right)dX_A$$

To integrate, then

$$mC_V(MW_A)\int_{T_0}^{T}\frac{dT}{(-\Delta H)} = m_{A_0}\int_{X_{A_0}}^{X_A}dX_A$$

$$-(0.62)(54)\int_{1000}^{900}\frac{dT}{(18{,}900 + 5.8T)} = m_{A_0}\int_{0.0}^{0.75}dX_A$$

$$-(0.62)(54)\left(\frac{1}{5.8}\ln(18{,}900 + 5.8T)\right)\Bigg|_{900}^{1000} = 0.75\,m_{A_0}$$

$$m_{A_0} = \frac{(0.62)(54)\ln\left(\dfrac{24{,}700}{24{,}120}\right)}{(5.8)(0.75)}$$

$$m_{A_0} = 0.1826$$

■

ILLUSTRATIVE EXAMPLE 12.6 Comment on the results of the two previous illustrative examples.

Solution. The assumption that ΔH is a constant is a reasonable one since a difference of about 1% is noted in the two calculations. ■

CSTR REACTORS

If the conservation law for energy is applied to a CSTR, information on temperature changes and variations within and across the reactor can be obtained. If the enthalpy of reaction is not negligible, or if it is required to supply or remove thermal energy in the form of heat from the reactor, an energy balance around the system must be obtained in order to determine the temperature T_1 within the reactor (see Figure 12.1). The terms in this equation are derived by applying the conservation law for energy provided in Equation (12.11) to a tank flow reactor.

Terms 1–2: This term is given by

$$\dot{Q} = -UA(T_1 - T_a) \tag{12.34}$$

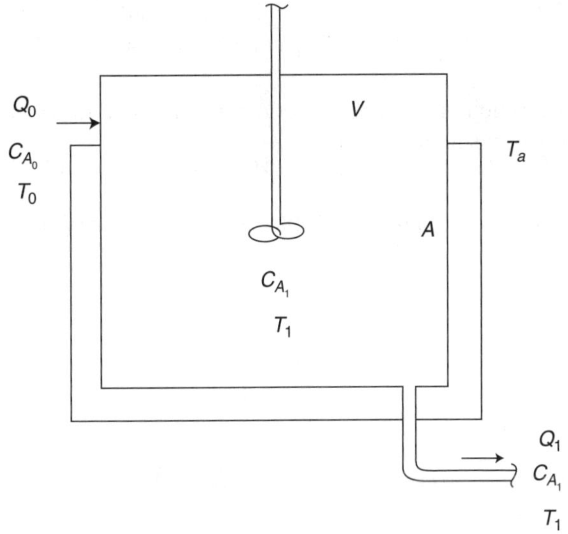

Figure 12.1 Line diagram: tank flow reactor, thermal effects.

Term 3: This is the first of the two convective terms and arises due to the energy a fluid possesses on flowing into or out of the reactor. This is given by

$$\sum_{i=1}^{n} F_i C_{p_i}(T_0 - T_b) \tag{12.35}$$

where F_i = inlet molar (or mass) flow rate of component i
C_{p_i} = molar (or mass) heat capacity of component i
T_b = a reference or base temperature

Term 4: This convective term is

$$\sum_{i=1}^{n} F_i^* C_{p_i}(T_1 - T_b) \tag{12.36}$$

where F_i^* is the outlet flow rate of i.

Term 5: This term is once again

$$|r_A|V(-\Delta H_A) \tag{12.37}$$

with ΔH_A evaluated at T_b.

Term 6: This term is zero for steady-state conditions.

Substituting into the conservation law for energy equation yields

$$-UA(T_1 - T_a) + \sum_{i=1}^{n} F_i C_{p_i}(T_0 - T_b) + \sum_{i=1}^{n} F_i^* C_{p_i}(T_1 - T_b) + |r_A|V(-\Delta H_{\text{A}}) = 0 \tag{12.38}$$

Since F_i^* is usually not known, set

$$T_b = T_1$$

Equation (12.38) then becomes

$$-UA(T_1 - T_a) + \Sigma F_i C_{p_i}(T_0 - T_1) + |r_A|V(-\Delta H_A) = 0 \tag{12.39}$$

If average heat capacity data for the mixture are employed over the temperature range T_o and T_1, then

$$-UA(T_1 - T_a) + F\langle C_p\rangle(T_0 - T_1) + |r_A|V(-\Delta H_A) = 0 \tag{12.40}$$

where the first term on the LHS may be denoted by $\dot{Q}$. Once again, the reader is reminded that the units of F and C_p must be consistent, and the terms in the equation must be dimensionally the same.

For adiabatic conditions, a rather simple situation develops since

$$UA(T_1 - T_a) = 0 \tag{12.41}$$

and

$$|r_A|V = F_A X_A \tag{12.42}$$

where

$$F_A = \text{inlet molar (or mass) flow rate of } A$$

$$X_A = \frac{\text{moles of } A \text{ converted}}{\text{moles (or mass) of } A \text{ entering the reactor}}$$

Equation (12.40) becomes

$$F\langle C_p\rangle(T_0 - T_1) = F_A X_A(\Delta H_A) \tag{12.43}$$

and

$$T_1 = T_0 + \left[\frac{F_A(\Delta H_A)}{F\langle C_p\rangle}\right] X_A \tag{12.44}$$

Thus, the need for the simultaneous solution of the mass and energy transfer equations is removed.

For transient (unsteady-state) conditions, Term 6 is nonzero. The time rate of change of the energy content of the reactor is

$$\frac{d[mC_p(T_1 - T_b)]}{dt} = \frac{d(mC_pT_1)}{dt} \tag{12.45}$$

where m = total moles (or mass) in the reactor at time t. If m and C_p are constant, this term is given by $mC_p(dT_1/dt)$ so that Equation (12.40) becomes

$$-UA(T_1 - T_a) + F\langle C_p\rangle(T_0 - T_1) + |r_A|V(-\Delta H_A) = mC_p\frac{dT_1}{dt} \tag{12.46}$$

Under these conditions, the describing equations for both mass and energy transfer are first order ordinary differential equations. These must be solved simultaneously using a suitable analytical or numerical procedure.

Fogler[2] has accounted for enthalpy of reaction variation for CSTRs with temperature through the equation

$$\dot{Q} - [\Delta H_R^0(T_R) + \Delta C_p(T - T_R)]F_{Ao}X_A = F_{Ao}\Sigma\theta_i C_{p_i}(T - T_0) \tag{12.47}$$

where $\Delta H_R^0\,(T_R)$ = enthalpy of reaction at temperature T_R

ΔC_p = heat capacity difference between products and reactants

$\theta_i C_{pi}$ = applies only to the species in the feed stream

Note again that for the reaction

$$aA + bB \longrightarrow cC + dD$$

$$\Delta C_p = cC_{pC} + dC_{pD} - aC_{pA} - bC_{pB} \tag{12.48}$$

Finally, note that for a non-adiabatic reactor, the term $\dot{Q}$ must be retained in the energy equation.

There is a great variety in the modes of operation of semibatch reactors and transient systems. An energy balance general enough to satisfy all these possibilities becomes unwieldy and often difficult to solve. Several authors have recommended that the energy balance be derived on a case-by-case, system-by-system basis.

ILLUSTRATIVE EXAMPLE 12.7 A continuous bench-scale stirred-tank reactor was used to study the reaction

$$A \xrightarrow{k_A} \text{products}$$

The process was run adiabatically with an inlet temperature of 200°F. If the exit temperature was 350°F, determine the volume required to handle 1.2 lbmol/h of pure A. Assume the conversion is small enough so that the heat capacity is constant and the total number of moles in and out of the system is also constant.

Data:

$$C_{PA} = 17.5 \text{ Btu/lbmol} \cdot {}^\circ\text{R}$$
$$\Delta H_A = -33{,}000 \text{ Btu/lbmol}$$
$$C_{A_0} = 0.75 \text{ lbmol/ft}^3$$
$$k_A = 0.18\ (\text{h})^{-1}$$

Solution. The describing energy balance equation is

$$F\langle C_p\rangle(T_0 - T_1) - FX_A^*(\Delta H_A) = 0;\ \text{note that } F = F_A \tag{12.43}$$

For this system $X_A^* = X_A$. Solving for X_A gives

$$X_A = \frac{\langle C_p \rangle (T_0 - T_1)}{\Delta H_A}$$

$$= \frac{(17.5)(200 - 350)}{-33{,}000}$$

$$= 0.0795$$

Therefore,

$$C_{A_1} = C_{A_0}(1.0 - X_A)$$

$$= 0.75(1.0 - 0.0795)$$

$$= 0.69\ \text{lbmol/ft}^3$$

Now substitute these values into the CSTR design mass transfer equation (noting that q, and not Q, represents the volumetric flow rate)

$$\frac{V}{q} = \frac{C_{A_1} - C_{A_0}}{-k_A C_A}$$

and solve for V.

$$\frac{V}{1.2} = \frac{0.69 - 0.75}{-(0.18)(0.69)}$$

$$V = 0.48\ \text{ft}^3$$

■

ILLUSTRATIVE EXAMPLE 12.8 A 2 : 1 molar mixture of ethylene oxide (A) and water was fed to a 10 liter adiabatic tank flow reactor. The flow rate of the solution was 1000 L/h. The initial concentration of ethylene oxide was measured to be 38.3 gmol/L. The temperature at the reactor outlet was found to be 375°F. The heat of reaction, activation energy, average heat capacity and reaction velocity constant for this system are known:

$\Delta H_A = 23{,}000$ Btu/lbmol
C_p (mixture) $= 10.0$ Btu/lbmol · °R
$E = E_a = 32{,}525$ Btu/lbmol
$k_A = 0.00343(\text{s})^{-1}$ at 400°F

The first three values can be considered independent of temperature and conversion. What is the inlet temperature of this mixture?

Solution. Using a simple energy balance,

$$FC_p(T_0 - T) + F_{A_0}X_A(-\Delta H_A) = 0 \tag{12.43}$$

$$T_0 = T_1 + \left(\frac{X_A\,\Delta H_A}{C_p}\right)\frac{F_{A_0}}{F}$$

The design equation for the CSTR is

$$\frac{V}{Q_0} = \frac{C_{A_1} - C_{A_0}}{-k_A C_{A_1}}$$

Use the Arrhenius equation to obtain k at 375°F.

$$\ln\left(\frac{k_2}{k_1}\right) = \frac{E}{k}\left[\frac{T_2 - T_1}{T_2 T_1}\right]; \qquad \begin{aligned} k_1 &= 0.00343\,\text{s}^{-1} \\ T_1 &= 860\,^\circ\text{R} \\ T_2 &= 835\,^\circ\text{R} \end{aligned}$$

Solving

$$k_2\,(375\,^\circ\text{F}) = 0.00194\,(\text{s})^{-1}$$

Thus,

$$\frac{10}{1000} = \frac{C_{A_1} - 38.3}{(0.00194)(3600)C_{A_1}}$$

$$C_{A_1} = 35.8\,\text{gmol L}; \qquad X = \frac{38.3 - 35.8}{38.3} = 0.0653$$

Returning to the energy balance equation,

$$\begin{aligned} T_0 &= 375 + (0.0653)(23{,}000)(2/3)/10 \\ &= 375 + 100 = 475\,^\circ\text{F} = 935\,^\circ\text{R} \end{aligned}$$

Once again, the reader should check that the terms in the equation are dimensionally consistent. ■

ILLUSTRATIVE EXAMPLE 12.9 Consider the following elementary liquid phase irreversible reaction:

$$A + 2B \longrightarrow C$$

The reaction takes place in a CSTR at 170°F. Calculate the conversion and volume required for adiabatic conditions. The feed enters at 20°F. Data are provided below.

$$C_{A_0} = 0.1 \text{ lbmol/ft}^3; \qquad C_{pA} = 30 \text{ Btu/lbmol} \cdot {}^\circ\text{F}$$

$$C_{B_0} = 0.2 \text{ lbmol/ft}^3; \qquad C_{pB} = 20 \text{ Btu/lbmol} \cdot {}^\circ\text{F}$$

$$F_{A_0} = 40 \text{ lbmol/h}; \qquad C_{pC} = 45 \text{ Btu/lbmol} \cdot {}^\circ\text{F}$$

At 70°F,

$$\begin{aligned} \Delta H^\circ_{fA} &= -60{,}000 \text{ Btu/lbmol} \\ \Delta H^\circ_{fB} &= -120{,}000 \text{ Btu/lbmol} \\ \Delta H^\circ_{fC} &= -395{,}000 \text{ Btu/lbmol} \end{aligned}$$

$$k\,(170^\circ\text{F}) = 50.0 \text{ (units consistent with concentration and feed rate)}$$

Solution. Calculate the standard enthalpy of reaction at 70°F.

$$\begin{aligned} \Delta H^\circ(70^\circ\text{F}) &= -395{,}000 - [-60{,}000 + (2)(-120{,}000)] \\ &= -95{,}000 \text{ Btu/lbmol} \end{aligned}$$

Determine ΔC_p.

$$\begin{aligned} \Delta C_p &= C_{pC} - 2C_{pB} - C_{pA} \\ &= 45 - (2)(20) - 30 \\ &= -25 \text{ Btu/lbmol.}^\circ\text{F} \end{aligned}$$

Calculate θ_A and θ_B.

$$\begin{aligned} \theta_A &= 1. \\ \theta_B &= 2. \end{aligned}$$

Therefore,

$$\begin{aligned} \Sigma\theta_i C_{pi} &= C_{pA} + \theta_B C_{pB} \\ &= 30 + (2)(20) \\ &= 70 \text{ Btu/lbmol} \cdot {}^\circ\text{F} \end{aligned}$$

Write the equation describing the conversion in terms of the (outlet) temperature in the CSTR. Employ Equation (11.48) with $\dot{Q} = 0$.

$$X = -\Sigma\theta_i C_{pi}(T - T_0)/[\Delta H^\circ(T_R) + \Delta C_p(T - T_R)]$$

Calculate the conversion if the discharge temperature is 170°F.

$$\begin{aligned} X &= -70(170 - 20)/[-95{,}000 + (-25)(170 - 70)] \\ &= 0.108 \end{aligned}$$

The design equation for a CSTR is once again

$$V = \frac{F_{A_0}X}{-r_A}$$

and the rate of reaction, $-r_A$, in terms of the conversion X is

$$\begin{aligned} -r_A &= kC_A C_B^2 \\ &= kC_{A_0}^3(1-X)(2-2X)^2 \end{aligned}$$

The rate at the operating conditions ($X = 0.108$) in the reaction is therefore

$$\begin{aligned} -r_A &= (50)(0.1)^3(1-0.108)[2-(2)(0.108)]^2 \\ &= 0.1419; \quad \text{consistent units} \end{aligned}$$

Finally, the volume, V, of the reactor in ft^3 is

$$\begin{aligned} V &= (40)(0.108)/(0.1419) \\ &= 30.43\ \text{ft}^3 \end{aligned}$$

■

ILLUSTRATIVE EXAMPLE 12.10 The elementary irreversible liquid phase reaction

$$A + B \longrightarrow 2P$$

occurs in a CSTR reactor. Feed at 100°F consists of a 2 : 1 molar ratio of B to A and is at a total rate of 8 lbmol/min.

Data is provided below

$$\begin{aligned} \Delta H_A^\circ(70^\circ\text{F}) &= +40\ \text{kJ/mol}; \quad C_{pA} = 15\ \text{J/mol}\cdot{}^\circ\text{R} \\ \Delta H_B^\circ(70^\circ\text{F}) &= -50\ \text{kJ/mol}; \quad C_{pB} = 25\ \text{J/mol}\cdot{}^\circ\text{R} \\ \Delta H_P^\circ(70^\circ\text{F}) &= -10\ \text{kJ/mol}; \quad C_{pP} = 20\ \text{J/mol}\cdot{}^\circ\text{R} \end{aligned}$$

$$\begin{aligned} C_{A_0} &= 0.2\ \text{lbmol/ft}^3 \\ k\,(100^\circ\text{F}) &= 0.1\ \text{ft}^3/\text{lbmol}\cdot\text{s} \\ E_a &= 30{,}000\ \text{J/mol} \end{aligned}$$

1. Obtain the temperature as a function of conversion.
2. Calculate the volume of the CSTR at 90% conversion when operating adiabatically.

Solution First note that

$$y_{A_0} = \frac{1}{3} = 0.333\,;\ y_{B_0} = \frac{2}{3} = 0.667$$

$$T_R = 70°\text{F} = 530°\text{R}$$

$$T_0 = 100°\text{F} = 560°\text{R}$$

Apply Equation (12.47) under adiabatic conditions:

$$\dot{Q} - [\Delta H°(T_R) + \Delta C_p(T - T_R)]F_{A_0}X_A = F_{A_0}\Sigma\theta_i C_{pi}(T - T_0)$$

$$\Delta H°(T_R) = [m_P \cdot \Delta H_P^\circ] - [(m_A \cdot \Delta H_A^\circ) + (m_B \cdot \Delta H_B^\circ)]$$

$$\Delta H°(T_R) = [2(-10\text{kJ/mol}) - (+40\text{kJ/mol} - 50\text{kJ/mol})]$$

$$\Delta H°(T_R) = -10\text{kJ/mol}$$

$$\Delta C_p = [(n_P)(C_{pP})] - [(n_A)(C_{pA}) + (n_B)(C_{pB})]$$

$$\Delta C_p = 2(20\text{ J/mol}\cdot°\text{R}) - (15 + 25)\text{ J/mol}\cdot°\text{R}$$

$$\Delta C_p = 0$$

$$\theta_A = \frac{y_{A_0}}{y_{A_0}} = \frac{0.333}{0.333} = 1.0$$

$$\theta_B = \frac{y_{B_0}}{y_{A_0}} = \frac{0.667}{0.333} = 2.0$$

$$\Sigma\theta_i C_{pi} = [(\theta_A C_{pA} + \theta_B C_{pB})]$$

$$\Sigma\theta_i C_{pi} = [(1)(15\text{ J/mol}\cdot°\text{R}) + (1)(15\text{ J/mol}\cdot°\text{R})]$$

$$\Sigma\theta_i C_{pi} = 45\text{ J/mol}\cdot°\text{R}$$

Solve for $T(X_A)$.

$$[-\Delta H(T_{Ref}) - \Delta C_p(T - T_R)]F_{A_0}X_A = F_{A_0}\Sigma\theta_i C_{pi}(T - T_{lo})$$

$$(10{,}000)(X_A) = (45)(T - 560)$$

$$222.22\,X_A = T - 560$$

$$T = 222.2\,X_A + 560$$

The CSTR design equation is once again

$$V = \frac{F_{A_0}X_A}{-r_A}\ ;\ X_A = 0.9$$

The rate law is

$$-r_A = kC_AC_B$$
$$C_A = C_{A_0}(1 - X)$$
$$C_B = C_{A_0}(\theta_B - X)$$
$$-r_A = kC_{A_0}^2(1 - X)(\theta_B - X); \;\; X = X_A$$

Calculate k at $X = 0.9$

$$k = Ae^{-E_{Act}/RT}$$

with

$$T(X_A) = T(X) = 560 + 222.22(X)$$
$$T(0.9) = 560 + 222.22(0.9)$$
$$T = 760\,\text{R}$$

Substituting with $T = 560°\text{R}$,

$$A = 1.089 \times 10^4$$

At $T = 760°\text{R}$

$$k\,(760°\text{F}) = 1.089 \times 10^4\, e^{\frac{-30{,}000}{(8.314)(760)(1/1.8)}} = 2.116\ \text{ft}^3/\text{lbmol} \cdot \text{s}$$

The design equation is employed to calculate the volume.

$$V = \frac{F_{A_0}(X)}{kC_{A_0}^2(1 - X)(\theta_B - X)}$$

$$V = \frac{8(0.9)}{(2.116)(60/1)(0.2)^2(1 - 0.9)(2 - 0.9)}$$

$$V = \frac{8(0.9)}{(2.116)(60)(0.2)^2(0.1)(1.1)}$$

$$= 12.88\,\text{ft}^3$$

■

ILLUSTRATIVE EXAMPLE 12.11 The elementary gas phase reaction (with kinetics implied as written)

$$A \longrightarrow 2B$$

is to be carried out in one of the reactors below. The feed, which is at a temperature of 27°C, consists of 30 mol% of A and the remainder inerts. The volumetric flow rate entering the reactor at this temperature is 1000 ft^3/min. The concentration of A in the feed at 27°C is 0.5 lbmol/ft^3.

Data:

$$C_{pA} = 25\ \text{Btu/lbmol}\cdot{}^{\circ}\text{C}$$
$$C_{pB} = 20\ \text{Btu/lbmol}\cdot{}^{\circ}\text{C}$$
$$C_{pI} = 30\ \text{Btu/lbmol}\cdot{}^{\circ}\text{C}$$

The heat of reaction is a function of temperature, and its value at 300°K is –15,000 Btu/lbmol of A. k_1 varies with temperature as follows:

$$k_1 = 0.217\ (\text{min})^{-1} \quad \text{at} \quad 300^{\circ}\text{K}$$

$$k_1 = 0.325\ (\text{min})^{-1} \quad \text{at} \quad 340^{\circ}\text{K}$$

1. Generate the two coefficients in the Arrhenius expression.
2. For adiabatic operation, develop an equation describing the temperature in terms of the conversion. What is the temperature for 40% conversion?
3. *Outline* a method to calculate the volume for a CSTR reactor which is operated adiabatically at a given conversion.

Solution

1. Applying the equation $k = Ae^{-E/RT}$. Using the 2 data points gives

$$A = 6.54\ (\text{min})^{-1};\ E_a = 2030\ \text{cal/gmol}\cdot\text{A}$$

2. Since $\dot{Q} = 0$, Equation (12.48) becomes

$$X[\Delta H_R(T_R) + (2C_{pB} - C_{pA})(T - T_0)] = (C_{pA} + 2.333C_{pI})(T - T_0)$$

Solving for T,

$$T = T_0 + \frac{(-\Delta H_A)(T_R)X}{C_{pA} + 2.33C_{pI} + X(2C_{pB} - C_{pA})}$$

$$= 300 + \frac{15{,}000X}{[25 + (2.33)30] + X(2 \times 20 - 25)}$$

$$T = 300 + \frac{15{,}000X}{95 + 15X}$$

If $X = 0.4$,

$$T = 300 + 59.4 = 359.4^{\circ}\text{K} = 86.4^{\circ}\text{C}$$

3. Since this is a gaseous reaction,

$$-r_A = kC_A = kC_{A_0}\left(\frac{1 - X}{1 + \varepsilon X}\right)\frac{T_0}{T}$$

$$= Ae^{-E/RT}C_{A_0}\left(\frac{1 - X}{1 + \varepsilon X}\right)\frac{T_0}{T}$$

and

$$V = \frac{F_{A_0}X}{-r_A};\ \theta_A = 1.0,\ \theta_B = 0.0;\ \theta_I = 2.33,\ \varepsilon = 0.3$$

Substitute the rate expression into the above equation, and one can then calculate the volume. ■

ILLUSTRATIVE EXAMPLE 12.12 The following elementary liquid phase irreversible reaction:

$$A + B \xrightarrow{k} 3C$$

is carried out in a CSTR equipped with a heat exchanger. The reactants consist of equimolar concentrations (5 gmol/L) of A and B at 50°C. The reacting system must produce C at a concentration of 13 gmol/L. The total volumetric flow rate of the feed is known to be 10 L/min.

If 20,000 KJ/min of heat are added to the reactor, what is the temperature of the exit stream and what size volume reactor is necessary to complete the reaction as specified? The mixer in the CSTR also requires an input of 2000 KJ/min.

Information provided:

$$R = 8.314\ \text{KJ/gmol}\cdot\text{K}$$
$$E_a = E = 6000\ \text{KJ/gmol}\cdot\text{K}$$
$$k = 0.054\ \text{KJ/gmol}\cdot\text{min at } 50^\circ\text{C}$$
$$\Delta H_R^\circ = 75\ \text{KJ/gmol of } A \text{ at } 50^\circ\text{C}$$
$$C_{pA} = 3\ \text{KJ/gmol}\cdot\text{K}$$
$$C_{pB} = 2\ \text{KJ/gmol}\cdot\text{K}$$
$$C_{pC} = 6\ \text{KJ/gmol}\cdot\text{K}$$

Solution. This is an elementary, liquid, irreversible, equimolar nonadiabatic reaction with work input.

Preliminary calculations

$$F_{A_0} = (5)(10) = 50\ \text{gmol/min}$$
$$\theta_A = 1;\ \theta_B = 1;\ \theta_C = 0.$$
$$\Delta C_p = (3)(6) - (3)(1) - (2)(1) = 13\ \text{kJ/gmol}\cdot\text{K}$$
$$T_R = T_0 = 50^\circ\text{C} = 323\text{K}$$
$$-r_A = kC_AC_B = kC_{A_0}^2(1 - X)^2;\ X = X_A$$
$$C_C = C_{A_0}(\theta_c + 3X)$$
$$X_A = X = 13/(5)(3) = 0.867$$

Apply Equation (12.44) with a work term included.

$$\dot{Q} - \dot{W}_S - [\Delta H_R^\circ(T_R) + F\Delta C_p(T - T_R)]F_{A_0}X = F_{A_0}\Sigma\theta_i C_{pi}(T - T_0)$$

Substituting and solving,

$$20{,}000 - (-2000) - [75 + 13(T - 323)](50)(0.867) = 50[(3 + 2 + 0)(T - 323)]$$

Solving,

$$T = 346^\circ\text{K}$$

Evaluate k at $T = 346^\circ$K.

$$\ln\left(\frac{k_2}{k_1}\right) = -\frac{E}{R}\left[\frac{1}{T_2} - \frac{1}{T_1}\right]$$

$$\ln(k_2/0.05) = \left(\frac{-6000}{8.314}\right)\left(\frac{1}{346} - \frac{1}{323}\right)$$

$$k_2 = 5.8 \times 10^{-2}\,\text{L/gmol}\cdot\text{min}$$

Apply the design equation

$$V = F_{A_0}X/-r_A$$

Substituting,

$$V = (50)(0.867)/(5.8\times10^{-2})(5.0)^2(1 - 0.867)^2$$
$$= 1690\,\text{L}$$

■

ILLUSTRATIVE EXAMPLE 12.13 The following elementary, liquid phase reaction is carried out in a CSTR:

$$A + 2B \longrightarrow 3C + 2D$$

The reactants are fed to the reactor at a rate of 10 L/min. The inlet concentrations of A and B are 0.333 lbmol/L and 0.667 lbmol/L, respectively. The inlet temperature to the reactor is 100°F. Because of safety considerations, the temperature in the reactor must not exceed 150°F.

1. What is the conversion (based on component A) if the reactor is operated adiabatically at 150°F?
2. Show that the reactor volume necessary to achieve this conversion is approximately 14.40 liters.

Suppose an 80% degree of conversion is desired.

3. What must the volumetric flowrate be in order to achieve this conversion if the reactor volume given in (2) is used?
4. What is the rate of heat removal in (3) from the reactor such that the temperature does not exceed 150°F?

Additional data:

$$C_{pA} = 5.0\,\frac{\text{Btu}}{\text{lbmol}\cdot{}^\circ\text{F}};\quad C_{pC} = 7.0\,\frac{\text{Btu}}{\text{lbmol}\cdot{}^\circ\text{F}}$$

$$C_{pB} = 6.0\,\frac{\text{Btu}}{\text{lbmol}^\circ\cdot\text{F}};\quad C_{pD} = 8.0\,\frac{\text{Btu}}{\text{lbmol}\cdot{}^\circ\text{F}}$$

$$k(100^\circ\text{F}) = 1.0\times 10^{-7}\,\frac{\text{L}^2}{\text{gmol}\cdot\text{s}}$$

$$\Delta H_R(100^\circ\text{F}) = -3500\,\frac{\text{Btu}}{\text{lbmol A}}$$

$$E = 3000\,\frac{\text{cal}}{\text{gmol A}}$$

Solution. Preliminary calculations

$$\begin{aligned} F_{A_0} &= C_{A_0}Q_0 = 0.333\times 10 = 3.33\ \text{lbmol/min}\\ F_{B_0} &= C_{B_0}Q_0 = 0.667\times 10 = 6.66\ \text{lbmol/min}\\ \theta_A &= 1.0,\ \theta_B = 2.0 \end{aligned}$$

1. If adiabatic, $\dot{Q} = 0$, so that

$$-X[\Delta H_R(T_R) + \Delta C_p(T - T_R)] = \Sigma\theta_i C_{pi}(T - T_0)$$

Additional calculations

$$\begin{aligned} \Delta C_p &= (3)(7) + (2)(8) - (1)(5) - (2)(6)\\ &= 21 + 16 - 5 - 12 = 20\ \text{Btu/lbmol}\cdot{}^\circ\text{F}\\ T - T_R &= 150^\circ\text{F} - 100 = 50^\circ\text{F}\\ T - T_i &= 150 - 100 = 50^\circ\text{F}\\ \sum\theta_i C_{pi} &= (1)(5) + (2)(6) = 17\ \text{Btu/lbmol}\cdot{}^\circ\text{F} \end{aligned}$$

Substituting,

$$X = -\frac{(17)(50)}{3500 + (20)(50)} = \frac{850}{2500} = 0.34$$

2. Use the following equation to calculate k at 150°F

$$k_2 = k_1 e^{-\frac{E}{R}\left[\frac{1}{T_2} - \frac{1}{T_1}\right]}$$

$$T_1 = 100°\text{F},\ T_1 = 311°\text{K}$$

$$T_2 = 150°\text{F},\ T_2 = 339°\text{K}$$

$$k_2 = 1.07 \times 10^{-7} e^{-\frac{3000}{1.987}\left[\frac{1}{339} - \frac{1}{311}\right]}$$

$$k\,(150°\text{F}) = 1.493 \times 10^{-7}\ \text{L}^2/(\text{gmol} \cdot \text{s})$$

Apply the CSTR equation

$$V = \frac{F_{A_0}X}{kC_{A_0}^3(1-X)(\theta_B - 2X)^2};\quad \theta_B = 2.0$$

Substituting,

$$V = \frac{(3.33)(0.34)}{(1.493 \times 10^{-7})(60)(454)^2(0.333)^3(1-0.34)(2-0.68)^2}$$

$$= 14.40\ \text{L}$$

3. Rewrite the CSTR equation as

$$V = \frac{C_{A_0}q_0X}{kC_{A_0}^3(1-X)(\theta_B - 2X)^2};\quad X = 0.8$$

and solve for $q = q_0$.

$$Q_0 = 0.0745\ \text{L/min}$$

This result is reasonable since a higher X lowers q_0 and increases τ.

4. Since there is a higher X, more heat is generated. In order to retain the outlet temperature at 150°F, one must *remove* heat from the system. Since q_0 is different in (3), the F_{io}'s are also different. Thus,

$$F_{A_0} = C_{A_0}q_0 = (0.333)(0.0745) = 0.0248\ \text{lbmol/min}$$

$$F_{B_0} = 2F_{A_0} = 0.049\ \text{lbmol/min}$$

The energy equation is again applied.

$$\dot{Q} - F_{A_0}X[\Delta H_R(T_R) + \Delta C_p(T - T_R)] = F_{A_0}\Sigma\theta_i C_{pi}(T - T_0) \qquad (12.48)$$

For this case,

$$[\Delta H_R(T_R) + \Delta C_p(T - T_R)] = -2500$$

$$\Sigma\theta_i C_{pi}(T - T_0) = 850$$

Therefore

$$\begin{aligned}\dot{Q} &= F_{A_0}[(X)(-2500) + 850] \\ &= 0.0248[-2000 + 850] \\ &= -28.72 \text{ Btu/min}\end{aligned}$$

The negative sign requires heat be *removed.* ■

TUBULAR FLOW REACTORS

The temperature in a TF reactor can also vary with position (volume) due to enthalpy of reaction effects or transfer of energy in the form of heat across the walls of the reactor. The earlier TF equations must then include temperature effects and temperature variations before they are solved. In order to obtain information on the temperature at every point in the reactor, the conservation law for energy is applied to the system (see Figure 12.2). This analysis assumes the following:

1. Steady conditions.
2. Plug flow.
3. Radial (perpendicular to the direction of flow) temperature effects are negligible so that the reacting mass and products at any r for a given z are at an average temperature.
4. Diffusion effects[3] in the axial (z) direction are neglected.

The six terms in the conservation law for energy in Equation (12.11) are now evaluated.

Terms 1–2: This term is given by and accounts for energy transfer between the reactor contents and the surroundings.

$$d\dot{Q} = -U(T - T_a)\, dA \tag{12.49}$$

A is the effective surface area available for heat transfer in dV.

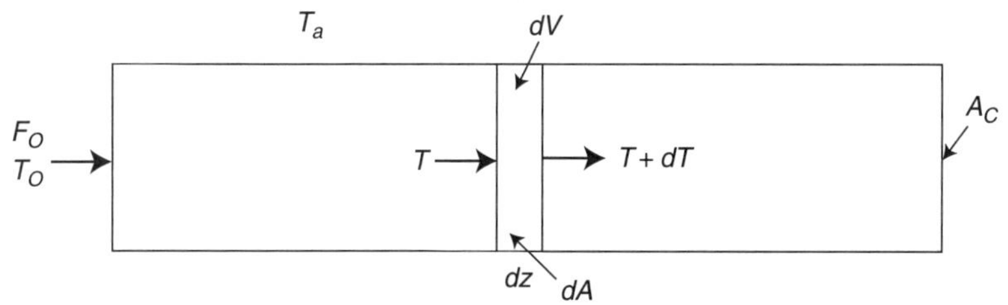

Figure 12.2 Energy balance on differential element.

Terms 3–4: This reduces to

$$-F\langle C_p\rangle\, dT \tag{12.50}$$

where F = total molar (or mass) flow rate at dV

$\langle C_p\rangle$ = average molar (or mass) heat capacity for the fluid at dV.

and

A_c = cross-sectional area of conduit.

Since the molar flow will in general vary with position, it is more convenient to use mass units for F.

Term 5: This given by

$$|r_A|(\Delta H_A)\, dV \tag{12.51}$$

Term 6: For steady conditions, this term is zero.

Inserting the above terms into the energy equation gives

$$-U(T - T_a)\, dA_e - F\langle C_p\rangle\, dT + |r_A|(-\Delta H_A)\, dV = 0 \tag{12.52}$$

To obtain this equation in terms of V or z, substitute

$$dA = P\, dz;\ P = \text{perimeter} \tag{12.53}$$

$$dV = A_C\, dz \tag{12.54}$$

and obtain either

$$\frac{dT}{dV} = \frac{-UP(T - T_a)/A_C + |r_A|(-\Delta H_A)A_C}{F\langle C_p\rangle} \tag{12.55}$$

or

$$\frac{dT}{dz} = \frac{-UP(T - T_a) + |r_A|(-\Delta H_A)}{F\langle C_p\rangle} \tag{12.56}$$

This equation must be solved with the appropriate form of the mass transfer design equation. The reaction velocity constant in the rate of reaction term is expressed in the normal Arrhenius form:

$$k_A = A_e^{-E_a/RT} \tag{12.57}$$

Thus, the mass and energy equations are interdependent (not independent) since both contain T and X_A, and must therefore be solved simultaneously.

For adiabatic operation, heat transfer with the surroundings is zero. The energy equation reduces to

$$FC_p\, dT = |-r_A|(-\Delta H_A)\, dV \tag{12.58}$$

Since $|-r_A|\, dV = F_{A_0}\, dX_A$ for a TF reaction, the above equation becomes

$$FC_p\, dT = F_{A_0}(-\Delta H_A)\, dX_A \tag{12.59}$$

which may be integrated to give

$$T = T_0 + [F_{A_0}(-\Delta H_A)/FC_p]X_A \tag{12.60}$$

The term in brackets is a constant if the enthalpy of reaction and the average heat capacity are assumed independent of temperature. The temperature is then a *linear* function of conversion.

Fogler's equation(2) takes the same form as that provided in the previous section for CSTRs, i.e.,

$$\dot{Q} - \left[\Delta H_R^\circ(T_R) + \Delta C_p(T - T_R)\right]F_{A_0} X_A = F_{A_0}\Sigma\theta_i C_{pi}(T - T_{i0}) \tag{12.47}$$

■

ILLUSTRATIVE EXAMPLE 12.14 The gaseous reaction

$$C_2H_4 + H_2O \longrightarrow C_2H_5OH$$

is studied in a TF reactor. If the reactants enter at 47°C with a feed rate of 1600 lb/h consisting of 60 wt% C_2H_4, determine the exit temperature for which 56% of C_2H_4 is converted to products. The reactor operates adiabatically and the enthalpy of the reaction and heat of capacity of the reaction mixture can be assumed constant and equal to −11,000 cal/gmol and 0.436 cal/gmol · °C, respectively.

Solution. The energy equation for the adiabatic case is

$$FC_p\frac{dT}{dV} = F_A\left(\frac{-\Delta H_A}{M_A}\right)\frac{dX_A}{dV};\ F_A = \text{mass basis} \tag{12.59}$$

in consistent units. Rearranging the above

$$dT = \left(\frac{-\Delta H_A}{M_A C_p}\right)\left(\frac{F_A}{F}\right)dX_A$$

and integrating gives

$$T - T_0 = \left(\frac{-\Delta H_A}{M_A C_p}\right)\left(\frac{F_A}{F}\right)X_A$$

where F_A/F is the wt% A in the feed. Substituting,

$$T = 47 + \left[\frac{11{,}000}{(28)(0.436)}\right](0.6)(0.56) = 350^\circ\text{C}$$

■

ILLUSTRATIVE EXAMPLE 12.15 It is desired to use a 250 ft³ tubular flow reactor for a new liquid process to be run isothermally at 685°R.

$$2A \longrightarrow B;\ k_A = 200\ \text{ft}^3/\text{lbmol}\cdot\text{s} \quad \text{at } 685^\circ\text{R}$$

1. If a conversion of 80% is desired for an inlet concentration of A equal to 2.0×10^{-2} lbmol/ft³, calculate the feed rate of pure A in lbmol/s.

2. If the above reactor is run at 885°R, calculate the reaction velocity constant and new conversion at the same operating conditions. The energy of activation for the above reaction is 10,000 Btu/lbmol.

Solution.

1. The TF equation is

$$V = F_{A_0}\int\frac{dX}{(-r_A)}$$

$$-r_A = k_A C_A^2 = k_A C_{A_0}^2(1-X)^2$$

$$\frac{V}{F_{A_0}} = \frac{1}{k_A C_{A_0}^2}\int_0^X \frac{dX}{(1-X)^2}$$

Substituting and integrating gives

$$\frac{250}{F_{A_0}} = \frac{1}{(200)(10^{-4})(4.0)}\int_0^{0.8}\frac{dX}{(1-X)^2}$$

$$= \frac{1}{8\times 10^{-2}}\left[\frac{1}{1-X}\right]_0^{0.8}$$

$$= \frac{1}{8\times 10^{-2}}[5-1]$$

$$F_{A_0} = (250)(8\times 10^{-2})(4) = 5.0\,\text{lbmol/s}$$

2. If operated at 885°R, a new k is required. Apply the Arrhenius equation

$$k\,(885°\text{R}) = 200\,e^{\frac{10{,}000}{1.987}\left[\frac{1}{685}-\frac{1}{885}\right]}$$

$$= 1055\,\text{ft}^3/\text{lbmol}\cdot\text{s}$$

The CSTR equation is now

$$\frac{V}{F_{A_0}} = \frac{1}{k_A C_{A_0}^2}\left[\frac{1}{1-X}\right]_0^X$$

Substituting leads to

$$\frac{250}{5} = \frac{50^2}{1055}\left[\frac{1}{1-X}-1\right]21 = \frac{1}{1-X} - 1X = 0.955$$

■

ILLUSTRATIVE EXAMPLE 12.16 The elementary irreversible organic liquid-phase reaction

$$A + B \longrightarrow C$$

is carried out adiabatically in a tubular flow reactor. An equal molar feed of A and B enters at 27°C, and volumetric flow rate is $2\,\text{dm}^3/\text{s}$.

1. Obtain the temperature, T, as a function of the conversion, X.
2. What is the maximum *inlet* temperature one could have so that the boiling point of the liquid (550 K) will not be exceeded, even for complete conversion?
3. *Outline* how to calculate the volume necessary to achieve 85% conversion when the reaction is carried out adiabatically.

Additional data:

$$\Delta H_A^\circ(273\text{K}) = -20\,\text{kcal/gmol}$$
$$\Delta H_B^\circ(273\text{K}) = -15\,\text{kcal/gmol}$$
$$\Delta H_C^\circ(273\text{K}) = -41\,\text{kcal/gmol}$$

$$C_{A_0} = C_{B_0} = 0.1\,\text{kgmol/m}^3$$
$$C_{pA} = C_{pB} = 15\,\text{cal/gmol}\cdot\text{K},\ C_{PC} = 30\,\text{cal/gmol}\cdot\text{K}$$
$$k = 0.01\,\text{dm}^3/\text{gmol}\cdot\text{s at }300\,\text{K};\quad E = 10{,}000\,\text{cal/gmol}$$

Solution. Preliminary calculation

$$-r_A = kC_AC_B;\ C_A = C_{A_0}(1-X),\ C_B = C_{A_0}(1-X)$$
$$k_T = k(300)\exp\left[\frac{E}{R}\left(\frac{1}{300}-\frac{1}{T}\right)\right]$$

Since adiabatic, the describing energy equation is

$$T = T_0 + \frac{X[-\Delta H^\circ(T_R)]}{\Sigma\theta_i C_{pi} + X\Delta\tilde{C}_p} \tag{12.43}$$

1.

$$\begin{aligned}\Delta C_p &= C_{pC} - C_{pB} - C_{pA}\\ &= 30 - 15 - 15 = 0!\\ \Delta H^\circ(T_R) = \Delta H_R^\circ &= \Delta H_C^\circ - \Delta H_B^\circ - \Delta H_A^\circ\\ &= (-41{,}000) - (-15{,}000) - (-20{,}000)\\ &= -6{,}000\,\text{cal/gmol A}\quad\text{(and constant!)}\\ \Sigma\theta_i C_{pi} &= C_{pA} + \theta_B C_{pB}\\ &= 15 + 15 = 30\ \text{cal/gmol}\cdot\text{K}\end{aligned}$$

Substituting,

$$\begin{aligned}T &= 300 + \frac{6000X}{30}\\ &= 300 + 200X\end{aligned}$$

2. For this case,

$$T = T_o + \frac{(-\Delta H_R)X}{\Sigma \theta_i C_{pi}}$$

$$T = T_o + 200X$$

For maximum conversion ($X = 1$) and $T = 550$. Therefore

$$T_o = 550 - 200 = 350^\circ\text{K}$$

3. For a TF unit, use any suitable numerical method with the temperature expressed as a function of temperature in the rate term.

$$V = F_{A_0} \int \frac{dX}{-r_A} = \int \left(\frac{F_{A_0}}{-r_A} \right) dX; \qquad X = 0.85$$

■

ILLUSTRATIVE EXAMPLE 12.17 Refer to (3) in the previous illustrative example. Calculate the volume.

Solution. Apply the trapezoidal rule. (See also Chapter 21.) Results are provided in Table 12.2.

$$V = \frac{h}{3}\left[f(0) + 4f(1) + 2f(2) + 4f(3) + f(4) \right] + \frac{h}{2}\left[f(4) + f(5) \right]; \; X = 0.85$$

$$= \frac{0.3}{3}\left[2000 + (4)(435) + (2)(163) + (4)(103) + 146 \right] + \frac{0.05}{2}\left[146 + 206 \right]$$

$$= 308 + 8.8$$

$$= 317 \text{ dm}^3$$

■

ILLUSTRATIVE EXAMPLE 12.18 The elementary liquid-phase reaction

$$A + B \longrightarrow 2C + D$$

takes place in a 2-inch I.D. (inside diameter) tubular reactor with the following characteristics:

$$\dot{Q} = -45{,}000 \text{ Btu/h}$$

$$\Delta H^\circ(60^\circ\text{F}) = -1100 \text{ Btu/lbmol}$$

TABLE 12.2 Trapezoidal Results for Illustrative Example 12.17

X	T	k_A	$-r_A \times 10^4$	$F_{A_0}/-r_A$
0	300	0.01	1.0	$f(0) = 2000$
0.2	340	0.072	4.6	$f(1) = 435$
0.4	380	0.34	12.3	$f(2) = 163$
0.6	420	1.21	19.3	$f(3) = 103$
0.8	460	3.42	13.7	$f(4) = 146$
0.85	470	4.31	9.71	$f(5) = 206$

$$q_0 = 6.67\ \text{ft}^3/\text{min}$$
$$k_A = 0.172\ \text{ft}^3/\text{lbmol} \cdot \text{min}$$
$$C_{A_0} = C_{B_0} = 0.75\ \text{lbmol}/\ \text{ft}^3$$
$$C_{pA} = 4.0\ \text{Btu}/\text{lbmol} \cdot {}^\circ\text{F}$$
$$C_{pB} = 2.0\ \text{Btu}/\text{lbmol} \cdot {}^\circ\text{F}$$
$$C_{pC} = 8.0\ \text{Btu}/\text{lbmol} \cdot {}^\circ\text{F}$$
$$C_{pD} = 11.0\ \text{Btu}/\text{lbmol} \cdot {}^\circ\text{F}$$

Calculate the conversion if the inlet and outlet temperatures to the reactor are 70°F and 50°F, respectively. Assume k_A is constant. Also calculate the conversion if k_A is treated as a function of the temperature. Assume all other process variables remain the same.

Solution. The energy equation for a tubular flow reactor is

$$(\dot{Q}/F_{A_0}) - \left[\Delta H_R^\circ(T_R) + \Delta C_p(T - T_R)\right]X = \sum \theta_i C_{pi}(T - T_0) \qquad (12.43)$$

Calculate all the pertinent terms in the energy equation.

$$\dot{Q}/F_{A_0} = -45{,}000/(60)(6.67)(0.75)$$
$$= -150\ \text{Btu}/\text{lbmol}$$
$$\Delta H^\circ(T_R) = \Delta H^\circ(60^\circ\text{F})$$
$$= -1100\ \text{Btu}/\text{lbmol}$$
$$\Delta C_p = (8)(2) + 11 - 2 - 4$$
$$= 21\ \text{Btu}/\text{lbmol} \cdot {}^\circ\text{F}$$
$$T = 50^\circ\text{F}$$
$$T_R = 60^\circ\text{F}$$
$$T - T_R = -10^\circ\text{F}$$
$$\theta_A = \theta_B = 1.0$$
$$\sum \theta_i C_{pi} = 4.0 + 2.0$$
$$= 6.0\ \text{Btu}/\text{lbmol} \cdot {}^\circ\text{F}$$
$$T_0 = 70^\circ\text{F}$$
$$T - T_0 = 50^\circ\text{F} - 70^\circ\text{F}$$
$$= -20^\circ\text{F}$$

Calculate the conversion for the conditions specified by substituting into Equation (12.43)

$$-150 = -X[-1100 + (21)(-10)] = (6)(50 - 70)$$
$$-1310(-X) = 30$$
$$X = 30/1310 = 0.023 = 2.3\%$$

Repeat the above calculation if k is treated as a variable. However, since k does not appear in above equation(s), the conversion X remains the same. ■

ILLUSTRATIVE EXAMPLE 12.19 The elementary liquid-phase reaction

$$A + B \longrightarrow C + D$$

is carried out in a constant volume adiabatic tubular flow reactor. Using the following data, calculate the reaction temperature at $X_A = 0.8$ and the volume necessary to achieve this conversion.

$$\Delta H_R^\circ(T_R) = \Delta H_R = -10{,}000\,\text{cal/gmol of at } 25^\circ\text{C}$$
$$E = 5000\,\text{cal/gmol}$$
$$F_T = 100\,\text{gmol/h}$$
$$y_{A_0} = y_{B_0} = 0.5$$
$$C_{A_0} = C_{B_0} = 2\,\text{gmol/L}$$
$$C_{pA} = 30\,\text{cal/gmol}\cdot{}^\circ\text{K}$$
$$C_{pB} = 20\,\text{cal/gmol}\cdot{}^\circ\text{K}$$
$$C_{pC} = C_{pD} = 15\,\text{cal/gmol}\cdot{}^\circ\text{K}$$
$$k_0 = 0.51\,\text{gmol}\cdot\text{h at } 25^\circ\text{C}$$

Assume that the enthalpy of reaction is independent of temperature for the above calculation. Is this a reasonable assumption?

Solution. For a tubular flow adiabatic reactor

$$T = T_0 + \left[\frac{F_{A_0}(-\Delta H_R)}{\sum F_i C_{pi}}\right] X$$

Substituting

$$T = 298 + \left[\frac{(0.5)(100)(10{,}000)}{(30)(50) + (20)(50)}\right] X$$
$$= 298 + 200X$$

For $X = 0.8$

$$T = 458^\circ\text{K}$$

The TF equation is

$$F_{A_0}\frac{dX}{dV} = -r_A = kC_A C_B; \quad \theta_B = 1.$$

$$\frac{dX}{dV} = \frac{C_{A_0}^2}{F_{A_0}} k_o \cdot e^{-\frac{E}{R}\left[\frac{1}{T} - \frac{1}{T_0}\right]}(1 - X)^2$$

TABLE 12.3 Integration Details for Illustrative Example 12.19

X	I
0.0	25.00
0.1	18.16
0.2	14.38
0.3	12.39
0.4	11.63
0.5	11.98
0.6	13.84
0.7	18.69
0.8	32.71
0.9	103.98
1.0	large

Rearranging

$$V = \int_0^{0.8} \frac{dX}{\left\{ \frac{(4)(0.5)}{50} e^{\frac{-5000}{1.987}\left[\frac{1}{298+200X} - \frac{1}{298}\right]} (1-X)^2 \right\}}$$

$$= \int_0^{0.8} I\,dX; \quad \text{where} \quad I = \frac{1}{\{\quad\}}$$

Use the trapezoidal rule (or the equivalent). See Table 12.3 for details

$$V = 2.158 + 1.627 + 1.3385 + 1.201 + 1.1805 + 1.291 + 1.6265 + 2.570$$
$$= 12.9925\,\text{L} = 13\,\text{L}$$

■

ILLUSTRATIVE EXAMPLE 12.20 It is desired to carry out the following elementary liquid phase reaction in an adiabatic tubular flow reactor.

$$A + 2B \longrightarrow C$$

The following information is known:

$$C_{pA} = 15, \quad C_{pB} = 7.5, \quad C_{pC} = 30; \quad \text{cal/gmol}\cdot{}^\circ\text{C}$$
$$C_p\,(\text{mixture}) = 10\,\text{cal/gmol}\cdot{}^\circ\text{C}\,(\text{constant})$$
$$\Delta H_R^\circ = -6000\,\text{cal/gmol B at } 25^\circ\text{C}$$

Activation energy (E) = 6000 cal/gmol A

Arrhenius Frequency Factor (A) = 250 $(\text{cm}^3)^2/(\text{gmol})^2 \cdot \text{min}$

$$C_{A_0} = 0.05\,\text{gmol/cm}^3; \quad C_{B_0} = 0.15\,\text{gmol/cm}^3$$
$$F_{A_0} = 25\,\text{gmol/h}$$

Inlet reactor temperature $= 25\,^\circ\text{C}$

Determine the reactor volume needed for a 60% conversion of A. Also calculate the exit temperature.

Solution. The rate equation is

$$\begin{aligned} -r_A &= k_A C_A C_B^2 \\ &= k_A C_{A_0}^3 (1 - X)\left(\theta_B - \frac{b}{a}X\right)^2 \end{aligned}$$

$$\frac{b}{a} = \frac{2}{1} = 2; \quad \theta_B = \frac{C_{B_0}}{C_{A_0}} = \frac{0.15}{0.05} = 3.0$$

$$-r_A = k_A C_{A_0}^3 (1 - X)(3 - 2X)^2$$

The TF equation is then

$$V = F_{A_0}\int_0^X \frac{dX}{-r_A} = \frac{F_{A_0}}{C_{A_0}^3}\int_0^3 \frac{dX}{k_A(1 - X)(3 - 2X)^2}$$

Obtain the Arrhenius equation as a function of temperature

$$\begin{aligned} k &= Ae^{-E/RT} \\ &= 250\,e^{-\frac{6000}{(1.987)T}} \end{aligned}$$

Combining the above two equation yields

$$V = \frac{F_{A_0}}{C_{A_0}^3}\int_0^{0.6} \frac{dX}{250\,e^{-\frac{6000}{1.987T}}(1 - X)(3 - 2X)^2}$$

Since the temperature T is varying, T must be expressed in terms of X. By applying an energy balance (as before) one ultimately obtains

$$T = 298 + 300X$$

At $X = 0.6$, T is

$$\begin{aligned} T &= 298 + (300)(0.6) \\ &= 478\text{K} \end{aligned}$$

The above integral for V now becomes

$$V = \frac{F_{A_0}}{C_{A_0}^3} \int_{0.0}^{0.6} \frac{dX}{250\, e^{-\frac{6000}{1.987(298+300X)}}(1 - X)(3 - 2X)^2}$$

$$= \frac{F_{A_0}}{C_{A_0}^3} \int_{0.0}^{0.6} \frac{dX}{I}$$

$$= \frac{F_{A_0}}{C_{A_0}^3} \int Y\, dX; \quad Y = \frac{1.0}{I}$$

Using Simpson's rule (see Chapter 21 once again)

$$\int_0^X f(X) = \frac{h}{3}\left[Y(0) + 4Y(1) + 2Y(2) + 4Y(3) + 2Y(4) + 4Y(5) + Y(6)\right]; \quad h = 0.1$$

Results are provided in Table 12.4. Therefore

$$\int Y\, dX = \frac{0.1}{3}\left[11.18 + 4(5.65) + 2(3.41) + 4(2.38) + 2(1.89) + 4(1.69) + 1.71\right]$$

$$= 2.1$$

The volume is then

$$V = \left(\frac{F_{A_0}}{C_{A_0}^3}\right)(2.1); \quad F_{A_0} = 25/60 = 0.417\ \text{gmol/min}$$

$$= (0.417)(2.1)/(0.05)^3 = 7000\ \text{cm}^3 = 7.0\ \text{L}$$

■

TABLE 12.4 Simpson Rule Values for Illustrative Example 12.20

Point	X	Y
0	0	11.18
1	0.1	5.65
2	0.2	3.41
3	0.3	2.38
4	0.4	1.89
5	0.5	1.69
6	0.6	1.71

REFERENCES

1. L. Theodore, F. Ricci, and T. VanVliet, "*Thermodynamics for the Practicing Engineer*," John Wiley & Sons, Hoboken, NJ, 2009.
2. S. Fogler, "*Elements of Chemical Reactor Engineering*," 4th edition, Prentice-Hall, Upper Saddle River, NJ, 2006.
3. L. Theodore and F. Ricci, "*Mass Transfer Operations for the Practicing Engineer*," John Wiley & Sons, Hoboken, NJ, 2010.

CHAPTER 13

INTERPRETATION OF KINETIC DATA

INTRODUCTION

The application of chemical kinetic principles to the design of reactors for the chemical process industry is hindered in many cases by the lack of information on the rate of reaction. Therefore, one of the most important aspects of chemical kinetics is the prediction of the form of the reaction rate equation from experimental data. This is a very difficult task and is, for all intents and purposes, still a state of art. Yet, it is worthwhile to present and discuss some of the elementary methods employed in determining reaction mechanisms. Information on the reaction mechanism can provide a complete description of the behavior of a reacting system.

Kinetic data can be obtained experimentally in any one of several different ways. The choice is usually dictated by

1. convenience,
2. reaction system (species),
3. availability of equipment,
4. simplicity in experimental arrangement,
5. cost of experimentation, and
6. time of experimentation.

Not necessarily in that order. Some of the more common experimental methods of analysis include:

1. Refractive index
2. Colorimetric indicators—dyes, tracers, etc
3. Thermal conductivity—gas chromatograph
4. Dielectric constant
5. Electric conductance
6. Total pressure

All of these and other methods, excluding (6) require calibration charts or the equivalent: e.g., a plot of refractive index vs. concentration is needed with (1).

Chemical Reactor Analysis and Applications for the Practicing Engineer. By Louis Theodore

Laboratory data are usually generated by any one of the following experimental procedures.

1. In a batch reactor of constant volume and constant temperature.
2. In a variable-volume batch reactor operated at constant temperature and pressure.
3. In a tubular flow reactor at constant temperature and (for gas phase reactions) constant pressure.
4. In a CSTR at a constant temperature.

As one would expect, procedure (1) is the usual choice.

In addition to the introductory section, this chapter discusses the following topics:

Experimental Methods and Analysis of Kinetic Data

Method of Least Squares

Application to Specific Reactors

Reactions of Complex Mechanism

EXPERIMENTAL METHODS AND ANALYSIS OF KINETIC DATA

There are two experimental methods of operation and both are applicable to CSTRs and TF reactors, as well as batch reactors. In one, the data are obtained over a very small interval of reaction time—so small, in fact, that it is defined as a *differential reaction system*. Changes in concentration are usually so small that they may be assumed constant during the short period of reaction time. This permits a direct calculation of the rate. The rate calculated may be interpreted as an average rate since it is based on average concentration(s) and an average temperature. This represents an acceptable method of obtaining information on the rate equation. However, experimental problems can arise in the measurement of small concentration changes. This technique is used for systems in which it is difficult to maintain isothermal conditions. In the second method, concentration (or pressure) data are gathered over a long interval of time. This mode of operation is defined as an *integral reaction system*. The analytical methods of analyses to follow will focus on integral reaction systems.

Assume in the development below that concentration (or the equivalent) data are available as a function of time. There are two analytical techniques that can be used to analyze the experimental data. These are the:

1. differentiation method, and
2. integration method.

In the *differentiation method*, not to be confused with a differential reaction system, solutions have traditionally been obtained using graphical or equivalent means. Consider the elementary irreversible reaction

$$nA \longrightarrow \text{products} \tag{13.1}$$

The rate equation is

$$\frac{dC_A}{dt} = -k_A C_A^n \tag{13.2}$$

If C_A is provided as a function of time, one can plot C_A vs. t. The rate of change of C_A with respect to t on the LHS of Equation (13.2) may be obtained by drawing a tangent to the curve or calculating the slope at that point in time (see Figure 13.1). The rate, dC_A/dt, will then be available as a function of time. Equation (13.2) is now written as

$$-\frac{dC_A}{dt} = k_A C_A^n \tag{13.3}$$

Taking the logarithm of both sides of Equation (13.3) gives

$$\log\left(-\frac{dC_A}{dt}\right) = \log k_A + n \log C_A \tag{13.4}$$

Equation (13.4) indicates that a plot of $\log\left(-dC_A/dt\right)$ vs. $\log C_A$ [or $\ln(-dC_A/dt)$ vs. $\ln C_A$] will yield a straight line (if the assumed mechanism above is correct) of slope n and intercept $\log k_A$ (see Figure 13.2). Thus, one sees that this method, with the generation of two graphs, provides information on both the order of the reaction and the numerical value of the reaction velocity constant. An alternate, but not as accurate a method, is to arbitrarily choose two data points and calculate n and k_A from the equation

$$n = \frac{\log\left(-\frac{dC_A}{dt}\right)_2 - \log\left(-\frac{dC_A}{dt}\right)_1}{\log C_{A_1} - \log C_{A_2}} \tag{13.5}$$

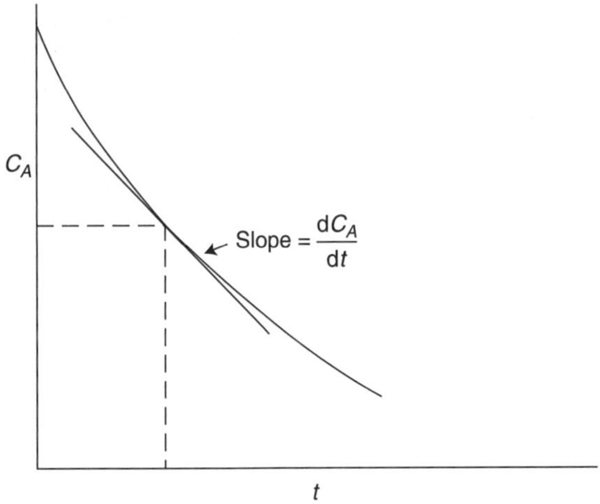

Figure 13.1 Change of concentration with respect to time.

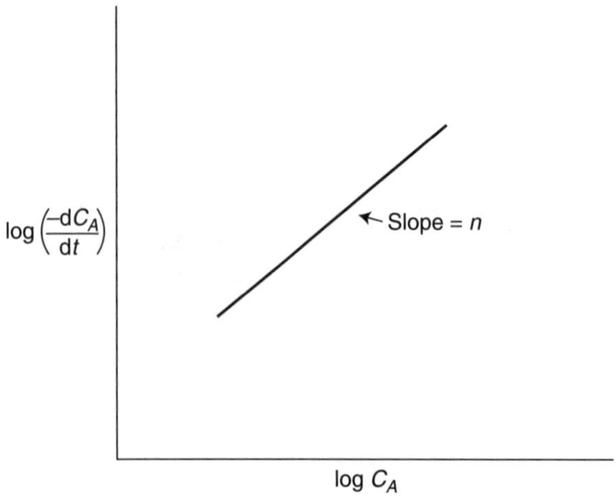

Figure 13.2 Order of reaction from rate–concentration plot.

and

$$k_A = \frac{\log\left(-\frac{dC_A}{dt}\right)_1}{C_{A_1}^n} \tag{13.6}$$

or

$$k_A = \frac{\log\left(-\frac{dC_A}{dt}\right)_2}{C_{A_2}^n}. \tag{13.7}$$

The *integration method* requires that the rate equation be integrated analytically for an assumed reaction mechanism or order. The integrated rate equation compares most favorably with the experimental data then corresponds to the correct equation. Once again, consider Equation (13.1):

$$nA \longrightarrow \text{products}$$

The rate equation is now written in the form

$$\frac{dC_A}{C_A^n} = -k_A dt \tag{13.8}$$

The above equation may be integrated to give

$$k_A = \frac{1}{t(n-1)}\left[\frac{1}{C_A^{n-1}} - \frac{1}{C_{A_0}^{n-1}}\right]; \quad n \neq 1 \tag{13.9}$$

If $n = 2$, then

$$k_A = \frac{1}{t}\left[\frac{1}{C_A} - \frac{1}{C_{A_0}}\right] \tag{13.10}$$

Values of k_A are calculated from Equation (13.10) for various values of C_A vs. t data. The assumed order is correct if relatively constant values of k_A are generated from the data. This procedure can be repeated for different orders until the above criterion is met. Thus, it can be seen that the integration method requires a trial-and-error calculation.

The application of either of the above two methods to reaction rate equations containing more than one concentration term can be somewhat complex. Consider the equation (with C_A, C_B vs. t data) available

$$iA + jB \longrightarrow \text{products} \tag{13.11}$$

For the differentiation technique, Equation (13.4) becomes

$$\log\left(-\frac{dC_A}{dt}\right) = \log k_A + i \log C_A + j \log C_B \tag{13.12}$$

Three unknowns appear in this equation; a trial-and-error graphical calculation is required. For example, one can assume values of j until a straight line log–log plot of $\left(-dC_A/dt\right)$ vs. C_A is obtained. The integrated form of the rate equation may also be employed but is also complex. If one assumes $i = 1$ and $j = 1$, then

$$\frac{dC_A}{dt} = -k_A C_A C_B \tag{13.13}$$

which may be integrated to give

$$k_A = \frac{1}{t(C_{A_0} - C_{B_0})} \ln \frac{C_{B_0} C_A}{C_{A_0} C_B} \tag{13.14}$$

A constant value of k_A for the data would indicate that the assumed order ($i = 1, j = 1$) is correct.

The *isolation method* has often been applied to a mixed reaction system as in Equation (13.11). The procedure employed here is to set the initial concentration of one of the reactants, say B, so large that the change in concentration of B during the reaction is vanishingly small. The rate equation may then be approximated by

$$\frac{dC_A}{dt} \approx k_A^* C_A^i \tag{13.15}$$

where

$$k_A^* = k_A C_B \approx \text{constant} \tag{13.16}$$

Either the differentiation or integration method may be used with Equation (13.16). For example, in the differentiation method, a log–log plot of $-dC_A/dt$ vs. C_A will yield a straight line of slope i. A similar procedure is employed to obtain j.

The method of *fractional* or *half-lives* has limited application. Here, the time required to react a certain fraction of a reactant is obtained for different initial conditions. A modified form of either the integration or differentiation techniques is then applied to the data.

ILLUSTRATIVE EXAMPLE 13.1 The half-life of a certain compound undergoing a first order irreversible reaction is 30 min under certain conditions. What is the reaction rate constant in both engineering and metric units?

Solution. For this reaction, write

$$A \longrightarrow \text{products}$$

with

$$t_{1/2} = 30\text{ min} = 1800\text{ s}$$

One can show that for half of A reacting,

$$t_{1/2} = \left(\frac{1}{k_A}\right)\ln(2)$$

Solving for k_A gives

$$k_A = \frac{\ln 2}{t_{1/2}} = \frac{0.693}{1800}(\text{s})^{-1} = 3.85 \times 10^{-4}(\text{s})^{-1}$$

The metric and engineering units for this reaction velocity constant are identical. ■

ILLUSTRATIVE EXAMPLE 13.2 In the very early stages of the *elementary* reversible reaction

$$A + B \underset{k'_A}{\overset{k_A}{\rightleftharpoons}} 2C$$

r_A is found to be -2.42×10^{-3} gmol/L · s, and at equilibrium there is an 18% conversion of A. What is the rate equation for this reaction? The initial concentrations of A and B are each 0.5 gmol/L.

Solution. For this reaction,

$$\frac{dC_A}{dt} = -k_A C_A C_B + k'_A C_C^2$$

In the early stages of this reaction, $C_C \approx 0$. Therefore,

$$-2.42 \times 10^{-3} = -k_A(0.5)(0.5)$$

and

$$k_A = 9.68 \times 10^{-3}$$

The stoichiometric equation (at equilibrium) is now written

$$\begin{array}{ccccc} 0.5 & & 0.5 & & 0 \\ A & + & B & \rightleftharpoons & 2C \\ 0.5 - 0.09 & & 0.5 - 0.09 & & 0.18 \end{array}$$

The equilibrium constant based on concentration is

$$K = \frac{(0.18)^2}{(0.41)(0.41)} = 0.193$$

since

$$K = \frac{k_A}{k'_A}$$

$$k'_A = \frac{9.68 \times 10^{-3}}{0.193} = 50.2 \times 10^{-3}$$

Therefore,

$$r_A = \frac{\mathrm{d}C_A}{\mathrm{d}t} = -9.68 \times 10^{-3} C_A C_B + 50.2 \times 10^{-3} C_C^2$$

■

ILLUSTRATIVE EXAMPLE 13.3 The following elementary reaction was studied experimentally:

$$A + 2B \longrightarrow \text{products}$$

The initial mole fractions (consistent units) of the reactants were $A = 0.99$, $B = 0.01$. A plot of $(C_B)^{-1}$ vs. time gave a slope of 1.0 $(\text{min})^{-1}$. Calculate the rate constant for B.

Solution. For the above reaction,

$$r_B = \frac{\mathrm{d}C_B}{\mathrm{d}t} = -k_B C_A C_B^2$$

Since the concentration of A is so much greater than that of B, it can be assumed that it remains constant during the reaction. Therefore,

$$\frac{\mathrm{d}C_B}{\mathrm{d}t} = -k_B^* C_B^2$$

where

$$k_B^* = C_A k_B \approx 0.99\, k_B$$

The integrated form of the above equation is

$$\frac{1}{C_B} - \frac{1}{C_{B_0}} = k_B^* t$$

Substituting gives

$$\frac{1}{C_B} - \frac{1}{0.01} = 0.99\, k_B t$$

or

$$\frac{1}{C_B} = 100 + (0.99\, k_B)t$$

Based on the graphical results,

$$1.0 = 0.99\, k_B$$

so that

$$k_B \approx 1.0\, (\text{min})^{-1}$$

■

Finally, when an engineer/scientist finds it necessary to perform statistical analyses of experimental data, he/she probably has little or no knowledge of the reaction mechanism. He/she is therefore more often interested in obtaining information on

1. how well the proposed reaction mechanism is, rather than,
2. how well the data fit the proposed model (reaction mechanism) or in obtaining the best fit of the data to the model by regressing the data.

A least-squares analysis,[1] or the equivalent, provides no information to (1). An analysis of variance (ANOVA)[1] must be performed to answer (1). The reader is referred to the statistics literature for details on ANOVA.

METHOD OF LEAST SQUARES

It is no secret that many statistical calculations are now performed with spreadsheets or packaged programs. This statement is particularly true with regression analysis. The result of this approach has been to reduce or eliminate one's fundamental understanding of this subject. This section attempts to correct this shortcoming.

Engineers and scientists often encounter applications that require the need to develop a mathematical relationship between data for two or more variables. For example, if Y (a dependent variable) is a function of or is dependent on X (an independent variable), i.e.,

$$Y = f(X) \tag{13.17}$$

one may be required to express this (X,Y) data in equation form. This process is referred to as *regression analysis*, and the regression method most often employed is the method of *least squares*.

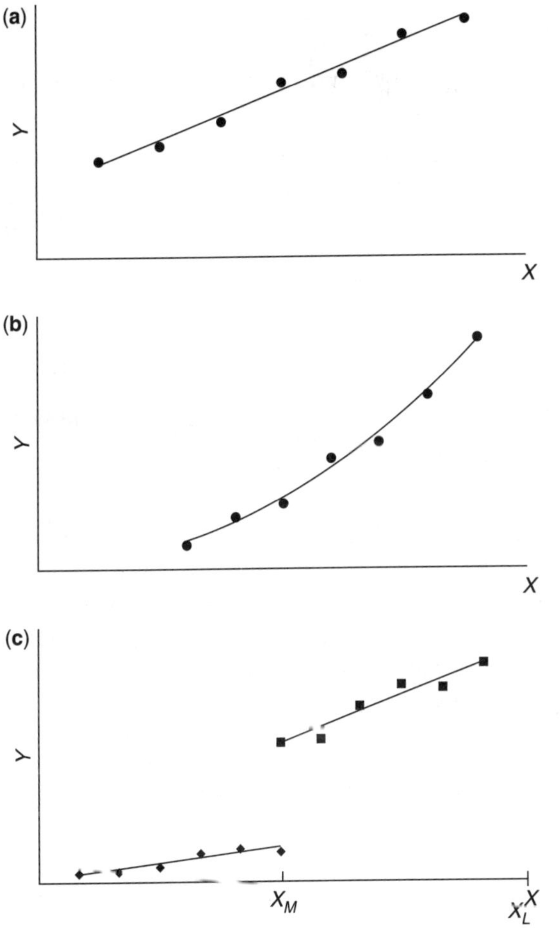

Figure 13.3 Scatter diagrams: (a) linear relationship, (b) parabolic relationship, and (c) dual-linear relationship.

An important step in this procedure—which is often omitted—is to prepare a plot of Y vs. X. The result referred to as a *scatter* diagram, could take on any form. Three such plots are provided in Figures 13.3(a)–(c). The first plot (a) suggests a linear relationship between X and Y, i.e.,

$$Y = a_0 + a_1 X \tag{13.18}$$

The second graph (b) appears to be best represented by a second order (or parabolic) relationship, i.e.,

$$Y = a_0 + a_1 X + a_2 X^2 \tag{13.19}$$

The third plot suggests a linear model that applies over two different ranges, i.e., it should represent the data

$$Y = a_0 + a_1 X; \quad X_0 < X < X_M \tag{13.20}$$

and

$$Y = a_0' + a_1'X; \quad X_M < X < X_L \tag{13.21}$$

This multiequation model finds application in representing adsorption equilibria, multiparticle size distributions and quantum energy relationships. In any event, a scatter diagram and individual judgment can suggest an appropriate model at an early stage in the analysis.

Some of the models often employed by technical individuals are as follows:[1]

1.	$Y = a_0 + a_1X$	Linear	(13.22)
2.	$Y = a_0 + a_1X + a_2X^2$	Parabolic	(13.23)
3.	$Y = a_0 + a_1X + a_2X^2 + a_3X^3$	Cubic	(13.24)
4.	$Y = a_0 + a_1X + a_2X^2 + a_3X^3 + a_4X^4$	Fourth order	(13.25)

Procedures to evaluate the regression coefficients a_0, a_1, a_2, etc., are provided below. The reader should note that the analysis is based on the method of least squares. This technique provides numerical values for the regression coefficients a_i such that the sum of the square of the difference (error) between the actual Y and the Y_e predicted by the equation or model is minimized. This is shown in Figure 13.4 with the dots (experimental value of Y) and triangles (equation or model value of Y, i.e., Y_e) represent the data and model values, respectively. On examining the two figures that apply to the same data, one can immediately conclude that the error $(Y - Y_e)$ squared and summed for the four points is less for the lower figure. Also note that a dashed line represents the error. The line that ultimately produces a minimum of the sum of the individual errors squared, i.e., has its smallest possible value, is the regression model (based on the method of least squares). The proof is left as an exercise.

To evaluate a_0 and a_1 for a linear model (see Equation 13.22) one employs the following least-squares algorithm for n data points of Y and X.

$$a_0 n + a_1 \Sigma X = \Sigma Y \tag{13.26}$$

$$a_0 \Sigma X + a_1 \Sigma X^2 = \Sigma XY \tag{13.27}$$

All the quantities given, except a_0 and a_1 can be easily calculated from the data. Because there are two equations and two unknowns, the set of equations can be solved for a_0 and a_1. For this case,

$$a_1 = \frac{n\Sigma XY - \Sigma X \Sigma Y}{n\Sigma X^2 - (\Sigma X)^2} \tag{13.28}$$

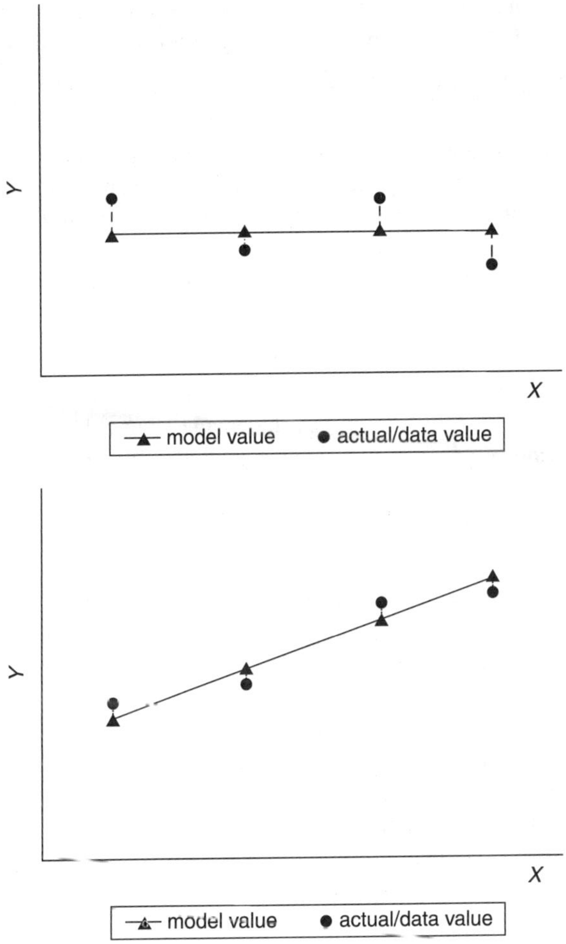

Figure 13.4 Error difference: actual and predicted values.

Dividing numerator and denominator by n, and defining $\overline{X} = \Sigma X/n$ and $\overline{Y} = \Sigma Y/n$, leads to

$$a_1 = \frac{\Sigma XY - \dfrac{\Sigma X \Sigma Y}{n}}{\Sigma X^2 - \dfrac{(\Sigma X)^2}{n}} = \frac{\Sigma XY - n\overline{X}\,\overline{Y}}{\Sigma X^2 - n\overline{X}^2} \tag{13.29}$$

Using this value of a_1 produces the following equation for a_0:

$$a_0 = \overline{Y} - a_1\overline{X} \tag{13.30}$$

If the model (or line of regression) is forced to fit through the origin, then the calculated value of $Y_e = 0$ at $X = 0$. For this case, the line of regression takes the form

$$Y_e = a_1 X; \quad a_0 = 0 \tag{13.31}$$

with

$$a_1 = \Sigma XY \big/ \Sigma X^2 \tag{13.32}$$

A cubic model takes the form

$$Y = a_0 + a_1 X + a_2 X^2 + a_3 X^3 \tag{13.24}$$

For n pairs of $X - Y$ values, the constants a_0, a_1, a_2, and a_3 can be obtained by the method of least squares so that $\Sigma(Y - Y_e)^2$ again has the smallest possible value, i.e., is minimized. The coefficients a_0, a_1, a_2, and a_3 are the solution of the following system of four linear equations:

$$\begin{aligned} a_0 n + a_1 \Sigma X + a_2 \Sigma X^2 + a_3 \Sigma X^3 &= \Sigma Y \\ a_0 \Sigma X + a_1 \Sigma X^2 + a_2 \Sigma X^3 + a_3 \Sigma X^4 &= \Sigma XY \\ a_0 \Sigma X^2 + a_1 \Sigma X^3 + a_2 \Sigma X^4 + a_3 \Sigma X^5 &= \Sigma X^2 Y \\ a_0 \Sigma X^3 + a_1 \Sigma X^4 + a_2 \Sigma X^5 + a_3 \Sigma X^6 &= \Sigma X^3 Y \end{aligned} \tag{13.33}$$

Because there are four equations and four unknowns, this set of equations can be solved for a_0, a_1, a_2, and a_3. This development can be extended to other regression equations, e.g., exponential, hyperbola, or higher order models.

The correlation coefficient provides information on how well the model, or line of regression, fits the data. It is denoted by r and is given by

$$r = \frac{\Sigma XY - \dfrac{\Sigma X \Sigma Y}{n}}{\sqrt{\left[\Sigma X^2 - \dfrac{(\Sigma X)^2}{n}\right]\left[\Sigma Y^2 - \dfrac{(\Sigma Y)^2}{n}\right]}} \tag{13.34}$$

or

$$r = \frac{n\Sigma XY - \Sigma X \Sigma Y}{\sqrt{n\left[\Sigma X^2 - (\Sigma X)^2\right]\left[n\Sigma Y^2 - (\Sigma Y)^2\right]}} \tag{13.35}$$

or

$$r = \frac{\Sigma XY - n\overline{X}\overline{Y}}{\sqrt{\left(\Sigma X^2 - n\overline{X}^2\right)\left(\Sigma Y^2 - n\overline{Y}^2\right)}} \tag{13.36}$$

This equation can also be shown to take the form

$$r = \pm\left[\frac{\Sigma(\overline{Y} - Y_e)^2}{\Sigma(Y - \overline{Y})^2}\right]^{0.5} \tag{13.37}$$

The correlation coefficient satisfies the following six properties.

1. If all points of a scatter diagram lie on a line, then $r = +1$ or -1. In addition, $r^2 = 1$. The square of the correlation coefficient is defined as the coefficient of determination.
2. If no linear relationship exists between the X's and Y's, then $r = 0$. Furthermore, $r^2 = 0$. It can be concluded that r is always between -1 and $+1$, and r^2 is always between 0 and 1.
3. Values of r close to $+1$ or -1 are indicative of a strong linear relationship.
4. Values of r close to 0 are indicative of a weak linear relationship.
5. The correlation coefficient is positive or negative depending on whether the linear relationship has a positive or negative slope. Thus, positive values of r indicate that Y increases as X increases; negative values indicate that Y decreases as X increases.
6. If $r = 0$, it only indicates the lack of linear correlation; X and Y might be strongly correlated by some nonlinear relation, as discussed earlier. Thus, r can only measure the strength of linear correlations; if the data are nonlinear, one should attempt to linearize before computing r.

It should be noted that the correlation coefficient only provides information on how well the model fits the data. It is emphasized that r provides no information on how good the model is or, to reword this, whether this is the *correct* or *best* model to describe the functional relationship of the data.

ILLUSTRATIVE EXAMPLE 13.4 Chemical A is undergoing a reaction in a specially controlled laboratory experiment. The data in Table 13.1 have been obtained for the reaction rate, $-r_A$, vs. concentration, C_A. Using the data, estimate the coefficient k_A and α in the equation below:

$$-r_A = k_A C_A^{\alpha}$$

Solution. For the problem at hand, linearize the equation by taking the natural logarithm (ln) of both sides of the equation.

$$-r_A = k_A C_A^{\alpha}$$
$$\ln(-r_A) = \ln(k_A) + \alpha \ln(C_A)$$

Change the variables to Y and X, so that

$$Y = A + BX$$

TABLE 13.1 Reaction Rate Data

Reaction Rate, $-r_A$ ($lbmol/ft^3 \cdot s$)	Concentration, C_A ($lbmol/ft^3$)
48	8
27	6
12	4
3	2

Regress the preceding four data points using the method of least squares, where $A = \ln(k_A)$ and $B = \alpha$. Once again, the method of least squares requires that the sum of the errors squared between the data and the model is minimized.

$$A = -0.2878$$
$$B = 2.0$$

Take the inverse natural logarithm of A to obtain k_A:

$$A = \ln k_A = 0.75$$
$$B = 2.0$$

The equation for the rate of reaction is therefore

$$-r_A = 0.75C_A^{2.0}$$ ■

APPLICATION TO SPECIFIC REACTORS

The purpose of this section is to apply the introductory statistics materials to each of the traditional three reactors. The emphasis is on batch reactors since most of the experimental work conducted for the purpose of generating kinetic models, employs batch reactors. As such, most of the illustrative examples to follow are concerned with this class of reactor. Examples on CSTRs and tubular reactors are also presented in two later subsections.

Batch Reactors

As noted in an earlier chapter, the variables affecting the rate of reaction in a reactor are the concentration of the species in the reacting system, temperature, pressure, and catalyst effects. Experimentation and the subsequent analysis of the data involves an attempt to isolate the variables and determine mathematical relationships between them.

In a very real sense, the development material provided in the section titled "Experimental Methods and Analysis of Kinetic Data" applies directly to batch reactors and will therefore not be repeated here. Developmental material for both CSTRs and TF reactors differs from that provided for batch reactors and receives treatment in the two subsections to follow.

To summarize, either *differential* or *integral* batch reactors are employed during the data gathering process. Differential units operate for short periods of time; there is usually one data point per run and the extent of the conversion is low with little or no temperature variation problems. The integral unit operates for long reaction times generating either one data point per run or many; the extent of conversion will usually be high with temperature variation a potential problem.

Ten batch reactor illustrative examples follow.

ILLUSTRATIVE EXAMPLE 13.5 Given a homogeneous-phase batch polymerization occurring at a constant temperature, with 20% of the monomer disappearing within 34 minutes for initial monomer concentrations of both 0.04 and 0.8 gmol/L. What can be said about the rate of disappearance of the monomer, i.e., what is the order of the reaction and the reaction velocity constant?

Solution. For this batch reactor,

$$t = C_{A_0} \int \frac{\mathrm{d}X}{-r_A}; \qquad -r_A = kC_A^n$$

If $n = 0$

$$t = \frac{C_{A_0}X}{k}; \quad \text{reject since dependent on } C_{A_0}$$

If $n = 1$

$$t = -\frac{1}{k}\ln(1 - X); \quad \text{independent of } C_{A_0}$$

Therefore

$$n = 1.0$$

Solving for k,

$$k = -\frac{1}{t}\ln(1 - X)$$

$$k = -\frac{1}{34}\ln(1 - 0.2)$$
$$= 6.56 \times 10^{-3}\,(\text{min})^{-1}$$

■

ILLUSTRATIVE EXAMPLE 13.6 The aqueous-phase reaction

$$A \underset{k'}{\overset{k}{\rightleftharpoons}} R + S$$

proceeds as follows,

Time, min	0	36	65	100	160	∞
C_A, gmol/L	0.1823	0.1453	0.1216	0.1025	0.0795	0.0494

with

$$C_{A_0} = 0.1823 \text{ gmol/L}$$
$$C_{R_0} = 0$$
$$C_{S_0} \approx 55 \text{ gmol/L}$$

Is the reverse reaction important?

Solution. For this reversible reaction *at equilibrium*,

$$K = K_{eq} = \frac{k}{k'} = \frac{C_R C_S}{C_A}$$
$$= \frac{(0.1823 - 0.0494)(55)}{0.0494}$$
$$= 148$$

This large number suggests that $k \ggg k'$, i.e., the reverse reaction proceeds slowly relative to the forward reaction. Therefore, the reverse reaction can be neglected in most analyses. ■

ILLUSTRATIVE EXAMPLE 13.7 A small reaction bomb, fitted with a sensitive pressure-measuring device is flushed out and then filled with a gaseous mixture of 76.94% reactant A and 23.06% inert at 1 atm pressure. The operation is carried out at 14°C, a temperature low enough that the reaction does not proceed to any appreciable extent.

The temperature is raised as rapidly as possible to 100°C by plunging the bomb into boiling water, and the readings below are obtained. The stoichiometric equation for the reaction is

$$A \rightleftharpoons 2R$$

After leaving the bomb in the bath over the weekend the contents were analyzed for A; none were found. Outline a method to obtain a rate equation in concentration units of lbmol/ft^3 which will satisfactorily fit the data provided in Table 13.2.

TABLE 13.2 Time–Total Pressure Data for Illustrative Example 13.7

t, min	P, atm
0.5	1.5
1	1.65
1.5	1.76
2	1.84
2.5	1.90
3	1.95
3.5	1.99
4	2.025
5	2.08
6	2.12
7	2.15
8	2.175

Solution. Since no reactant was found when $t \rightarrow \infty$, the reaction is irreversible. One need only consider the gaseous reaction

$$A \longrightarrow 2R$$

with

$$y_{A_0} = 0.7694$$
$$\delta = 2 - 1 = 1$$
$$\varepsilon = 0.7694$$

and proceed as suggested earlier. ■

ILLUSTRATIVE EXAMPLE 13.8 The reaction

$$A + B \longrightarrow C$$

takes place in a constant volume batch reactor. Using the time-concentration data below in Table 13.3, determine the order of the reaction and the specific reaction rate constant by the *integration* method. Also, obtain the reaction rate equation.

The initial concentration of B is 18.25 gmol/L.

Solution. Develop the solution in terms of X. For this case,

$$C_A = C_{A_0}(1 - X); \quad X = 1 - \frac{C_A}{C_{A_0}}; \qquad \theta_B = \frac{18.25}{35.35} = 0.52$$

Assume the following mechanism

$$-r_A = k_A C_A^{\alpha} C_B^{\beta}$$

TABLE 13.3 Time-Concentration Data for Illustrative Example 13.8

t (min)	C_A (gmol/L)
0	35.35
4.75	30.5
10	27.0
20	23.2
35	20.3
54	18.6
∞	17.1

where

$$C_A = C_{A_0}(1 - X)$$
$$C_B = C_{A_0}(\theta_B - X)$$

and

$$-r_A = k_A C_{A_0}^{\alpha+\beta}(1 - X)^{\alpha}(\theta_B - X)^{\beta}$$

For a batch reactor, one may write

$$C_{A_0}\frac{dX}{dt} = -r_A = k_A C_{A_0}^{\alpha+\beta}(1 - X)^{\alpha}(\theta_B - X)^{\beta}$$
$$\frac{dX}{dt} = k_A \frac{C_{A_0}^{\alpha+\beta}}{C_{A_0}}(1 - X)^{\alpha}(\theta_B - X)^{\beta}$$

Assume $\alpha = \beta = 1$ so that

$$\frac{dX}{dt} = k_A C_{A_0}(1 - X)^{1.0}(\theta_B - X)^{1.0}$$

and integrate,

$$k_A C_{A_0}\int_0^t dt = \int_0^X \frac{dX}{(1 - X)(\theta_B - X)}$$

See also Illustrative Example 21.18 in Part IV.

$$k_A C_{A_0} t = \frac{1}{(\theta_B - 1.0)}\ln\left[\frac{(\theta_B - X)}{\theta_B(1 - X)}\right]$$
$$k_A = \frac{1}{(35.35)(t)}\left(\frac{1}{0.52 - 1}\right)\ln\left[\frac{(0.52 - X)}{0.52(1 - X)}\right]$$

Calculate X from C_{A0}, C_A data, and substitute time into the above equation to calculate k. The results are provided in Table 13.4. Since k_A is relatively constant, the reaction is therefore second order with $\alpha = 1$, $\beta = 1$. The average value of k is

TABLE 13.4 Data and Calculations for Illustrative Example 13.8

t (min)	C_A (gmol/L)	X	k_A
0	35.35	0	–
4.75	30.5	0.137	0.001966
10	27.0	0.236	0.001978
20	23.2	0.344	0.001950
35	20.3	0.426	0.001946
54	18.6	0.474	0.001947
∞	17.1	0.516	–

$$k_A = 0.001957\,\text{L/gmol} \cdot \text{min}$$

so that

$$-r_A = 0.001957\,C_A C_B$$

■

ILLUSTRATIVE EXAMPLE 13.9 The stoichiometry of the liquid phase reaction between trimethylamine and n-propyl-bromide is

$$N(CH_3)_3 + CH_3CH_2CH_2Br \rightleftharpoons (CH_3)_3(CH_2CH_2CH_3)N^+ + Br^-$$

The kinetic data in Table 13.5 were obtained at 40°C in a batch reactor using an initial charge of 0.2 gmol/L for *each* reactant.

Use the data below to test the rate expression for *irreversible* kinetics by determining the rate constant at 40°C and the order of the reaction. Obtain the rate expression for this reaction.

Hint: The reaction is known to be of integer order.

Solution. For an irreversible reaction one need only consider

$$A + B \longrightarrow P; \qquad A = N(CH_3)_3, \quad B = CH_3CH_2CH_2Br$$

Assume an elementary reaction so that

$$-r_A = kC_AC_B = kC_A^2 = kC_{A_0}^2(1 - X)^2$$

TABLE 13.5 Time–Conversion Data Illustrative Example 13.9

Time (min)	13	34	59	120
Conversion (%)	11.2	25.7	36.7	55.2

As developed before for a batch reactor,

$$-r_A = -\frac{dC_A}{dt} = C_{A_0}\frac{dX}{dt}$$

$$C_{A_0}\frac{dX}{dt} = kC_{A_0}^2(1-X)^2$$

$$kC_{A_0}\int dt = \int\frac{dX}{(1-X)^2}$$

$$kC_{A_0}t = \left[\frac{1}{1-X} - 1\right] = \frac{X}{1-X}; \qquad C_{A_0} = 0.2\,\text{gmol/L}$$

One can now calculate k, or plot $[X/1-X]$ vs. t which will yield a slope of kC_{A_0}. Calculations are provided in Table 13.6.

From a plot (near linear)

$$\text{slope} = 0.0102\,(\text{min})^{-1}$$

so that the reaction is elementary (and second order) and

$$k = \frac{0.0102}{C_{A_0}} = \frac{0.0102}{0.2} = 0.05\,\text{L/gmol}\cdot\text{min}$$

Or, using the integration approach, calculate k using the above equation

$$k = \frac{1}{C_{A_0}t}\left[\frac{X}{1-X}\right]$$

For run#1; $\quad k\dfrac{1}{(0.2)(13)}\left[\dfrac{0.112}{0.888}\right] = 0.53$

For run#2; $\quad = 0.49$

etc.

Finally, one *cannot* write the rate expression since the reaction is second order overall. ■

TABLE 13.6 Calculations For Illustrative Example 13.9

t (min)	X	$X/(1-X)$
13	0.112	0.126
34	0.257	0.346
59	0.367	0.580
120	0.552	1.23

ILLUSTRATIVE EXAMPLE 13.10 Refer to the previous illustrative example. How else could the problem have been solved?

Solution. Using the differential approach,

$$\frac{dX}{dt} = kC_{A_0}^{n-1}(1 - X^n)$$

Plot X vs. t to obtain dX/dt. Then, taking the ln of both sides of the equation,

$$\underbrace{\ln\frac{dX}{dt}}_{} = \underbrace{\ln\left[kC_{A_0}^{n-1}\right]}_{} + \underbrace{n}_{}\ \underbrace{\ln(1 - X)}_{}$$

One may now plot Y vs. X to generate the coefficients k and n. ■

ILLUSTRATIVE EXAMPLE 13.11 The following gas phase reaction is carried out *isothermally* in a constant volume batch reactor:

$$2A \longrightarrow B + C + D$$

Initially there is pure A in the reactor at STP (32°F, 1 atm). The following pressures in Table 13.7 were recorded at subsequent times during the reaction.

1. Calculate δ and ε.
2. Develop an expression for the conversion in terms of the total pressure.
3. Is the reaction reversible? Explain.
4. Verify that the order of the reaction is approximately 1.4, and then calculate the specific reaction rate constant using any convenient method.

Solution. Rewrite the gas phase reaction

$$2A \longrightarrow B + C + D$$

as

$$A \longrightarrow (1/2)B + (1/2)C + (1/2)D$$

TABLE 13.7 Time–Pressure Data for Illustrative Example 13.11

Time (min)	0	1.20	1.95	2.90	4.14	5.70	8.10	∞
Total pressure (atm)	1	1.10	1.15	1.20	1.25	1.30	1.35	1.51

Preliminary calculations:

$$\begin{aligned}
y_{A_0} &= 1.0 \\
\delta &= (3/2) - 1 = 1/2 \\
\varepsilon &= (1.0)(1/2) = 1/2 = 0.5 \\
V &= V_0(1 + \varepsilon X)(T_0/T)(P/P_0);\ T_0/T = 1.0 \\
&= V_0(1 + \varepsilon X)(P/P_0)
\end{aligned}$$

Since the volume is constant, $V = V_0$, so that

$$\begin{aligned}
X &= \frac{P - P_0}{\varepsilon P_0}; \qquad P_0 = 1.0 \\
&= 2P - 2
\end{aligned}$$

For a batch reactor,

$$r_A = \frac{1}{V}\frac{dN_A}{dt} = \frac{dC_A}{dt}; \qquad \text{constant volume}$$

Assume the mechanism to be of the form

$$r_A = kC_A^n$$

Since

$$C_A = C_{A_0}(1 - X)$$

$$-C_{A_0}\frac{dX}{dt} = kC_A^n$$

One now needs both C_A and X expressed in terms of P.

$$C_A = C_{A_0}\left[1 - \left(\frac{P - P_0}{\varepsilon P_0}\right)\right] = \frac{C_{A_0}}{\varepsilon P_0}[(1 + \varepsilon)P_0 - P]$$

Therefore

$$-r_A = k\left(\frac{C_{A_0}}{\varepsilon P_0}\right)^n [(1 + \varepsilon)P_0 - P]^n = kC_A^n \tag{1}$$

Also note that since

$$X = \frac{P - P_0}{\varepsilon P_0}$$

the derivative is

$$\frac{dX}{dt} = \frac{1}{\varepsilon P_0}\frac{dP}{dt} \tag{2}$$

Combining (1) and (2) yields

$$\frac{\mathrm{d}P}{\mathrm{d}t} = k\left(\frac{C_{A0}}{\varepsilon P_0}\right)^{n-1}\left[(1+\varepsilon)P_0 - P\right]^n \tag{3}$$

Use any approach to generate $\mathrm{d}P/\mathrm{d}t$ information (see also Table 13.8).

Employing (3), one requires a log–log plot of $\mathrm{d}P/\mathrm{d}t$ vs. $[(1+\varepsilon)P_0 - P]$, i.e., $[1.5 - P]$. The slope is then n. For $n = 1.4$,

$$\text{intercept} = k\left(\frac{C_{A0}}{\varepsilon P_0}\right)^{n-1} = k\left(\frac{1}{\delta RT}\right)^{n-1}$$

and

$$k = 1.04\ \ (\text{gmol})^{0.4}(\text{min})^{-1}$$

For example, at

$$\frac{\mathrm{d}P}{\mathrm{d}t} = 0.01\ \text{atm/min}$$

$$(1+\varepsilon P_0) - P = 0.513\ \text{atm}$$

and

$$k = \frac{\left[(1.5)(0.082)(273)\right]^{0.4}(0.01)}{(0.513)^{1.4}} = 1.04$$

One may also numerically integrate (3) with $n = 1.4$ and calculate k and observe if the calculated values of k are constant. ■

TABLE 13.8 Generating $\mathrm{d}P/\mathrm{d}t$

t	P	$\Delta P/\Delta t$	$\mathrm{d}P/\mathrm{d}t$ (from plot)
0	1.0		0.0968
		0.0833	
1.2	1.1		0.0718
		0.0667	
1.95	1.15		0.0584
		0.0537	
2.90	1.20		0.0472
		0.0403	
4.14	1.25		0.0368
		0.0320	
5.70	1.30		0.0277
		0.0208	
8.10	1.35		0.0158

ILLUSTRATIVE EXAMPLE 13.12 Dimethyl ether decomposes by an irreversible first order reaction. A certain amount of this ether is placed in a batch reactor and allowed to decompose according to the reaction

$$CH_3\,OCH_3 \longrightarrow CH_4 + H_2 + CO$$

At the end of one week, a gaseous mixture is produced which analyzes 25 mol% methane. Calculate the reaction rate constant for this gas phase reaction. Use the conversion variable α rather than X or C in the solution.

Solution. Choose a basis of 1.0 mol of ether initially present.

$$\begin{array}{lllllllll} 1 & & 0 & & 0 & & 0\;; & & \Sigma = 1.0 \\ CH_3\,OCH_3 & \longrightarrow & CH_4 & + & H_2 & + & CO & & \\ (A) & & (B) & & (C) & & (D) & & \\ 1-\alpha_A & & \alpha_A & & \alpha_A & & \alpha_A\;; & & \Sigma = 1.0 + 2\alpha_A \end{array}$$

From the example statement

$$0.25 = y_B$$

$$0.25 = \frac{\alpha_A}{1 + 2\alpha_A}$$

$$\alpha_A = 0.5$$

For this first order reaction

$$\frac{dC_A}{dt} = -k_A C_A$$

Integrating and rearranging gives

$$k_A = -\frac{\ln(C_A/C_{A_0})}{t}$$

$$= -\frac{\ln(N_A/N_{A_0})}{t} = -\frac{\ln[(N_{A_0} - \alpha_A)N_{A_0}]}{t}$$

$$= -\frac{\ln[1.0 - 0.5)/1.0]}{168}\;; \qquad 24 \times 7 = 168\,\text{h/week}$$

$$k_A = 4.12 \times 10^{-3}(\text{h})^{-1}$$

Since $\alpha_A = 0.5$, the half-life formula

$$k_A = \frac{\ln 2}{t_{1/2}}$$

could also have been used, i.e.,

$$k_A = \frac{\ln 2}{168}$$
$$= 4.12 \times 10^{-3}$$

■

ILLUSTRATIVE EXAMPLE 13.13 The irreversible liquid phase reaction

$$A + B \xrightarrow{k} C + D$$

is carried out in a batch reactor. Reactants A and B are initially placed in the reactor in an equimolar ratio. The conversion of A is measured at various times during the course of the reaction; see Table 13.9. Determine the overall order of the reaction and the reaction velocity constant k. Use the integration approach since the overall order in known to be an integer value.

$$\text{Data:} \qquad C_{A_0} = C_{B_0} = 2.0 \text{ gmol/L}$$

Solution. Write the rate of reaction, $-r_A$, in terms of C_A and C_B.

$$-\mathrm{d}C_A/\mathrm{d}t = -r_A$$
$$= kC_A^{\alpha}C_B^{\beta}; \qquad \alpha + \beta = n = \text{overall order}$$

Express C_A and C_B in terms of X and rewrite the above equation.

$$C_A = C_{A_0}(1 - X)$$
$$C_B = C_{A_0}(\theta_B - X); \qquad \theta_B = 1.0$$
$$\mathrm{d}C_A/\mathrm{d}t = -kC_{A_0}^n(1 - X)^n$$

The rate of reaction equation may also be written in terms of $\mathrm{d}X/\mathrm{d}t$. Since

$$\mathrm{d}C_A/\mathrm{d}t = -C_{A_0}(\mathrm{d}X/\mathrm{d}t)$$
$$\mathrm{d}X/\mathrm{d}t = kC_{A_0}^{n-1}(1 - X)^n$$

TABLE 13.9 Time–Conversion Data for Illustrative Example 13.13

t (s)	X
15	0.473
30	0.626
45	0.696
60	0.736

TABLE 13.10 Calculated k Values; $n = 0$

X	t (s)	k (gmol/L · s)
0.473	15	0.063
0.627	30	0.042
0.696	45	0.031
0.736	60	0.025

Separate the variables in this equation.

$$dX/(1 - X)^n = kC_{A_0}^{n-1}\,dt$$

Assume, as a hypothesis, a zero order reaction and integrate the above equation.
For $n = 0$,

$$\int dX = (k/C_{A_0}) \int dt$$

$$X = (k/C_{A_0})t$$

$$k = C_{A_0}X/t$$

Calculated values of k are provided in Table 13.10. Trend! Therefore, reject the hypothesis that $n = 0$.

Assume, as a hypothesis, a first order reaction ($n = 1$) and integrate the above equation.

$$\int [1/(1 - X)dX = \int k\,dt$$

$$-\ln(1 - X) = kt$$

$$k = -\ln(1 - X)/t$$

Calculated values of k are provided in Table 13.11. Trend! Therefore, reject the hypothesis that $n = 1$.

TABLE 13.11 Calculated k Values: $n = 1$

X	t (s)	k (s)$^{-1}$
0.473	15	0.043
0.627	30	0.033
0.696	45	0.027
0.736	60	0.022

TABLE 13.12 Calculated k Values; $n = 2$

X	t (s)	k (L/gmol · s)
0.473	15	0.063
0.627	30	0.045
0.696	45	0.037
0.736	60	0.032

Assume, as a hypothesis, a second order ($n = 2$) reaction and integrate the equation as before.

$$\int [1/(1-X)^2]\mathrm{d}X = kC_{A_0}\int \mathrm{d}t$$

$$-[1/(1-X)] = kC_{A_0}t$$

$$k = (1/C_{A_0}t)[1/(1-X)]$$

Calculated values of k are provided in Table 13.12. Trend! Therefore, reject the hypothesis that $n = 2$.

Assume, as a hypothesis, a third order $n = 3$ reaction and integrate the equation.

$$\int [1/(1-X)^3]\mathrm{d}X = kC_{A_0}^2\int \mathrm{d}t$$

$$-[1/2(1-X)^2] = kC_{A_0}^2 t$$

$$k = (1/2C_{A_0}^2 t)[1/(1-X)^2]$$

Calculated values of k are provided in Table 13.13. Is the third order reaction hypothesis correct? No Trend! Accept the hypothesis that $n = 3$.

The results clearly show that this is a third order reaction overall. Because of the form of the data, i.e., $C_{A_0} = C_{B_0}$ and $C_A = C_B$ at all times, no conclusions can be drawn regarding the order of the reaction with respect to A and B individually. ■

ILLUSTRATIVE EXAMPLE 13.14 The following gas phase irreversible reaction:

$$A + B \longrightarrow 3C$$

TABLE 13.13 Calculated k Values; $n = 3$

X	t (s)	k (L^2/gmol^2 · s)
0.473	15	0.0600
0.627	30	0.0599
0.696	45	0.0601
0.736	60	0.0599

TABLE 13.14 Time-Pressure Data for Illustrative Example 13.14

Time (min)	Total pressure (atm)
0	12.20
4.163	14.03
12.0	15.25
96.0	17.08

occurs in a constant volume isothermal batch reactor. The reactor was initially charged with 0.5 gmol/L of A and 0.5 gmol/L of B. Prove that the overall order of the reaction is third order and determine the specific reaction rate constant by using the integration method. Also determine the time required to achieve a conversion of 90% in this reaction. Time-total pressure data is provided above in Table 13.14.

Assume a reaction mechanism of the form:

$$-r_A = kC_A^{\alpha}C_B^{\beta}$$

Solution. For this reaction,

$$-r_A = -\frac{dC_A}{dt} = kC_A^{\alpha}C_B^{\beta}$$

One can either convert this equation to "P" form, or convert the data to C_A or X form. For this example, convert the above equation to X form, i.e., convert the data from P to X. Use the earlier equation (See Illustrative Example 13.8).

$$C_{A_0}\frac{dX}{dt} = k_A C_{A_0}^{\alpha}(1-X)^{\alpha}C_{A_0}^{\beta}(\theta_B - X)^{\beta}; \qquad \theta_A = \theta_B = 1$$

$$\frac{dX}{dt} = k_A C_{A_0}^{\alpha+\beta-1}(1-X)^{\alpha+\beta} = k_A C_{A_0}^{n-1}(1-X)^{n}$$

Separate and integrate,

$$\int \frac{dX}{(1-X)^n} = k_A C_{A_0}^{n-1}\int dt$$

For $n = 3$

$$k = k_A = \int_0^X \frac{dX}{(1-X)^3} \Big/ tC_{A_0}^2 = \left[\frac{1}{2(1-X)^2} - \frac{1}{2}\right] tC_{A_0}$$

Since this is a gas, constant volume, and isothermal reaction,

$$1.0 = \frac{P_0}{P}1.0(1+\varepsilon X); \qquad \varepsilon = (3-1-1)(0.5) = 0.5$$

TABLE 13.15 Calculations for Illustrative Example 13.14

t, min	P, atm	X	k
4.613	14.03	0.3	0.5000
12.00	15.25	0.5	0.5000
96.00	17.08	0.8	0.5000

Solving for X,

$$X = (P - 12.2)/(12.2)(0.5); \qquad P_0 = 12.2$$
$$= (P - 12.2)/6.1$$

One can now calculate k as provided above in Table 13.15. Therefore,

$$k = k_A = 0.050001 \text{ L}^2/\text{gmol} \cdot \text{min}\,; \text{ 3rd order reaction}$$

Note that $\alpha + \beta = 3$, but values of α and β once again cannot be determined with the data provided.

For $X = 0.9$, $C_{A_0} = 0.5$, $k_A = 0.500$,

$$t = 395.95 = 396 \text{ min} = 6.6 \text{ h}$$ ■

CSTR Reactors

Tank flow reactors are occasionally used to obtain information on reaction velocity constant(s) and order(s) of reaction. These reactors are operated under conditions approaching perfect mixing and steady-state. The same two basic methods of analysis are available: the *integration* and *differentiation* methods, although there is no need to integrate or differentiate the describing equations.

The integration method is first presented. Consider the irreversible reaction

$$nA \longrightarrow \text{B (product)} \tag{13.38}$$

where

$$-r_A = k_A C_A^n \tag{13.39}$$

For a CSTR, the describing equation becomes

$$V/q = \tau = (C_{A_0} - C_{A_1})/k_A C_{A_1}^n \tag{13.40}$$

which may also be written

$$k_A = (C_{A_0} - C_{A_1})/\tau C_{A_1}^n \tag{13.41}$$

If τ (or q and V), C_{A_0}, and C_{A_1} experimental data are available for the system, one can assume an order n until the calculated values for k_A become "constant" and do not exhibit a trend in any direction.

In the differentiation method, the above equation is written in the following form:

$$k_A C_{A_1}^n = (C_{A_0} - C_{A_1})/\tau \tag{13.42}$$

Taking the log of both sides of this equation gives

$$\log k_A + n \log C_{A_1} = \log [(C_{A_0} - C_{A_1})/\tau] \tag{13.43}$$

This is a linear equation of the form

$$A + BX = Y \tag{13.44}$$

Therefore, a plot of $[(C_{A_0} - C_{A_1})/\tau]$ vs. C_{A_1} on log–log coordinates will provide direct information on both k_A and n.

To summarize, both *differential* and *integral* reactors are employed with CSTRs to generate experimental data.

The differential unit is generally small, operating with high flow rates and generating one data point (DP) per run so that both the residence time τ and conversion are low (see Figure 13.5). The describing equations remain.

$$-r_A = \frac{F}{V}\Delta X_A$$

or

$$-r_A = \frac{q_0(C_{A_0} - C_A)}{V}$$

One may now use the graphical or integration method of analysis.

Integration units are generally large with low flow rates and with both high residence time and high conversion; more than one data is generated with each run.

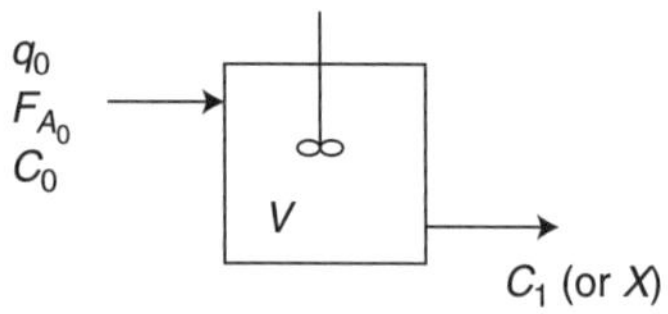

DP	q_0, F_{A_0}, C_0	C_1 (or X)
1	q_0'	C_1'
2	q_0''	C_1''
3	q_0'''	C_1'''
⋮	⋮	⋮
n	q_0^n	C_1^n

Figure 13.5 Differential CSTR unit.

TABLE 13.16 Data for Illustrative Example 13.15

$(C_{A_0} - C_{A_1})/\tau$, lbmol/ft^3 · min	C_{A_1}, lbmol/ft^3
3.0	2.0
12	4.0
27	6.0
48	8.0

ILLUSTRATIVE EXAMPLE 13.15 The irreversible liquid phase reaction

$$nA \longrightarrow \text{products}$$

is carried out in a CSTR. Using a modified differentiation approach (regressing data using the method of least squares), determine the order of the reaction, n. The data in Table 13.16 have been obtained for C_{A_0} (inlet concentration of A to the CSTR), C_{A_1} (outlet concentration), and τ (residence time).

Solution. Write the rate of reaction, $-r_A$, in terms of the concentration, C_{A_1}.

$$-r_A = k_A C_A^n; \qquad C_{A_1} = C_{A_0}(1 - X)$$

The design equation for a CSTR is

$$V/q = \tau = (C_{A_0} - C_{A_1})/k_A C_{A_1}^n$$

Rearrange this design equation in terms of $(C_{A_0} - C_{A_1})/\tau$.

$$kC_{A_1}^n = (C_{A_0} - C_{A_1})/\tau$$

Linearize the equation.

$$\ln(k) + n(\ln C_{A_1}) = \ln[(C_{A_0} - C_{A_1})/\tau]$$

Evaluate the coefficients in the linear equation by regressing the data. The reader should note that this calculation was performed previously in Illustrative Example 13.4. The values of n and k are

$$\ln k = -0.2878$$
$$k = 0.75$$
$$n = 2.0$$

■

ILLUSTRATIVE EXAMPLE 13.16 Refer to the previous illustrative example. Write the rate expression for this reaction.

Solution. This is a second-order reaction and

$$-r_A = 0.75\, C_A^2$$

■

Tubular Flow Reactors

Tubular flow reactors are also employed to obtain information on the reaction mechanism, particularly for gas phase reactions. Like batch reactors, experimental data can be obtained using a *differential* or *integral* reactor and the analysis performed using the *differentiation* or *integration* method.

In a differential reactor, inlet and outlet concentrations (or the equivalent) and flow rate data are obtained for a *small* reactor volume. A number of runs are made varying the inlet concentration. Changes in concentration must be extremely small so that both the concentration and rate may be assumed constant in the reactor. This permits a direct calculation of the rate assuming the CSTR equation applies.

$$-r_A = F_{A_0}X_A/V = \frac{q_0(C_{A_0} - C_A)}{V} \quad (13.45)$$

with

$$C_A = (C_{A_0} + C_A)/2 \quad (13.46)$$

If the differentiation method of analysis is employed for an assumed mechanism, e.g.,

$$-r_A = k_A C_A^n \quad (13.47)$$

then a plot of $-r_A$ vs. C_A on log–log coordinates can yield the reaction velocity constant k_A and order n. In the integration scheme, one assumes n and checks to see whether calculated values for k_A are constant. Data requirements are presented in Table 13.17.

If the small differences in concentration at the inlet to and outlet from the reactor cannot be experimentally determined accurately, the above differential method is not satisfactory and an integral reactor system is employed. Here, the concentration at various locations and flow rate data are gathered for a larger reactor. The flow and/or inlet concentration may be varied. This information is presented in Figure 13.6 and Table 13.18.

For the differentiation method of analysis, Equation (13.45) is written in differential form as

$$dX_A/d(V/F_{A_0}) = -r_A \quad (13.48)$$

TABLE 13.17 Required Data for a Differential Reactor

Data point	q_0, F_{A_0}, C_{A_0}	C_A (or X_A)
0		
1		
2		
⋮		
n		

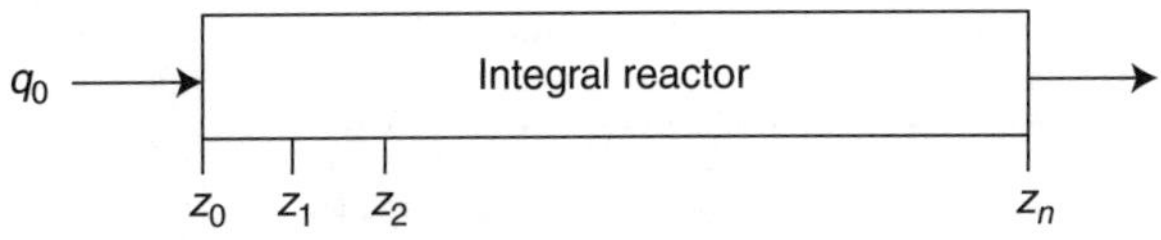

Figure 13.6 Integral tubular flow reactor.

TABLE 13.18 Required Data For an Integral Reactor

Data point	Position, z	C_A (or X_A)
0	z_0	C_{A_0}
1	z_1	C_{A_0}
2	z_2	C_{A_1}
⋮	⋮	⋮
n	z_n	C_{A_n}

and a plot of X_A vs. V/F_{A_0} yields information on $-r_A$. Refer to Figure 13.7. A log–log plot of $-r_A$ vs. C_A will then provide information on k_A and n. This plot was presented earlier in Figure 13.2. In the integration method of analysis, one integrates the design equation for an assumed n and checks to see if the calculated values for k_A are constant.

To summarize, both *differential* and *integral* reactors are employed with TF reactors to generate experimental data. The differential unit is generally small, operating with high flow rates, generating one data point per run, so that both the residence time τ and conversion are low. Integration units are generally large with low flow rates and with both high residence time and high conversion; more than one data is usually generated with each run.

ILLUSTRATIVE EXAMPLE 13.17 The following gaseous reaction

$$A \longrightarrow \tfrac{1}{2}P$$

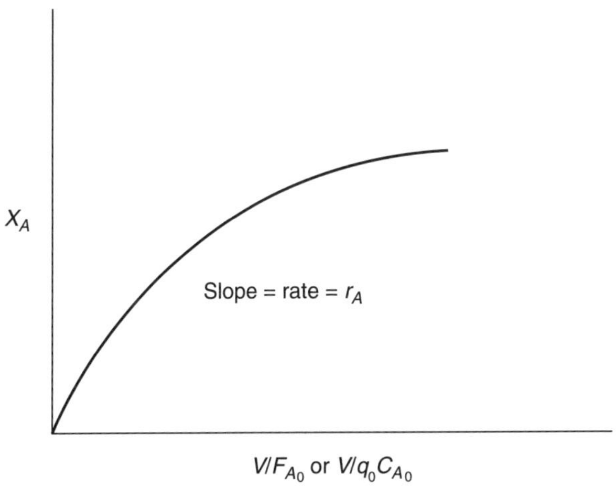

Figure 13.7 Integral reactor plot.

TABLE 13.19 Conversion–Residence Time Data for Illustrative Example 13.17

X	τ_o (s)
0.20	1120
0.40	2610
0.60	4915
0.80	10025

is carried out isothermally in a plug flow reactor. Only pure A is fed to the reactor. The conversion of A was measured at various values of the inlet space (or residence) time τ_0 (see Table 13.19). Calculate the order of the reaction α and the reaction velocity constant k_A using the integration approach. Assume a mechanism of the form

$$-r_A = k_A C_A^n$$

where n can have an integer value, including zero.

Data: $C_{A_0} = 0.02\,\text{gmol/L}$

Solution. Write the rate of reaction, $-r_A$, in terms of the conversion, X, for this gas phase reaction,

$$-r_A = k_A C_A^n; \qquad C_A = C_{A_0}(1 - X)/(1 + \varepsilon X)$$
$$= kC_{A_0}^n[(1 - X)/(1 + \varepsilon X)]$$

The design equation for a plug flow tubular reactor is once again,

$$V = F_{A_0}\int_0^X (-1/r_A)\,dX; \qquad F_{A_0} = C_{A_0}q_0$$

Assume, as a hypothesis, a zero-order reaction and integrate the design equation. Set $n = 0$.

$$V = C_{A_0}q_0\int_0^X (1/k_A)\,dX$$
$$= C_{A_0}q_0X/k$$
$$k = C_{A_0}q_0/V$$
$$= C_{A_0}X/\tau$$

Calculated values of k are provided in Table 13.20. Trend! Therefore, reject the zero-order hypothesis.

Assume, as a hypothesis, a first-order ($n = 1$) reaction and integrate the design equation.

TABLE 13.20 Calculated Values of k; $n = 0$

X	τ (s)	k (gmol/L · s)
0.2	1120	3.6×10^{-6}
0.4	2610	3.1×10^{-6}
0.6	4915	2.4×10^{-6}
0.8	10025	1.6×10^{-6}

For this gas phase reaction,

$$y_{A_0} = 1.$$
$$\delta = 0.5 - 1.0 = -0.5$$
$$\varepsilon = (1)(-0.5) = -0.5$$

The design equation becomes

$$\tau = C_{A_0} \int (1/kC_A)\,\mathrm{d}X$$
$$= C_{A_0} \int [(1 + \varepsilon X)/kC_{A_0}(1 - X)]\,\mathrm{d}X$$
$$= (1/k) \int_0^X [(1 + \varepsilon X)/(1 - X)]\,\mathrm{d}X$$
$$= (1/k)\{(1 + \varepsilon)\ln[1/(1 - X)] - \varepsilon X\}; \text{ see also Chapter 21}$$

Calculated values of k are provided in Table 13.21. Trend! Therefore, reject the first order hypothesis.

Assume, as a hypothesis, a second order ($n = 2$) reaction and integrate the design equation.

$$\theta = C_{A_0} \int (1/kC_A^2)\,\mathrm{d}X$$
$$= C_{A_0} \int \left[(1 + \varepsilon X)^2/kC_{A_0}^2(1 - X)^2\right] \mathrm{d}X$$
$$k = (1/\theta C_{A_0})[2\varepsilon(1 + \varepsilon)\ln(1 - X) + \varepsilon^2 X + (1 + \varepsilon)^2 X/(1 - X)]; \text{ see Chapter 21}$$

TABLE 13.21 Calculated Values of k; $n = 1$

X	τ (s)	k (s)$^{-1}$
0.2	1120	1.89×10^{-4}
0.4	2610	1.75×10^{-4}
0.6	4915	1.54×10^{-4}
0.8	10025	1.20×10^{-4}

TABLE 13.22 Calculated Values of k; $n = 2$

X	τ (s)	k (L/gmol · s)
0.2	1120	0.010
0.4	2610	0.010
0.6	4915	0.010
0.8	10025	0.010

Calculated values of k are provided in Table 13.22. There is no trend. Accept the second order hypothesis, and

$$-r_A = k_A C_A^2; \qquad k_A = 0.010 \text{ L/gmol} \cdot \text{s}$$ ■

ILLUSTRATIVE EXAMPLE 13.18 The following information in Table 13.23 were obtained for the reaction.

$$A + 2B \xrightarrow{k_A} P$$

$$C_{A_0} = 0.268 \text{ gmol/L}; \qquad C_{B_0} = 0.379 \text{ gmol/L}$$

Determine the reaction order and the rate constant using the integration method of analysis.

Solution. First assume a first-order reaction in A so that

$$r_A = -k_A C_A = -k_A C_{A_0}(1 - X_A)$$

The describing equation becomes

$$\frac{V}{F_A} = \int_0^{X_A} \frac{dX_A}{k_A C_{A_0}(1 - X_A)}$$

$$= -\frac{1}{k_A C_{A_0}} \ln(1 - X_A)$$

TABLE 13.23 V/F_A vs X_A Data Information for Illustrative Example 13.18

V/F_A (L · s/gmol)	C_A (gmol/L)	X_A (calc)
13.2	0.233	0.132
14.5	0.1995	0.257
15.8	0.1750	0.348
16.9	0.1575	0.413
19.6	0.1324	0.507

Solving for k_A gives

$$k_A = -\frac{\ln(1 - X_A)}{C_{A_0}(V/F_A)}$$

For the first data point

$$k_A = -\frac{\ln(1 - 0.132)}{(0.268)(13.2)} = 4.00 \times 10^{-2}\,(s)^{-1}$$

For the other data points in $(s)^{-1}$:

$$\begin{aligned} k_A &= 7.63 \times 10^{-3} \\ &= 9.43 \times 10^{-3} \\ &= 12.90 \times 10^{-3} \\ &= 13.45 \times 10^{-3} \end{aligned}$$

There is a trend. Since there is an increasing trend in k_A, the above assumption is false.

Assume a second order reaction of the form

$$r_A = -k_A C_A C_B$$

where

$$\begin{aligned} C_A &= C_{A_0}(1 - X_A) \\ C_B &= C_{B_0} - 2C_{A_0}X_A \end{aligned}$$

The describing equation is then

$$\frac{V}{F_A} = \int_0^{X_A} \frac{dX_A}{k_A C_{A_0}(1 - X_A)(C_{B_0} - 2C_{A_0}X_A)}$$

Integrating results in the following equation for k_A.

$$k_A = -\frac{1}{C_{A_0}(V/F_A)}\left[\frac{1}{(-0.536 + 0.379)}\ln\left(\frac{0.379 - 0.536X_A}{1 - X_A}\right)\right]\Bigg|_0^{X_A}$$

The calculated values for k_A are 1.86, 1.85, 1.83, 1.85, 1.85, with $\bar{k}_A = 1.85$ L/gmol · s. This assumed mechanism is correct. ■

REACTIONS OF COMPLEX MECHANISM

Stoichiometric reaction equations often produce rate equations that are not elementary but rather of a complex mechanism. The resulting rate expressions are frequently so complex that empirical equations are employed. These equations may be used satisfactorily when the reverse reaction is negligible, equilibrium is not a consideration, and in the absence of a chain and/or consecutive reactions.

For example, the mechanism of a reaction may be defined by a number of elementary reactions that together represent the overall reaction depicted in the stoichiometric equation. Consider the first order reaction

$$2N_2O_5 \longrightarrow 2N_2O_4 + O_2 \tag{13.49}$$

Based on the stoichiometric equation, this would appear to be a second order reaction. However, the actual mechanism is given by

$$N_2O_5 \longrightarrow NO + NO_2 + O_2 \tag{13.50}$$

$$NO + NO_2 + N_2O_5 \longrightarrow 2N_2O_4 \tag{13.51}$$

The second reaction is very rapid relative to the first; it may be assumed at equilibrium. The first reaction is, therefore, the rate controlling reaction, and this is a first order reaction. Although the actual mechanism is rarely described by the stoichiometric equation in industrial reactions, the stoichiometric equation is usually employed to describe the reaction mechanism in this text.

If a reaction has a complex reaction mechanism, the rate equation can become unwieldly and difficult to handle. These may contain fractional or even negative reaction order terms. It is sometimes near impossible or inconvenient to treat these systems in the manner outlined above. The application of statistical principles can provide quantitative solutions to the equation(s).[2]

The application of any interpretation of kinetic data approach to reaction rate equations containing more than one concentration term can be somewhat complex. As presented earlier in the chapter consider the reaction

$$iA + jB \longrightarrow \text{products}$$

where

$$\frac{dC_A}{dt} = -k_A C_A^i C_B^j$$

For the differentiation technique, this equation is written

$$\log\left(-\frac{dC_A}{dt}\right) = \log k_A + i \log C_A + j \log C_B$$

Three unknowns appear in this equation and can be solved by regressing the data; a trial-and-error graphical calculation is required. Using the integration technique is also complex. If one assumes $i = 1$ and $j = 1$, then (as shown earlier)

$$\frac{dC_A}{dt} = -k_A C_A C_B \tag{13.52}$$

which may be integrated to give

$$k_A = \frac{1}{t(C_{A_0} - C_{B_0})} \ln\left(\frac{C_{B_0} C_A}{C_{A_0} C_B}\right) \tag{13.53}$$

A constant value of k_A for the data would indicate that the assumed order ($i = 1, j = 1$) is correct.

The isolation method has often been applied to the above mixed reaction system. And, as noted earlier in the chapter, the procedure employed here is to set the initial concentration of one of the reactants, say B, so large that the change in concentration of B during the reaction is vanishingly small. The rate equation may then be approximated by

$$\frac{dC_A}{dt} \approx -k_A^* C_A^i \tag{13.54}$$

where

$$k_A^* = k_A C_B^j \approx \text{constant} \tag{13.55}$$

Either the differentiation or integration method may then be used. For example, in the differentiation method, a log–log plot of $(-dC_A/dt)$ vs. C_A will give a straight line of slope i. A similar procedure is employed to obtain j.

ILLUSTRATIVE EXAMPLE 13.19 The following liquid-phase reaction takes place in a constant volume, isothermal batch reactor:

$$A + B + C \longrightarrow P$$

The *initial* reaction rates were measured at various *initial* concentrations at 0°C. The following data were obtained[3] in Table 13.24.

TABLE 13.24 Information for Illustrative Example 13.19

Run	$C_{A_0} \times 10^4$ (gmol/L)	$C_{B_0} \times 10^2$ (gmol/L)	$C_{C_0} \times 10^2$ (gmol/L)	Initial Reaction Rate $\times 10^7$ gmol/(L·min)
1	0.712	2.06	3.0	4.05
2	2.40	2.06	3.0	14.60
3	7.20	2.06	3.0	44.60
4	0.712	2.06	1.8	0.93
5	0.712	2.06	3.0	4.05
6	0.712	2.06	9.0	102.00
7	0.712	2.06	15.0	508.00
8	0.712	2.06	3.0	4.05
9	0.712	5.18	3.0	28.0
10	0.712	12.50	3.0	173.0

Assuming a reaction mechanism of the form

$$-r_A = k_A C_A^{\alpha} C_B^{\beta} C_C^{\gamma}$$

1. Estimate the order of the reaction with respect to A, α.
2. Estimate the order of the reaction with respect to B, β.
3. Estimate the order of the reaction with respect to C, γ.
4. Estimate the overall order of the reaction.

Solution. The initial rate of reaction, $-r_{A_0}$, in terms of the initial concentrations C_{A_0}, C_{B_0}, and C_{C_0} for the reaction at hand is

$$-r_{A_0} = kC_{A_0}^{\alpha} C_{B_0}^{\beta} C_{C_0}^{\gamma} \tag{13.56}$$

Linearizing the above equation by taking logs,

$$\log(-r_{A_0}) = \underset{(1)}{\log(k)} + \underset{(2)}{\alpha \log(C_{A_0})} + \underset{(3)}{\beta \log(C_{B_0})} + \underset{(4)}{\gamma \log(C_{C_0})} \tag{13.57}$$

One may also use ln instead of log in the above equation.

The above can also be expanded to other more complicated systems. It is suggested to use initial rates if

$$aA \underset{k'}{\overset{k}{\rightleftharpoons}} cC \tag{13.58}$$

Here,

$$-r = kC_A^i - k'C_c^j \tag{13.59}$$

For the forward reaction, set

$$C_c \approx 0 \qquad \text{so that} \qquad -r_A = kC_A^i \tag{13.60}$$

For the reverse reaction, set

$$C_A \approx 0 \qquad \text{so that} \qquad -r_A = k'C_C^j \tag{13.61}$$

Also, allow the system to come to equilibrium which can provide additional information

$$K = \frac{k}{k'}$$

As noted above, the application of any interpretation of kinetic data to reaction rate equations containing more than one concentration term can be somewhat complex. For runs (1)–(3), the concentrations of B and C are constant (for each run at the initial conditions specified). Therefore, columns (3) and (4) in Table 13.24

TABLE 13.25 Information for α

$\log(-r_{A_0})$	$\log(C_{A_0})$
−6.3925	−4.1475
−5.8356	−3.6197
−5.3506	−3.1426

above may be treated as constants. Thus,

$$\log(-r_{A_0}) = \log B + \alpha \log(C_{A_0})$$

where

$$B = kC_{B_0}^{\beta} C_{C_0}^{\gamma}$$

To obtain α graphically, plot the three values provided in Table 13.25 or note that $\alpha \approx 1.0$.

Use runs (8)–(10) in order to establish the order of the reaction with respect to β in Table 13.26. For this set of runs, $\beta \approx 2.0$.

Use runs (4)–(7) to estimate the order the reaction with respect to C in Table 13.27. For this case, $\gamma \approx 3.0$.

The overall order of the reaction, n, is therefore

$$n = \alpha + \beta + \gamma$$
$$\approx 1.0 + 2.0 + 3.0 = 6.0$$

■

ILLUSTRATIVE EXAMPLE 13.20 Refer to the previous example. Outline how to calculate k_A.

Solution. To calculate k_A, use the α, β, and γ values above and calculate 10 values of k_A. Sum the results and average. The units are $L^5/(\text{gmol}^5 \cdot \text{min})$. ■

ILLUSTRATIVE EXAMPLE 13.21 Refer to the two previous illustrative examples. Outline how to calculate the results numerically by regressing the data.

TABLE 13.26 Information for β

$\log(-r_{A_0})$	$\log(C_{B_0})$
−6.3925	−1.6861
−5.5528	−1.2856
−4.7619	−0.9039

TABLE 13.27 Information for γ

$\log(-r_{A_0})$	$\log(C_{C_0})$
−7.0315	−1.7440
−6.3925	−1.5228
−4.9913	−1.0457
−4.2941	−0.8239

Solution. In order to calculate α, β, γ, and k by regressing the data, the equation (with logs) is of the form

$$Y = A_0 + A_1X_1 + A_2X_2 + A_3X_3$$

where $Y = \log(-r_0)$

$$A_0 = \log k$$

$$A_1 = \alpha$$

$$A_2 = \beta$$

$$A_3 = \gamma$$

$$X_1 = \log(C_{A_0})$$

$$X_2 = \log(C_{B_0})$$

$$X_3 = \log(C_{C_0})$$

The method of least squares may now be used to generate numerical results. ■

REFERENCES

1. S. Shaefer and L. Theodore "*Probability and Statistics Calculations for Environmental Science*", CRC Press/Taylor & Francis, Boca Raton, FL, 2007.
2. L. Theodore, *personal notes*, East Williston, NY, 1969.
3. J. Castellar, "*Physical Chemistry*", Addison-Wesley, Boston MA, 1964.

CHAPTER 14

NONIDEAL REACTORS

INTRODUCTION

Three types of reactors were reviewed in Part II: batch, CSTR, and TF. Each was assumed to operate in an "ideal" mode. The first two types of reactors are vessels with a length-to-diameter ratio of approximate unity, usually defined as a tank. The third type logically is a cylindrical vessel with large length-to-diameter ratio, usually defined as a tube or pipe. The flow pattern for these two main types of "vessel" represents two extreme cases. The first two assume perfect mixing or backmix flow; there is complete mixing of either the reactor contents or the feed due in part to a mixing process. There is no variation in concentration, temperature, and pressure thoughout the unit. Any discharge stream has the same properties as those inside the unit. The TF reactor assumes plug flow for which the velocity is uniform (constant) throughout the entire cross section. Each element of fluid that enters the reactor travels through the unit without mixing (of any form) with other fluid parcels that are within the system. These two classes of reactors are ideal. A real reactor employed in industry lies betweeen these extremes. In effect, real reactors never satisfy the above idealized assumption. However, most reactors closely approximate this condition so that one can often assume them to be ideal with little error from an engineering perspective.

The problem of nonideal flow is directly related to scale-up. For example, there is no reason to pilot-plant if the reactor operates in an ideal mode. However, the magnitude or degree of the nonideality of flow can significantly impact design and performance equations. This effect should not be ignored by the practicing engineer in some applications; it is discussed in more detail in a later section.

Of course, nonideal flow problems exist with other equipment in the chemical process industries. Units that can be significantly impacted include:

1. packed towers,[1,2]
2. plate towers,[1,2]
3. double pipe head exchangers,[3]
4. shell and tube heat exchangers,[3]
5. many environment control equipment,[4,5]
   etc.

Chemical Reactor Analysis and Applications for the Practicing Engineer. By Louis Theodore

Thus, it is fair to say that chemical reactors are not alone in experiencing this nonideal phenomena.

Chapter contents following this Introduction include:

Nonideal Approaches

Definitions

Mean and Variance

Residence Time Distributions

Residence Time Distribution Functions

Experimental Tracer Techniques

NONIDEAL APPROACHES

Nonideal approaches can be divided into three main categories: imperfect mixing, bypassing/short circuiting, or both. Each is considered in the development to follow.

Pockets of stagnant fluid or "dead spots" can exist such as those in Figures 14.1 and 14.2. The conversion in these regions is higher but the fluid at these locations will not leave the reactor. Thus, the overall conversion in the discharge stream is lower because the remainder of the fluid resides in the reactor for a shorter period of time. The insertion of baffles can either hinder or (usually) enhance the conversion process.

Consider the imperfect mixing situations pictured in Figure 14.1 for a batch reactor. A similar condition could exist with a CSTR, as demonstrated in Figure 14.2.

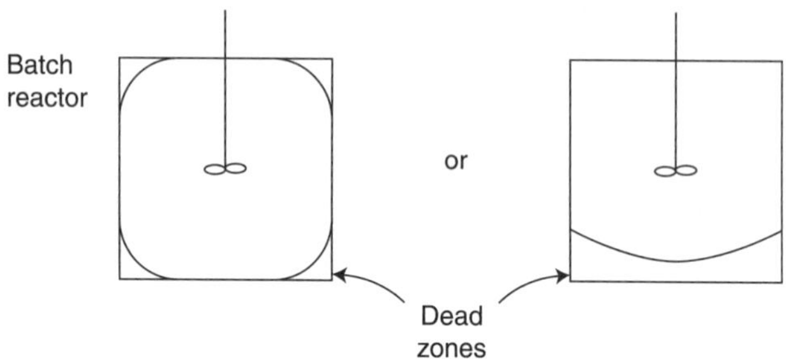

Figure 14.1 Imperfect mixing: batch reactor.

Consider now the nonideal tubular flow reactor pictured in Figures 14.3(a)–(c). As with the imperfect mixing case above in bypassing and/or short circuiting, the feed effectively takes a path through the reactor in which it does not mix with the remaining fluid in the reactor. The conversion in the discharge stream is therefore lower than the ideal plug flow case. In addition, the velocity flow profile may take the form shown in Figures 14.4(a)–(d). Obviously, the residence time of the fluid varies across the cross-sectional area of the reactor. This variation produces varying degrees of conversion for the fluid elements in the reactor, with lower conversions for those elements passing through the unit at high(er) velocities.

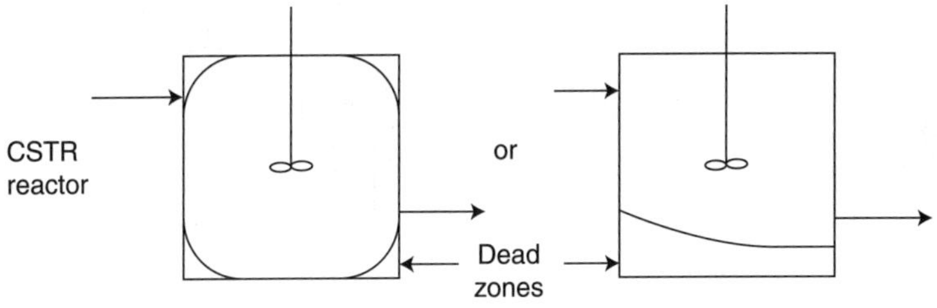

Figure 14.2 Imperfect mixing: CSTR.

Fogler[6] has classified these as

1. segregated flow in which the elements of fluid do not mix but follow separate paths through the reactor so that they have different residence time, and
2. micromixing in which the adjacent elements of fluid are partially mixed.

Unfortunately, much of this information is not available for a real reactor, and other approaches must be employed since data gathering on smaller (or even similar units) rarely produce meaningful results. (Experimental methods are briefly discussed in the last section.)

So what options are available to the practicing engineer? The author[7] has employed a method that involves employing a combination of reactors. For example,

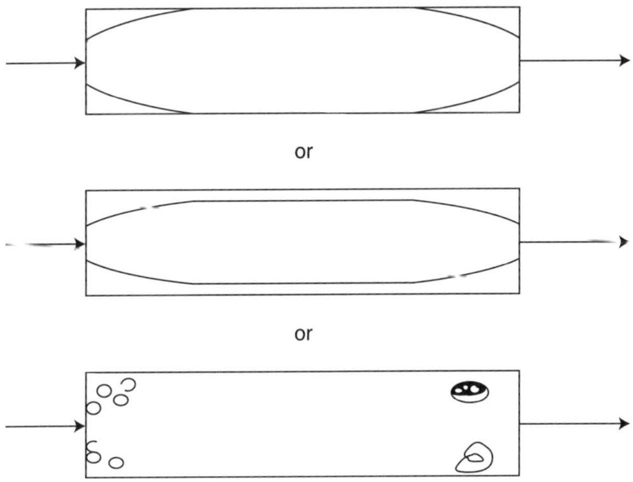

Figure 14.3 Nonideal tubular flow reactor.

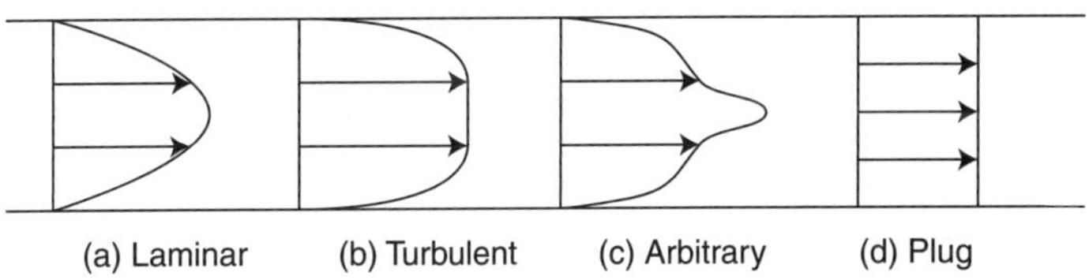

Figure 14.4 Tubular flow velocity profile.

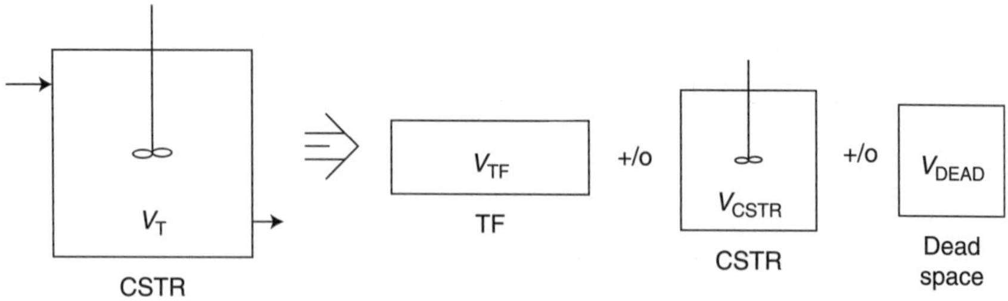

Figure 14.5 Nonideal reactor representation.

the CSTR pictured in Figure 14.5 can be represented by either two or three units. For this case,

$$V_T = V_{TF} + V_{CSTR} + V_{DEAD} \tag{14.1}$$

Thus, a combination of reactor types might be used to simulate an actual reactor. In addition, the flow rates and/or volumes provide adjustable parameters which can be chosen to fit response data, i.e., modifications in the geometric model and physical/chemical system can be made to better fit conversion data. Bypassing (the equivalent in some cases of dead space) can also be included in the analyses.

Perhaps, the simplest approach is to introduce a "fudge" or "fineagle" factor (f) in one of the design or performance equation(s). For example, the volume requirement for a given conversion would be given by $V_{act} = (V_{ideal})\, f_V : f_V > 1$. For the conversion, one could employ $X_{act} = (X_{ideal})\, f_X : f_X < 1$.[8]

The important fact for the practicing engineer to remember is that deviations from ideal behavior negatively impacts the performance of a reactor. Every attempt should be made to minimize or eliminate this effect.

DEFINITIONS[9]

There are several terms that should be defined before proceeding to calculational details. Some of these terms will receive only superficial treatment to follow: The two key terms: cumulative distribution function (cdf) and probability distribution function (pdf), are treated less superficially.

1. **Age:** The age of a fluid element at a given instant of time is the time between a fluid element's entrance into the reactor and the time it exits.
2. **Cumulative distribution function, cdf, of a random variable:** A function F defined for all real numbers such that $F(x) = P(X \leq x)$. $F(x)$ is the cumulative sum of all probabilities assigned to real numbers less than or equal to x.

   Example 1: See Figure 14.6. X has the pdf defined by

$$\begin{aligned} f(x) &= 0; & x &< -2 \\ &= 0.25; & -2 \leq x &\leq 2 \\ &= 0; & x &> 2 \end{aligned}$$

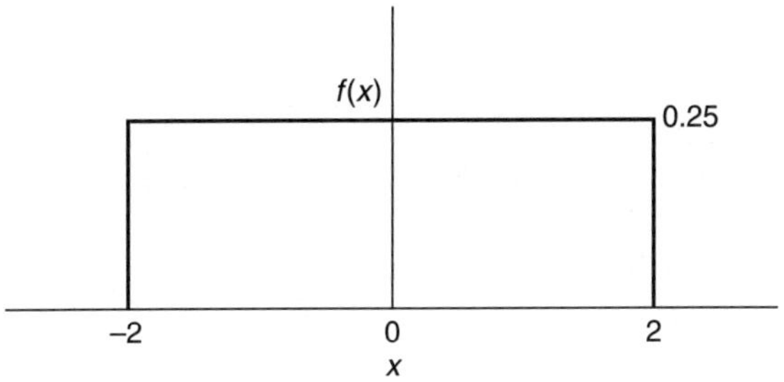

Figure 14.6 A cumulative distribution function.

Then the cdf of X is defined by

$$\begin{aligned} F(x) &= 0; & x &< -2 \\ &= 0.25(x+2); & -2 \le x &\le 2 \\ &= 1; & x &> 2 \end{aligned}$$

Example 2: X has the pdf defined by

$$f(x) = e^{-x}; \ \ x > 0$$

Then the cdf of X is defined by

$$F(x) = \int_{-\infty}^{x} f(x)\, dx$$

Therefore,

$$\begin{aligned} F(x) &- 0; \ \ x < 0 \\ &= \int_{0}^{x} e^{-x}\, dx = 1 - e^{-x}; \ \ x \ge 0 \end{aligned}$$

3. **Discrete random variable**: A random variable whose range of possible values consists of a finite or countable number of values. The random variable X described earlier is a discrete random variable. X has a finite number of values, namely, 0 and 1.
4. **Distribution**: A term to describe the variation of a term or property.
5. **Event**: A subset of the sample space.

   Example: Let A be the event of at least one head appearing when a coin is tossed twice. Then $A = \{$HH, TH, HT$\}$ and the sample space is $\{$HH, TH, HT, TT$\}$.
6. **Expected value of a random variable**: The average value of the random variable. If X is the random variable, its expected value is denoted by $E(X)$.

7. **External Age Distribution**: The external age distribution is simply the residence time distribution. The age-distribution frequency provides the fraction of ages per unit time.

8. **Internal Age Distribution**: The internal age distribution is related to the fraction of the fluid elements between t and $t + \Delta t$ in the reactor.

9. **Parameter**: A constant describing the population, e.g., the population mean and the population variance.

10. **Population**: The totality of elements under consideration in a statistical investigation. These elements may be people, objects, molecules, or measurements associated with them. Frequently, the population will be the set of possible values of a random variable.

11. **Probability distribution function, pdf, of a continuous random variable**: A function when integrated over an interval gives the probability that the random variable values assume in the interval. If f is the pdf of a continuous random variable X then

$$\int_a^b f(x)\,dx = P(a < X < b)$$

Example: For the random variable T, a possible pdf might be defined as follows:

$$f(t) = e^{-t}; \quad t > 0$$

Then,

$$P(1 < T < 3) = \int_1^3 e^{-t}\,dt = e^{-1} - e^{-3} = 0.367 - 0.050 = 0.317$$

12. **Probability distribution function, pdf, of a discrete random variable**: A function that assigns probability to each of the possible values of the discrete random variable. If f is the pdf of a discrete random variable X then $f(x) = P(X = x)$ is the probability assigned to the value x of the random variable X.

Example 1: For the random variable X defined earlier a possible pdf might be defined as follows:

$$\begin{aligned} f(x) &= 0.01; \quad x = 1 \\ &= 0.99; \quad x = 0 \end{aligned}$$

This pdf would assign probability 0.01 to drawing a defective transistor, and 0.99 to drawing a nondefective transistor.

Example 2: For the random variable Y, a possible pdf might be defined as follows:

$$g(y) = (0.001)(0.999)^{y-1}; \quad y = 1, 2, \ldots, n, \ldots 3$$

Then $P(Y = 1) = 0.0001$, $P(Y = 2) = (0.001)\,(0.999)$, $P(Y = 3) = (0.001)$ $(0.999)^2$, etc.

13. **Probability of event A, $P(A)$**: A number about which the relative frequency of event A tends to cluster when a random experiment is repeated indefinitely. This interpretation of $P(A)$ is the relative frequency interpretation. Another interpretation relates $P(A)$ to the degree of belief, measured on a scale from 0 (low) to 1 (high), that the event A occurs.

14. **Random experiment**: An experiment whose outcome is uncertain.

15. **Random variable**: A variable describing the outcome of a random experiment. The values of a random variable are real numbers associated with the elements of the sample space.

    Example 1: Let X denote the result of classifying a transistor drawn at random from a lot as defective or non-defective. Associate 1 with the drawing of a defective and 0 with the drawing of a non-defective.

    Example 2: Let T denote the time of failure of a CSTR in a series of CSTRs. Then T is a random variable whose values consist of the positive real numbers representing the possible times to failure.

    Example 3: Let Y denote the number of throws of a switch prior to the first failure. Then Y is a random variable with values $1, 2, \ldots, n, \ldots$.

    Note that in Example 1 the outcomes of the random experiment had to be assigned a numerical code to define the random variable. In Example 2 and Example 3 the outcome of the random experiment was described numerically, and the numerical values thus introduced became the values of the random variable.

16. **Relative frequency of an event**: The ratio m/n where n is the number of times the random experiment is performed and m is the number of times the event occurs.

    Example: Toss a coin 100 times. Let A be the event of heads on a single toss. Suppose A occurs 40 times. Then $m = 40$, $n = 100$, and the relative frequency of A is $40/100$ or 0.40.

17. **Residence time**: The residence time of a fluid element is the time a fluid element enters a reactor to the time it leaves the reactor. As defined earlier, if a fluid enters and leaves a tubular reactor of volume V at a volumetric flow rate Q, then the residence time τ is V/Q.

18. **Sample**: A subset of a population.

19. **Sample space**: The set of possible outcomes of a random experiment.

    Example: Toss a coin twice. Sample space = {HH, HT, TH, TT} or {0, 1, 2 heads} or {0, 1, 2 tails}. Note that the description of a sample space is not unique.

20. **Standard deviation of a random variable**: The positive square root of the variance of the random variable. The standard deviation is expressed in the same units as the random variable.

21. **Statistic**: A variable describing the sample, e.g., the sample mean and the sample variance. A statistic is a variable in the sense that its value varies

from sample to sample from the same population. A parameter, on the other hand, is constant for any given population.

Example: Consider a lot of 10,000 catalyst pellets each of which is either active or non-active. Suppose a sample of 100 pellets is drawn at random from the lot. The proportion of pellets in the lot that are active is a population parameter. The proportion of pellets in the sample that are active is a statistic whose value varies from sample to sample drawn from the lot.

22. **Variance of a random variable**: The expected value of the squared difference between a random variable and its mean. If X is a random variable with mean μ, then the variance of X is $E(X - \mu)^2$ and is denoted by σ^2. The variance *and* the mean are discussed in the next section.

ESTIMATION OF MEAN AND VARIANCE[9]

The mean μ and the variance σ^2 of a random variable are constants characterizing the random variable's average value and dispersion, respectively, about its mean. The mean and variance can also be derived from the probability distribution function (pdf)—to be discussed shortly—of the random variable. If the pdf is unknown, however, the mean and the variance can be estimated on the basis of a random sample of some, but not all, observations on the random variable.

Let $X_1, X_2, \ldots, X_n$ denote a random sample of n observations on X. Then the sample mean $\overline{X}$ is defined by

$$\overline{X} = \sum_{i=1}^{n} \frac{X_i}{n} \tag{14.2}$$

and the *sample* variance s^2 is defined by

$$s^2 = \sum_{i=1}^{n} \frac{(X_i - \overline{X})^2}{n-1} \tag{14.3}$$

where $\overline{X}$ and s^2 are random variables in the sense that their values vary from the sample to sample of observations on X. It can be shown that expected value of $\overline{X}$ is μ and that the expected value of s^2 is σ^2 (the population variance). Because of this, $\overline{X}$ and s^2 are called unbiased estimators of μ and σ^2, respectively.

The calculation of s^2 can be facilitated by use of the equation.

$$s^2 = \frac{n\sum_{i=1}^{n} X_i^2 - \left(\sum_{i=1}^{n} X_i\right)^2}{n(n-1)} \tag{14.4}$$

For example, given the sample 5, 3, 6, 4, 7,

$$\sum_{i=1}^{5} X_i^2 = 135$$

$$\sum_{i=1}^{5} X_i = 25$$

$$n = 5$$

Substituting in Equation (14.4) yields

$$s^2 = \frac{(5)(135) - (25)^2}{(5)(4)} = 2.5$$

In the case of a random sample of observations on a continuous random variable assumed to have a so-called normal pdf, the graph of which is a bell-shaped curve, the following statements give a more precise interpretation of the sample standard deviation s as a measure of spread or dispersion.

1. $\overline{X} \pm s$ includes approximately 68% of the sample observations.
2. $\overline{X} \pm 2s$ includes approximately 95% of the sample observations.
3. $\overline{X} \pm 3s$ includes approximately 99.7% of the sample observations.

The source of these percentages is the normal probability distribution, which is discussed in more detail later in this chapter.

Chebyshev's theorem provides an interpretation of the sample standard deviation (the positive square root of the sample variance) as a measure of the spread (dispersion) of sample observations about their mean. Chebyshev's theorem states that with $k > 1$, at least $(1 - 1/k^2)$, of the sample observations lie in the interval $\overline{X}$ (ks, $\overline{X} + ks$). For $k = 2$, e.g., this means that at least 75% of the sample observations lie in the interval $\overline{X}$ (2s, $\overline{X}$ + 2s). The smaller the value of s, the greater the concentration of observations in the vicinity of $\overline{X}$.

ILLUSTRATIVE EXAMPLE 14.1 Define the following "central tendency" terms in equation form:

1. Arithmetic mean
2. Geometric mean
3. Median
4. Mode

Solution. Let $X_1, X_2, \ldots, X_n$ represented a set of n numbers.

1. For the arithmetic mean $\overline{X}$,

$$\overline{X} = \frac{X_1 + X_2 + \cdots + X_n}{n} \tag{14.5}$$

If the numbers X_i have weighing factors W_i associated with them,

$$\begin{aligned} \overline{X} &= \frac{W_1X_1 + W_2X_2 + \cdots + W_nX_n}{n} \\ &= \frac{\sum_{i=1}^{n} W_iX_i}{n} \end{aligned} \tag{14.6}$$

Weighing factors are often normalized, i.e., $\sum W_i = 1.0$. For this condition,

$$\overline{X} = \sum_{i=1}^{n} W_i X_i;\ W_i = \text{fractional value} \tag{14.7}$$

2. The geometric mean, $\overline{X}_G$ is given by

$$\overline{X}_G = (X_1 X_2, \ldots X_n)^{1/n}$$

$$\overline{X}_G = \left(\prod_{i=1}^{n} x_i\right)^{1/n} \tag{14.8}$$

3. The median is defined as the middle value (or arithmetic mean of the two middle values) of a set of the numbers. Thus, the median of 4, 5, 9, 10, and 15 is 9. It is also occasionally defined as the distribution midpoint. Further, the median of a continuous probability distribution function $f(x)$ is that value of c so that

$$\int_{-\infty}^{c} f(x)\,dx = 0.5 \tag{14.9}$$

4. The mode is the value that occurs with the greatest frequency in a set of numbers. Thus, it is the typical or most common value in a set.

Other definitions also exist in the literature for the aforementioned four terms. ■

ILLUSTRATIVE EXAMPLE 14.2 Qualitatively describe the mean.

Solution. One basic way of summarizing data is by the computation of a central value. The most commonly used central value statistic is the arithmetic average, or the mean discussed in the previous illustrative example. This statistic is particularly useful when applied to a set of data having a fairly symmetrical distribution. The mean is an efficient statistic because it summarizes all the data in the set and each piece of data is taken into account in its computation. However, the arithmetic mean is not a perfect measure of the true central value of a given data set because arithmetic means can overemphasize the importance of one or two extreme data points.

When a distribution of data is asymmetrical, it is sometimes desirable to compute a different measure of central value. This second measure, known as the *median*, is simply the middle value of a distribution, or the quantity above which half the data lie and below which the other half lie. If n data points are listed in their order of magnitude, the median is the $[(n+1)/2]$th value. If the number of data is even, then the numerical value of the median is the value midway between the two data points nearest the middle. Since the median is a positional value, it is less influenced by extreme values in a distribution than the mean. However, the median alone is usually not a good measure of central tendency. To obtain the median, the data provided must first be arranged in order of magnitude, such as

8, 10, 13, 15, 18, 22

Thus, the median is 14, or the value halfway between 13 and 15 because this data set has an even number of measurements.

Another measure of central tendency used in specialized applications is the aforementioned *geometric mean*, $\overline{X}_G$.

Generally, the mean falls near the "middle" of the distribution. Actually, the mean may be thought of as the *center of gravity* of the distribution. The mean has another important property. If each measurement is subtracted from the mean, one obtains n "discrepancies" or differences; some of these are positive and some are negative, but the algebraic sum of all the differences is equal to zero. ■

ILLUSTRATIVE EXAMPLE 14.3 Qualitatively describe the standard deviation.

Solution. The mean of a set of measurements provides some information about the location of the "middle" or "center of gravity" of the set of measurements, but it gives no information about the *scatter* (or *dispersion* or amount of concentration) of the measurements. For example, the five measurements 14.0, 24.5, 25.0, 25.5, and 36.0 have the same mean as the five measurements 24.0, 24.5, 25.0, 25.5, and 26.0, but the two sets of measurements have different amounts of scatter.

One simple indication of the scatter of a set of measurements is the *range*, i.e., the largest measurement minus the smallest. In the two sets of measurements mentioned, the ranges are 22 and 2, respectively. With fairly small sample sizes one would find the range to be very convenient. It is difficult, however, to compare a range for one sample size with that for a different sample size. For this and other reasons, the range, in spite of its simplicity, convenience, and importance, is used only in rather restricted situations.

One clearly needs a measure of scatter, which can be used in samples of any size and in some sense makes use of all the measurements in the sample. There are several measures of scatter that can be used for this purpose, and the most common of these is the *standard deviation*. The standard deviation may be thought of as the "natural" measure of scatter.

Calculation details for the variance and deviation are provided in this chapter and Chapter 8. The examples that follow will illustrate the procedure to obtain these values. ■

ILLUSTRATIVE EXAMPLE 14.4 The average weekly inlet reactor feed temperatures (°C) for six consecutive weeks are

22, 10, 8, 15, 13, 18

Find the median, the arithmetic mean, the geometric mean, and the standard deviation.

Solution. See Equations (14.2) and (14.8). For the above reactor temperatures (substituting T for X),

$$\overline{T}_G = [(8)(10)(13)(15)(18)(22)]^{1/6} = 13.54^\circ\text{C}$$

and the arithmetic mean, $\overline{T}$, is

$$\overline{T} = \frac{(8 + 10 + 13 + 15 + 18 + 22)}{6}$$
$$= 14.33^\circ\text{C}$$

As noted earlier, the most commonly used measure of dispersion, or variability, of sets of data is the standard deviation, s. Its defining formula is given by the expression in Equation (14.3).

$$s = \sqrt{\frac{\sum (X_i - \overline{X})^2}{n - 1}}$$

The expression $(X_i - \overline{X})$ shows that the deviation of each piece of data from the mean is taken into account by the standard deviation. Although the defining formula for the standard deviation gives insight into its meaning, the following algebraically equivalent formula (see Equation 14.4) makes computation much easier (now applied to the temperature, T):

$$s = \sqrt{\frac{\sum (T_i - \overline{T})^2}{n - 1}} = \sqrt{\frac{n \sum T_i^2 - \left(\sum T_i\right)^2}{n(n - 1)}}$$

The standard deviation may be calculated for the data at hand:

$$\sum T_i^2 = (8)^2 + (10)^2 + (13)^2 + (15)^2 + (18)^2 + (22)^2 = 1366$$

and

$$\left(\sum T_i\right)^2 = (8 + 10 + 13 + 15 + 18 + 22)^2 = 7396$$

Thus,

$$s = \sqrt{\frac{6(1366) - 7396}{(6)(5)}} = 5.16^\circ\text{C}$$

■

ILLUSTRATIVE EXAMPLE 14.5 Sulfur dioxide discharging from utility combustion devices are a concern to society. The following are SO_2 concentrations (μg/m^3) for March, April, and May of 2011 at Floral Park, NY—home of the Belmont Park racetrack (which the author regularly visits):

29 103 27 14 24 63 24

Calculate the following terms:

1. Mean
2. Median
3. Mode
4. Range
5. Standard deviation
6. Variance

Solution. In a very real sense, this is a repeat of the previous illustrative example.

1. For the mean,

$$\overline{X} = \frac{\sum_{i=1}^{8} X_i}{n}; \quad n = 7$$

$$\overline{X} = \frac{29 + 103 + 27 + 14 + 24 + 63 + 24}{7}$$

$$= 40.57 \ \mu\text{g/m}^3$$

2. For the median, the middle value is 27 μg/m^3 since,

$$14 \ \ 24 \ \ 24 \ \ \textcircled{27} \ \ 29 \ \ 63 \ \ 103$$

3. For the mode, the most frequently occurring value is 24 μg/m^3.
4. The range is given by the maximum value minus the minimum value. Thus, the range, R, is

$$R = 103 - 14 = 89 \ \mu\text{g/m}^3$$

5. For the standard deviation, apply Equation (14.4)

$$s = \sqrt{\frac{\sum X_i^2 - \frac{\left(\sum X_i\right)^2}{n}}{n - 1}}$$

The calculations are performed as follows based on the results provided in Table 14.1. From the calculation of $\sum X_i$ determine $\left(\sum X_i\right)^2$:

$$\left(\sum X_i\right)^2 = (284)^2 = 80{,}656$$

TABLE 14.1 Utility SO_2 Calculations

X_i	X_i^2
14	196
24	576
24	576
27	729
29	841
63	3969
103	10,609
$\sum X_i = 284$	$\sum X_i^2 = 17{,}496$

Substituting gives

$$s = \sqrt{\frac{17{,}496 - \dfrac{80{,}656}{7}}{6}}$$

$$= \sqrt{995.62}$$

$$= 31.55\ \mu g/m^3$$

6. Finally, the variance is defined as the square of the standard deviation

$$s^2 = (\text{standard deviation})^2$$

$$= (31.55)^2$$

$$= 995\ (\mu g/m^3)^2$$

■

ILLUSTRATIVE EXAMPLE 14.6 Continuous random variable X and Y have pdfs specified by $f(x)$ and $g(y)$, respectively, as follows:

$$f(x) = \frac{1}{2}; \quad 1 < x < 1$$

$$g(y) = \frac{1}{4}; \quad 2 < y < 2$$

Compute the mean and variance of X and Y and compare the results.

Solution. Note that for a continuous random variable, integration replaces summation in the calculation of the mean and variance. Furthermore, the probability that a continuous random variable lies in a certain interval is obtained by integrating the pdf over that interval.

First compute μ_x, the mean of X, and μ_y, the mean of Y. Use the following formulas:[9]

$$\mu_x = \int_{-\infty}^{\infty} xf(x)\,dx \tag{14.10}$$

$$\mu_y = \int_{-\infty}^{\infty} yg(y)\,dy \tag{14.11}$$

Substituting $f(x)$ and $g(y)$ gives

$$\mu_x = \int_{-1}^{1} x\left(\frac{1}{2}\right)dx = 0$$

$$\mu_y = \int_{-2}^{2} y\left(\frac{1}{4}\right)dy = 0$$

The terms σ_x^2, the variance of X, and σ_y^2, the variance of Y, are calculated as follows:[9]

$$\sigma^2 = E(X^2) - \mu_x^2 \tag{14.12}$$

$$= \int_{-1}^{1} x^2\left(\frac{1}{2}\right)dx = \frac{1}{3} = 0.333$$

$$\sigma^2 = E(Y^2) - \mu_y^2$$

$$= \int_{-2}^{2} y^2\left(\frac{1}{4}\right)dy = \frac{4}{3} = 1.333$$

■

ILLUSTRATIVE EXAMPLE 14.7 A continuous random variable X has a pdf given by

$$f(x) = 4x^3; \qquad 0 < x < 1$$

Calculate μ and σ^2 for x.

Solution. By definition,

$$\mu = \int xf(x)\,dx \tag{14.13}$$

Substituting gives

$$\mu = \int_0^1 x(4x^3)\,dx = \int_0^1 4x^4\,dx$$

Integrating gives

$$\mu = 4\left(\frac{x^5}{5}\right)_0^1$$
$$= \frac{4}{5}$$

For σ^2, one notes that

$$\sigma^2 = E(X^2) - \mu^2 \tag{14.14}$$
$$= \int x^2 f(x)\,dx - \mu^2$$

Substituting gives

$$\sigma^2 = \int_0^1 x^2(4x^3)\,dx - \mu^2$$

Integrating,

$$\sigma^2 = \frac{4x^6}{6} - \left(\frac{4}{5}\right)^2$$
$$= \frac{2}{3} - \frac{16}{25} = \frac{50 - 48}{75}$$
$$= \frac{2}{75} = 0.0267$$

■

Finally, the sample mean, $\overline{X}$, constitutes a so-called point estimate of the mean, μ, of the population from which the sample was selected at random. Instead of a *point* estimate, an *interval* estimate of μ may be required along with an indication of the confidence that can be associated with the interval estimate. Such an interval estimate is called a *confidence interval*, and the associated confidence is indicated by a *confidence coefficient*. The length of the confidence interval varies directly with the confidence coefficient for fixed values of n, the sample size; the larger the value of n, the shorter the confidence interval. Thus, for fixed values of the confidence coefficient, the limits that contain a parameter with a probability of 95% (or some other stated percentage) are defined as the 95% (or that other percentage) *confidence limits* for the parameter; the interval between the confidence limits is referred to as the aforementioned confidence interval.(9)

RESIDENCE TIME DISTRIBUTIONS

What follows is an attempt to describe the performance of nonideal reactors using some simple statistical models that can be represented by analytical solutions. This information can be considerably beneficial. The presence of the residence time distribution is, as noted previously, an undesirable phenomena for reactors.

One engineering design area, i.e., impacted by nonideal reactor behavior and the accompanying fluid residence time distribution(s) is scaleup. Suppose that a reactor study is conducted at the pilot scale level and that the conversion (or the equivalent) associated with volume flow rate Q_S are judged to be acceptable. The classical scaleup problem is to then design a larger process with flow rate $q_D \gg q_S$ which results in the same conversion. The scaleup factor SF is.

$$SF = \frac{q_D}{q_S} \tag{14.15}$$

The reactor would then be scaled by this factor. For example, the reactor volume would be

$$SF = \frac{V_2}{V_1} \tag{14.16}$$

It is difficult to maintain similarity (geometric, dynamic, thermal, etc.) during scaleup. The practicing engineer is concerned with the decision of which similarities to maintain and which to sacrifice.

The reader should note that to scaleup (or scaledown) a process, it is necessary to establish geometric and dynamic similarities between the model and the prototype. These two similarities are discussed below.

Geometric similarity implies using the same geometry of equipment. A circular pipe prototype should be modeled by a tube in the model. Geometric similarity establishes the scale of the model/prototype design. A 1/10th scale model means that the characteristic dimension of the model is 1/10th that of the prototype.

Dynamic similarity implies that the important dimensionless numbers must be the same in the model and the prototype. For a particle settling in a fluid. It has been shown that the drag coefficient, C_D, is a function of the dimensionless Reynolds number, Re. By selecting the operating conditions such that Re in the model equals the Re in the prototype, then the drag coefficient (*or friction factor*) in the prototype equals the friction factor in the model.[10]

RESIDENCE TIME DISTRIBUTION FUNCTIONS

There are several residence time distribution functions of concern that find application in nonideal reactors. The four distributions of main concern are:

1. normal,
2. lognormal,
3. Weibull, and
4. experimental.

These are continuous distributions. Discontinuous (or discrete) distributions—such as binomial, Poisson, hypergeometric, etc.—find extremely limited use. The

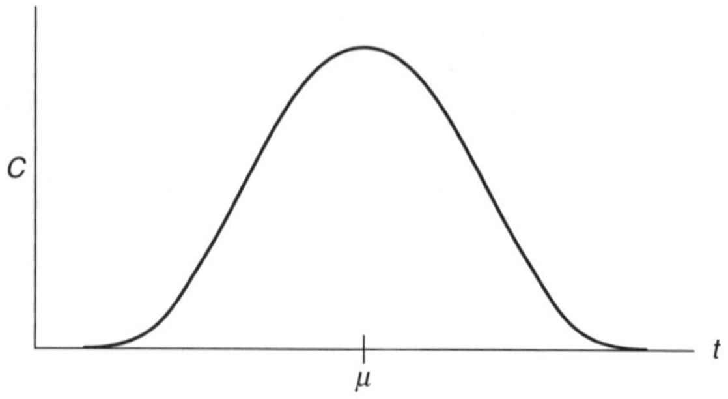

Figure 14.7 Normal pdf of concentration (*C*)–time (*t*).

development to follow will key on the normal and lognormal distributions since both find application in this field.

Normal Distributions

The initial presentation in this section will focus on normal distributions. When time, *t*, has a normal distribution, its probability distribution function (pdf) is given by

$$f(t) = \frac{1}{\sqrt{2\pi}\sigma}\exp\left[-\frac{1}{2}\left(\frac{t-\mu}{\sigma}\right)^2\right]; \quad -\infty < t < \infty \tag{14.17}$$

where μ is the mean value of t and σ is its standard deviation. The graph of $f(t)$ is the familiar bell-shaped curve shown in Figure 14.7.

It C is normally distributed with mean μ and standard deviation σ, then the random variable $(C - \mu)/\sigma$ is normally distributed with mean 0 and standard deviation 1. The term $(C - \mu)/\sigma$ is called a *standard normal variable*, i.e., often represented by Z. This is discussed in more detail later.

Table 14.2 is a tabulation of areas under a standard normal curve to the right of z_0 for non-negative values of z_0. Probabilities about a standard normal variable Z can be determined from this table. For example,

$$P(Z > 1.54) = 0.062$$

is obtained directly from Table 14.2 as the area to the right of 1.54. As presented in Figure 14.8, the symmetry of the standard normal curve about zero implies that the area to the right of zero is 0.5, and the area to the left of zero is 0.5.

Plots demonstrating the effect of μ and σ on the bell-shaped curve are available in the literature.[(9)]

Log-Normal Distribution

A non-negative random variable C has a log-normal distribution whenever ln C, i.e., the natural logarithm of C, has a normal distribution. The probability distribution function (pdf) of a random variable C having a log-normal distribution is

TABLE 14.2 Areas Under a Standard Normal Curve Between 0 and Z

The Standard Normal Distribution

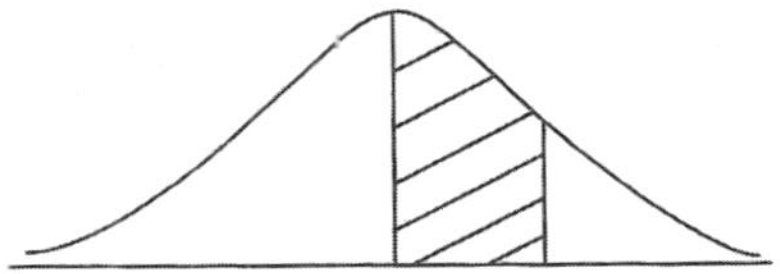

z	0.00	0.01	0.02	0.03	0.04	0.05	0.06	0.07	0.08	0.09
0.0	0.0000	0.0040	0.0080	0.0120	0.0160	0.0199	0.0239	0.0279	0.0319	0.0359
0.1	0.0398	0.0438	0.0478	0.0517	0.0557	0.0596	0.0636	0.0675	0.0714	0.0753
0.2	0.0793	0.0832	0.0871	0.0910	0.0948	0.0987	0.1026	0.1064	0.1103	0.1141
0.3	0.1179	0.1217	0.1255	0.1293	0.1331	0.1368	0.1406	0.1443	0.1480	0.1517
0.4	0.1554	0.1591	0.1628	0.1664	0.1700	0.1736	0.1772	0.1808	0.1844	0.1879
0.5	0.1915	0.1950	0.1985	0.2019	0.2054	0.2088	0.2123	0.2157	0.2190	0.2224
0.6	0.2257	0.2291	0.2324	0.2357	0.2389	0.2422	0.2454	0.2486	0.2517	0.2549
0.7	0.2580	0.2611	0.2642	0.2673	0.2704	0.2734	0.2764	0.2794	0.2823	0.2852
0.8	0.2881	0.2910	0.2939	0.2967	0.2995	0.3023	0.3051	0.3078	0.3106	0.3133
0.9	0.3159	0.3186	0.3212	0.3238	0.3264	0.3289	0.3315	0.3340	0.3365	0.3389
1.0	0.3413	0.3438	0.3461	0.3485	0.3508	0.3531	0.3554	0.3577	0.3599	0.3621
1.1	0.3643	0.3665	0.3686	0.3708	0.3729	0.3749	0.3770	0.3790	0.3810	0.3830
1.2	0.3849	0.3869	0.3888	0.3907	0.3925	0.3944	0.3962	0.3980	0.3997	0.4015
1.3	0.4032	0.4049	0.4066	0.4082	0.4099	0.4115	0.4131	0.4147	0.4162	0.4177
1.4	0.4192	0.4207	0.4222	0.4236	0.4251	0.4265	0.4279	0.4292	0.4306	0.4319
1.5	0.4332	0.4345	0.4357	0.4370	0.4382	0.4394	0.4406	0.4418	0.4429	0.4441
1.6	0.4452	0.4463	0.4474	0.4484	0.4495	0.4505	0.4515	0.4525	0.4535	0.4545

(Continued)

TABLE 14.2 *Continued*

The Standard Normal Distribution										
z	0.00	0.01	0.02	0.03	0.04	0.05	0.06	0.07	0.08	0.09
1.7	0.4554	0.4564	0.4573	0.4582	0.4591	0.4599	0.4608	0.4616	0.4625	0.4633
1.8	0.4641	0.4649	0.4656	0.4664	0.4671	0.4678	0.4686	0.4693	0.4699	0.4706
1.9	0.4713	0.4719	0.4726	0.4732	0.4738	0.4744	0.4750	0.4756	0.4761	0.4767
2.0	0.4772	0.4778	0.4783	0.4788	0.4793	0.4798	0.4803	0.4808	0.4812	0.4817
2.1	0.4821	0.4826	0.4830	0.4834	0.4838	0.4842	0.4846	0.4850	0.4854	0.4857
2.2	0.4861	0.4864	0.4868	0.4871	0.4875	0.4878	0.4881	0.4884	0.4887	0.4890
2.3	0.4893	0.4896	0.4898	0.4901	0.4904	0.4906	0.4909	0.4911	0.4913	0.4916
2.4	0.4918	0.4920	0.4922	0.4925	0.4927	0.4929	0.4931	0.4932	0.4934	0.4936
2.5	0.4938	0.4940	0.4941	0.4943	0.4945	0.4946	0.4948	0.4949	0.4951	0.4952
2.6	0.4953	0.4955	0.4956	0.4957	0.4959	0.4960	0.4961	0.4962	0.4963	0.4964
2.7	0.4965	0.4966	0.4967	0.4968	0.4969	0.4970	0.4971	0.4972	0.4973	0.4974
2.8	0.4974	0.4975	0.4976	0.4977	0.4977	0.4978	0.4979	0.4979	0.4980	0.4981
2.9	0.4981	0.4982	0.4982	0.4983	0.4984	0.4984	0.4985	0.4985	0.4986	0.4986
3.0	0.4987	0.4987	0.4987	0.4988	0.4988	0.4989	0.4989	0.4989	0.4990	0.4990

Adapted from: http://www.statsoft.com/textbook/sttable.html

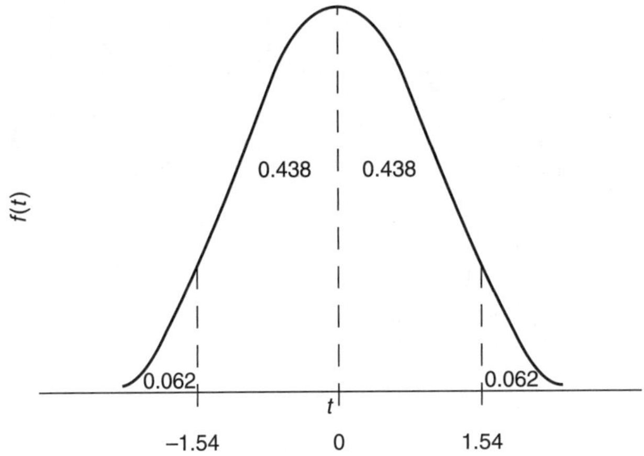

Figure 14.8 Areas under a standard normal curve.

specified by

$$f(x) = \frac{1}{\sqrt{2\pi\beta}} x^{-1} \exp\left[-\frac{(\ln C - \alpha)^2}{2\beta^2}\right]; \quad x > 0$$
$$= 0; \quad \text{elsewhere} \tag{14.18}$$

The mean and variance of a random variable C having a log-normal distribution are given by

$$\mu = e^{\alpha+\beta^2/2} \tag{14.19}$$

$$\sigma^2 = e^{2\alpha+\beta^2}(e^{\beta^2} - 1) \tag{14.20}$$

Figure 14.9 plots the pdf of the log-normal distribution for $\alpha = 0$ and $\beta = 1$. Probabilities concerning random variables having a log-normal distribution can be calculated from the previously employed table of the normal distribution. If X has a log-normal distribution with parameters α and β, then the ln C has a normal distribution with $\mu = \alpha$ and $\sigma = \beta$. Probabilities concerning C can therefore be converted into equivalent probabilities concerning ln C.

For example, suppose that C has a log-normal distribution with $\alpha = 2$ and $\beta = 0.1$. Then

$$\begin{aligned}
P(6 < C < 8) &= P(\ln 6 < \ln C < \ln 8) \\
&= \left[\frac{\ln 6 - 2}{0.1} < \frac{\ln C - 2}{0.1} < \frac{\ln 8 - 2}{0.1}\right] \\
&= P(-2.08 < Z < 0.79) \\
&= (0.5 - 0.019) + (0.5 - 0.215) \\
&= 0.481 + 0.285 \\
&= 0.78
\end{aligned}$$

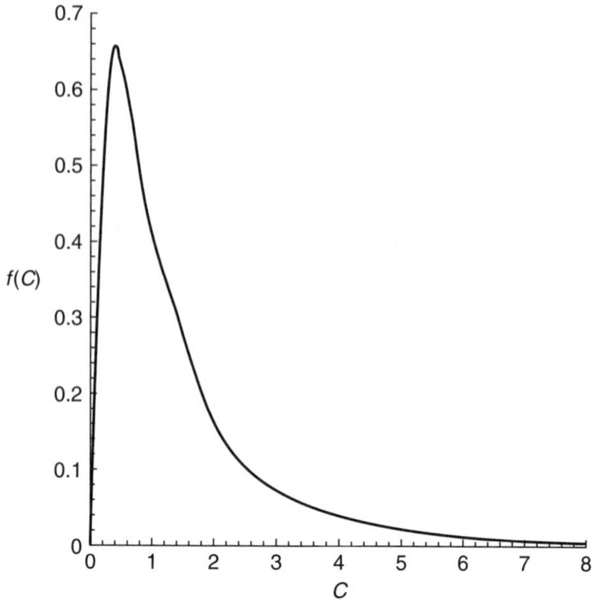

Figure 14.9 Log-normal pdf for $\alpha = 0$, $\beta = 1$.

Estimates of the parameters α and β in the pdf of a random variable C having a log-normal distribution can be obtained from a sample of observations on C by making use of the fact that $\ln C$ is normally distributed with mean α and standard deviation β. Therefore, the mean and standard deviation of the natural logarithms of the sample observations on C furnish estimates of α and β.

The log-normal distribution has been employed as an appropriate model in a wide variety of situations from nonideal reactor studies to environmental management to biology to economics. Additional applications include the distributions of personal incomes, inheritances, bank deposits, and also the distribution of organism growth subject to many small concentrations of impurities. Perhaps the primary application of the log-normal distribution has been to represent the distribution for particle sizes in gaseous emissions from many industrial processes.[2,4]

Exponential Distribution

The exponential distribution is another important distribution. The probability that an "event" will occur during $(0, t)$ is

$$F(t) = 1 - e^{-\lambda t} \tag{14.21}$$

This represents the cumulative distribution function (cdf) of t. One can therefore show that the pdf is[7]

$$f(t) = e^{-\lambda t} \tag{14.22}$$

Note that the parameter $1/\lambda$ (sometimes denoted as μ) is the expected value. Normally, the reciprocal of this value is specified and represents the expected value of $f(t)$.

Because the exponential function appears in the expression for both the pdf and cdf, the distribution is understandably referred to as the *exponential distribution.* A typical pdf of t plot is provided in Figure 14.10.

Alternatively, the cumulative exponential distribution can be obtained from the pdf (x):

$$F(t) = \int_0^t \lambda e^{-\lambda t}\, dt = 1 - e^{-\lambda t} \tag{14.23}$$

All that remains is a simple evaluation of the negative exponent in Equation (14.23). One can show[(7)] that

$$\begin{aligned}\lambda &= \frac{\lambda e^{-\lambda t}}{1-(1-e^{-\lambda t})} \\ &= \frac{\lambda e^{-\lambda t}}{e^{-\lambda t}}\end{aligned} \tag{14.24}$$

This distribution finds a wide application in life testing. Equation (14.24) indicates that the probability of concern is constant, irrespective of time. It implies that the probability of a reactor or a "component" whose time-to-failure distribution is exponential fails in an instant during the first hour of its life and is the same as its failure probability during an instant in the thousandth hour—presuming it has survived up to that instant. It is for this reason that the parameter λ is usually referred to on life-test applications as the *failure rate*. This definition generally has meaning only with an exponential distribution. This natural association with life-testing and the fact that it is very tractable mathematically makes the exponential distribution attractive as representing the life distribution of a complex system or several complex systems. In fact, the exponential distribution is as prominent in reliability analysis as the normal distribution is in other branches of science, including residence time distribution in nonideal reactors.

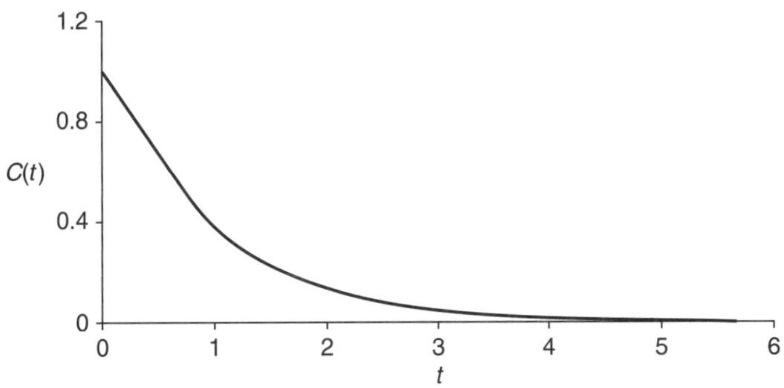

Figure 14.10 Exponential distribution.

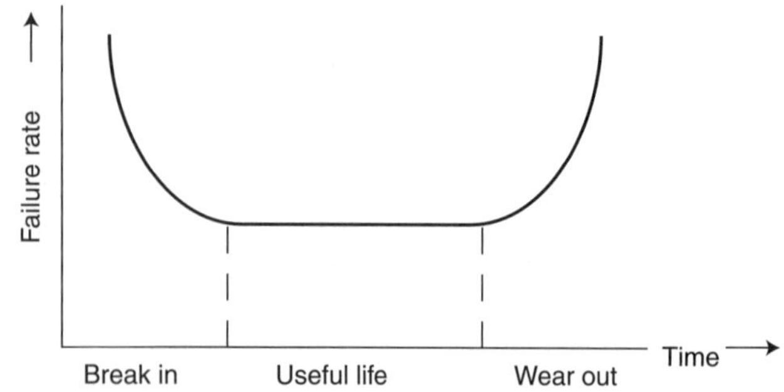

Figure 14.11 Bathtub curve.

Weibull Distribution

Unlike the exponential distribution, the failure rate of equipment such as reactors frequently exhibits three stages: a break-in stage with a declining failure rate, a useful life stage characterized by a fairly constant failure rate, and a wear out period characterized by an increasing failure rate. Many industrial parts and components also follow this path. This occurs only occasionally in representing residence time distribution(s) in nonideal reactors. A failure rate curve exhibiting these three phases (see Figure 14.11 is called a *bathtub* curve.[9,11]

In the case of the bathtub curve, failure rate during useful life is constant. Letting this constant be α leads to[9]

$$F(t) = \alpha \exp\left(-\int_0^t \alpha\, dt\right) \tag{14.25}$$

$$= \alpha \exp(-\alpha t); \quad t > 0$$

$$= \alpha e^{-\alpha t}$$

as the pdf of time to failure during the useful life stage of the bathtub curve. Equation (14.25) defines an exponential pdf, i.e., a special case of the pdf defining the Weibull distribution.

Weibull introduced the distribution, which bears his name, principally on empirical grounds, to represent certain life-test data. Note that Weibull distribution provides a mathematical model of all three stages of the bathtub curve.

Shaefer and Theodore[9] provide additional details on this and the three previous distributions plus a host of other distributions.

EXPERIMENTAL TRACER TECHNIQUES

Tracer techniques have long been employed in an attempt to describe nonideal reactor behavior by quantifying residence time information and data. As noted in an earlier section, some of the elements of a fluid remain in a reactor longer than other elements, with the result that the conversion may vary from element to element depending on its residence time.

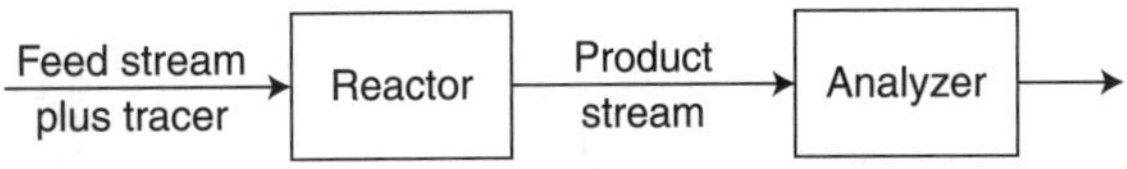

Figure 14.12 Tracer experiment setup.

Analytical instrumentation is required for any tracer experiment to be successful. The measuring device is installed at the outlet of the reactor as shown in Figure 14.12. Typically, but not always, the concentration of the tracer can be linearly related to the signal. Different analytical equipments are noted in Table 14.3.

The two most common tracers can be classified as *radioactive* and *colored*. The two most common methods of injection are *pulse input* and *step input*, with the former favored in practice. Details on the pulse and step method follow.

In the method known as pulse, an amount of tracer is injected into the feed entering the reactor over a period of time approaching zero. The discharge concentration (or equivalent) is then measured as a function of time. Typical concentration curves at the outlet of the reactor can take the form of any of the residence time distribution plots discussed earlier. The most usual response takes the form similar of that in Figure 14.13. Generally, the response approaches a normal or log-normal distribution curve.

TABLE 14.3 Analytical Measurement Equipment for Tracer Experiments

Equipment	Key measurement
Conductometry	Electrical conductivity
Gas chromatography	Adsorption of a compound on a carrier material (gas phase)
Liquid chromatography	Adsorption of a compound on a carrier material (liquid phase)
Mass spectroscopy	Different mass numbers of components
Paramagnetic analysis	Paramagnetic properties of compounds
Photometry	Light absorbance (visible or UV light)
Radioactivity	Radioactive radiation

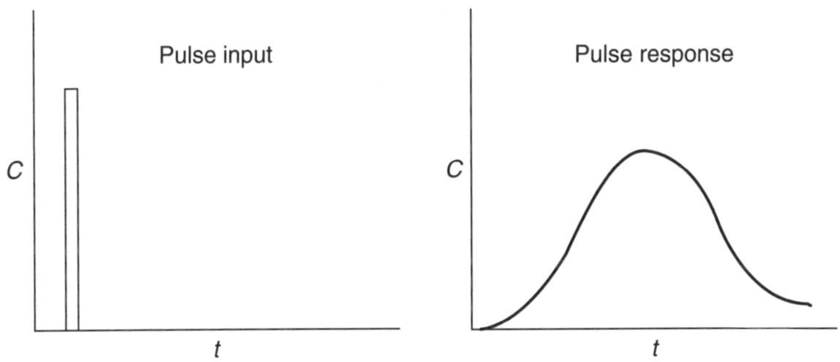

Figure 14.13 Pulse response beta.

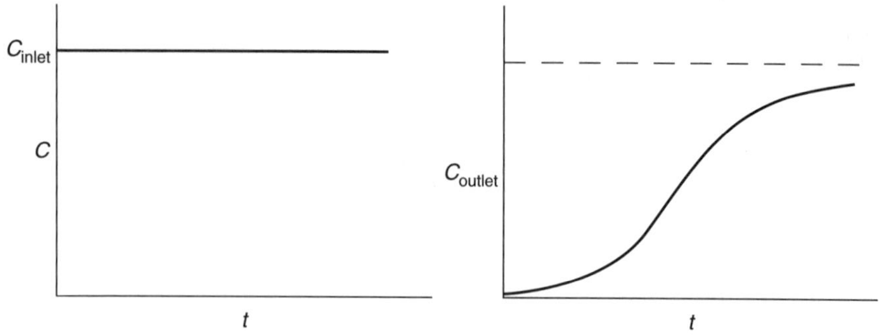

Figure 14.14 Step injection behavior.

In the step injection method, the reactor receives a step input of a fixed and constant tracer concentration for a reactor with a constant volumetric flow rate. The concentration of the tracer is kept constant until the concentration at the outlet is equal to that at the feed; the rest may then be discontinued. A typical concentration–time curve for this type of injection is presented in Figure 14.14. An analytical description of the outlet concentration–time relationship is usually difficult to represent. However this injection method is usually easier to carry out.

To summarize, this experimental method involves perturbing the inlet flow to the reactor and noting the effect on the output from the reactor. Analyzing the response often provides information about the residence time distribution.

REFERENCES

1. L. Theodore and F. Ricci, "*Mass Transfer Operations for the Practicing Engineer,*" John Wiley & Sons, Hoboken, NJ, 2010.
2. J. Reynolds, J. Jeris, and L. Theodore, "*Handbook of Chemical and Environmental Engineering Calculations,*" John Wiley & Sons, Hoboken, NJ, 2004.
3. L. Theodore, "*Heat Transfer for the Practicing Engineer,*" John Wiley & Sons, Hoboken, NJ, 2011.
4. L. Theodore, "*Air Pollution Control Equipment Calculations,*" John Wiley & Sons, Hoboken, NJ, 2008.
5. M. K. Theodore and L. Theodore, "*Introduction to Environmental Management,*" CRC Press/Taylor & Francis, Boca Raton, FL, 2010.
6. S. Fogler, "*Elements of Chemical Reaction Engineering,*" 4th edition, Prentice-Hall, Upper Saddle River, NJ, 2006.
7. L. Theodore, personal notes, East Williston, NY, 1976.
8. L. Theodore, personal notes, East Williston, NY, 1990.
9. S. Shaefer and L. Theodore, (adapted from), "*Probability and Statistics Calculations for Environmental Science,*" CRC Press/Taylor & Francis, Boca Raton, FL, 2007.
10. P. Abulencia and L. Theodore, "*Fluid Flow for the Practicing Engineer,*" John Wiley & Sons, Hoboken, NJ, 2009.
11. L. Theodore and R. Dupont, "*Environmental Health Risk and Hazard Risk Assessment Calculations*" CRC Press/Taylor & Francis, Boca Raton, FL, 2012.

CHAPTER 15

REACTOR DESIGN CONSIDERATIONS

INTRODUCTION

Industrial processes for which reactors are employed include: petroleum units, coffee roasters, fat rendering, wire enameling ovens, packing house effluents, pulp and paper mills, rubber curing, paint and varnish cookers, charcoal broilers, incinerators, textile finishing, nitric acid plants, sulfuric acid plants, and so on. For example, in order to control the discharge of sulfur dioxide, it may be reacted with oxygen in a catalytic reactor at 830°F. The sulfur trioxide formed is then reacted with water vapor to form sulfuric acid, a valuable end product.

Before proceeding to the remainder of the chapter including the illustrative examples, the reader is reminded that reactors can be used for both homogeneous and heterogeneous systems.

Variations and unknowns in many of the parameters in the describing equation(s) make it nearly impossible to provide a generalized procedure for the design of reactors. However, design considerations must definitely include:

1. Reactor volume or mass of catalyst for a given conversion at a particular flow rate.
2. Energy (fuel) requirements.

The analysis should also provide for:

1. provision for energy recovery (if applicable),
2. means for ensuring adequate turbulence for complete reaction,
3. specification of construction materials,
4. controls for optimum operation,
5. specification of auxiliary equipment, e.g., feed lines, fans, pumps, etc.,
6. safety provisions.

If the process is examined using a forward marching approach, one realizes that energy requirements and degree of mixing often dictate the reaction temperature. This in turn sets the reaction rate. A kinetic analysis then provides process design information.

Chemical Reactor Analysis and Applications for the Practicing Engineer. By Louis Theodore

The chapter is concluded by presenting several illustrative examples. As mentioned earlier, attention is focused on both homogeneous and heterogeneous reactions. The last illustrative example sets forth several design calculations employed by the practicing engineer.

Following this Introduction, the remaining chapter sections include:

Design Principles
Specific Design Considerations
Operation and Maintenance, and Improving Performance
Reactor Selection
Applications

DESIGN PRINCIPLES

Current design practices for chemical reactors usually fall into the category of state of the art and pure empiricism. Past experience with similar applications is commonly used as the sole basis for the design procedure. The vendor maintains proprietary files on past installations; these files are periodically revised and expanded as new installations are evaluated. In designing a new unit, the files are consulted for similar applications and old designs are heavily relied on.

By contrast, the engineering profession in general, and the chemical engineering profession in particular, has developed well defined procedures for the design, construction, and operation of chemical plants. These techniques, tested and refined for nearly a century, are routinely used by today's engineers. These same procedures may also be used in the design of chemical reactors.

The purpose of this section is to introduce the reader to some of these process design fundamentals. Such an introduction to design principles, however sketchy, can provide the reader with a better understanding of the major engineering aspects of a chemical reactor, including some of the operational, economic, controls and instrumentation for safety and regulatory requirements, and environmental factors associated with the unit. No attempt is made in the sections that follow to provide extensive coverage of this topic; only general procedures and concepts are presented and discussed.

Preliminary Studies

A process design engineer is usually involved in one of two activities: building a plant or deciding whether to do so. The skills required in both cases are quite similar, but the money, time, and details involved are not as great in the latter situation. It has been estimated that only one out of 15 proposed new processes ever reaches the construction stage. Thus, knowledge at the preliminary stage is vital to prevent financial loss on one hand and provide opportunity for success on the other. In well managed process organizations, the engineer's evalution is a critical activity that usually involves considerable preliminary research on the proposed process. Successful process development consists of a series of actions and decisions, the most significant of which takes place well before plant construction.

It is important to determine whether a project has promise as early in the development stage as possible. This section discusses some of the preparatory work required before the design of a chemical reactor can be formally initiated. In the chemical process industry, there may be an extended period of preparatory work required if the reactor is a unique or first-time application. This can involve bench-scale work by chemists to develop and better understand the process chemistry. This is often followed by pilot experimentation by process and/or development engineers to obtain scale-up and equipment performance information. However, these two steps are usually not required in the design of an "established" reactor. This is presently the situation with most designs although some bench-scale or pilot work may be necessary and deemed appropriate by management.

In general design practice, there are usually five levels of sophistication for evaluating and estimating. Each is discussed in the following list, as it applies to a chemical plant.

1. The first level requires little more than identification of products, raw materials, and utilities. This is what is known as an *order of magnitude estimate* and is often made by extrapolating or interpolating from data on similar existing processes. The evaluation can be done quickly and at minimum cost, but with a probable error exceeding $\pm 50\%$.
2. The next level of sophistication is called a *study estimate* and requires a preliminary process flow sheet (to be discussed in the next sub-section) and a first attempt at identification of utilities, materials of construction, and other ancillary equipment. Estimation accuracy improves to within $\pm 30\%$ probable error, but more time is required and the cost of the evaluation can escalate to over \$30,000 for a \$5 million plant. Evaluation at this level usually precedes expenditures for site selection, market evaluation, pilot plant work, and detailed equipment design. If a positive evaluation results, pilot plant and other activities may also begin.
3. A *scope* or *budget authorization*, the next level of economic evaluation, requires a more defined process definition, detailed process flow sheets, and prefinal equipment design (discussed in a later section). The information required is usually obtained from pilot plant, marketing, and other studies. The scope authorization estimate could cost upwards of \$80,000 for a \$5 million facility with a probable error exceeding $\pm 20\%$.
4. If the evaluation is positive at this stage, a *project control estimate* is then prepared. Final flow sheets, site analyses, equipment specifications, and any architectural and engineering sketches are employed to prepare this estimate. The accuracy of this estimate is about $\pm 10\%$ probable error. A project control estimate can serve as the basis for a corporate appropriation, for judging contractor bids, and for determining construction expenses. Due to increased intricacy and precision, the cost for preparing such as estimate for the process can approach \$150,000.
5. The final economic analysis is called a *firm* or *contractor's estimate*. It is based on detailed specifications and actual equipment bids. It is employed by the contractor to establish a project cost and has the highest level of

accuracy, ±5% probable error. A cost of preparation results from engineering, drafting, support, and management/labor expenses. Because of unforeseen contingencies, inflation, and changing political and economic trends, it is impossible to assure actual costs for even the most precise estimates.

In the case of a chemical reactor design, data on similar existing processes are normally available and economic estimates or process feasibility are determined from these data. It should be pointed out again that most processes in real practice are designed by duplicating or *mimicking* similar existing systems. Simple algebraic correlations that are based on past experience are the rule rather than the exception. This stark reality is often disappointing and depressing to students and novice engineers involved in design. For chemical reactors, the only preparatory work normally required is the gathering of all existing physical and chemical property data of the feed, and auxiliary materials and chemicals. Process chemistry information may also be needed, but this too may be obtained directly from the literature, company files, or a similar type of application.

Process Schematics

To the practicing engineer, particularly the chemical engineer, the process flow sheet is the key instrument for defining, refining, and documenting not only a reactor but also a chemical process. The process flow diagram is the authorized process blueprint, the framework for specifications used in equipment designation and design; it is the single, authoritative document employed to define, construct, and operate the chemical process.[(1)]

Beyond equipment symbols and process stream flow lines, there are several essential constituents contributing to a detailed process flow sheet. These include equipment identification numbers and names, temperature and pressure designations, utility designations, volumetric or molar flow rates for each process stream, and a material balance table pertaining to process flow lines. The process flow diagram may also contain additional information such as energy requirements, major instrumentation, and physical properties of the process streams. When properly assembled and employed, a process schematic provides a coherent picture of the process. It can point out some deficiencies in a chemical reactor that may have been overlooked earlier in the study, e.g., by-products (undesirable or otherwise) and recycle needs, safety considerations, etc. But basically, the flow sheet symbolically and pictorially represents the interrelation between the various flow streams and equipment, and permits easy calculations of material and energy balances. Controls and instrumentation must also be considered in the overall requirements of a reactor system. These concerns are briefly covered later in this chapter. As one might expect, a process flow diagram for a chemical or petroleum plant is usually significantly more complex than that for a reactor.

There are a number of symbols that are universally employed to represent equipment, equipment parts, valves, piping, etc. These are available in the literature.[(1)] Although a significant number of these symbols are used to describe some of the chemical and petrochemical processes, only a few are needed for even the most complex chemical reactors. These symbols obviously reduce, and in some instances, replace detailed written descriptions of the process. Note that many of the symbols are pictorial, which helps in better describing other process components, units, and equipment.

The degree of sophistication and details of a flow sheet usually vary with time. The flow sheet may initially consist of a simple free-hand block diagram with limited information that includes only the equipment; later versions may include line drawings with pertinent process data such as overall and componential flow rates, utility and energy requirements, and instrumentation. During the later stages of a design project, the flow sheet will consist of a highly detailed P&I (piping and instrumentation) diagram; this aspect of the design procedure is beyond the scope of this text; for information on P&I diagrams, the reader is referred to the literature.(2)

In a sense, flow sheets are the international language of the engineer, particularly the chemical engineer. Chemical engineers conceptually view a (chemical) plant as consisting of a series of interrelated building blocks that are defined as *units* or *unit operations*. The plant essentially ties together the various pieces of equipment that make up the process. Flow schematics follow the successive steps of a process by indicating where the pieces of equipment are located and the material streams entering and leaving each unit.

Instrumentation and Controls

The control of a chemical reactor requires careful consideration of all operational and (potential) regulatory requirements. The reactor is designed to process feed material. Design characteristics were covered in earlier chapters. Safety should also be a concern of all individuals involved with reactors.

The safe operation of a reactor requires that the controls keep the system operating within a safe operating envelope (see also Chapter 21). The envelope is based on many of the design and process constraints. These are placed on the unit to ensure the proper operation. Additional controls may be installed to operate additional equipment needed for energy recovery, neutralization, etc.

The control system should also be designed to vary one or more of the process variables to maintain the appropriate conditions within the reactor unit. These variations are programmed into the system based on the past experience of the unit manufactured. The operational parameters that may vary include the flow rate, temperatures in the reactor, system pressure, and any auxiliary equipment.

The proper control system has also been subjected to extensive analysis on operational problems and items that could go wrong. A hazard operability (*HAZOP*) analysis should be conducted on the control system to examine and identify all possible failure mechanisms (see Chapter 21). It is important that all of the failure mechanisms have appropriate response reactions by the control system. Several of the failure mechanisms that must be addressed within the appropriate control system response are excess or minimal temperature, excessive or subnormal flow rate, equipment failure, component failure, sticking or inoperable components, and broken or failed circuits. All should be examined, including the response time of the control system to a problem and the appropriate response of the system to the problem.

The system must identify the problem and integrate necessary actions and alarms into the system.

A complete review of the maximum and minimum process rate of change and the equipment rate of change must be identified before the control system can be defined. The system must be reviewed to define the primary control parameters for operation and safety. The limitations that are imposed on the unit must be identified and monitored to ensure that they are not exceeded.

Scale-Up

The flow sheet referred to in the previous section may also be altered from *batch* to *continuous* operation by scaling. In batch operation, streams are defined in terms of amounts, whereas in continuous operation, streams are defined by flow rates. For example, one could multiply all streams by a constant, e.g., (1000 lb/h)/lb. A word of warning regarding the word "constant": the multiplier constant used for scale-up must be truly independent of all streams and the material (A) in those streams. Determining whether the "constant" is a valid one is not always obvious. Suppose, one used (1000 lbmol/h)/lb as a multiplier. There may be a problem here since this multiplier usually varies from stream to stream.

The fact that balanced flow sheets can always be scaled allows the practicing engineer, at the start of calculations, to assume, for the sake of convenience, any amount of material for a given stream. After applying the material balances to balance the flow sheet, the process can very easily be scaled up or down, or changed from batch to continuous operation. The assumed amount of material is called a *basis*. If the amount or flow rate of a stream is given in a particular design or problem statement, it is usually convenient (but not necessary) to choose this quantity as a basis. If the amount or flow rate is not given, any convenient basis, such as 100 lb or 100 lbmol/h of one of the streams or one of the components of one of the streams, should be chosen.

The design of commercial-sized equipment and plants are based on experience, test work, fundamental principles, and correlations of data. When available, experience is fine; test work properly done is invaluable, though expensive; fundamental principles and data correlations are highly desirable and of course inexpensive if already existing. In such fields of chemical engineering interest as fluid flow or heat and mass transfer, many correlations of the pertinent variables have been developed, permitting the design of large-scale equipment with a limited amount of direct data. Though much interest exists and much effort has been expended in developing similar generalizations of data for the design of chemical reactors, somewhat less success has been achieved because of the greater complexity of this field. Consequently, greater dependence must usually be placed on direct experimental work. Since the smaller the test scale the less expensive the work, the question arises, how far upward is it possible to extrapolate reliably small-scale laboratory or pilot-plant results? No really simple or firm answer exists, but it seems to be fairly commonly agreed that scale-up may be performed satisfactorily by stages in which the test equipment is successively five to ten times the previous size, with the degree of satisfaction depending, of course, on the thoroughness of the test work and its correlations.[3]

Application of chemical similitude to reactions leads to a number of interesting conclusions. It appears that for all types of reactions, temperatures or the temperature-time profile must be the same in the large and small units. In homogeneous systems the reaction times and the initial concentrations must be the same; in heterogeneous ones, the product of the reaction time and the interfacial area per unit volume of the reactor must be the same for both.[3] Both dynamic and chemical similarity ordinarily are rarely achieved simultaneously; e.g., if reaction times are preserved, the Reynolds numbers which also involve linear or mass velocities, cannot also be preserved.[4]

ILLUSTRATIVE EXAMPLE 15.1 The following process is being used to produce sodium hydroxide crystals from a sodium hydroxide solution reactor discharge.

A feed NaOH solution first enters a mixer (M) where a recycle stream, consisting of a more concentrated NaOH solution, is added to it. The mixed stream, which has a mass flow rate of 15,000 lb/h and a NaOH mass fraction of 0.531, moves to a concentrator (C) where pure water is removed. The stream exiting the concentrator has a mass flow rate of 11,850 lb/h. This stream enters a cooling unit (CL) in which crystallization occurs. After the crystals are formed, the solution is decanted and recycled back to the mixer. This recycle stream has a mass flow rate of 4200 lb/h and consists of a saturated NaOH solution (i.e., saturated at the temperature of the cooling unit) with a NaOH mass fraction of 0.471. The product stream (wet crystals) is a mixture of pure NaOH crystals and entrained solution; the NaOH mass fraction of the entrained solution is also 0.471. Refer to Figure 15.1. Which flow diagram best describes this system?

Suggestion: The product stream consists of two phases, solid (dry) crystals and liquid solution. Even though the two phases are not physically separated when leaving the crystallizer, the problem can be more easily solved by considering each phase as a separate product stream.

Solutions. Consider the flow diagram in (A). Since two flow streams enter unit CL, (A) is incorrect.

Consider the flow diagram in (B). Since recycle occurs from C to CL, (B) is incorrect.

Consider the flow diagram in (C). Since recycle occurs form CL to C, (C) is incorrect.

Consider the flow diagram in (D). Since recycle occurs from CL to M, (D) is correct.

The correct answer is therefore (D). ■

ILLUSTRATIVE EXAMPLE 15.2 Refer to the previous example. Calculate the mole fraction NaOH leaving the concentrator (and entering the cooling unit)

Solution. Draw a flow diagram of the process and include all pertinent data. See Figure 15.2. The desired mass fraction NaOH is represented by x_4.

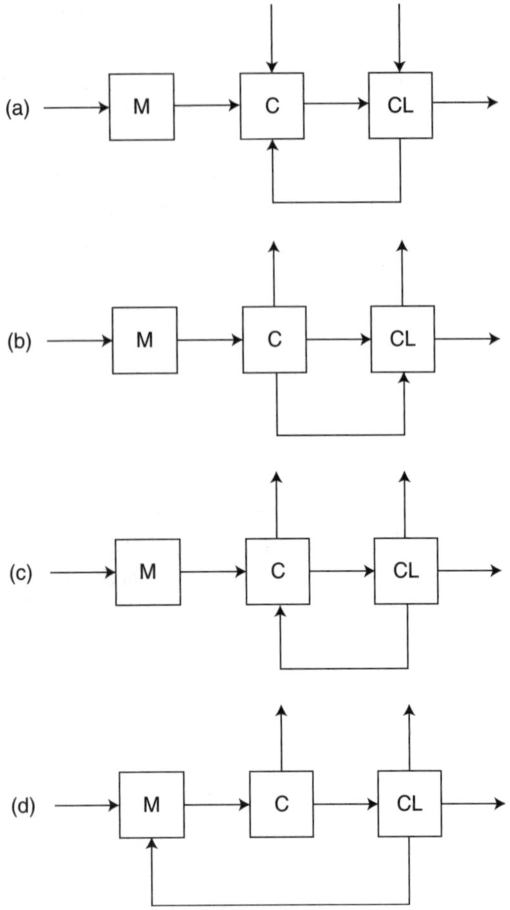

Figure 15.1 Flow diagrams for Illustrative Example 15.1.

Calculate the flow rate of the feed solution by performing a total material balance around the mixer (M).

$$15{,}000 = N_1 + 4200$$
$$N_1 = 10{,}800 \text{ lb/h}$$

Calculate the flow rate of the pure water stream exiting the concentrator by performing a total material balance around the concentrator (C).

$$15{,}000 = N_3 + 11{,}850$$
$$N_3 = 3150 \text{ lb/h}$$

The NaOH mass fraction of the solution leaving the concentrator may now be obtained.

$$(0.531)(15{,}000) = (0)(3150) + (x_4)(11{,}850)$$
$$x_4 = 0.672$$

■

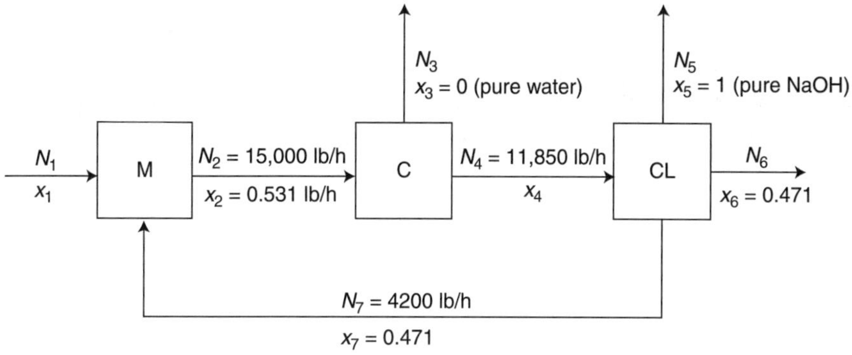

Figure 15.2 Flow diagram for Illustrative Example 15.2.

ILLUSTRATIVE EXAMPLE 15.3 A hot process stream is mixed with recycled gas from an absorber (A), and the mixture passes through a reactor (R), which uses water as the cooling medium. It then passes through a water spray quencher (Q) in which the temperature of the mixture is further decreased and, finally, through an absorber (A) in which water is the absorbing agent (solvent) for one of the species in the product gas stream. Prepare a simplified flow diagram for this process.

Solution. Before attempting to calculate the raw material or energy requirements of a process, it is desirable to attain a clear picture of the process. As noted above, the best way to do this is to draw a flow diagram. A flow diagram is a line diagram showing the successive steps of a process by indicating the pieces of equipment in which the steps occur and the material streams entering and leaving each piece of equipment. Flow diagrams are used to conceptually represent the overall process.

Lines are usually used to represent streams, and boxes may be used to represent equipment. A line diagram of the process is first prepared (Figure 15.3). The equipment in Figure 15.3 may now be labeled as in Figure 15.4. The flow streams labels are shown in Figure 15.5. ■

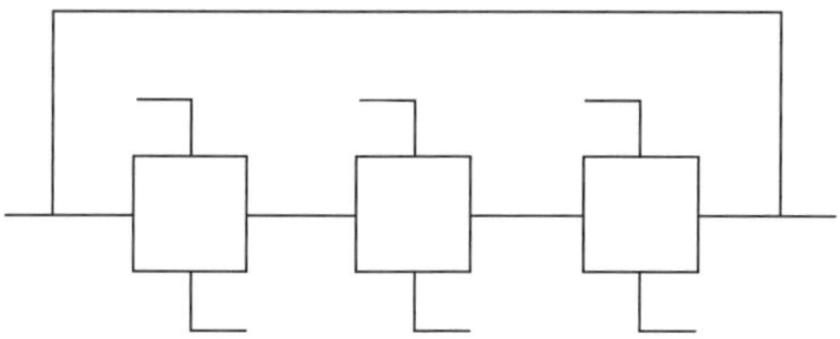

Figure 15.3 Line diagram.

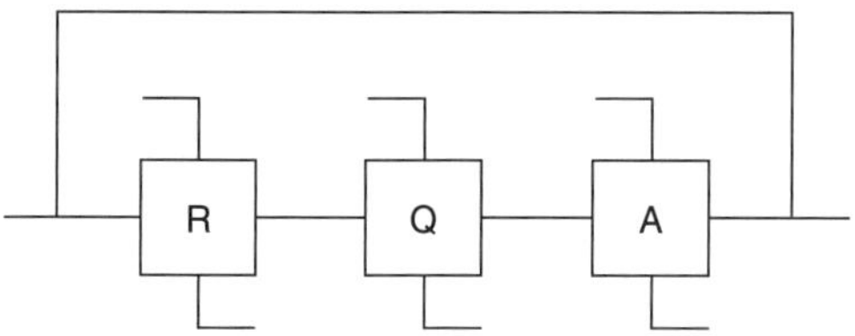

Figure 15.4 Line diagram with equipment labels.

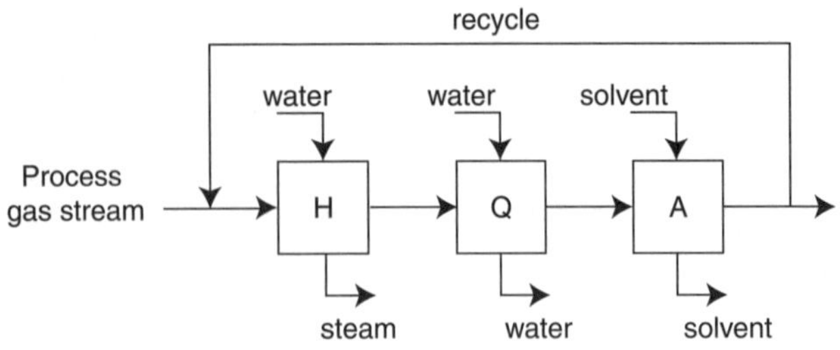

Figure 15.5 Line diagram with stream labels.

SPECIFIC DESIGN CONSIDERATIONS

Theodore[5] has summarized six specific design considerations that the practicing engineer should be aware of. These are detailed below and includes information on catalytic reactors to be discussed in Part IV.

1. Adiabatic operation is most ideal. Generally, do not employ heat exchangers if at all possible. The practicing engineer should first examine the temperature (T) vs. conversion (X) relationship for the reactor. Refer to Figure 15.6. If the relationship is too steep, one should consider the possibility of using intercoolers as shown in Figure 15.7. If the T vs. X slope is low, consider proceeding adiabatically.
2. Once a decision has been made on a heat exchange for this particular reaction try not to go below 1 inch or $1\frac{1}{2}$ inch tubes for catalytic or non-catalytic reactor systems.
3. One can control heat transfer primarily through a heating medium. Then apply the general heat transfer equation.[6]

$$\dot{Q} = UA\,\Delta T_{\mathrm{LM}} \tag{15.1}$$

   The area A can then be calculated.
4. The engineer must also decide on the number and the diameter of tubes. Care should be exercised here because of:
   a. velocity constraints, and
   b. pressure drop considerations.[4]

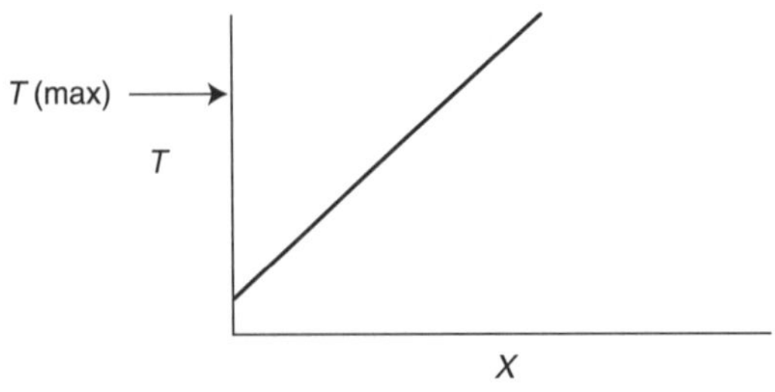

Figure 15.6 T vs. X relations.

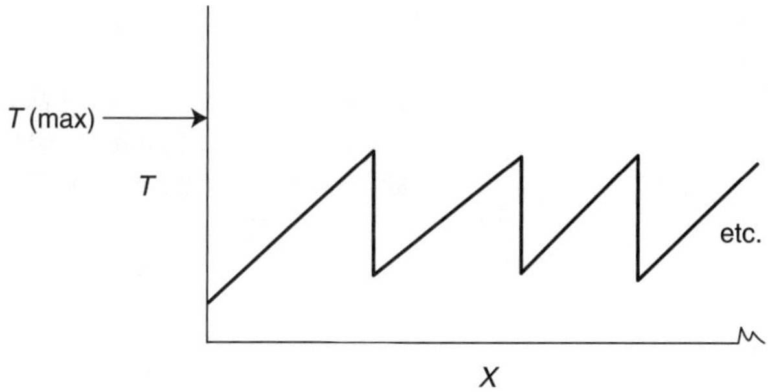

Figure 15.7 Reactor with intercoolers.

5. Fluidized beds usually operate adiabatically. Therefore, heat exchangers are often not necessary. However, velocity constraints still apply. As will be shown in Chapter 18, Part IV, the pressure drop is given by the weight of the bed.
6. For heterogeneous/catalytic reactors/reactions (to be discussed in Chapters 17–19, Part IV) remember the following.
   a The same basic forms of equations apply.
   b The volume *V* is replaced by the catalytic mass *W*.
   c A reaction velocity constant *k* and the corresponding rate equation applies to a particular catalyst.
   d For fixed bed units employ the TF model.
   e For fluidized bed units use the CSTR model.

OPERATION AND MAINTENANCE AND IMPROVING PERFORMANCE

Little to no information has been published on the section title, as it applies to chemical reactors. However, Santoleri et al.[1] and Theodore[6,7] have provided information on incinerators, which, in a very real sense, can in some aspects also be applied to chemical reactors. This information follows.

Normal operation of an incinerator should be quite simple. A controller should be incorporated into the design to maintain the outlet temperature at a fixed value by varying the auxiliary fuel input. Combustion air (assuming limited air in the fume stream) is usually controlled by the average amount of fumes to be incinerated. This adjustment is normally set manually if the fume flow rate and fume heat content are fairly constant. If the fume heat content or flow rate are likely to be highly variable, a more sophisticated control system may be appropriate—one that analyzes combustion efficiency at the outlet as well as fume flow and outlet temperature, and varies both auxiliary fuel input and combustion air accordingly. But, normal operation of even a complicated control system usually requires nothing more than an occasional monitoring of the instruments.

Most incinerators are custom-designed within certain basic parameters. Therefore, they are likely to be accompanied by a very complete instruction manual that should include the manufacturer's basic maintenance instructions from all the sub-suppliers, i.e., the incinerator manufacturer will have purchased equipment, such as the refractory, valves, and controls, from other suppliers. The operating and maintenance instructions for this equipment will be quite extensive and complete because it has been written by the original manufacturer, who is concerned only with particular items. The instruction manual is therefore a very useful document from a maintenance viewpoint and should be followed explicitly. (The instruction manual may not be so useful from a system operations viewpoint because incinerator systems are usually custom-designed and system problems cannot always be anticipated, which often results in some field modifications to the operating instructions.)

There are some general maintenance guidelines that can be discussed. The insulation or refractory should be inspected on a regular basis. Cracks may develop, especially in brick joints, and thermal shock damage (spalling) should be repaired with a suitable plastic of the same thermal properties. The burner (if applicable) should be inspected at regular intervals for signs of warpage or corrosion. Moving parts should be lubricated with graphite or a similar high-temperature lubricant. Lubricants that carbonize should not be used under any circumstances. Also dirt, mortar, carbon, or other foreign matter should be cleaned from the burner area. The pressure seals around any parts projecting through the burner or incinerator shell should be inspected. Usually, these are asbestos rope packing glands and should be fairly tight after the adjusting/retaining screws have been loosened. These seals should only be lubricated with flake or powder graphite. This should never be mixed with oil, as the oil will carbonize. The gas jets should be free of corrosion and should be cleared of any deposits on burners using gas as an auxiliary fuel.

The outer shell of the incinerator should be inspected, especially when a new lining has been installed, for signs of thermal shock, i.e., welds, especially at the outlet should be checked for hairline cracks, which are the first signs of poor thermal design.

The auxiliary fuel piping train should be inspected in accordance with the manufacturer's instructions. Electrically operated valves and interlock switches should also be inspected frequently for conditions that might cause "shorting" (short-circuiting), e.g., dirty contacts, moisture leaks, deteriorating insulation, etc. Air supply lines and filters (to air-operated valves) should be inspected for dirt or blockages. The valves themselves are usually provided with air signal and air supply pressure gauges, and these should be checked occasionally for accuracy. If there are shutoff dampers in the ductwork to or from the incinerator, their seals should be checked frequently.

Maintenance procedures for catalytic incinerators should include catalyst cleaning every 3 months to a year. Cleaning is usually accomplished by blowing clean compressed air through the catalyst elements, by vacuuming, or by washing with water or a mild detergent not containing phosphates. Iron oxide deposits can be removed by soaking with a mild organic acid followed by a water rinse.

Improving the performance of a thermal incinerator basically involves the optimization of the combustion process. Ideally, no more combustion air should be used than is required for complete combustion of the fumes and the auxiliary fuel. The auxiliary fuel should be used only in amounts required to maintain the design furnace temperature.

An incinerator operating efficiently should normally have only 1–2% O_2 and 0–1% combustibles in the outlet gases. Monitors are available that can indicate these parameters to the operator as well as provide automatic control of the incinerator when required.

The following questions often require answers. If odors are involved, has their concentration been reduced or eliminated? Has there been adequate reduction in the emissions of all types of reactive hydrocarbons? Has there been a reduction in any possible explosion or public health hazard that might exist in the manufacturing process? If so, the air pollution control system should be well designed, and operated and maintained properly.

REACTOR SELECTION

There are a number of factors to be considered prior to selecting a reactor.[8] In general, they can be grouped into three categories: engineering, economic, and environmental.

Engineering

1. Contaminant characteristics (i.e., physical and chemical properties, concentration, particulate shape and size distribution—in the case of particulates, chemical reactivity, corrosivity, abrasiveness, toxicity, etc.)
2. Feed stream characteristics (i.e., volumetric flow rate, temperature, pressure, humidity, composition, viscosity, density, reactivity, combustibility, corrosivity, toxicity, etc.)
3. Design and performance characteristics of the particular reactor (i.e., size, weight, pressure drop, reliability and dependability, turndown capability, power requirements, utility requirements, temperature limitations, maintenance requirements, flexibility, etc.)

Economics (see also Chapter 21)

1. Capital cost (equipment, installation, engineering, etc.)
2. Operating cost (utilities, maintenance, etc.)
3. Expected reactor lifetime and salvage value
4. Profitability

Environmental (if applicable, see also Chapter 21)[9]

1. Equipment location (see Illustrative Example 15.4)
2. Available space
3. Ambient conditions
4. Availability of adequate utilities (i.e., power, water, etc.) and ancillary system facilities (i.e., waste treatment and disposal, air pollution control equipment, etc.)
5. Maximum allowable emissions (environmental codes)
6. Aesthetic considerations (i.e., visible steam or water vapor plume, etc.)
7. Noise levels

Proper selection of a particular chemical reactor for a specific application at times can be extremely difficult and complicated. In view of the multitude of complex and variety of reactors, it is in the best interest of the practicing engineer to work closely with experienced engineer and operators as early in the selection process as possible. Obviously, previous experience on a similar application cannot be overemphasized.

If should be noted that a multitude of points have been raised above. Economics is almost always the ultimate driving force in the selection of the type of reactor for a particular application. When the reactor required is very large, the cost of the reactor for the process can be controlling. However, the major issue is the concern that the reactor will be both efficient and dependable, and operate safely during the operation of the process.

Finally, the optimum reactor design for a given application is one that results in the manufacture of a desired product(s) at maximum profit and the lowest cost. The optimum design must consider all fixed and operating charges, raw-material and feed costs plus by-product credits and costs. A perturbation technique can often be employed where the basic design factors are systematically perturbed (varied) and the overall economics from each design are generated. The combination which produces the maximum profit is then selected.[10]

ILLUSTRATIVE EXAMPLE 15.4 List several guidelines that should be followed when selecting an "optimum" site for a chemical reactor.

Solution. The following guidelines should be followed when selecting a site for either a reactor or a plant.

1. *Topography.* A fairly level site is needed to contain spills and prevent spills from migrating and creating more of a hazard. Firm soil above water level is recommended.
2. *Utilities and water supply.* The water supply must be adequate for cooling. The source of electricity should be reliable to prevent unplanned shutdowns.
3. *Roadways.* Roadways should allow access to trucks entering and exiting the site plus emergency vehicles such as ambulances and fire engines in the event of an emergency.

4. *Neighboring communities and plants.* Population density and proximity to the plant should be considered for the initial site and in anticipation of a possible future expansion.
5. *Waste disposal.* Waste disposal systems containing flammable, corrosive, or toxic materials should be a minimum distance of 250 ft from the reactor.
6. *Climate and natural hazards.* Lightning arrestors should be installed to reduce/eliminate ignition sources in any flammable areas.
7. *Emergency services.* Emergency services should be readily available and appropriately equipped. ■

ILLUSTRATIVE EXAMPLE 15.5 Outline the key safety factors that should be considered in the layout of a new chemical reactor.

Solution. The key safety factors that should be considered in the layout of a new reactor are:

1. site selection (see previous example),
2. water supply,
3. utilities,
4. offices and ancillary equipment,
5. storage and loading areas,
6. effect of other process equipment,
7. safety equipment,
8. access in and out of the plant. ■

ILLUSTRATIVE EXAMPLE 15.6 Provide suggestions for purchasing a *new* chemical reactor.[11]

Solution. Prior to the purchase of any reactor, experience has shown that the following 12 points should be emphasized.

1. Refrain from purchasing any chemical reactor without reviewing *certified independent test data* on its performance under a similar application. Request the manufacturer to provide performance information and design specifications.
2. In the event that sufficient performance data are unavailable, request that the equipment supplier provide a small pilot model for evaluation under existing conditions.
3. Request participation of important administration in the decision-making process.
4. Prepare a good set of design specifications.
5. Make a careful material balance study before purchasing the reactor.

6. Refrain from purchasing any reactor until *firm* installation cost estimates have been added to the equipment cost. *Escalating installation costs are the rule rather than the exception.*
7. Give operation and maintenance costs high priority on the list of reactor selection factors.
8. Refrain from purchasing any reactor until assured that the new reactor will utilize any fuel, controllers, filters, motors, etc., that are compatible with those already available at the plant.
9. The specification should include written assurance of *prompt* technical assistance from the reactor supplier. This, together with a complete operating manual (with parts list and full schematics), is essential and is too often forgotten in the rush to get the reactor operating.
10. Meeting schedules can be critical. In such cases, *delivery guarantees* should be obtained from the manufacturers and penalties identified.
11. Include a *strong performance guarantee* from the manufacturer to ensure that the reactor will meet all design specifications.
12. Finally, withhold 10–15% of the purchase price until design operating performance is clearly demonstrated.

■

ILLUSTRATIVE EXAMPLE 15.7 Discuss new reactor problems.

Solution. The usual design/procurement/construction/startup problems can be compounded by any one or combination of the following.

1. Unfamiliarity of process engineers with a chemical reactor.
2. New suppliers with frequently unproven equipment.
3. Lack of industry standards for some reactors.
4. Compliance schedules that are too tight.
5. Vague specifications.
6. Weak guarantees.
7. Unreliable delivery schedules.
8. Effect of process unreliability problems on the new reactor.

■

APPLICATIONS

This last section provides the reader with examples dealing with not only reactor design and predictive calculations but also with several ancillary equipment that often are part of the reactor "process". A total of ten illustrative examples follow. The last illustrative example examines a formaldehyde process from a conversion, yield, material balance, etc., point of view.

ILLUSTRATIVE EXAMPLE 15.8 Determine the hydraulic, brake and installed horsepower for a centrifugal pump to move 95,000 lb/hr of kerosene with a specific gravity of 0.95 from a tank at ground level and atmospheric pressure to the top of a 35-foot-tall location of a reactor operating at 58 psig. Assume losses across valves, pipes, and fittings to be 25 psig. Comment on pipe sizing.

Typical centrifugal pump efficiencies are provided in Table 15.1. Reasonable pipe velocities for low-viscosity liquids range from 4 to 6 feet per second.(4)

Solution. Calculate the volumetric flow rate of the liquid in ft^3/s and gal/min.

$$\begin{aligned} \text{Density, } \rho &= (0.95)(62.4) = 59.3 \text{ lb/ft}^3 \\ \text{Volumetric Flow, } q &= \dot{m}/\rho \\ &= (95{,}000)/[(59.3)(3600)] \\ &= 0.445 \text{ ft}^3/\text{s} \\ &= (0.445)(60)(7.481) \\ &= 199.7 \text{ gal/min} \end{aligned}$$

Estimate the required pressure increase in psi.(4)

$$\begin{aligned} \Delta P &= \rho(h_2 - h_1) + (P_2 - P_1) + \text{loss} \qquad (15.2) \\ &= (59.3)(35)/(144) + 58 + 25 \\ &= 97.4 \text{ psi} \end{aligned}$$

Calculate the hydraulic horsepower (HHP).(4)

$$\begin{aligned} \text{HHP} &= 0.000583(q)(\Delta P) \qquad (15.3) \\ &= 0.000583(199.7)(97.4) \\ &= 11.3 \text{ hp} \end{aligned}$$

TABLE 15.1 Horsepower-Efficiency Data

Horsepower	Efficiency
1	25%
2	35%
5	50%
10	55%
20	65%
50	70%
100	75%
200	80%
500	82%

Calculate the brake horsepower (BHP). Initially select an overall pump efficiency of 60%

$$\begin{aligned} \text{BHP} &= \text{HHP/E} \\ &= 11.3/0.60 \\ &= 19.2\ \text{hp} \end{aligned} \tag{15.4}$$

Select the installed horsepower (IHP).

$$\text{IHP} = 20\ \text{hp}$$

Typical velocities for liquids flowing in pipes are:

Water	3–10 ft/s
Liquid hydrocarbons	4–6 ft/s

For highly viscous fluids, lower pipe velocities are more economical. Like pumps, pipes do not come in all possible sizes. To select an available size, use a table of standard pipe sizes and choose one close to the diameter estimated from the assumed velocity. Refer to the literature for additional details.[12] ■

ILLUSTRATIVE EXAMPLE 15.9 Determine the hydraulic, brake and installed horsepower required to compress a reactor feed of 5000 lb/h of ethane vapor at 20 psia and 75°F, having a density of 0.105 lb/ft^3, to 200 psia. Large compressors typically have efficiencies in the range of 75% to 85%; employ a median value of 80% in the calculation.

Solution. Calculate the volumetric flow rate of the inlet gas in ft^3/s.

$$\begin{aligned} q &= \dot{m}/\rho \\ &= (5000)/[(0.105)(3600)] \\ &= 13.2\ \text{ft}^3/\text{s} \end{aligned}$$

The molar flow rate is

$$\begin{aligned} \dot{n} &= 5000/30.07;\ \text{MW(ethane)} = 30.07 \\ &= 166.3\ \text{lbmol/h} \end{aligned}$$

Calculate the energy required per mole for isothermal compression[13] and the accompanying horsepower

$$\begin{aligned} W &= RT \ln (P_2/P_1) \\ &= (1.987)(535) \ln (200/20) \\ &= 2448\ \text{Btu/lbmol} \end{aligned} \tag{15.5}$$

$$\begin{aligned} \text{HP} &= (2448)(166.3) \\ &= 407{,}102\ \text{Btu/h} \\ \text{HP} &= 407{,}102/2545 \\ &= 160\ \text{hp} \end{aligned}$$

By definition, the brake horsepower is

$$\begin{aligned} \text{BHP} &= \text{HP}/E \\ &= 160/0.80 \\ &= 200 \text{ hp} \end{aligned}$$

Select the installed horsepower

$$\text{IHP} = 200\text{–}220 \text{ hp (range)}$$

Typical velocities for gases flowing in pipes are:

Air	50–100 ft/s
Hydrocarbon vapors	30–80 ft/s
Steam $\leq$50 psig	50–100 ft/s
Steam $>$50 psig	100–200 ft/s

Like compressors, pipes do not come in all possible sizes. To select an available size, use a table of standard pipe sizes and choose one close to the diameter estimated for the assumed velocity.[4,12] The pipe Schedule Number is an indicator of wall thickness. ■

ILLUSTRATIVE EXAMPLE 15.10 Estimate the heat transfer area required to preheat a reactor feed of 40,000 lbs per hour of cyclohexane liquid from 75°F to 220°F using saturated steam at 300°F as the heat source. The heat capacity of cyclohexane is 0.57 Btu/lb · °F. The overall heat transfer coefficient for the exchanger is 138 Btu/ft^2 · °F · h.

Solution. Calculate the total heat duty.

$$\begin{aligned} \dot{Q} &= (40{,}000)(0.57)(220 - 75) \\ &= 3{,}306{,}000 \text{ Btu/h} \end{aligned}$$

Calculate the "log-mean temperature difference" (LMTD).

$$\begin{aligned} \text{LMTD} &= (\Delta T_2 - \Delta T_1)/\ln(\Delta T_2/\Delta T_1) \\ &= (225 - 80)/\ln(225/80) \qquad (15.6) \\ &= 140.2°\text{F} \end{aligned}$$

Calculate the required heat exchanger area. Employ Equation (15.1).

$$\begin{aligned} A &= \dot{Q}/[(U)(\text{LMTD})] \\ &= (3{,}306{,}000)/[(138)(140.2)] \\ &= 171 \text{ ft}^2 \end{aligned}$$

The reader is referred to the literature for more extensive details on heat exchangers and heat exchanger design.[14,15] ■

ILLUSTRATIVE EXAMPLE 15.11 Consider the following reaction

$$A \longrightarrow 2B;\ k = 3.75\ (\min)^{-1}$$

The reaction occurs in the gaseous phase, and is elementary and irreversible. If the inlet flow rate is 5 lbmol/min, and the reaction occurs at 100°F and 10 atm, determine the required reactor volume and space time needed to convert 90% of A in a CSTR and TF.

Solution. Preliminary calculations:

$$\delta = 2 - 1 = 1; \quad \varepsilon = y_{A_0}\delta = 1(1) = 1$$

$$C_A = C_{A_0}(1 - X); \quad X = X_A$$

$$q_0 = \frac{F_{A_0}RT}{P} - \frac{(5\,\text{lbmol/min})\left(0.73\dfrac{\text{ft}^3\cdot\text{atm}}{\text{lbmol}\cdot{}^\circ\text{R}}\right)(100+460)^\circ\text{R}}{10\,\text{atm}}$$

$$q_0 = 204.4\,\text{ft}^3/\text{min}$$

$$-r_A = kC_A = kC_{A_0}\left(\frac{1-X}{1+\varepsilon X}\right) = \frac{kF_{A_0}}{q_0}\left(\frac{1-X}{1+\varepsilon X}\right); \quad X = X_A$$

For a CSTR:

$$V = \frac{F_{A_0}X}{-r_A} = \frac{q_0X(1+\varepsilon X)}{k(1-X)}$$

Substituting

$$V = \frac{204.4[0.9(1+0.9)]}{3.76(1-0.9)} = 930\,\text{ft}^3$$

and

$$\tau = \frac{V}{q_0} = \frac{930}{204.4} = 4.55\ \text{min}$$

For a TF:

$$V = F_{A_0}\int_0^X \frac{dX}{(-r_A)} = \frac{F_{A_0}}{kC_{A_0}}\int_0^X \frac{(1+\varepsilon X)\,dX}{(1-X)}$$

$$= \frac{q_0}{k}\int_0^X \frac{(1+\varepsilon X)\,dX}{(1-X)}$$

From Illustrative Example 21.18,

$$V = \frac{q_0}{k}\left\{(1+\varepsilon)\ln\left(\frac{1}{1-X}\right) - \varepsilon X\Big|_0^X\right\}$$

Substituting

$$V_{\text{TF}} = \left(\frac{204.4}{3.76}\right)\left[(1+1)\ln\left(\frac{1}{1-0.9}\right) - 1(0.9)\right]$$

$$= 54.36\,\text{ft}^3[2\ \ln 10 - 0.9] = 201.4\,\text{ft}^3$$

$$\tau = \frac{V_{\text{TF}}}{q_0} = \frac{201.4}{204.4} = 0.985\,\text{min} = 60\,\text{s}$$

As expected, from an "economical" viewpoint, a TF is more feasible to accomplish a 90% conversion of A. ■

ILLUSTRATIVE EXAMPLE 15.12 Refer to the previous example. Determine the time it takes for a batch reactor to achieve a 80% conversion of A under the same (now initial) conditions as the previous example.

Solution. For a batch reactor

$$t = N_{A_0}\int_0^X \frac{\text{d}X}{-r_A}$$

Since $V = V_0(1 + \varepsilon X)$

$$t = \frac{N_{A_0}}{kC_{A_0}}\int_0^X \frac{(1+\varepsilon X)\,\text{d}X}{(1-X)(1+\varepsilon X)}$$

which reduces to

$$t = \frac{C_{A_0}}{kC_{A_0}}\int_0^X \frac{\text{d}X}{(1-X)} = \frac{1}{k}\ln\left(\frac{1}{1-X}\right)\Bigg|_0^X$$

Substituting

$$t = t_R = \frac{1}{3.76}\ln\left(\frac{1}{1-0.8}\right) = 0.428\ \text{min}$$

■

ILLUSTRATIVE EXAMPLE 15.13 Refer to the previous illustrative example. What is the production rate of B if it takes 20 minutes to empty, 2 hours to clean, and 15 minutes to fill the reactor. The reactor initially contains 250 lbmol of A.

Solution. The production rate of B is given by

$$P_B = \frac{\text{moles of } B \text{ produced}}{\text{total cycle time}} = \frac{N_B}{\text{tct}}$$

$$= \frac{2N_{Ao}}{t_R + t_{\text{fill}} + t_{\text{clean}} + t_{\text{empty}}}$$

Substituting

$$P_B = \frac{(2)(250)(0.8)}{0.428 + 20 + 15 + 120}$$
$$= 2.57\ \text{lbmol/min}$$ ■

ILLUSTRATIVE EXAMPLE 15.14 Comments on the results of the previous illustrative example

Solution. The reaction time is so low that it may be neglected in the calculations. ■

ILLUSTRATIVE EXAMPLE 15.15 Determine the total flowrate of cooling water required for the services listed below in Table 15.2 for a chemical reactor if a cooling tower system supplies the water at 90°F with a return temperature of 115°F. How much fresh water makeup is required if 5% of the return water is sent to "blow-down"? Note that the cooling water heat capacity is 1.00 Btu/lb · °F, the heat of vaporization at cooling tower operating conditions is approximately 1030 Btu/lb, and the density of water at cooling tower operating conditions is 62.0 lb/ft^3.

Solution. The required cooling water flowrate, q_{CW}, is given by the following equation:

$$q_{\text{CW}} = \dot{Q}_{\text{HL}}/[(\Delta T)(c_p)(\rho)] \tag{15.7}$$

where

$$\dot{Q}_{\text{HL}} = \text{heat load, Btu/min}$$
$$\Delta T = \text{change in temperature} = 115°\text{F} - 90°\text{F} = 25°\text{F}$$
$$c_p = \text{heat capacity} = 1.00\ \text{Btu/lb} \cdot °\text{F}$$
$$\rho = \text{density of water} = (62.0\ \text{lb/ft}^3)(0.1337\ \text{ft}^3/\text{gal})$$
$$= 8.289\ \text{lb/gal}$$

The heat load is

$$\dot{Q}_{\text{HL}} = (12 + 6 + 23.5 + 17 + 31.5)(10^6\ \text{Btu/h})/60\ \text{min/h}$$
$$= 1{,}500{,}000\ \text{Btu/min}$$

Thus,

$$q_{\text{CW}} = \frac{1{,}500{,}000\ \text{Btu/min}}{(25°\text{F})(1.00\ \text{Btu/lb} \cdot °\text{F})(8.289\ \text{lb/gal})} = 7250\ \text{gpm}$$

The blow-down flow, q_{BD}, is given by the following:

$$q_{\text{BD}} = (\text{BDR})(q_{\text{CW}}) \tag{15.8}$$

where BDR is the blow-down rate = 5% = 0.05.

TABLE 15.2 Reactor Cooling Water Required

Process unit	Heat duty (Btu/h)	Required temperature (°F)
1	12,000,000	250
2	6,000,000	200–276
3	23,500,000	130–175
4	17,000,000	300
5	31,500,000	150–225

Thus,

$$q_{\mathrm{BD}} = (0.05)(7250\,\mathrm{gpm}) = 362.5\,\mathrm{gpm}$$

The rate of water vaporized by the cooling tower, ΔH_v is given by:

$$q_v = (q_{\mathrm{HL}})/[(\rho)(\Delta H_v)]$$

where ΔH_v is the heat of vaporization = 1030 Btu/lb.
Substitution yields

$$\begin{aligned} q_v &= (1{,}500{,}000\ \mathrm{Btu/min})/[(8.289\ \mathrm{lb/gal})(1030\ \mathrm{Btu/lb})] \\ &= 175.7\,\mathrm{gpm} \end{aligned}$$

■

ILLUSTRATIVE EXAMPLE 15.16 Determine how many pounds per hour of steam are required for the following two applications involving four reactors, if:

1. steam is provided at 500 psig, and
2. steam is provided at both 500 and 75 psig pressures.

The reactors have the following heating requirements (see Table 15.3). Note the properties of saturated steam are provided in Table 15.4.

Solution. **1.** The total required flow-rate of 500 psig steam, $\dot{m}_{\mathrm{BT}}$, is given by:

$$\dot{m}_{\mathrm{BT}} = \dot{m}_{\mathrm{B1}} + \dot{m}_{\mathrm{B2}} + \dot{m}_{\mathrm{B3}} + \dot{m}_{\mathrm{B4}}$$

TABLE 15.3 Heating Requirements

Reactors	Unit heat duty (UHD; Btu/h)	Required temperature (°F)
1	10,000,000	250
2	8,000,000	450
3	12,000,000	400
4	20,000,000	300

TABLE 15.4 Steam properties

Pressure provided (psig)	Saturation temperature (°F)	Enthalpy of vaporization (ΔH_v) (Btu/lb)
75	320	894
500	470	751

For the above equation:

$\dot{m}_{\text{B1}}$ (mass flow-rate of 500 psig steam through reactor 1)
$= \text{UHD}/\Delta H_v = 13{,}320$ lb/h
$\dot{m}_{\text{B2}}$ (mass flow-rate of 500 psig steam through reactor 2)
$= \text{UHD}/\Delta H_v = 10{,}655$ lb/h
$\dot{m}_{\text{B3}}$ (mass flow-rate of 500 psig steam through reactor 3)
$= \text{UHD}/\Delta H_v = 15{,}980$ lb/h
$\dot{m}_{\text{B4}}$ (mass flow-rate of 500 psig steam through reactor 4)
$= \text{UHD}/\Delta H_v = 26{,}635$ lb/h

Thus,

$$\sum_{i=1}^{4} \dot{m}_{\text{Bi}} = \dot{m}_{\text{BT}}$$

$$\sum_{i=1}^{4} \dot{m}_{\text{BT}} = 66{,}590 \text{ lb/h}$$

2. The required combined total flowrate of 500 and 75 psig steam, $\dot{m}_{\text{CT}}$, is given by

$$\dot{m}_{\text{CT}} = \dot{m}_{75,1} + \dot{m}_{\text{B2}} + \dot{m}_{\text{B3}} + \dot{m}_{75,4}$$

For this situation:

$\dot{m}_{75,1}$ (mass flow-rate of 75 psig steam through reactor 1)
$= \text{UHD}/\Delta H_v = 11{,}185.7$ lb/h
$\dot{m}_{75,4}$(mass flow-rate of 75 psig steam through reactor 4)
$= \text{UHD}/\Delta H_v = 22{,}371.4$ lb/h

Thus,

$$\begin{aligned}\dot{m}_{\text{CT}} &= 11{,}186 + 10{,}655 + 15{,}980 + 22{,}371 \\ &= 60{,}192 \text{ lb/h}\end{aligned}$$

Note that since the saturation temperature of 75 psig steam is lower than two of the four reactors, the 500 psig steam must be used for reactors 2 and 3. ■

ILLUSTRATIVE EXAMPLE 15.17 The following information and data is to be employed for Questions (1–10).
Methane and oxygen can react to produce formaldehyde (HCHO) according to the following reaction:

$$CH_4 + O_2 \rightarrow HCHO + H_2O$$

The following undesirable reaction also occurs;

$$CH_4 + 2O_2 \rightarrow CO_2 + 2H_2O$$

These reactions take place in the reactor RX (see Figure 15.8). MX is a mixer, SEP a separations system, and SPL a splitter. The molar composition of stream 2 is 40% methane, 40% oxygen, and 20% nitrogen. The HCHO molar flow rate (stream 4) is 8% of the stream 2 flow rate. The splitter flow ratio (stream 8: stream 7) is 3:1, i.e., Q_8 is three times Q_7. The molar ratio of oxygen to carbon dioxide in stream 3 is 2:1.

Selecting 100 gmol/h as a basis for stream 2, answer the following questions.

1. Determine the componential flow rate of HCHO for stream 3.
2. Determine the componential flow rates of O_2, N_2, CO_2 and H_2O for stream 3.
3. Determine the total flow rates for streams 4, 5, and 6.
4. Determine the total flow rates for streams 7 and 8.
5. Determine the composition (mode percent basis) of stream 6.
6. Determine the composition (mode percent basis) of stream 8.
7. Determine the total flow-rate of the process feed stream 1.

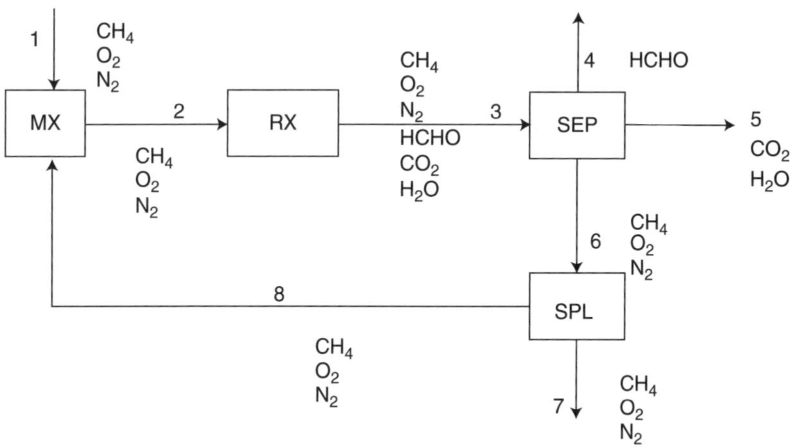

Figure 15.8 Formaldehyde process.

8. What are the single pass conversion and the overall conversion of the methane?
9. What are the yield of formaldehyde, and the selectivity of the formaldehyde (over carbon dioxide)?
10. If a formaldehyde production rate of 10,000 lbmol/month is desired, what is the required feed rate (stream 1)?

Solution

1. From the balanced equation in the Illustrative Example statement and data, the inlet and outlet flow for each species around the reactor (assuming that x is the number of moles of CH_4 reacted in the first reaction and y is the number of moles of CH_4 reacted in the second reaction) are:

Inlet	40		40		0		0
	CH_4	+	O_2	$\rightarrow$	HCHO	+	H_2O
Outlet	$40 - x - y$		$40 - x - 2y$		x		$x + 2y$

	CH_4	+	$2O_2$	$\rightarrow$	CO_2	$2H_2O$
Outlet	$40 - x - y$		$40 - x - 2y$		y	$x + 2y$

From the problem statement, and assuming all of the HCHO exits in stream 4, $x = (0.08)100 = 8$ gmol/h.

2. The ratio of oxygen to carbon dioxide in the product gas (stream 3) is 2:1. The following equation may therefore be written:

$$\frac{40 - x - 2y}{y} = 2$$

Solve for y in terms of x and note that $x = 8$.

$$y = \frac{40 - x}{4} = \frac{40 - 8}{4}$$

$$y = 8 \text{ gmol/h}$$

The outlet stream of the reactor (stream 3) consists of

O_2 : $40 - x - 2y = 40 - 8 - 2(8) = 16$ gmol/h
N_2 : 20 gmol/h (problem statement)
CO_2 : $y = 8$ gmol/h
H_2O : $x + 2y = 8 + 2(8) = 24$ gmol/h

3. The total flow rate for stream 4 is equal to the flowrate of HCHO in stream 3. Therefore,

$$q_4 = 8 \text{ gmol/h}$$

All of the carbon dioxide and water in stream 3 enters stream 5. The total flow rate for stream 5 is then

$$q_5 = CO_{2,\text{stream 3}} + H_2O_{\text{stream 3}} = 8 + 24 = 32 \text{ gmol/h}$$

All CH_4, O_2 and N_2 in stream 3 enter stream 6. The total flowrate for stream 6 is

$$q_6 = CH_{4,\text{ stream 3}} + O_{2,\text{stream 3}} + N_{2,\text{stream 3}} = 24 + 16 + 20 = 60 \text{ gmol/h}$$

4. A material balance around the splitter yields

$$q_7 + q_8 = Q_6 = 60$$

The ratio of q_8 to q_7 is:

$$\frac{q_8}{q_7} = \frac{3}{1} = 3$$

One may solve both equations above simultaneously to obtain:

$$\begin{aligned} q_7 + 3q_7 &= 60 \\ q_7 &= 15 \text{ gmol/h} \\ q_8 &= 45 \text{ gmol/h} \end{aligned}$$

5. The compositions for stream 6 are:

$$\begin{aligned} CH_{4,\text{ stream 6}} &= \frac{24}{60} = 0.4 = 40\% \\ O_{2,\text{ stream 6}} &= \frac{16}{60} = 0.267 = 26.7\% \\ N_{2,\text{ stream 6}} &= \frac{20}{60} = 0.333 = 33.3\% \end{aligned}$$

6. The composition of stream 8 is the same as the composition of stream 6, i.e., 40% CH_4, 26.7% O_2, and 33.3% N_2.

7. An overall balance around the mixer gives:

$$45 + q_1 = 100$$

Thus,

$$q_1 = 100 - 45 = 55 \text{ gmol/h}$$

8. Conversions are calculated as follows:

$$\text{Single pass conversion} = \frac{CH_{4,\text{ stream 2}} - CH_{4,\text{ stream 3}}}{CH_{4,\text{ stream 2}}}$$

$$\text{Single pass conversion} = \frac{40 - 24}{40} = 0.4 = 40\%$$

$$\text{Overall conversion} = \frac{CH_{4,\text{ stream 1}} - CH_{4,\text{ stream 7}}}{CH_{4,\text{ stream 1}}}$$

$$\text{Overall conversion} = \frac{(0.4)(55\text{ gmol/h}) - (0.4)(15\text{ gmol/h})}{(0.4)(55\text{ gmol/h})}$$

$$= 0.73 = 73\%$$

9. The yield and selectivity are given by:

$$\text{Yield} = \frac{8}{40} = 0.2 = 20\%$$

$$\text{Selectivity} = \frac{8}{8} = 1.0 = 100\%$$

10. Scaling up the feed from the product ratio yields:

$$\left(\frac{55\text{ feed}}{8\text{ HCHO product}}\right)10{,}000\text{ lbmol/month} = 68{,}750\text{ lbmol/month}$$

■

REFERENCES

1. J. Santoleri, J. Reynolds, and L. Theodore, "*Introduction to Hazardous Waste Incineration*," 2nd edition, John Wiley & Sons, Hoboken, NJ, 2000.
2. G. Ulrich, "*A Guide to Chemical Engineering Process and Economics*," John Wiley & Sons, Hoboken, NJ, 1984.
3. Source, date, and location unknown.
4. P. Abulencia, and L. Theodore, "*Fluid Flow for the Practicing Engineer*," John Wiley & Sons, Hoboken, NJ, 2009.
5. L. Theodore, personal notes, East Williston, NY, 1974.
6. L. Theodore, "*Ask the Experts: Designing Thermal Afterburners*," Chem. Eng. Progress, New York City, NY, April 2005.
7. L. Theodore, personal notes, East Williston, NY, 1975.
8. L. Theodore and A. J. Buonicore, "*Industrial Air Pollution Control Equipment for Particulates*," CRC Press, Boca Raton, FL, 1976.
9. M. K. Theodore and L. Theodore, "*Introduction to Environmental Management*," CRC Press/Taylor & Francis, Boca Raton, FL, 2010.
10. J. Reynolds, J. Jeris, and L. Theodore, "*Handbook of Chemical and Environmental Engineering Calculations*," John Wiley & Sons, Hoboken, NJ, 2004.
11. L. Theodore, personal notes, 1978.
12. D. Green and R. Perry (editors), "*Perry's Chemical Engineers' Handbook*," 8th edition, McGraw-Hill, New York City, NY, 2008.

13. L. THEODORE, F. RICCI, and T. VANVLIET, "*Thermodynamics for the Practicing Engineer*," John Wiley & Sons, Hoboken, NJ, 2009.
14. L. THEODORE, "*Heat Transfer for the Practicing Engineer*," John Wiley & Sons, Hoboken, NJ, 2011.
15. D. KERN, "*Heat Transfer*," McGraw-Hill, New York City, NY, 1950.

PART **IV**

OTHER REACTOR TOPICS

Elegance in prose composition is mainly this: a just admission of other topics and of words; neither too many nor too few of either

—*Imaginary Conversations, ibid. Chesterfield* and *Chatham*, Walter Savage Landor, 1775–1864

This book is divided into four Parts, and as noted in the Preface and Introduction, this last Part, Part (IV) concerned with what the author has defined as "other reactor topics". The six chapter titles are:

1. Catalysts (Chapter 16)
2. Catalytic Reactions (Chapter 17)
3. Fluidized and Fixed Bed Reactors (Chapter 18)
4. Biochemical Reactors (Chapter 19)
5. Open-Ended Questions (Chapter 20)
6. Abet-Related Topics (Chapter 21)

This Part is also supplemented with a number of illustrative examples, particularly in the latter four chapters (18–21).

Chemical Reactor Analysis and Applications for the Practicing Engineer. By Louis Theodore

CHAPTER 16

CATALYSTS*

INTRODUCTION

One of the more important fields/considerations in kinetics and chemical reactor design is catalysis (Greek word: loosen), and this chapter serves to introduce the reader to this subject. Chemists recognized in the early 1800s that small quantities of (primarily) solid material could profoundly influence both the rate and mechanism of a reaction.

There is some question regarding the "birth" of catalysts. However, in the early 1800s, it was observed that the nature of some reactions was influenced by the presence of a substance which was in itself unchanged in the process. This phenomenon was observed in numerous applications. The opinion at that time was that a catalytic force was present in these reactions and the reaction brought about under the influence of this force was referred to as catalysis. Although this earlier concept has now been discarded, the term catalysis is used today to describe all processes in which the rate of a reaction is influenced by a substance called a **catalyst** that remains chemically unaffected. If the reaction is hindered by the added substance it is said to be a **negative catalyst**; however the word catalyst, when used alone, is almost invariably taken to imply acceleration of the chemical reaction process.

There are three general characteristics common to all catalysts, as detailed below.

1. The catalyst is unchanged chemically at the end of the reaction; however it does undergo a physical change.
2. A small amount of catalyst is often sufficient to bring about a considerable extent of reaction; generally, the *rate* of a catalytic reaction is proportional to the concentration of the catalyst.
3. The catalyst does not affect the equilibrium state in a reversible reaction; since the catalyst is unchanged, it contributes no energy to the system so that the second law of thermodynamics applies and that the same equilibrium state is reached whether a catalyst is employed or not.
4. Increases the rate of a reaction.

*Adapted from "Harshaw Catalysts for Industry", Harshaw Chemical Company, Cleveland, Ohio, date unknown.

Chemical Reactor Analysis and Applications for the Practicing Engineer. By Louis Theodore

5. Directs the reaction toward a particular product.
6. Eliminates undesired side reactions.
7. Initiates a new reaction.

Metals in the platinum family are recognized for their ability to promote reactions at low temperatures. Other catalysts include various oxides of copper, chromium, vanadium, nickel, and cobalt. These catalysts are subject to poisoning, particularly from halogens, sulfur compounds, zinc, arsenic, lead, mercury, and particulates. High temperatures can also reduce catalyst activity. It is therefore important that catalyst surfaces be clean and active to insure optimum performance. Catalysts can usually be regenerated with superheated steam.

Catalysts may be porous pellets, usually cylindrical or spherical in shape, ranging from $1/16$ to $1/2$ inch in diameter. Small sizes are recommended, but the pressure drop through the reactor increases. Other shapes include honeycombs, ribbons, wire mesh, etc. Since catalysis is a surface phenomena, an important physical property of these particles is that the internal pore surface may be many magnitudes greater than the outside surface. The reader is referred to the remainder of this chapter for more information on catalyst preparation, properties, comparisons, costs, and impurities.

Some of the advantages of catalytic reactors are.

1. low fuel requirements,
2. lower operating temperatures,
3. little or no insulation requirements,
4. reduced fire hazards,
5. reduced flashback problems.

The disadvantages include:

1. high initial cost,
2. catalyst poisoning,
3. (large) particles must first be removed,
4. some liquid droplets must first be removed,
5. catalyst regeneration problems.

It should also be noted once again that a catalyst does not, in general, alter the position of equilibrium. For a reversible process it must therefore affect both the forward and reverse reactions to the same extent, since $K = k/k'$; thus, catalysts will accelerate the forward reaction at the same degree as the reverse change.

The remainder of this chapter reviews the following topics:

Key Definitions and Testing Procedures

Chemical and Formulated Catalysts

Catalytic Processes

Catalyst Selection and Evaluation

KEY DEFINITIONS AND TESTING PROCEDURES

Some definitions that relate to catalyst and to catalytic reactions (see next chapter) are provided below. Testing procedures are included with some of the definitions.

1. Absorption—The transfer of a substance through a phase boundary followed by a solution with the receiving phase.
2. Activity—The ability of a catalyst to alter the rate of a chemical reaction.
3. Adsorption—The adhesion of the molecules of gases or of dissolved substances or liquids to the surfaces of solid bodies with which they are in contact.
4. Carrier—A physical support for a catalyst deposited in or on it. A carrier may be catalytically active or inert. Support, mount, and base are also synonyms for carrier.
5. Catalyst Preformed—A solid catalyst made in a prescribed geometric shape.
6. Inhibitor—A substance which retards, delays or prevents a chemical reaction.
7. Poison—A material which reduces or destroys the effectiveness of a catalyst under normal conditions of use.
8. Promoter—A substance which is added to a catalyst that disproportionately benefits its activity or selectivity.
9. Selectivity–The preferential activity of a catalyst for a specified reaction (see also Chapter 5).
10. Space Velocity—The volume of gas or liquid measured under specified conditions passing through a unit volume of catalyst bed per unit time (see also Chapter 10).
11. Surface Area—The available internal and external area of a catalyst as measured by gas absorption, usually expressed as m^2/g of catalyst.
12. Apparent Bulk Density (ρ_B)—Abbreviated ABD, this is a measured weight per unit volume. Two values may be reported: packed and loose. Packed density is calculated from the weight obtained by filling a graduated cylinder in small increments, tapping the cylinder on a soft surface after each addition until a constant volume results. Loose density is calculated from the weight obtained by gently filling a graduated cylinder from a funnel without shocking nor tapping.
13. Crushing Strength—An arbitrary measurement of the force needed to collapse a single tablet; the force is usually described in pounds (force).
14. Erosion or Abrasion—A stainless steel tube is partially filled with a weighed sample. The tube is mounted on a revolving drum and rotated for a given time. The fines generated are reported as weight percent.
15. Micro Mesh Sieve Analysis—A method to determine particle size distribution of powders, usually finer than 100 microns (μm) in diameter. A

series of micro mesh sieves are employed, with screen openings of 20, 45, 60, and 90 microns, nested in a Rotap. The dry method is used whenever possible but in some instances, it may be necessary to use a compatible liquid for complete separation on the finer screens. At times, full wet screening is necessary.

16. Porosity—The ratio of the internal void space *within* a solid particle to the total external volume of the solid particle is the volume porosity of that material. Porosity does *not* include the external void space among particles.

17. Pore Volume—The simplest procedure is to calculate the volume of water absorbed per unit weight of dry sample from the porosity sample above[(16)], This includes all pore volumes larger than about 10 Å (Angstroms) diameter and usually is expressed as cc per gram. Immersion liquids other than water may also be used, necessarilly applying the specific gravity of the particular liquid to the calculations.

18. Pore Diameter, Average—This value assumes all pores are cylindrical and that surface is attributed to the cylinder walls. The pore volume divided by surface area gives an average pore diameter.

19. Pore Volume Distribution—A range of pore sizes and their respective volumes in a catalyst.

20. Surface Area—This method involves the measurement of a quantity of adsorbed gas which would be required to cover the entire surface of the pore walls, including roughness, with a one molecule thick layer.

21. Voids—It is the free volume within the catalyst bed external to the catalyst (does not include pore volume). It is expressed in per cent or a fractional value and is usually determined along with the porosity.

ILLUSTRATIVE EXAMPLE 16.1 Discuss procedures usually employed to measure the apparent bulk density.

Solution. The diameter of the cylinder employed should be at least eight times that of the largest dimension of the catalyst particle. In large diameter reactors (low ratio of wall area to volume), the filling weight would be approximately 90% of the packed apparent bulk density, ABD. In small diameter, multi-tube reactors (high ratio of wall area to volume), the filling weight should be between the packed and loose ABD. ■

ILLUSTRATIVE EXAMPLE 16.2 Discuss procedures available to measure the crushing strength of a catalyst.

Solution. Many types of apparatus can be used such as pneumatic lever crushers, spring operated hand crushers or lead shot crushers. Also, the force can be applied to either side or ends of the catalyst. The Harshaw average side crushing strength is determined by automatically feeding shot into a pan at one end of a lever with a knife-edge fulcrum; at the other end, the increasing load is applied to the cylindrical surface of the test pellet between two parallel plates until it collapses, which automatically cuts off the shot feed. The load in pounds force required to produce collapse is the crushing strength and is determined from the weight of shot and the mechanical advantage of the lever. At least 50 determinations are usually required to obtain a reliable average value. ■

ILLUSTRATIVE EXAMPLE 16.3 Outline a procedure to measure porosity.

Solution. Volume porosity may be determined by completely immersing a dry sample of known weight in water and boiling to eliminate air. After boiling, the water is adjusted to a measured volume. The free water is then poured off and measured. The difference between the total volume and the volume of the free water is the total external volume of the solid. The internal pore space is equivalent to the difference between the wet and dry weight of the sample (assume a specific gravity of unity for water). This difference in weight divided by the total external volume of the solid (as determined above) is the volume porosity ratio. This value is approximate and does not include porosities of less than 10 Å diameter. ■

ILLUSTRATIVE EXAMPLE 16.4 An outdated catalyst size analysis is provided in Table 16.1. Discuss the essence of this data.

Solution. These numbers mean that 40% of the particles by weight are greater than 5 μm in size, 27% are less the 5 μm but greater than 2.5 μm, 20% are less than 2.5 μm but greater than 1.5 μm, and the remainder (13%) are less than 1.5 μm. ■

ILLUSTRATIVE EXAMPLE 16.5 Outline how to calculate the geometric mean of various catalyst particle sizes.

Solution. The reader is referred to Chapter 14 for additional details. The geometric mean can be calculated using the following equation:

$$\bar{X}_G = n\sqrt{(X_1)(X_2)\cdots(X_n)} \tag{16.1}$$

■

ILLUSTRATIVE EXAMPLE 16.6 Outline how to calculate the standard deviation of a group of catalyst particle sizes.

TABLE 16.1 Early Notation of Particle Size Distribution

–	>5.0	μm	40%
<5	>2.5	μm	27%
<2.5	>1.5	μm	20%
<1.5	–	μm	13%
–	–	–	100%

Solution. The most commonly used measure of dispersion, or variability, of sets of data is the *standard deviation* σ. Its defining formula is given by expression (see also Chapter 14)

$$\sigma = \sqrt{\frac{\sum (X_i - \bar{X})^2}{n - 1}} \tag{16.2}$$

where $\sigma =$ standard deviation (always positive)

$X_i =$ value of ith data point

$\overline{X} =$ mean of data sample

$n =$ number of observations

If applied to the particle diameter d:

$$\sigma = \sqrt{\frac{\sum (d_i - \bar{d})^2}{n - 1}} = \sqrt{\frac{n \sum d_i - \left(\sum d_i\right)^2}{n(n - 1)}} \tag{16.3}$$

■

ILLUSTRATIVE EXAMPLE 16.7 Discuss catalyst fouling and deactivation.

Solution. Many reaction rates decline due to an accumulation of a foreign *deposit* on the catalyst, even if the reactor is operating under steady conditions. This is defined as *fouling* of the catalyst and there is a need to remove this deposit since it can cause a serious problem during operation as well as in the design of a reactor.

The effect of fouling can be determined by noting the effect of processing time on conversion during operation. ■

ILLUSTRATIVE EXAMPLE 16.8 Briefly describe catalyst activity.

Solution. Catalyst activity for a specific reaction refers to an expression of the reactant conversion or product yield relative to that obtained with a reference catalyst under similar conditions, or by comparing the residence time required to achieve the same conversion with both. ■

CHEMICAL AND FORMULATED CATALYSTS

The various physical forms of catalysts include:

1. Extrusion
2. Flake
3. Granule
4. Liquid

5. Powder
6. Ring
7. Sphere
8. Tablet

A list of the various catalysts employed in practice is provided below. Note that the list is alphabetized by the chemical symbol for the catalyst in question, i.e., they are identified by the elemental symbols of the principal atom.

Catalysts are also listed below by their industrial names.

1. Silver, Ag
2. Alumina, Al
3. Barium Fluoride, Ba
4. Chromia Alumina, Cr
5. Cobalt, Co
6. Cobalt Molybdate, CoMo
7. Copper, Cu
8. Copper Chromite, Cr_2O_3 and CuO
9. Iron, Fe
10. Magnesia, Mg
11. Manganese, Mn
12. Molybdena Alumina, Mo
13. Nickel, Ni
14. Nickel Molybdenum, Ni
15. Nickel Tungsten, Ni
16. Palladium, Pd
17. Titanium, Ti
18. Tungsten, W
19. Vanadia, V
20. Zinc, Zn
21. Zinc Chromite
22. Zinc Oxide
23. Zinc Acetate catalyst
24. Zirconia, Zr
25. Boron Trifluoride Dihydrate
26. Boron Fluoride Monoethylamine
27. Boron Trifluoride Phosphoric Acid
28. Cadmium Acetate
29. Chromium Chloride

30. Cobalt Acetate
31. Cobalt Sulfate
32. Copper Chloride
33. Hydrofluoric Acid, Anhydrous
34. Manganese Acetate
35. Magnesium Fluoborate
36. Nickel Acetate
37. Nickel Nitrate
38. Platinum Monometallic Reforming
39. Platinum Bimetallic Reforming
40. Sodium Methylate
41. Selenium Powder
42. Zinc Acetate
43. Zinc Fluoborate
44. Zinc Nitrate

The 44 catalysts listed above is but a sampling of the major ones employed by industry. As one might suppose, there are numerous other catalysts. Finally, activity in the nanotechnology[1] area has produced catalysts that have outperformed catalysts employed in the past.

CATALYTIC PROCESSES

This final section provides a very brief description of the catalytic processes that are currently found in industry.

1. *Alkylation*—The alkylation reaction commonly used in the petroleum industry involves condensation of an isoparaffin (e.g., isobutane) with olefins—usually propylene, butylene, or amylene—to produce highly branched gas components with desirable stability properties and high octane number. It is said to be second only to reforming as a means to economical octanes.
2. *Ammonia Dissociation*—Ammonia is dissociated yielding a gas consisting of 75% hydrogen and 25% nitrogen by volume and usually less than 0.1% residual free ammonia. The cracking occurs at about 1000°C with catalyst. The catalyst is activated by the ammonia feed when the exit gas no longer contains moisture. Other catalysts can also be used to crack ammonia.
3. *Chromatography*—This is a method of separation based upon selective adsorption of components from a gas or liquid sample. As the sample is carried through the chromatographic column by the flow of a carrier gas, individual molecules of the sample gas are continuously adsorbed and then

released (desorbed) from the surface of the adsorbent. The adsorbent delays the passage of the individual components through the column, with the amount of delay being dependent upon each component's adsorption characteristics. The time delay among appearances of the individual components in the column effluent permits identification.

In a liquid system, solution mixtures are separated by passage through a chromatographic column. Different substances flow with varying speeds down the column and become separated into distribution zones. The column core can then be "pushed" out and the zones of material cut apart, or the zones can be eluted by passing more solvent down the column and collecting material in small fractions.

4. *Cyclization*—The conversion of aliphatic hydrocarbons containing six or more carbon atoms in a chain to the corresponding aromatic hydrocarbon is known as dehydrocyclization. The reaction sequence is believed to involve dehydrogenation of a paraffin to an olefin followed by ring closure and subsequent dehydrogenation of the ring compound to an aromatic. In many cases, isomerization (see 18) reactions also take place.

5. *Dehydration and Hydration*—Addition or removal of water from a compound may be accomplished by a catalyst. Dehydration may occur as an independent reaction or as a heterogeneous reaction such as the condensation of an alcohol with ammonia, or the formation of esters from alcohols and acids. Alcohols can be dehydrated to form olefins or ethers. Other dehydration reactions are glycerine to acrolein and acetic acid to acid anhydride. Olefins can be condensed with water to form ethers, aldehydes, ketones and alcohols. The hydration reactions are governed by pressure, temperature, and the mole ratio of reactants. High pressures are occasionally necessary to counteract dehydration tendency and promote hydration.

6. *Dehydrogenation*—In general, hydrogenation catalysts can also remove hydrogen from organic compounds. The production of olefins and diolefins from paraffins is an important commercial process. Formation of butene from butane is favored at about atmospheric pressures while butadiene is favored at reduced pressures.

7. *Esterification*—Favorable equilibrium can be obtained in ester formation by using dehydration catalysts. Alumina may be used, but normally sulfuric acid is employed. Methyl esters are produced from glycerides and excess methanol at moderate temperatures and atmospheric pressure.

8. *Halogenation*—This reaction incorporates one of the halogen elements, usually chlorine or fluorine into a compound. Halogen gas can be used, as in the addition of chlorine to benzene to form chlorobenzene. Copper chloride on various supports are also used to add Cl_2 or HC to olefins. Fluorine or HF can be added to olefins.

9. *Hydrodealkylation*—This reaction removes alkyl side chains from alkyl-aromatic hydrocarbons, and adds hydrogen—used in the production of naphthalene from methyl naphthalene and benzene from toluene.

10. *Hydrocracking*—Hydrocracking in a broad sense is defined as a process in which cracking reactions take place in the presence of hydrogen over an acidic catalyst containing a hydrogenating component. A large number of simultaneous reactions occur, including cracking of higher molecular weight hydrocarbons to lower molecular weights, isomerization (see 18 on next page), hydrogen transfer and hydrogenation. Many variations and combinations of these reactions are feasible depending upon the operating conditions of temperature and pressure and upon the nature of the feedstock. Hydrocracking can be carried out at any pressure if the temperature is sufficient to induce cracking. For long cycles, however, the pressure must be high enough to prevent coking which tends to increase with temperature. For coking polycyclic aromatics, pressures must be high enough to perform ring saturation since unsaturated mixtures will not crack.

11. *Hydrogen Treating*—Hydrogen treating, as applied to the petroleum refining industry where it has its most common application, is defined as the contacting of petroleum feedstocks with hydrogen. The treated products from this process are usually intermediates that eventually see further conversion via fluid cat cracking or reforming. Hydrogen treating is most commonly used in the treatment of:

    a. Naphtha

    b. Lube Oil Wax and Solvent

    c. Gas Oil for Catalytic Cracking

    d. Furnace Oil, Diesel Fuel, and Kerosene

    e. Residual Stocks.

12. *Hydrogenation of Benzene*—Benzene is readily hydrogenated. The reaction, converting benzene to cyclohexane, proceeds even at ambient temperature and moderate pressure, undiluted or in the presence of hexane or cyclohexane.

13. *Hydrogenation of Unsaturated Carbon–Carbon and Carbon–Oxygen Bonds*—In general, nickel catalysts are employed for hydrogenating unsaturated carbon to carbon bonds. Conversely, copper chromite catalysts hydrogenate carbonyl bonds in preference to carbon to carbon bonds. This selective activity, along with specific pressure and temperature conditions, can be favorably applied to preferentially hydrogenate one of two possible sites.

14. *Hydrogenation of Carbon Oxides*—Methanation is the hydrogenation to methane of carbon oxides present as impurities in hydrogen or synthesis gas streams. The process requires a highly active, rugged, supported nickel catalyst in a fixed bed. Several reactions (all exothermic) are possible depending on quality of feed gas and operating conditions.

15. *Hydrogenation of Fats, Oils, and Fatty Acids*—These products include:

    a. Margarine

    b. Shortening

c. Salad and Cooking Oils

d. Fatty Acids

e. Castor Wax.

16. *Hydrogenation of Nitrogen Compounds*—Amines are derivatives of ammonia in which one or more hydrogen atoms are replaced by alkyl or aryl groups. Amines, because of an unshared pair of electrons on the nitrogen atom, are bases in that they are proton acceptors or electron donors. However, they are relatively weak bases because they are found in water solution largely as hydrated molecules in which association is caused by hydrogen bonding. Lower molecular weight amines are gases with an odor resembling ammonia. They are soluble in water and form salt similar to ammonia salts. The solubility of amines in water decreases with increasing molecular weight, but because of salt formation, they are all soluble in dilute acids. General methods for amine preparations are the alkylation of ammonia by alkyl halides, alcohols, or carbonyl compounds in the presence of hydrogen, and the hydrogenation of nitro compounds, nitriles, and amides. In most cases, mixtures of primary, secondary, and tertiary amines are obtained.

17. *Interesterification (rearrangement)*—Random interesterification of triglyceride fats, particularly lard, produces a refined crystal structure, thus enhancing texture and cake baking properties in shortening formulations.

18. *Isomerization*—Isomerization converts hydrocarbons or other organic compounds into compounds whose molecules have a different arrangement of atoms, but the same number and kind of atoms. An example is the conversion of normal butane to iso-butane in the production of iso-octane and other high-grade motor fuels.

19. *Methanol Synthesis*—The methanol synthesis is represented by the following equations:

$$CO + 2H_2 \longrightarrow CH_3OH$$
$$CO_2 + 3H_2 \longrightarrow CH_3OH + H_2O$$

These reactions are accelerated by zinc chromite catalyst. The rate of conversion and heat of formation increase with pressure. Typical optimum conditions are about 250 atmospheres and 300°C.

20. *Oxidation*—In general, the metals-silver, vanadium, copper, molybdenum, platinum, palladium, cobalt, nickel, manganese, tin, and lead-in either oxide or metallic form, mounted on suitable supports, are used in the oxidation of hydrocarbons. For complete oxidation, active supports such as gamma aluminas have been used. For selective oxidation, alpha alumina or silicon carbide are employed. Sometimes bi-metals, tin–vanadium, iron–molybdenum, and vanadium–molybdenum are mounted on the carrier.

21. *Oxo Process*—The Oxo Process involves two separate catalytic reactions. The first is the addition of carbon monoxide and hydrogen to olefins (hydroformylation) to produce aldehydes. The second reaction is hydrogenation of the aldehydes to the corresponding alcohols.

22. *Purification of Gas Streams*—Reduced copper or nickel catalysts are used to remove oxygen from most gas streams. If hydrogen is present in the gas stream, the oxygen is removed catalytically; water is the product. In hydrogen-free gas streams, either reduced nickel or reduced copper will remove oxygen by simply converting the active metals to their oxides. These beds must be reactivated periodically to re-reduce the catalyst. Hydrogen fluoride removal is also possible.

23. *Reforming*—Catalytic reforming, as applied to the petroleum refining industry, is defined as the conversion of either straight chain or cyclic hydrocarbons that boil substantially within the gasoline range, in the presence of hydrogen and a precious metal catalyst, to produce motor fuels and/or aromatic hydrocarbons. Motor fuel production via catalytic reforming usually involves the transformation of low octane naphthas. Aromatic hydrocarbon manufactured by catalytic reforming commonly converts paraffinic hydrocarbons to benzene, toluene, and xylene. The catalytic reforming process for both motor fuels and aromatics, incorporates the following simultaneous reactions:

    a. Dehydrogenation of C_6-ring napthenes
    b. Isomerization of other napthenes and their dehydrogenation to aromatics
    c. Dehydrocyclization of paraffins
    d. Hydrocracking
    e. Isomerization of paraffins.

24. *Saturation of Aromatics*—Petroleum recycle stocks, high in polycyclic aromatics, make poor feed for catalytic cracking. Pre-treatment hydrogenation is feasible. Polycyclics can be substantially saturated at pressures in the 2000–3000 psi range but with moderate temperatures, not exceeding 750°F. Under these conditions, some desulfurization and denitrogenation also result. The heavy recycle stocks thus treated make excellent catalytic cracking feeds.

25. *Sulfur Removal*—Small amounts of sulfur from natural gas and other gases can be removed with zinc sulfide.

26. *Vinyl Acetate Synthesis*—Acetylene and acetic acid are combined over zinc acetate on carbon catalyst at 170–210°C to form vinyl acetate monomer. The reaction is highly exothermic in the 200°C temperature range and catalytic converters must be designed with cooling facilities for heat removal to prevent catalyst destruction. With fresh catalyst, the temperature should be limited to the 170°C range, gradually increasing to 210°C maximum to maintain the conversion rate as catalytic activity decreases.

Finally, it should also be noted that molecular sieves have become a major new class of catalysts in recent years that can be synthesized with custom-built activity and selectivity. These sieves, also defined as zeolites, are a particular class of aluminosilicate compounds with well-defined crystal structures. They can be employed in a variety of chemical processes, but have had the greatest impact in catalytic cracking, hydrocracking, alkylation, and isomerization processes.

CATALYST SELECTION AND EVALUATION

The job of selecting, evaluating and developing a catalyst is certainly difficult and complex. The factors that determine catalyst activity is still in a state of flux despite the enormous amount of research and development work performed during and immediately following World War II.

The selection of a particular catalyst for a process is often a major problem for the practicing engineer. Furthermore, the development of a new catalyst is a formidable task that involves significant time, manpower, and money. It is not uncommon for thousands of trials and experiments to be performed in developing an acceptable catalyst for a new process.

The search for new and/or improved catalysts appears to be a never-ending process that has recently received impetus because of developments in the nanotechnology field(1). This predicament is further exacerbated because small changes in the operating temperature and/or pressure, or small change in yields or product distribution can have significant economic impact on the process of concern.

The general properties of catalysts were discussed earlier. Some of this material can be used as a guide in the selection of the appropriate catalyst for a specific operation. However, the selection process is still more art than science.

The ultimate goal of the practicing engineer is to develop a rational method of selecting the best catalyst for a specific reaction. Naturally, any study for a new catalyst usually begins with substances that can be expected to positively interact with the reactants, producing product quantity and quality of a desirable nature. Although the engineer is aware of what general class of catalyst is suitable for a particular reaction/process, their individuality differs greatly in activity, selectivity, resistance to degradation, economics, etc. Even small differences in a catalyst's properties can ultimately impact the economics at the commercial scale.

If a practicing engineer has been assigned a catalyst development project, there are four procedures that he/she should investigate before entering into a potentially extensive, time consuming, and expensive experimentation program.

1. Literature search
2. Patent search
3. Economic feasibility
4. Thermodynamics

The next step involves the development of the aforementioned experimental program. The nature of this aspect of the program will depend on whether it involves the development of:

1. a new catalyst,
2. a new catalyst for a new or novel process,

3. a catalyst modification, or
4. a catalyst modification for an existing process.

The reaction in question can then be categorized as oxidation, polymerization, isomerization, etc. The engineer/scientist at this point is then somewhat constrained to the use of certain catalysts for the reaction under study.

De Maio and Naglieri[2] have provided the following regarding the catalyst evaluation's last step:

> "Rapid screening of catalysts can be performed in a battery of microreactors maintained at the same operating conditions of flow rate, temperature, and pressure. The optimum data obtained during the initial screening can serve as an arbitrary standard for comparison until such time as better and more extensive results are obtained.
>
> Since commercial processes require catalysts with physical and chemical stability, high activity, high selectivity, catalyst regenerability, etc., the evaluation effort may be time consuming and costly. It may be justified since the overall economics and eventual profitability are at stake.
>
> Finally, a program for selection and evaluation of catalysts can be outlined as follows:
>
> 1. Vary the flow rate, catalyst volume and size, to test for mass transfer limitations.
> 2. If a supported catalyst is used, check the carrier for catalytic activity. Vary the properties of both the carrier and the catalyst, and attempt to relate the effects to activity and selectivity.
> 3. Conduct a study to determine the effect of varying the preparation technique on catalyst properties.
> 4. Develop a mathematical index (or the equivalent), that defines catalyst quality in terms of temperature, reactant concentration and product formation.
> 5. Compare physical and chemical properties of fresh and spent/used catalysts to determine factors contributing to catalyst deactivation/decay.
> 6. Evaluate the regenerability of catalyst (if applicable)."

REFERENCES

1. R. Kunz, and L. Theodore, "*Nanotechnology: Environmental Implications and Solutions*," John Wiley & Sons, Hoboken, NJ, 2006.
2. D. De Maio and A. Naglieri, "*Catalyst Selection and Evolution* (adapted from)," Chemical Engineering, New York City, NY, July 29, 1968.

CHAPTER 17

CATALYTIC REACTIONS

INTRODUCTION

As already mentioned on several occasions, a catalyst is a substance which influences the rate of a reaction; during the process it may or may not become altered itself. Catalysts are most often employed to speed up a reaction, i.e., either those that are slow or will not otherwise proceed at all. Also, they may change the operating temperature level, or influence the product distribution, or more rarely, slow down a reaction. Thus, when the reaction proceeds by more than one irreversible path, the catalyst may favor one path over another and thus can lead to a product distribution different from that of the uncatalyzed reaction.

Catalytic processes provide excellent examples of complex reactions. The catalyst is viewed as combining with some of the reactants to form an intermediate species which subsequently reacts to form products with the return of the catalyst to its original state so that it can continue to further participate in the reaction. Scientists and physical chemists have concluded that the catalyzed path requires a lower activation energy and thus can proceed more rapidly.

Two broad classes of reaction are recognized: homogeneous and heterogeneous. The former were discussed in Parts II and III. Solid catalysts are widely employed for fluid-phase reactions. These are the most important heterogeneous types and are considered at length in the next chapter, as opposed to the general characteristics of catalysis and catalytic reactions.

These catalytic reactions are influenced by the same kinds of variables as homogeneous reactions. However, the quantity and chemical nature of the catalyst are additional variables, with some reactions approaching first-order behavior with respect to catalyst concentration.

This introduction is followed by six sections that will primarily address the mechanism(s) associated with catalytic reactions.

The Overall Process

Convective Transfer

Molecular Diffusion

Adsorption/Desorption

Chemical Reactor Analysis and Applications for the Practicing Engineer. By Louis Theodore

Chemical Reaction

The Controlling Step(s)

More time is spent on adsorption since it is often the controlling step in a catalytic reaction.

THE OVERALL PROCESS

Fluid–solid phase catalytic reactions are generally recognized to proceed according to the following seven (7) steps:

1. Convective transfer of the reacting molecule(s) from the bulk fluid to the outer surface of the catalyst
2. Molecular diffusion of the reacting molecule(s) to the pore surface within the catalyst
3. Adsorption of the reacting molecule(s) on the pore surface
4. Reaction on the catalyst surface
5. Desorption of the product molecule(s) from the pore surface
6. Molecular diffusion of the product molecule(s) from the pore surface to the outer surface of the catalyst
7. Convective transfer of the product molecules from the outer surface of the catalyst to the bulk fluid.

The above process is schematically represented in Figure 17.1 for the reaction A → B. Note that the figure presents steps (1)–(4); steps (5)–(7) are simply the reverse of steps (1)–(3) and are on the upper RHS of the figure.

With reference to Figure 17.1, steps (1)–(4) can be seen on the LHS of the figure. The RHS provides the reverse steps, i.e., steps (4)–(7). Also note that step (4), i.e., the reaction itself, constitutes only *one* of the seven steps. An equation attempting to represent the reaction mechanism for a catalytic reaction could possibly include the effects of each of these steps.

The rate at which each of these steps occurs ultimately determines the distribution of the participating species (reactants and products) in the system; in addition, it plays a major role in determining the over-all rate of heterogeneous catalytic reactions. The factors and effects influencing each of these rates is considered in the next four sections. The final section is concerned with analyzing and determining the controlling step or steps for the reaction in question. This topic will be reviewed qualitatively since much of this material is both difficult and complex; a qualitative presentation is beyond the scope of this text.

CONVECTIVE TRANSFER

The first step in attempting to describe the mechanism of a catalytic reaction is defined as convective transfer. It is during this step that a molecule that will ultimately react is transported from the bulk fluid (motion) to the outer surface of the catalyst. When the

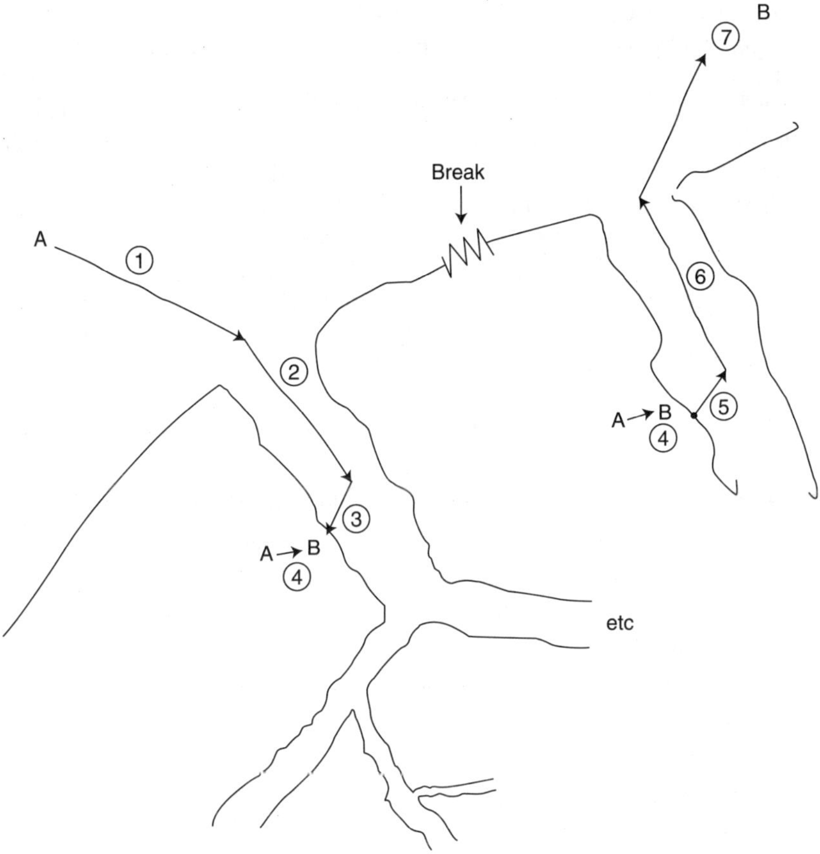

Figure 17.1 Heterogeneous catalyst mechanism for A → B;

fluid flows past the catalyst surface under such conditions that the fluid is in turbulent flow, the actual velocity of small parcels or lumps of fluid cannot be described as simply as in laminar flow. Since fluid flows in smooth streamlines in laminar flow, its behavior can usually be described mathematically. However, there are no orderly streamlines or equations to describe fluid behavior in turbulent motion. However, there are large eddies or "chunks" of fluid which move rapidly in a seemingly random fashion. This transfer process is usually very rapid in comparison to the relatively slow process of molecular diffusion (to be discussed in the next section), where each molecule to be reacted must move by random motion through the fluid.

Theodore and Ricci[1] provide the following: "When a fluid flows past a surface under conditions such that turbulence generally prevails, a thin laminar-type sublayer film exists adjacent to the surface. The mass transfer in this region occurs by molecular diffusion since little or no eddies are present. Since this is a slow process, a large concentration gradient or decrease in concentration across this laminar film occurs. Adjacent to this is the transition or buffer region. Here, some eddy activity exists and the transfer occurs by the sum of molecular and turbulent diffusion. In this region, there is a gradual and non-abrupt transition from the total transfer occurring by almost pure molecular diffusion at one end to mainly turbulent transfer at the other end. The concentration decrease is much less in this region. Although most of the transfer is by turbulent or eddy diffusion, molecular diffusion still occurs, but it contributes little to the overall transfer. The concentration decrease is very small here since the rapid eddy movement evens out any gradients tending to exist.

Many approaches to the turbulent (convective) mass transfer problem exist: film theory, combined film-surface-renewal theory, boundary layer theory, empirical approaches, etc. One former theory has somehow managed to survive the test of time, having been successful in interpreting the results of some two-phase mass transfer operations of industrial importance. Film theory[2] (as applied by Whitman) postulates the existence of an imaginary stagnant film next to the interface whose resistance to mass transfer is equal to the total mass transfer resistance of the system. The difficulty with this theory is in the calculation of the effective film thickness. Some of the other theories are briefly detailed below.

1. Surface-renewal theory assumes that a clump of fluid far from the interface:
   - **a.** moves to the interface without transferring mass,
   - **b.** sits there stagnant, transferring mass by molecular diffusion for a time short enough such that little change in the concentration profile occurs in the clump, and
   - **c.** then moves away from the interface without transferring mass en route and mixes with the bulk fluid instantly.

   This theory is somewhat more satisfactory in general than film theory.
2. Boundary layer theory rests on the solution of a set of simplified differential equations which are approximations to a more nearly correct set of differential equations.[3]
3. Empirical approaches, which are merely data correlations, serve for specific cases, but give little information about extrapolation.

Further details regarding any of these approaches are available in the literature.[3]"

For certain simplified cases, equations can be derived to determine precisely the rate at which mass is being transferred, e.g.,

$$r = [ka](C_{AO} - C_{AS}) \tag{17.1}$$

where r = rate of transfer by convection

ka = individual mass transfer coefficient

C_{AO} = concentration of reacting species A in the bulk fluid

C_{AS} = concentration of reacting species A at the outer catalyst surface site

The bracketed term above is an exact definition for the individual mass transfer coefficient corresponding to the steady-state situation of one component diffusing through a non-diffusing second component. The two most frequently encountered situations involve equimolar counter transfer and the transfer of one component through another non-diffusing component.[4,5] Theodore and Ricci[1] provide additional details.

Once a product species, e.g., B, has been formed, step (7) discussed in the previous section, provides the rate process in the reverse direction. For this step

$$r = [ka][C_{BS} - C_{BO}] \tag{17.2}$$

MOLECULAR DIFFUSION

The second and sixth steps involve the transport of the reacting molecule(s) from the outer surface to the active site of the catalyst and the transport of the product molecule(s) from the active site to the outer surface, respectively. Fick's law[5] describes this process and applies to both steps (2) and (6).

As noted earlier, molecular diffusion results from the motion of molecules. At any instant, the individual molecules in a fluid are moving in random directions at speeds varying from low to high values. The molecules move at random, frequently colliding with one another. Because of the frequent collisions, the molecular velocities are continually changing in both direction and magnitude. Diffusion is more rapid at higher temperatures due to greater molecular velocities. For gases, it is more rapid at low pressures because the average distance between the molecules is greater and the collisions are less frequent.

If a fluid is not uniform in concentration, the solution is gradually brought into uniformity by diffusion; the molecules move from an area of high concentration to one of low concentration. The rate at which travel occurs depends on the concentration gradient which exists in the pore. This gradient applies across adjacent regions of high and low concentrations. However, a quantitative measure of rate is needed to describe what is occurring.

The rate of diffusion can be described in terms of a molar flux term, with units of moles/area · time, and with the area measured as that which the molecule diffuses through. The diffusivity, or diffusion coefficient, D_{AF} of component A in fluid F which is a measure of its diffusive mobility, is defined as the ratio of its flux, J_A, to its concentration gradient and is given by

$$J_A = -D_{AF}\frac{\partial C_A}{\partial z} \tag{17.3}$$

This is Fick's first law[6] written for the z direction. The concentration gradient term represents the variation of the concentration, C_A, in the z direction. The negative sign accounts for diffusion occurring from high to low concentrations. The diffusivity is a characteristic of the component and its environment (temperature, pressure, concentration, etc.). The diffusivity is usually expressed with units of (length)^2/time or moles/time · length.

ILLUSTRATIVE EXAMPLE 17.1 Express the diffusivity in English units.

Solution. Based on its definition, the units of the diffusivity may be expressed as either

$$\text{ft}^2/\text{h}$$

or

$$\text{lbmol/h} \cdot \text{ft}$$

The latter units are derived by simply multiplying ft^2/h by the molar density in $lbmol/ft^3$.

Interestingly, there are three different cases of steady-state molecular diffusion in gases.[5,7]

1. Diffusion of A through non-diffusing B.
2. Steady state equimolar counter-diffusion.
3. Steady-state diffusion in multicomponent mixtures.

No purely theoretical generalized correlation of liquid phase diffusivities has yet been found, but certain empirical equations are available. This probably reflects the inadequacy of kinetic theory when applied to liquids. It is therefore preferable to use experimental data for liquid phase diffusivities. Liquid diffusivities are available in the literature.[2,8] The two principal situations for diffusion in liquids are as follows.

1. Steady-state diffusion of A through non-diffusing B.
2. Steady-state equimolar counterdiffusion. ■

ADSORPTION/DESORPTION

The third and fifth steps involve the adsorption of the reacting molecule, e.g., A, onto the active site of the catalysts and the desorbtion of the product molecule (following reactor), e.g., B, away from the active site. Note that the development to follow applies to both steps (3) and (5). Also note that since the adsorption/desorption process(es) are often the controlling step in the overall reaction mechanism, it receives more extensive treatment.

Catalysts are highly porous particles and adsorption occurs primarily on the internal surface of the particles. The attractive forces that hold the molecule to the surface of the solid are the same that cause vapors to condense (*van der Waals* forces). The adsorption process is classified as either *physical* or *chemical*. The basic difference between physical and chemical adsorption is the manner in which the molecule is bonded to the catalyst. In physical adsorption the molecule is bonded to the catalyst by weak forces of intermolecular cohesion. The chemical nature of the adsorbed species remains unchanged; therefore, physical adsorption is a readily reversible process. In chemical adsorption a much stronger bond is formed between the molecule and the catalyst surface. A sharing or exchange of electrons takes place—as happens in a chemical bond. Chemical adsorption is not easily reversible.

The forces active in physical adsorption are electrostatic in nature. These forces are present in all states of matter: gas, liquid, and solid. They are the same forces of attraction that cause gases to condense and real gases to deviate from ideal behavior. Physical adsorption is sometimes referred to as *van der Waals* adsorption. The electrostatic effect that produces the van der Waals forces depends on the polarity of both the gas (or liquid) and solid molecules.

Physical adsorption can occur from three different effects: an orientation effect, a dispersion effect, or an induction effect. For polar molecules, attraction to each other occurs because of an orientation effect. The orientation effect describes the attraction that occurs between the dipoles of two polar molecules. The negative area of one is attracted to the positive area of the other. Chemical adsorption, or chemisorption, results from the chemical interaction between a molecule and a solid. The molecule is held to the surface of the adsorbate by the formation of a chemical bond. Adsorbents used in chemisorption can be either pure substances or chemicals deposited on an inert carrier material.[(9)]

All known adsorption processes are *exothermic*, whether adsorption occurs from chemical or physical forces. In adsorption, molecules are transferred from the molecule to the surface of a solid. The fast-moving molecules lose their kinetic energy of motion to the adsorbent in the form of heat. In chemisorption, the heat of adsorption is comparable to the heat evolved from a chemical reaction, usually over 10 kcal/gmol. The heat given off by physical adsorption is much lower, approximately 100 cal/gmol, which is comparable to the heat of condensation (or vaporization). Since adsorption is an exothermic process, the concentration of adsorbed gas decreases with increased temperature at a given operating pressure.[(9)]

The relation between the amount of substance adsorbed by an adsorbent and the equilibrium partial pressure or concentration at constant temperature is called an adsorption *isotherm*. The adsorption isotherm is the most important and by far the most often used of the various equilibria data that can be measured. Five general types of isotherms have been observed in the adsorption of gases on solids. These are shown in Figure 17.2. In cases of chemisorption, only isotherms of type I are encountered, while in physical adsorption, all five types occur. Also note that the development to follow will primarily be concerned with gas–solid adsorption.

Freundlich Equation

In isotherms of Type I, the amount of gas adsorbed per given quantity of adsorbent increases relatively rapidly with partial pressure and then much more slowly as the surface becomes covered with gas molecules. To represent the variation of the amount of adsorption per unit area or unit mass with partial pressure, Freundlich proposed the following equation

$$Y = kp^{1/n} \tag{17.4}$$

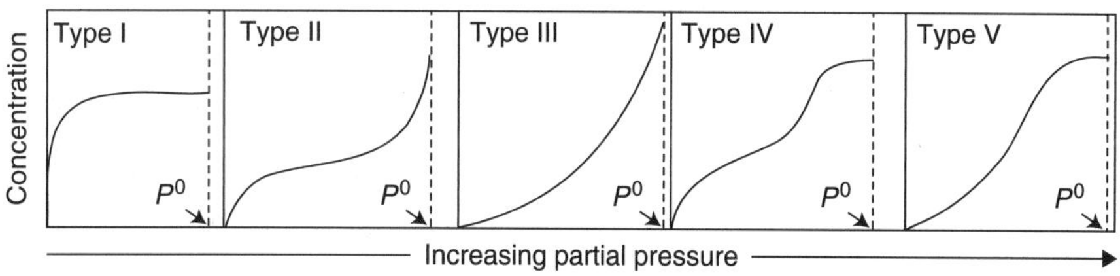

Figure 17.2 Types of adsorption isotherms. P^0 represents the saturation pressure.

where Y is the weight or volume of gas adsorbed per unit area or unit mass adsorbent, p is the equilibrium partial pressure of the species being adsorbed while k and n are empirical constants dependent on the nature of the solid and reaction species, and on the temperature. Equation (17.4) may be rewritten as follows. Taking logarithms of both sides,

$$\log(Y) = \log(k) + \frac{1}{n}\log(p) \tag{17.5}$$

If the $\log(Y)$ is now plotted against $\log(p)$, a straight line should result with the slope equal to $(1/n)$ and an ordinate intercept equal to $\log(k)$. Although the requirements of the equation are met satisfactorily at lower pressures, the experimental points curve away from the straight line at higher pressures, indicating that this equation does not have general applicability in reproducing adsorption of gases by solids.[9]

Langmuir Isotherms

A much better equation for Type I isotherms was deduced by Langmuir from theoretical considerations. Langmuir postulated that gases, on being adsorbed by a solid surface, cannot form a layer more than a single molecule in depth. Further, he visualized the adsorption process as consisting of two opposing actions, a condensation of molecules from the gas phase onto the surface and an evaporation of molecules from the surface back into the body of the gas. When adsorption first begins, every molecule colliding with the surface may condense on it. However, as adsorption proceeds, only those molecules that occupy a part of the surface not already covered by adsorbed molecules may be expected to be adsorbed. The result is that the initial rate of condensation of molecules on a surface is high and then falls off as the surface area available for adsorption is decreased. On the other hand, a molecule adsorbed on a surface may, by thermal agitation, become detached from the surface and escape into the gas. The rate at which desorption will occur will depend, in turn, on the amount of surface covered by molecules and will increase as the surface becomes more fully saturated. These two rates, condensation (adsorption) and evaporation (desorption) will eventually become equal and when this happens, an adsorption equilibrium will be established.

If θ is the fraction of the total catalyst surface covered by adsorbed molecules at any instant, then the fraction of bare catalyst surface available for adsorption is $(1 - \theta)$. According to kinetic theory, since the rate at which molecules strike a unit area of a surface is proportional to the pressure of the gas, the rate of condensation of molecules should be determined both by the partial pressure and the fraction of bare catalyst surface or $k_1(1 - \theta)p$, where k_1 is a proportionality constant. If k_2 is the rate at which molecules evaporate from a unit surface when the surface is fully covered, then for a fraction θ of a fully covered surface, the rate of evaporation will be $k_2\theta$. For adsorption equilibrium, these rates must be equal. Therefore,

$$k_1(1 - \theta)p = k_2\theta$$

or

$$\theta = \frac{k_1 p}{k_2 + k_1 p} = \frac{bp}{1 + bp} \tag{17.6}$$

where $b = k_1/k_2$. Now, the amount of gas adsorbed per unit area or per unit mass of catalyst, Y, must obviously be proportional to the fraction of surface covered: hence

$$Y = k\theta = \frac{kbp}{1+bp} = \frac{ap}{1+bp} \tag{17.7}$$

where the constant a has been written for the product kb. Equation (17.7) is the Langmuir adsorption isotherm. The constants a and b are characteristic of the system under consideration and are evaluated from experimental data. Their magnitude also depends on the temperature. The validity of the Langmuir adsorption equation at any one temperature can be verified most conveniently by first dividing both sides of Equation (17.7) by p and then taking reciprocals. The result is

$$\frac{p}{Y} = \frac{1}{a} + \frac{b}{a}p \tag{17.8}$$

Since a and b are constants, a plot of (p/Y) vs p should yield a straight line with slope equal to (b/a) and an ordinate intercept equal to $(1/a)$.[9]

ILLUSTRATIVE EXAMPLE 17.2 Gas–solid equilibrium adsorption data (including calculations) are presented in Table 17.1. Determine the constants of the Freundlich equation for this system at a temperature of 50°C.

Solution. Apply the Freundlich equation

$$Y = kp^{1/n} \tag{17.4}$$

The calculations presented in Table 17.1 are presented on the LHS of Figure 17.3. For the Freundlich equation (1), the plot of (log Y) vs (log p) in Figure 17.3 leads to the equation:

$$Y = 30p^{0.7}$$

■

ILLUSTRATIVE EXAMPLE 17.3 Refer to Illustrative Example 17.2. Determine the constants of the Langmuir equation.

TABLE 17.1 Calculations for the Freundlich Equation at 50°C

Y (cm^3/g)	p (atm)	Log Y	Log p
30	1	1.477	0.000
51	2	1.708	0.301
67	3	1.826	0.477
81	4	1.909	0.602
93	5	1.969	0.699
104	6	2.017	0.778

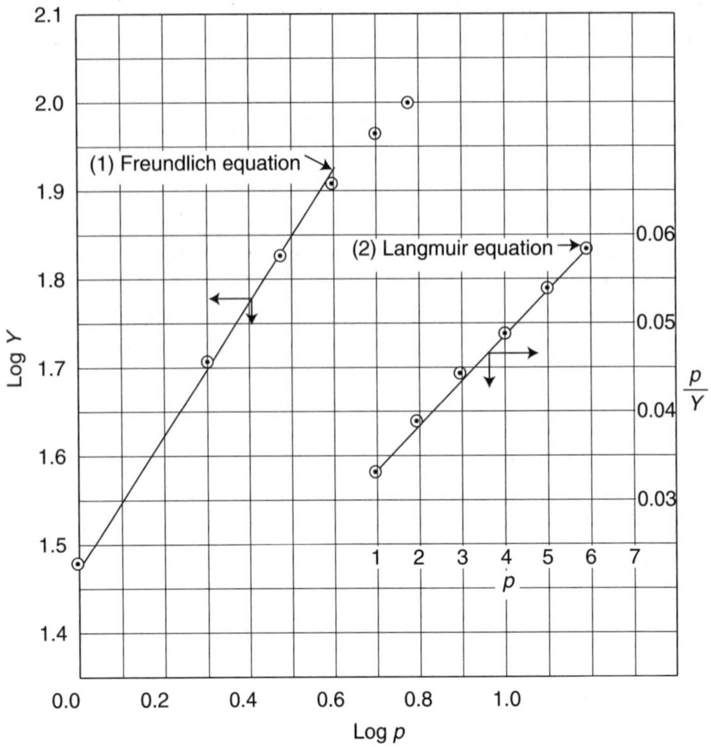

Figure 17.3 Adsorption isotherms from Illustrative Example 17.2 and 17.3.

Solution. For the Langmuir equation, employ

$$\frac{p}{Y} = \frac{1}{a} + \frac{b}{a}p \tag{17.8}$$

Transfer the information generated in Table 17.2 to the RHS of Figure 17.3. For the Langmuir equation (2) the plot of (p/Y) vs p leads to the equation

$$Y = \frac{35.7p}{1 + 0.186p}$$

TABLE 17.2 Calculations for the Langmuir Equation at 50°C

p/Y	p
0.033	1
0.039	2
0.045	3
0.049	4
0.054	5
0.058	6

■

ILLUSTRATIVE EXAMPLE 17.4 Which equation provides a better fit to the experimental data provided in the two previous illustrative examples.

Solution. As is evident from the lines drawn in Figure 17.3, the Langmuir equation provides the better fit. Strictly speaking, a correlation coefficient should be calculated for both results to provide a better basis for comparison. ■

The adsorption/desorption process can be expressed in another form. The following notation applies to the development that follows for the A → B reaction.

p_i = partial pressure of species i in the gas, atm

C_v = surface concentration of vacant catalyst sites, gmol/gcat

$C_{i\cdot s}$ = surface concentration of sites occupied by species i, gmol$i$/gcat

S_a = surface area per unit mass of catalyst, cm^2/gcat

$C'_{i\cdot s}$ = surface concentration of species i based on area, gmol$i$/cm^2

Thus, one may write

$$C'_{i\cdot s} = C_{i\cdot s}/S_a \tag{17.9}$$

and

$$C_t = C_v + C_{A\cdot S} + C_{B\cdot S} \quad \text{(for the adsorption of } A \text{ and } B\text{)} \tag{17.10}$$

For the overall adsorption process

$$A + S \underset{k'_A}{\overset{k_A}{\rightleftharpoons}} A \cdot S \tag{17.11}$$

The rate of adsorption is

$$r_{ADS} = k_A p_A C_v - k'_A C_{A\cdot S} \tag{17.12}$$

By definition, the equilibrium constant is given by

$$K_A = k_A/k'_A \tag{17.13}$$

Therefore

$$r_{ADS} = k_A\left(p_A C_v - \frac{C_{A.S}}{K_A}\right) \tag{17.14}$$

At equilibrium, one may write

$$r_{ADS} = O = k_A\left(p_A C_v - \frac{C_{A.S}}{K_A}\right) \tag{17.15}$$

so that

$$C_{A\cdot S} = K_A p_A C_v \tag{17.16}$$

The term C_v may be replaced with that provided in Equation (17.10). Rearrangement leads to

$$C_{A\cdot S} = \frac{K_A p_A C_t}{1 + K_A p_A} \tag{17.17}$$

The essence of this development is to provide the concentration of reactant A on the catalyst site. This information is required in order to determine the *rate of chemical reaction* at the site, i.e., step (4) in the overall process. Step (4) is considered in the next section.

Equation (17.18) describes the desorption process for product B in equation form,

$$B \cdot S \rightleftharpoons B + S \tag{17.18}$$

Fogler[10] provides an expanded treatment of the above material.

CHEMICAL REACTION

Much of the material for this section has already been presented in the text. In a very real sense, the kinetics of homogeneous reactions still apply. The describing equation for the kinetics as it relates to catalytic reactions is now examined. Since the reactions occur on the catalyst surface, it would be more appropriate to describe this step as a chemical surface reaction. Four possible surface reactions are provided in Equations (17.19)–(17.22)

$$A \cdot S \longrightarrow B \cdot S \tag{17.19}$$

$$A \cdot S \cdot S \rightleftharpoons B \cdot S + S \tag{17.20}$$

$$A \cdot S + B \cdot S \longrightarrow C \cdot S + D \cdot S \tag{17.21}$$

$$A \cdot S + B(\text{g}) \longrightarrow C \cdot S + D(\text{g}) \tag{17.22}$$

Obviously, there are a near infinite number of chemical surface reaction possibilities.

Finally, it should be noted that it is usually assumed that only *one* of seven steps reviewed in this and the previous three sections is the slowest controlling step. The three controlling steps that most often come into play in determining the catalytic reaction mechanisms are:

step (3)—adsorption

step (4)—chemical reaction

step (5)—desorption

It is somewhat surprising that both convective transfer and molecular diffusion resistances rarely contribute to the overall process. Note that some studies include more than one step in the analysis.

THE CONTROLLING STEPS(S)

Developing an equation describing the mechanism and rate of a catalytic reaction is an extremely difficult task if all seven steps in the process are included in the analysis. It is almost always impossible to accurately predict the coefficient(s) in the rate equation, even with extensive laboratory data. The practicing engineer usually hopes that only one of the aforementioned steps contributes significantly to the resistance associated with the reaction. The factors influencing the rates of the seven individual steps have already been developed in this chapter. What remains is to present some qualitative ways of predicting the controlling step(s).

Some of the "simpler" catalytic reactions have fairly "simple" rate expressions. In other instances, the rate equation may be so complex due to a multitude of mechanisms involved that purely empirical correlations of the kinetic variables may have to be employed using (perhaps) sophisticated statistics and numerical methods.

As noted in the previous section, either the adsorption (step 3) or the chemical reaction (step 4) or the desorption (step 4) process primarily contributes to the rate of reaction. For example, consider the case where steps (1) and (4) are the controlling resistances. Refer to Figure 17.4. Then,

Step (1): $r = [ka](C_{AO} - C_{AS})$ (17.23)

Step (4): $r_A = k_A C_{AS}$ (17.24)

As with an earlier heat exchanger analogy[11], set

$$r = r_A \tag{17.25}$$

so

$$k_A C_{AS} = ka(C_{AO} - C_{AS}) \tag{17.26}$$

Solving for C_{AS} (a term that cannot be measured)

$$C_{AS} = \frac{C_{AO}}{1 + \left(\frac{k_A}{ka}\right)} \tag{17.27}$$

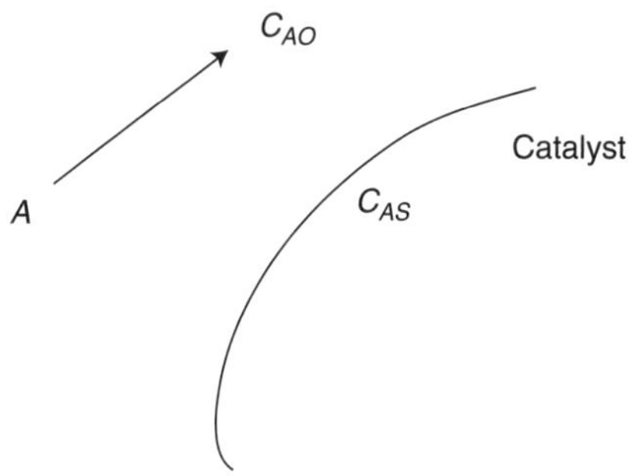

Figure 17.4 Controlling resistances: steps (1) and (4).

Since the practicing engineer generally prefers to work with measurable quantities, one may then write:

$$-r_A = k_O C_{AO} \tag{17.28}$$

Substituting for C_{AS} from Equation (17.27) into Equation (17.28) gives

$$-r_A = -r = \left[\frac{(k_A)(ka)}{k_A + ka}\right] C_{AO} \tag{17.29}$$

What is k_O? It is simply

$$k_O = \frac{1}{\frac{1}{k_A} + \frac{1}{ka}} = \frac{(k_A)(ka)}{k_A + ka} \tag{17.30}$$

The equations look familiar to the overall heat transfer coefficient, U_o, employed in heat exchanger studies.[11] Assuming this model is correct, one can obtain a numerical value for k_A from kinetic experimental data and ka can be estimated using mass transfer theory.[1–5]

In any event, the practicing engineer may be confronted with the need to analyze experimental data for the purpose of developing a reaction rate equation for a catalytic reaction. These rate equations rarely follow elementary power law kinetics so that the describing equation may be anything but elementary.

Fogler[10] has examined the catalytic reversible decomposition of cumene (C) to benzene (B) and pentene (P). Neglecting the resistance to the overall reaction rate process associated with convection and diffusion, i.e., steps (1) plus (7) and (2) and (6), respectively, one is left with the following (remaining) three resistances:

step (3): adsorption (A)

step (4): reaction (R)

step (5): desorption (D)

The above three reaction step sequence (as before) become

$$C + S \underset{k'_A}{\overset{k_A}{\rightleftharpoons}} C \cdot S \tag{17.31}$$

$$C \cdot S \underset{k'_R}{\overset{k_R}{\rightleftharpoons}} B \cdot S + P(g) \tag{17.32}$$

$$B \cdot S \underset{k'_D}{\overset{k_D}{\rightleftharpoons}} B + S \tag{17.33}$$

Based on this reaction sequence, i.e., steps (3)–(5), one may write

For (A): $r_A = k_A p_C C_v - k'_A C_{C \cdot S}$ (17.34)

For (R): $r_R = k_R C_{C \cdot S} - k'_R p_P C_{B \cdot S}$ (17.35)

For (D): $r_D = k_D C_{B \cdot S} - k'_D p_D C_v$ (17.36)

At steady-state, the rates of each step, of necessarity, must be equal, as with the heat transfer analogy associated with the overall heat transfer coefficient[11], i.e.,

$$r_A = r_R = r_D \tag{17.37}$$

What usually remains is to determine which step is controlling. Data is then gathered under a variety of conditions which in some instances leads to a rate expression suitable for design or predictive purposes. Fogler[10] provides the development for this reaction assuming the adsorption step (A), i.e., Equation (17.34) is the controlling resistance. Surprisingly, as noted earlier, it is the adsorption step, i.e., often found to be the controlling resistance.

REFERENCES

1. L. Theodore and F. Ricci "*Mass Transfer Operations for the Practicing Engineer*," John Wiley & Sons, Hoboken, NJ, 2010.
2. W. Whitman, *G. Chem. Met. Eng*., **29**, 147, 1923.
3. R. Bird, W. Stewart, and E. Lightfoot, "*Transport Phenomena,*" 2nd edition, John Wiley & Sons, Hoboken, NJ, 2004.
4. C. Geankopolis, "*Transport Processes and Unit Operations,*" Allyn and Bacon, Boston, MA, 1978.
5. R. Treybal, "*Mass Transfer Operations,*" 1st edition, McGraw-Hill, New York City, NY, 1955.
6. A. Fick, *Pogg. Ann*., **XCIV**, 59, 1855.
7. C. Bennett and J. Meyers, "*Momentum, Heat, and Mass Transfer,*" McGraw-Hill, New York City, NY, 1962.
8. D. Green and R. Perry (editors), "*Perry's Chemical Engineers', Handbook,*" 8th edition, McGraw-Hill, New York City, NY, 2008.
9. L. Theodore, "*Air Pollution Control Equipment Calculations*," John Wiley & Sons, Hoboken, NJ, 2008.
10. S. Fogler, "*Elements of Chemical Reaction Engineering,*" 4th edition, Prentice Hall, Upper Saddle River, NJ, 2006.
11. L. Theodore, "*Heat Transfer for the Practicing Engineer,*" John Wiley & Sons, Hoboken, NJ, 2011.

CHAPTER **18**

FLUIDIZED AND FIXED BED REACTORS

INTRODUCTION

Catalytic or heterogeneous reactors are an alternative to homogenous reactors. If a solid catalyst is added to the reactor, the reaction is said to be heterogeneous. For simple reactions, the effect of the presence of a catalyst is to

1. increase the rate of reaction,
2. permit the reaction to occur at a lower temperature,
3. permit the reaction to occur at a more favorable pressure,
4. reduce the reactor volume,
5. increase the yield of a product(s).

The basic problem in the design of a heterogeneous reactor is to determine the quantity of catalyst and/or reactor size required for a given conversion and flow rate. In order to obtain this, information on the rate equation(s) and their parameter(s) must be made available. A rigorous approach to the evaluation of reaction velocity constants has yet to be accomplished for catalytic reactions; at this time, industry still relies on the procedures set forth in the previous chapter. For example, in catalytic combustion reactions, the rate equation is extremely complex and cannot be obtained either analytically or numerically. A number of equations may result and some simplification is often warranted. As mentioned earlier, in many cases it is safe to assume that the expression may be satisfactorily expressed by the rate equation of a single step.

It is common practice to write the describing equations for mass and energy transfer for homogeneous and heterogeneous flow reactors in the same way. However, the (units of the) rate of reaction may be expressed as either

(moles reacted)/(time)(volume of reactor)

or

(moles reacted)/(time)(mass of catalyst)

Chemical Reactor Analysis and Applications for the Practicing Engineer. By Louis Theodore

The latter is normally the preferred method employed in industry since it is the mass of catalyst present in the reactor that significantly impacts the reactor design. Since the rate expression is often more complex for a catalytic reaction than for a non-catalytic (homogeneous) reactor, the design equation may be difficult to solve analytically. Numerical solution of the reactor design equation is usually required when designing tubular flow reactors for catalytic reactions.

As indicated above, the principal difference between reactor design calculations involving homogeneous reactions and those involving catalytic (fluid–solid) heterogeneous reactions is that for the latter the reaction rate is based on the mass of solid, W, rather than on reactor volume V. For a fluid–solid heterogeneous system the rate of reaction of species A is then defined as $-r_A'$ with the aforementioned units of (mols A reacted)/(mass catalyst) (time).

In addition to this introduction, the chapter primarily addresses two types of catalytic reactors. Although there are numerous types of catalytic units, this chapter solely reviews fluidized bed reactors and fixed bed reactors. Details on other reactors are available in the literature[1,2]. The remaining sections of this chapter include:

Fluidized Bed Reactors

Fixed Bed Reactors

Pressure Drop Calculations

Catalytic Reactor Design Considerations

FLUIDIZED BED REACTORS

Fluidization is the process in which fine solid particles are transformed into a fluid-like state through contact with either a gas or liquid, or both. Fluidization is normally carried out in a vessel filled with catalyst solids. The fluid is introduced through the bottom of the vessel and forced up through the bed. At a low flow rate, the fluid (liquid or gas) moves through the void spaces between the stationary and solid catalyst particles and the bed is referred to as *fixed*. (This topic is treated in the development that follows in the next section.) As the flow rate increases, the catalysts begin to vibrate and move about slightly, resulting in the onset of an *expanded bed*. When the flow of fluid reaches a certain velocity, the solid catalysts become suspended because the upward frictional force between the catalyst and the fluid balances the gravity force associated with the weight of the catalyst. This point is termed *minimum fluidization* or *incipient fluidization* and the velocity at this point is defined as the *minimum* or *incipient fluidization velocity*. Beyond this stage, the bed enters the fluidization state where bubbles of fluid rise through the solid catalysts, thereby producing a circulatory and/or mixing pattern.[3]

From a force balance perspective, as the flow rate upward through a packed bed is increased, a point is reached at which the frictional drag and buoyant force is enough to overcome the downward force exerted on the bed by gravity. Although the bed is supported at the bottom by a screen, it is free to expand upward, as it will if the velocity is increased above the aforementioned minimum fluidization velocity. At this point, the catalysts are no longer supported by the screen, but rather are suspended in the fluid in equilibrium and act and behave as the fluid. The bed is then said to be *fluidized*. From a momentum or force balance perspective, the sum of the drag, buoyancy, and gravity forces must be equal to zero.

The terminal settling velocity can be evaluated for the case of flow past one catalyst particle in the bed. By superimposition, this case is equivalent to that of the terminal velocity that a catalyst particle would attain flowing through a fluid.[4] Once again, a force balance can be applied and empirical data used to evaluate a friction (drag) coefficient.

At intermediate velocities between the minimum fluidization velocity and the terminal velocity, the bed is expanded above the volume that it would occupy at the minimum value. Note also that above the minimum fluidization velocity, the pressure drop stays essentially constant.

One of the novel characteristics of fluidized beds is the uniformity of temperature found throughout the system. Essentially constant conditions are known to exist in both the horizontal and vertical directions in both short and long beds. This homogeneity is due to the turbulent motion and rapid circulation rate of the solid catalyst particles within the fluid stream described above. In effect, excellent fluid-particle contact results. Temperature variations can occur in some beds in regions where quantities of relatively hot or cold catalyst particles are present but these effects can generally be neglected. Consequently, fluidized beds find wide application in industry, e.g., oil cracking, zinc coating, coal combustion, gas desulfurization, heat exchangers, plastics cooling, and fine powder granulation.

Minimum Fluidization Velocity

Figure 18.1 is a photograph of the fluidization experimental unit at Manhattan College. Figure 18.2 shows the kinds of contact between solids and a fluid, starting from a packed bed and ending with "pneumatic" transport. At a low fluid velocity, one observes a fixed bed configuration of height L_m, a term i.e., referred to as the *slumped* bed height. As the velocity increases, fluidization starts, and this is termed the *onset of fluidization*. The superficial velocity (that velocity which would occur if the actual flow rate passed through an empty vessel) of the fluid at the onset of fluidization is noted again as the *minimum fluidization velocity*, v_{mf}, and the bed height is L_{mf}. As the fluid velocity increases beyond v_{mf}, the bed expands and the bed void volume increases. At low fluidization velocities (fluid velocity $> v_{\text{mf}}$), the operation is termed *dense phase fluidization*. As noted in the previous section, the onset of fluidization (or *minimum fluidization condition*) in the packed bed occurs when drag forces due to friction by the upward moving gas equal the gravity force of the catalyst particles minus the buoyancy force on the particles.

There are two modes of fluidization. When the fluid and solid catalyst densities are not too different, or the particles are very small, the bed is fluidized evenly. This is called *smooth* fluidization, and is typical of liquid–solid systems. If the fluid and solid densities are significantly different, or the catalyst particles are large, the velocity of the flow must be relatively high. In this case, fluidization is uneven, and the fluid passes through the bed mainly in large bubbles. These bubbles burst at the surface, spraying the solid catalyst above the bed. Here, the bed has many of the characteristics

Figure 18.1 Fluidization experiment.

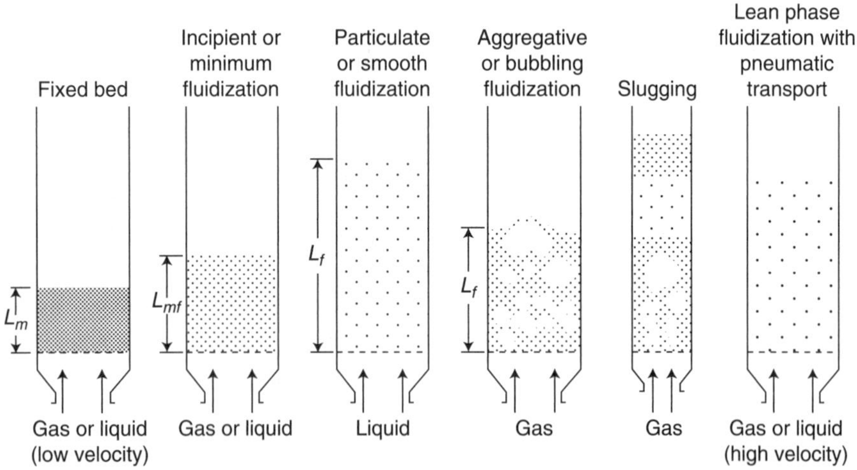

Figure 18.2 Types of particle-fluid contact in a bed.

of a liquid with the fluid phase acting as a gas bubbling through it. This is called *bubbling* (or aggregative) fluidization; it is typical of gas–solid systems and is due to the large density difference between the solid and gas.

The approximate criterion to estimate the transition from bubbling to smooth fluidization is expressed in terms of the dimensionless Froude number at minimum fluidization. This is presented in Equation (18.1) in terms of the minimum fluidization velocity, v_{mf}, the catalyst diameter, d_p, and the acceleration due to gravity, g, as:

$$Fr = \frac{v_{\mathrm{mf}}^2}{g d_p} \tag{18.1}$$

If $Fr < 0.13$, the fluidization mode is smooth; if $Fr > 0.13$, then the fluidization is bubbling.

Some of the advantages of employing a fluidized bed unit are as follows.

1. Easy to control the heat transfer within the fluid system.
2. Easy to control the flow within the system.
3. Savings in operating expenses due to (potential) heat recovery.
4. Rapid mixing of reactants–catalysts; little to no radial and axial gradients.
5. High heat transfer rates.

Disadvantages include the following

1. Backmixing due to catalyst particle distribution.
2. Inefficient contacting due to solids movement.
3. Bypassing of catalyst by bubbles.
4. Possible channeling, slugging, and attrition of catalyst.
5. Possible agglomeration and sintering of fine catalyst particles.
6. A major operating problem is that of entrainments; this is minimized with cyclone separators,[5] frequently mounted within the fluidized fed reactor.
7. Short circuiting.

The type of the catalytic reactor in common use is the fluidized-bed. The fluidized-bed reactor is analogous to the CSTR in that its contents, though heterogeneous, are well mixed, resulting in a uniform concentration and temperature distribution throughout the bed. The fluidized-bed reactor can therefore be modeled, as a first approximation, as a CSTR. For the ideal CSTR, the reactor design equation based on volume is

$$V = F_{A_0} X_A / -r_A \tag{18.2}$$

The companion equation for a catalytic or fluid–solid reactor, with the rate based on the mass of solid, is

$$W = F_{A_0} X_A / -r_A' \tag{18.3}$$

The reactor volume is simply the catalyst weight, W, divided by the fluidized bed density, ρ_f, of the catalyst in the reactor.

$$V = W/\rho_f \tag{18.4}$$

The fluid bed catalyst density is normally expressed as some fraction of the catalyst bulk density, ρ_B. The volume of the reactor is

$$V = \frac{W}{f\rho_B} \tag{18.5}$$

where ρ_B is the bulk density of the bed for the fixed bed unit to be discussed in the next section.

A distinction should be made between *true* density and *bulk* density. The true density is the actual density of a discrete particle or solid, whereas the bulk density is the density with the void volume between the particles included in the determination. In lieu of data, the bulk density may be assumed, for purposes of engineering calculations to be approximately 60% of the true density.[(6)]

ILLUSTRATIVE EXAMPLE 18.1 The elementary gas-phase reaction

$$A + 2B \longrightarrow C + D; \qquad k_A = 100.0\,\mathrm{L^3/min \cdot gmol^2 \cdot kg\,cat}$$

takes place at constant temperature in a fluidized-bed catalytic reactor. An equimolar ration of A and B is fed to the reactor at a rate of 500 L/min and a temperature of 627°C. The feed is pressurized and maintained at 20 atm. If the bulk density of the catalyst before fluidization is 0.9 g/cm^3, and the bed can be expected to expand by 125%, what minimum volume of vessel should be used to achieve a conversion of 90% of the limiting reactant.

Solution. Since this is a gas phase reaction, first calculate y_{A_0}, δ and

$$y_{A_0} = 0.5$$
$$\delta = -1$$
$$\varepsilon = -0.5$$

and express C_A and C_B in terms C_{A_0}, ε and X_A.

$$\theta_A = \theta_B = 1.$$
$$b/a = 2.0$$

Therefore (with $X = X_A$)

$$C_A = C_{A_0}[(1 - X)/(1 + \varepsilon X)]; \quad C_B = C_{A_0}[(1 - 2X)/(1 + \varepsilon X)]$$

The rate of reaction, $-r'_A$, in terms C_{A_0}, and X, is

$$-r'_A = kC_{A_0}^3(1 - X)(1 - 2X)^2/(1 - 0.5X)^3; \quad k = k_A$$

Calculate C_{A_0} in gmol/L using the ideal gas law

$$\begin{aligned} C_{A_0} &= (y_{A_0})(P)/RT \\ &= (0.5)(20)/(0.082)(900) \\ &= 0.1355 \text{ gmol/L} \end{aligned}$$

Calculate F_{A_0} in gmol/min.

$$\begin{aligned} F_{A_0} &= C_{A_0} q \\ &= (0.1355)(500) \\ &= 67.75 \text{ gmol/min} \end{aligned}$$

The design equation for W for this fluidized bed catalytic reaction is

$$W = F_{A_0} X_A / (-r'_A) \qquad (18.3)$$

Since B is the limiting reactant, $X_B = 0.9$ and $X_A = X = 0.45$. Substituting into the above equation gives

$$\begin{aligned} W &= (67.75)(0.45)(1 - 0.225)^3/(100)(0.1355)^3(1 - 0.45)(1 - 0.9)^2 \\ &= 10{,}372 \text{ kg cat} \end{aligned}$$

The volume of the reacting mixture is then

$$V = W/\rho_f; \quad \rho_f = 0.9/2.25 \qquad (18.5)$$

$$\begin{aligned} &= 10{,}372(2.25/0.9) \\ &= 25{,}931 \text{ L} \end{aligned}$$

Comment on the applicability of the ideal gas law at the operating pressure. ■

ILLUSTRATIVE EXAMPLE 18.2 Consider the hydrodemethylation of toluene.

$$C_6H_5CH_3 + H_2 \longrightarrow C_6H_6 + CH_4$$

Feed conditions are: 5.0 atm, 800°C, 25% toluene, 40% hydrogen, remainder inerts, total volumetric flow rate is 200 L/min. The rate equation for the catalyst employed is given by:

$$-r'_T = 5.0 \times 10^{-2} \frac{(1 - X)^2}{1 - 0.2X}; \quad \text{gmol T/s} \cdot \text{kg cat}$$

Calculate the mass of catalyst required to achieve a conversion of 80% if a fluidized bed reactor is employed.

Solution. The reaction equation is

$$T + H_2 \longrightarrow B + M$$

with

$$-r'_T = 5 \times 10^{-2}\left[\frac{(1-X)^2}{1-0.2X}\right]; \quad \text{gmol T/s} \cdot \text{kg cat}$$

Substituting,

$$-r'_T = 5 \times 10^{-2}\left[\frac{(1-0.8)^2}{1-(0.2)(0.8)}\right]; \quad X = 0.8$$

$$= 2.38 \times 10^{-3}$$

Also calculate F_{T_0} using the ideal gas law.

$$F_{T_0} = \left(\frac{P_{T_0}}{RT}\right)Q_0$$

$$= \left[\frac{(5.0)(0.25)}{(0.082)(1073)}\right]200 = 2.84\,\text{gmol T/ min}$$

Applying the fluid bed design equation gives

$$W = \frac{F_{A_0}X}{-r_A} \tag{18.3}$$

$$W = \frac{(2.84)(0.8)}{(60)2.38 \times 10^{-3}}$$

$$= \frac{955}{60}\,\text{kg} = 15.9\,\text{kg}$$

■

ILLUSTRATIVE EXAMPLE 18.3 Refer to the previous illustrative example. Calculate the volume requirement if the catalyst density in the fluidized bed is 0.38 g/(cm)^3.

Solution. Apply the fluid density term to calculate the volume

$$V = W/\rho_f$$

$$= 15.9/0.38 \tag{18.5}$$

$$= 41.9\,\text{L}$$

■

ILLUSTRATIVE EXAMPLE 18.4 The rate equation for the formation of methane from hydrogen and carbon monoxide over a nickel catalyst is given by:

$$-r'_A = \frac{0.0183\sqrt{\theta_B}\left[\dfrac{y_{A_0}P(1-X)}{1+\varepsilon X}\right]^{3/2}}{1.0+1.5\left[\dfrac{y_{A_0}P(1-X)}{1+\varepsilon X}\right]}; \quad \text{gmole CO} \cdot \text{min/g cat}$$

$$CO + 3H_2 \longrightarrow CH_4 + H_2O; \quad A = CO,\ C = CH_4$$

It is desired to produce one ton per day of CH_4. The feed consists of 75% H_2 and 25% CO (by volume) at a temperature of 500°F and a pressure of 10 atm. Calculate the catalyst weight necessary to achieve 80% conversion in a fluidized bed reactor.

Solution. For a CSTR, substitute $X = 0.8$, $\delta = -2$, $y_{A_0} = 0.25$, $P = 10$ atm, $\varepsilon = 0.5$, and $\theta_B = 3$ into the rate expression provided in the statement. The result is

$$-r'_A = 0.0107 \text{ gmol CO} \cdot \text{min/kg cat}$$

Apply the CSTR equation to calculate the mass of catalyst

$$W = \frac{F_{A_0} X}{-r'_A}; \quad X = X_A \tag{18.3}$$

$$= \frac{(49.25)(0.8)}{0.0107}$$

$$= 3676 \text{ g cat}$$

■

ILLUSTRATIVE EXAMPLE 18.5 Rework the previous example if the reaction equation is (incorrectly) given by:

$$CO + 3H_2 \longrightarrow CH_4 + 2H_2O$$

Solution. For this case, $\delta = -1$, $y_{A_0} = 0.25$, $\varepsilon = -0.25$ and $\theta_B = 3$. Apply the same equations and substitute.

$$W = F_{A_0} X / -r'_A \tag{18.3}$$

$$W = \frac{F_{A_0} X}{0.0183\sqrt{3}(2.5)^{3/2}} \left[\left(\frac{1 - (0.25)(0.8)}{1 - 0.8} \right)^{3/2} + 3.75 \left(\frac{1 - (0.25)(0.8)}{1 - 0.8} \right)^{1/2} \right]$$

$$= 4875 \text{ g cat}$$

■

ILLUSTRATIVE EXAMPLE 18.6 Happel[7] once investigated the catalytic reaction:

$$CO + SO_2 \longrightarrow CO_2 + S$$

Note that this reaction eliminates two air pollutants while yielding a valuable end product.

Outline several options that are available to the researcher at this point to obtain a rate expression.

Solution. In order to obtain a rate expression for the above reaction, one could perform n tests in a *differential* reactor. The study would require p_{CO}, p_{SO_2} inlet (entrance) data along with p_{CO_2} at the outlet. The differential reactor would produce little change in p_{CO} and p_{SO_2} and a very low p_{CO_2}. This would enable one to calculate the rate of the reaction from the differential reactor equation given below.

$$r'_{CO_2} = \frac{q_0(C_{CO_2} - 0)}{W} = \frac{q_0 \Delta X_{CO}}{W} \tag{18.6}$$

This procedure is presented in Table 18.1, with the calculation for the rate provided in the last column. To obtain the reaction mechanism, or at a minimum, some form of the rate expression, one would employ a procedure similar to that outlined in this Table.

In addition to the information provided in the table (or the equivalent), one would employ

1. theory,
2. intuition,
3. raw guesswork,
4. guts, and
5. luck.

For example, it would not be uncommon to assume a rate equation that takes the functional form involving the product of the two reactants, i.e.,

$$r'_{CO_2} = f(\text{CO})g(\text{SO}_2)$$

Where does one proceed at this point? A tough question, but applying the type of information in Table 18.1 would be a good start. ■

ILLUSTRATIVE EXAMPLE 18.7 A fluidized-bed incinerator is to be designed to destroy 99.99% of a unique hazardous waste. Based on laboratory and pilot plant studies, researchers have described the waste reaction by a first-order reversible mechanism. Their preliminary findings are given here.

$$A \underset{k'}{\overset{k}{\longleftrightarrow}} B$$

$$k = 1.0\exp(-10{,}000/T) \quad T = \text{Rankine}$$
$$k' = 9.89\exp(-35{,}000/T) \quad T = \text{Rankine}$$

TABLE 18.1 Differential Reactor Data and Calculations

Run #	p_{CO}	p_{SO_2}	p_{CO_2}		r_{CO_2}
1					
2					
3				⇒	
⋮					
n					

Calculate a fluidized-bed incinerator operating temperature that will minimize the volume of the incinerator and achieve the desired degree of waste conversion (destruction). The operating temperature must be in the 1400–1700°F range.[8]

Solution. As noted above, a fluidized-bed incinerator is best described by a CSTR. According to Equation (18.6)

$$\tau = \frac{V}{q_a} = \frac{C_{A_0} - C_{A_1}}{-r'_{A_1}} = \frac{C_{A_0} - C_{A_1}}{kC_{A_1} - k'C_{B_1}}; q = q_{act} = q_a$$

For 99.99% destruction of the waste A, the conversion variable X_A becomes

$$X = X_A = 0 - 9999$$

So that

$$C_{A_1} = 0.0001C_{A_0} \quad C_{B_1} = 0.9999C_{A_0}$$

Therefore,

$$\frac{V}{q_a} = \frac{C_{A_0} - 0.0001C_{A_0}}{(k)(0.0001C_{A_0}) - (k')(0.9999C_{A_0})} = \frac{0.9999}{0.0001k - 0.9999k'}$$

To minimize the incinerator volume, $\mathrm{d}V/\mathrm{d}T$ is set to zero. Since q_a is assumed to be constant, $\mathrm{d}(V/q_a)/\mathrm{d}T$ can also be set to 0:

$$\frac{\mathrm{d}(V/q_a)}{\mathrm{d}T} - \frac{-9999[(\mathrm{d}k/\mathrm{d}T) - 9999(\mathrm{d}k'/\mathrm{d}T)]}{(k - 9999k')^2} = 0$$

Thus,

$$\frac{\mathrm{d}k}{\mathrm{d}T} = \frac{10^4\mathrm{d}k'}{\mathrm{d}T}$$

Substituting in the values of the reaction velocity constants,

$$k = 1.0 \exp\left(\frac{-10{,}000}{T}\right)$$

$$\frac{\mathrm{d}k}{\mathrm{d}T} = \left(\frac{-10{,}000}{T^2}\right)\exp\left(\frac{-10{,}000}{T}\right)$$

$$k' = 9.89 \exp\left(\frac{-35{,}000}{T}\right)$$

$$\frac{\mathrm{d}k'}{\mathrm{d}T} = \left(\frac{3.4615 \times 10^5}{T^2}\right)\exp\left(\frac{-35{,}000}{T}\right)$$

Therefore,

$$(10{,}000)\exp\left(\frac{-10{,}000}{T}\right) = (3.4165 \times 10^9)\exp\left(\frac{-35{,}000}{T}\right)$$

Solving for T,

$$T = 1960^\circ\mathrm{R} = 1500^\circ\mathrm{F}$$

There are other approaches that could have been used to solve this problem. ■

FIXED BED REACTORS

The shape of a fixed bed catalytic reactor is either cylindrical, or occasionally spherical and the flow throughout the bed may be upward or downward. Most of these reactors are designed with an internal lining and a distributor to control the gas passage through the catalyst bed. The internal reactor volume is occupied by the catalyst and any inert fill above and/or below the bed. The shell of the reactor is usually made of either steel alloy or carbon steel which is resistant to hydrogen and/or hydrogen sulfide. Steel alloy is employed for reactors that are in hydrogen and hydrogen sulfide service. It is not uncommon for these units to operate at or near 1000°F; temperature measurements are normally obtained with thermocouples rather than thermometers.

A fixed bed reactor contains small catalyst particles that are approximately 0.1 to 0.2 inches in diameter through which the reaction mixture is passed. Fill material is often used for support and to moderate temperature changes. The packing itself may be arranged in any of several different ways:

1. A single large bed.
2. Several horizontal beds supported on trays.
3. Several parallel packed tubes in a single shell.
4. Several beds, each in its own shell.

Variations from the large single bed (1) are often employed (see Figure 18.3) because of the need to either control the temperature or to improve gas distribution.

A fixed bed (packed-bed) reactor may be physically viewed as a tubular flow reactor which is packed with solid catalyst particles. This type of heterogeneous reaction system is most frequently used to catalyze gaseous reactions. The design equation

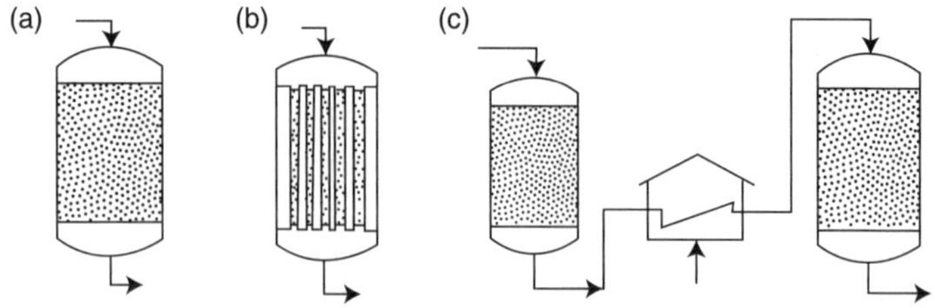

Figure 18.3 Fixed bed reactors, (a) single bed; (b) parallel packed tubes; (c) several connected in series.

for a TF reactor was previously shown to be (see Chapter 9)

$$V = F_{A_0} \int_0^X (-1/r_A)\,dX;\; X = X_A \tag{18.7}$$

The companion equation based on the mass of catalyst for a fixed-bed reactor is (see previous section)

$$W = F_{A_0} \int_0^X (-1/r_A')\,dX \tag{18.8}$$

The volume of the reactor, V, is then

$$V = W/\rho = W/\rho_B \tag{18.9}$$

where ρ_B = bulk density of the catalyst. See also Equation (18.5).

Fixed catalytic bed reactors have significant advantages relative to other types of heterogeneous catalytic reactors (particularly the fluid bed unit) and, consequently, are employed much more widely in the chemical industry. This section provides details on the procedures involved in the design of those reactors. As with fluidized bed reactors, global reactions rate information are required to predict the extent of conversion for a given set of operating conditions.

ILLUSTRATIVE EXAMPLE 18.8 Refer to Illustrative Equation 18.2. Calculate the mass of catalyst required to achieve a conversion of 80% if a fixed bed reactor is employed.

Solution. The describing equation for a fixed bed reactor in given by Equation (18.8)

$$W = F_{A_0} \int \frac{dX}{-r_A'}$$

Substituting for the rate gives

$$\begin{aligned}
W &= \frac{2.85}{(60)(5)\times 10^{-2}} \int_0^{0.8} \left(\frac{1-0.2X}{1-X}\right)^2 dX \\
&= \frac{2.84}{(60)(5)\times 10^{-2}} \left[\frac{(1-0.2)X}{1-X} + 0.2\ln\left(\frac{1}{1-X}\right)\right]; \quad X = 0.8, \\
&= \frac{2.84}{(60)(0.05)} \left[\frac{(1-0.2)0.8}{1-0.8} + 0.2\ln\left(\frac{1}{1-0.8}\right)\right] \\
&= \frac{2.84}{3.0}[3.2 + 0.322] = 3.34\,\text{kg}
\end{aligned}$$

■

ILLUSTRATIVE EXAMPLE 18.9 With reference to the previous illustrative example, calculate the volume requirement if the bulk density of the catalyst is $2.5\ \text{g}/(\text{cm})^3$.

Solution. The volume is given by Equation (18.9)

$$V = W/\rho_B$$

Substituting,

$$V = 3.34/2.5 = 1.33\,\text{L}$$

■

ILLUSTRATIVE EXAMPLE 18.10 Consider the following gas phase catalytic reaction:

$$A + B \overset{\text{cat}}{\longleftrightarrow} 3R$$

The volumetric flow rate to a small packed-bed reactor is 80 L/min at 400 K and 5 atm. The feed consists of 50% A, 25% B, and 25% inserts. The desired conversion is 60%. The bulk density of the catalyst in the tubular packed bed is $2.5\ \text{g/cm}^3$. The rate of reaction is given by

$$-r'_A = \frac{(0.05)\{1.0 + [(1 - X)/(1 + \varepsilon X)] + [3.125X/(1 + \varepsilon X)]\}}{[(1 - X)/(1 + \varepsilon X)][(2 - X)/(1 + \varepsilon X)] - [X/(0.86)(1 + \varepsilon X)]}$$

where r'_A has units of gmol A/g catalyst · h. Calculate the mass of catalyst required in grams if the reactor is to be designed in the fixed bed mode.

Solution. Determine y_{A_0}, δ, and ε for this reaction.

$$\begin{aligned} y_{A_0} &= 0.5 \\ \delta &= 3 - 1 - 1 = 1 \\ \varepsilon &= (y_{A_0})(\delta) \\ &= (0.5)(+1) \\ &= 0.50 \end{aligned}$$

Calculate C_{A_0} once again by applying the ideal gas law.

$$\begin{aligned} C_{A_0} &= y_{A_0}P/RT \\ &= (0.5)(5.0)/(0.082)(400) \\ &= 0.0762\ \text{gmol/L} \end{aligned}$$

Calculate F_{A_0} in gmol/h.

$$\begin{aligned} F_{A_0} &= qC_{A_0} \\ &= (80)(0.0762) \\ &= 6.098\ \text{gmol/min} \\ &= 365.85\ \text{gmol/h} \end{aligned}$$

Write the design equation for a fixed-bed reactor and calculate the mass of catalyst required to achieve a conversion of 60%.

$$W = F_{A_0} \int_0^X (-1/r'_A)\,dX \tag{18.8}$$

$$W = F_{A_0} \int_0^X f(X)\,dX; \qquad f(X) = -1/r'_A$$

Use Simpson's 5-point rule to evaluate the integral (see also Chapter 21)

$$\int_{0.0}^{0.6} f(X)dX = (h/3)[f(X_0) + 4f(X_1) + 2f(X_2) + 4f(X_3) + f(X_4)]$$

with

$$h = (0.6/4) = 0.15$$

Calculate $f(x)$ for various values of X.

X	$f(X)$
0.0	20.0
0.15	13.679
0.30	9.927
0.45	7.665
0.60	6.313

Substituting,

$$\begin{aligned} \int_{0.0}^{0.6} f(X)\,dX &= (0.15/3)[20.0 + 4(13.679) + 2(9.927) + 4(7.665) + 6.313] \\ &= 6.577\ \text{gmol/h} \end{aligned}$$

Thus,

$$\begin{aligned} W &= (365.85)(6.577) \\ &= 2407\ \text{g cat} = 2407\ \text{kg cat} \end{aligned}$$

■

ILLUSTRATIVE EXAMPLE 18.11 Refer to the previous example. Calculate the volume of a catalyst required in L.

Solution. Calculate the volume occupied by the catalyst bed in liters by employing Equation (18.9).

$$\begin{aligned} V &= 2407/2.5 = 962\,\mathrm{cm}^3 \\ &= 0.962\,\mathrm{L} \end{aligned}$$

■

ILLUSTRATIVE EXAMPLE 18.12 Refer to Illustrative Example 18.4. Perform the following two calculations

1. *Outline* the calculations to determine the catalyst weight necessary to achieve 80% conversion in a fixed bed reactor.
2. *Calculate* the catalyst weight necessary to achieve 80% conversion in a fixed bed reactor.

Solution. Apply the fixed-bed design equation

$$\frac{W}{F_{A_0}} = \int \frac{dX}{-r'_A}; \qquad X = 0.8$$

for the reaction

$$CO + 3H_2 \longrightarrow CH_4 + 2H_2O$$

Preliminary calculations:

$$\begin{aligned} 1.0 \text{ Ton } CH_4/\text{day} &\equiv 39.4\,\text{gmol/min} = F_C \\ F_C &= F_{A_0}X; \quad X = 0.9 \\ F_{A_0} &= 49.25\,\text{gmol/min} \\ \theta_B &= 3, \quad y_{A_0} = 0.25, \quad \delta = 1 + 1 - 1 - 3 = -2 \\ \varepsilon &= (0.25)(-2) = -0.5 \end{aligned}$$

Thus,

$$W = 49.25 \int \left[\frac{1 + 1.5\left[\frac{2.5(1-X)}{1-0.5X}\right]}{0.0317\left[\frac{2.5(1-X)}{(1-0.5X)}\right]} \right] dX = \int f(X)\,dX = \int I dX; \quad I = \text{integral}$$

Use Simpson's Rule (or equivalent) — see Chapter 21.

$$\int_{X_0}^{X_4} f(X)\,dX = \frac{h}{3}[f(X_0) + 4f(X_1) + 2f(X_2) + 4f(X_3) + f(X_4)]$$

Set $h = 0.2$ and once again calculate the integral at various values of X.

$$f(X_0) = f(0{,}0) = 37.91$$
$$f(X_1) = f(0{,}2) = 41.26$$
$$f(X_2) = f(0{,}4) = 46.81$$
$$f(X_3) = f(0{,}6) = 58.10$$
$$f(X_4) = f(0{,}8) = 93.3$$

Substituting into Simpson's rule gives

$$W = (49.25)\left(\frac{0.2}{3}\right)[37.91 + 4(41.26) + 2(46.81) + 4(58.10) + 93.3]$$
$$= 2043\ \text{g cat}$$

■

ILLUSTRATIVE EXAMPLE 18.13 Rework the previous example if the reaction equation is given by:

$$CO + 3H_2 \longrightarrow CH_4 + 2H_2O$$

(Note: Not realizing that the above equation is not (correctly) balanced stoichiometrically, the author once again, as in Illustrative Example (18.5), proceeded to solve the problem.) Perform the calculation using English units.

Solution. Apply the same design equation, i.e., Equation (18.8).

$$\frac{W}{F_{A_0}} = \int_0^X \frac{dX}{-r'_A}; \quad X = 0.8, \quad CO + 3H_2 \longrightarrow CH_4 + 2H_2O$$

Preliminary calculations:

$$y_{A_0} = 0.25, \quad y_{B_0} = 0.75; \quad \theta_B = 3$$
$$\delta = 1 + 2 - 1 - 3 = -1; \quad \varepsilon = y_{A_0}\delta = -0.25$$

$$F_{CH_4} = 2000\frac{\text{lb}}{\text{day}}\left(\frac{1}{16}\right) = 125\ \text{lbmol/day}$$

$$F_{A_0} = 125/0.8 = 156.25\ \text{lbmol/day}$$

$$\frac{W}{F_{A_0}} = \int_0^{0.8}\left[\frac{1 + 1.5\left(\frac{y_{A_0}P(1-X)}{1+\varepsilon X}\right)}{0.0183\sqrt{3}\left(\frac{y_{A_0}P(1-X)}{1+\varepsilon X}\right)^{3/2}}\right]dX$$

$$= \frac{1}{0.0183\sqrt{3}(y_{A_0}P)^{3/2}}\int_0^{0.8}\left[\left(\frac{1+\varepsilon X}{1-X}\right)^{3/2} + 1.5\,y_{A_0}P\left(\frac{1+\varepsilon X}{1-X}\right)^{1/2}\right]dX$$

$$= \frac{1}{0.0183\sqrt{3}(0.25\times 10)^{3/2}}\int_0^{0.8}\left[\left(\frac{1-0.25X}{1-X}\right)^{3/2} + 3.75\left(\frac{1-0.25X}{1-X}\right)^{1/2}\right]dX$$

TABLE 18.2 Calculations for Illustrate Example 18.13

X	$f(X)$	Simpson Factor, n	$n \times f(X)$
0.0	4.750	1.0	4.750
0.2	5.420	4.0	21.679
0.4	6.430	2.0	12.860
0.6	8.564	4.0	34.257
0.8	15.50	1.0	15.500

Let

$$f(X) = \int_0^X [(\mathrm{J})^{3/2} + 3.75(\mathrm{J})^{1/2}]\mathrm{d}X; \quad J = \left(\frac{1 - 0.25X}{1 - x}\right)$$

Apply Simpson's rule,

$$\int_0^{0.8} f(X)\,\mathrm{d}X = \frac{\Delta X}{3}[f(0) + 4f(0.2) + 2f(0.4) + 4f(0.6) + f(0.8)]$$

Results are provided in Table 18.2.
Substituting,

$$\int_0^{0.8} f(X)\,\mathrm{d}X = \frac{0.2}{3}[4.750 + 21.679 + 12.860 + 34.257 + 15.50]$$
$$= 5.936$$

Therefore,

$$W = \frac{5.936}{0.0183\sqrt{3}(2.5)^{3/2}}(156.25)\left(\frac{454\text{ g}}{\text{lb}}\right)\left(\frac{\text{day}}{24\text{ h}}\right)\left(\frac{h}{60\text{ min}}\right)$$
$$= 2331\text{ g cat}$$

■

ILLUSTRATIVE EXAMPLE 18.14 It is now January of 1990. A new Clean Air Law was passed by the President of the United States. Your company has been contacted by Frankel Motors to design a new catalytic converter. It is assumed that the catalytic converter acts as a fixed bed. It is known that the emissions the car exhaust gives off before reaching the converter has 10% toxic (T) gaseous materials while the rest of the exhaust can be considered as inerts. The Clean Air Law specifies that the toxic emissions should not exceed 1%. Calculate the weight of catalyst needed in grams. Data is provided below

$$\text{Pressure}, P = 1\,\text{atm}$$
$$\text{Temperature}, T = 400\,\text{K}$$
$$\text{Catalyst density}, \rho_C = 2.0\,\text{g/cm}^3$$
$$\text{Converter diameter} = D = 0.9\,\text{cm}$$
$$\text{Emission rate}, q = 100\,\text{L/s}$$

The stiochiometric reaction equation is given by

$$2T \longrightarrow C$$

The rate of reaction has been determined experimentally to be:

$$r'_A = 100\left[\frac{1+\varepsilon \cdot X}{1-X}\right]; \ \frac{\text{gmol T}}{\text{g cat} \cdot \text{s}}$$

Solution. Preliminary calculations, F_{A_0}, δ, X

$$\begin{aligned} y_{T_0} &= 0.1 \\ C_{T_0} &= \frac{(y_{T_0})(P)}{(R)T} \\ &= \frac{(0.1)(1)}{(0.0821)(400)} \\ &= 0.003045\,\text{gmol/L} \end{aligned}$$

Calculate the initial flow rate

$$\begin{aligned} F_{T_0} &= C_{T_0}q \\ &= (0.003045)(100) \\ &= 0.3045\,\text{gmol/s} \end{aligned}$$

From stoichiometry,

$$\begin{aligned} \delta &= 1 - 2 \\ &= -1 \end{aligned}$$

Calculate ε,

$$\begin{aligned} \varepsilon &= y_{A_0}\delta; \qquad \varepsilon = -0.1 \\ &= (0.1)(-1) = -0.1 \\ X &= \frac{y_{T_0} - y_{T_F}}{y_{T_0}}; \qquad X = 0.9 \\ &= \frac{0.1 - 0.01}{0.1} = 0.9 \end{aligned}$$

Assuming once again that a fixed-bed reactor has the same design equation as a plug-flow reactor,

$$W = F_{T_0} \int_0^X \frac{1}{-r'_T} dX \tag{18.8}$$

This integration can be performed analytical (see Illustrate Example 21.18).

$$W = F_{T_0}\left[(1+\varepsilon)\ln\left[\frac{1}{1-X}\right] - \varepsilon X\right]$$

Substituting,

$$W = 0.6584 \text{ gram cat}$$

Calculate the volume necessary for the reaction to occur

$$V = \frac{W}{\rho_C} \tag{18.9}$$

$$= \frac{0.6584}{2.0} = 0.3292 \text{ cm}^3$$

■

ILLUSTRATIVE EXAMPLE 18.15 With reference to the previous illustrative example, calculate the length of the converter assuming the converter has the shape of a cylinder.

Solution. The volume of the cylinder is;

$$V = 2\pi R^2 L; \quad R = \frac{D}{2}, \quad R = 0.45 \text{ cm}$$

Rearranging and solving

$$L = \frac{V}{2\pi R^2} = \frac{0.3292}{(2\pi)(0.45)^2}$$

$$L = 0.26 \text{ cm}$$

The converter appears to be awkwardly sized. ■

PRESSURE DROP CALCULATIONS

The calculation of pressure drop and pressure variation in reactors can be a difficult problem, particularly if a change of phase occurs or more than one phase is present. Although this effect is often small and/or can be neglected (particularly if the reactor is already under pressure), the reader should become familiar with the methods of calculating pressure drop in reactors. The literature[9,10] provides information for these calculations: Fanning's equation may be used to calculate pressure drop for homogenous reactions, while Ergun's equation is normally employed for fixed-bed reactors.[11] The mass of the catalyst divided by the cross-sectional area of the reactor provides a reasonable estimate of the pressure drop in fluidized-bed reactors. Fogler[1] provides equations and additional details for calculating these effects.

The following notations and (engineering) units are employed in the development to follow for fixed bed reactors. However, the reader should note that any consistent (from a dimensional perspective) set of units may be employed.

ΔP = pressure drop, ft of following fluid

L = packed bed length, ft

d_p = catalyst particle diameter, ft

ε = catalyst bed void volume, dimensionless

μ = fluid viscosity lb/ft · s

D = bed diameter, ft

v_s = superficial (empty bed) fluid velocity, ft/s

ρ = gas density, lb/ft^3

G_s = gas mass flux = ρv_s, lb/ft^2 · s

Extensive experimental measurements led to the following equation

$$v_s = \frac{\Delta P}{L}\frac{d_p^2}{150\mu}\frac{\varepsilon^3}{(1-\varepsilon)^2} \tag{18.10}$$

which some have defined as the Blake–Kozeny equation. This result is generally good for void fractions less than 0.5 and is valid only in the laminar region where the particle (catalyst) Reynolds number is given by $d_p G_s/\mu(1-\varepsilon) < 10$.[2] Data for highly turbulent flow in packed columns, which some have defined as the Burke–Plummer equation, are valid for $(d_p G_s/\mu)(1-\varepsilon) > 1000$.

$$\begin{aligned}\frac{\Delta P}{L} &= 3.50\frac{1}{d_p}\frac{\rho v_s^2}{2}\frac{1-\varepsilon}{\varepsilon^3}\\ &= 1.75\frac{\rho v_s^2}{d_p}\left(\frac{1-\varepsilon}{\varepsilon^3}\right)\end{aligned} \tag{18.11}$$

When the Blake–Kozeny equation for laminar flow and the Burke–Plummer equation for turbulent flow are simply added together, the result is

$$\frac{\Delta P}{L} = \frac{150\mu v_s}{d_p^2}\frac{(1-\varepsilon)^2}{\varepsilon^3} + \frac{1.75\rho v_s^2}{d_p}\frac{(1-\varepsilon)}{\varepsilon^3} \tag{18.12}$$

This may be rewritten in terms of dimensionless groups (numbers):

$$\left(\frac{\Delta P\rho}{G_0^2}\right)\left(\frac{d_p}{L}\right)\left(\frac{\varepsilon^3}{1-\varepsilon}\right) = 150\frac{1-\varepsilon}{d_p G_0/\mu} + 1.75 \tag{18.13}$$

This is the Ergun equation. It has been applied with success to gases by using the density of the gas at the arithmetic average of the inlet and outlet pressures; for large pressure drops, however, errors can arise because of the variation of ρ with both pressure and molecular weight and it is no longer valid. Ordinarily, it is satisfactory to use average values. Theodore[13] provides a procedure for addressing this situation.

Pressure drop information on different mesh catalyst sizes is presented in Figure 18.4 (EPA chart). These data are also often used in pressure drop calculations.

For fluidized beds the pressure drop is simply the mass of catalyst divided by the column area but may also be represented by:

$$\Delta P = (1 - \varepsilon_{\mathrm{mf}})\left(\frac{\rho_s - \rho}{\rho}\right)L_{\mathrm{mf}} \tag{18.14}$$

The equations for minimum fluidization are similar to those presented for a fixed bed, i.e., Equation (18.13). For laminar flow conditions ($\mathrm{Re}_p < 10$), the Blake–Kozeny equation is used to express the pressure in terms of the superficial gas velocity at minimum fluidization, v_{mf}, and other fluid and bed properties. This equation is obtained from Equations (18.10) and (18.14)

$$\Delta P = (1 - \varepsilon_{\mathrm{mf}})\left(\frac{\rho_s - \rho}{\rho}\right)L_{\mathrm{mf}} = 150\frac{v_{\mathrm{mf}}\mu}{\rho g}\frac{(1 - \varepsilon_{\mathrm{mf}})^2}{\varepsilon_{\mathrm{mf}}^3}\frac{L_{\mathrm{mf}}}{d_p^2} \tag{18.15}$$

Rearranging, one obtains the minimum fluidization velocity:

$$v_{\mathrm{mf}} = \frac{1}{150}\left(\frac{\varepsilon_{\mathrm{mf}}^3}{1 - \varepsilon_{\mathrm{mf}}}\right)\frac{g(\rho_s - \rho)d_p^2}{\mu}\ (\mathrm{Re}_p < 10) \tag{18.16}$$

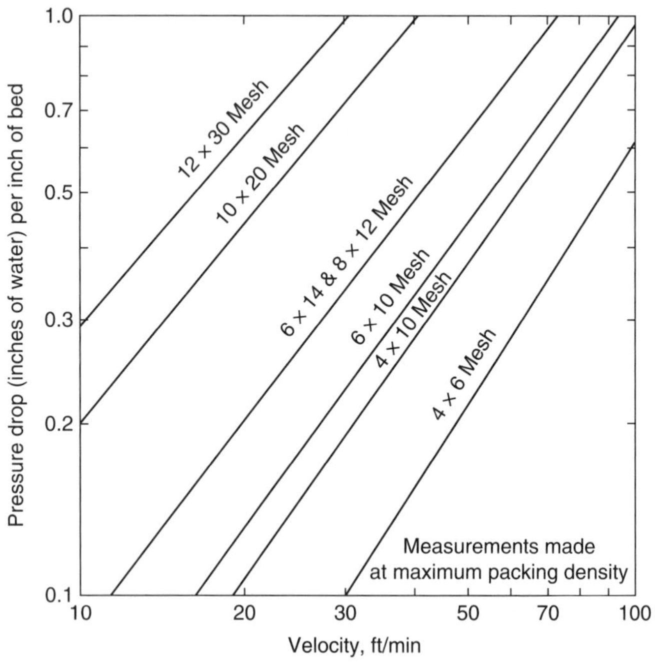

Figure 18.4 Pressure drop-velocity graphs.

with (once again)

$$\text{Re}_p = \frac{d_p v_{\text{mf}} \rho}{\mu(1 - \varepsilon_{\text{mf}})} \tag{18.17}$$

The Burke–Plummer equation is used to express the pressure drop for turbulent flow conditions ($\text{Re}_p > 1000$). For this condition, the result is:

$$v_{\text{mf}} = \sqrt{\frac{1}{1.75}\varepsilon_{\text{mf}}^3\left(\frac{\rho_s - \rho}{\rho}\right)gd_p} \tag{18.18}$$

In the absence of ε_{mf} data, the above equations can be approximated as

$$v_{\text{mf}} = \frac{1}{1650}\frac{g(\rho_s - \rho)d_p^2}{\mu} \quad \text{Re}_p < 10 \tag{18.19}$$

$$v_{\text{mf}} = \sqrt{\frac{1}{24.5}\left(\frac{\rho_s - \rho}{\rho}\right)gd_p} \quad \text{Re}_p > 1000 \tag{18.20}$$

where Re_p is the particle Reynolds number at minimum fluidization and is equal to:

$$\text{Re}_p = \frac{d_p v_{\text{mf}} \rho}{\mu} \tag{18.21}$$

Another approach that may be employed to calculate the minimum fluidization velocity is to employ Happel's equation.[13] Happel's equation is given by:

$$\frac{v_{\text{mf}}}{v_t} = \frac{3 - 4.5(1 - \varepsilon_{\text{mf}})^{1/3} + 4.5(1 - \varepsilon_{\text{mf}})^{5/3} - 3(1 - \varepsilon_{\text{mf}})^2}{3 + 2(1 - \varepsilon_{\text{mf}})^{5/3}} \tag{18.22}$$

where v_{mf} is the minimum fluidization velocity, v_t the terminal velocity, and ε_{mf} the bed porosity at minimum fluidization.

The selection of an optimum column diameter is also not a simple procedure. Generally speaking it is desirable to operate at high velocities in order to minimize both the convective mass transfer resistance and the undesirable effects of longitudinal and radial diffusion. However, high velocities result in high pressure drops which may be undesirable because of three concerns.

1. A high pressure drop may increase the rates of undesirable reactions and compromise the selectivity.
2. Excessive pressure drops may result in catalyst breakup.
3. A high pressure drop increases energy cost.

These three factors should be considered in determining the velocity and pressure drop which result in the most economical over-all operation.

ILLUSTRATIVE EXAMPLE 18.16 Comment on the relationship between the Ergun equation and the Burke–Plummer and Blake–Kozeny equations.

Solution. Note that for high rates of flow, the first term on the right-hand side of the Ergun equation drops out and the equation reduces to the Burke–Plummer equation. At low rates of flow, the second term on the right-hand side drops out and the Blake–Kozeny equation is obtained. It should be emphasized that the Ergun equation is but one of many that have been proposed for describing the pressure drop across packed columns. For example, Theodore[14] provide the following simple equation.

$$\Delta P = 0.37Z\left(\frac{v}{100}\right)^{1.56} \tag{18.14}$$

where ΔP = in H_2O

Z = bed depth, in

v = super fixed velocity, ft/s ■

ILLUSTRATIVE EXAMPLE 18.17 Determine the pressure drop of 60°F air flowing through a 3-inch diameter 10-ft packed bed with 0.24-inch protruded catalyst packing made of 316 stainless steel. The superficial velocity is 4.65 ft/s. The protruded packing has a fraction void volume, effective particle diameter and to surface area per unit packing of 0.89, 0.0078.5 ft, and 3305 ft^{-1}, respectively.

Solution. Use the Ergun equation

$$\Delta P = \left[\left(\frac{150 v_{s_0}\mu_g}{g_c(\Phi_s d_p)^2}\right)\left(\frac{(1-\varepsilon)^2}{\varepsilon^3}\right) + \left(\frac{1.75\rho_g v_{s_0}^2}{g_c(\Phi_s d_p)}\right)\left(\frac{(1-\varepsilon)^2}{\varepsilon^3}\right)\right]L \tag{18.13}$$

For air at 60°F,

$$\mu = 1.3 \times 10^{-5}\ \text{lb/ft}\cdot\text{s}$$

$$\rho = 0.067\ \text{lb/ft}^3$$

Plugging in values from the problem statement, one obtains

$$\begin{aligned}\Delta P &= \left[\left(\frac{(150)(4.65)(1.3E^{-5})}{(32.2)(0.007815)^2}\right)\left(\frac{(1-0.89)^2}{(0.89)^3}\right)\right.\\ &\quad \left.+\left(\frac{(1.75)(0.67)(4.65)^2}{32.2(0.007815)}\right)\left(\frac{(1-0.89)^2}{(0.89)^3}\right)\right]10\\ &= 10.25\ \text{lb/ft}^2\end{aligned}$$

■

ILLUSTRATIVE EXAMPLE 18.18 Air is used to fluidize a bed of spherical catalyst particles. The particles are 200 mesh uniform spheres; the bed diameter,

D_b, is 0.2 m; the ultimate (or true) solids density, ρ_s, is 2200 kg/m^3; the voidage at minimum fluidization, ε_{mf}, equals 0.45; and, bed length (height) at minimum fluidization, L_{mf}, is 0.3 mg. The air properties are $\rho = 1.2$ kg/m^3 and $\mu = 1.89 \times 10^{-5}$ kg/m · s. The diameter of a 200 mesh particle is approximately 74 μ m = 7.4×10^{-5} m.

Calculate the minimum fluidization mass flow rate of air and the pressure drop of air across the bed at minimum fluidization.

Solution. Assume laminar flow, with $\rho_s - \rho \cong \rho_s$, and employ Equation (18.16)

$$v_{mf} = \frac{1}{150}\frac{(1-0.45)}{(0.45)^3}\frac{9.807(2200)(7.4\times10^{-5})^2}{1.89\times10^{-5}} = 0.25\ \text{m/s}$$

Check the flow regime

$$\text{Re}_p = \frac{v_{mf}\,d_p}{\mu(1-\varepsilon_{mf})} = \frac{(0.25)(7.4\times10^{-5})}{(1.89\times10^{-5})(1-0.45)}$$
$$= 1.79 < 10$$

The flow is indeed laminar.

The mass flow rate of air is

$$\dot{m} = \frac{\pi D_b^2}{4} v_{mf}\rho = \frac{\pi(0.2)^2}{4}(0.25)(1.2) = 9.40^{-3}\ \text{kg/s}$$

Calculate the gas pressure drop across the bed. Use Equation (18.14).

$$\Delta P_{fr} = (1-0.45)(2200)(9.807)(0.3) = 3560\ \text{Pa} = 0.0351\ \text{atm}$$ ■

CATALYTIC REACTOR DESIGN CONSIDERATIONS

The process of a species undergoing a chemical reaction in the presence of a solid or catalyst can occur in a variety of ways. For reasons that should now be obvious, this chapter addresses fluidized bed reactors and fixed bed reactors. The material presented in the three previous sections attempted to provide some simple design procedures.

The practicing engineer can often apply much of this directly with confidence. However, good engineering judgement often plays an important rule in the final analysis, particularly when the ideal and/or simplified procedures presented earlier cannot be employed. Use of a heterogenous reaction rate, which is often of questionable value, further complicates the problem.

The design of an industrial scale fluidized bed reactor for a new or existing process can be challenging. Any study of an existing operating unit will immediately justify the basis of this statement.

The accurate design and prediction of the performance of a fixed bed reactor for a given set of operating conditions is also a problem that confronts the practicing engineer. The material presented earlier can be applied with some confidence for some applications. But, once again, good engineering judgment is often required.

The design of a fixed bed reactor is similar in principle to that of an empty tubular reactors discussed in Chapter 9. The principle differences are that in a fixed bed unit, temperature and concentration gradients occur radially as well as longitudinally whereas radial gradients are uncommon in fluid bed reactors. These variations adversely affect the fluidized reactors and these possibilities make the transition from pilot to commercial scale problematic. However, the calculational procedure for a fixed bed unit is similar to that presented earlier, with a heterogeneous rate equation employed.

In any event, the reader should remember that for heterogeneous catalytic reactors/reactions:

1. The same basic forms of equations apply.
2. The reactor volume V is replaced by the catalyst mass W.
3. The rate equation applies to a particular catalyst.
4. For fixed bed units use the tubular (plug) flow model.
5. For fluidized bed units use CSTR model.

The chapter concludes with four illustrative applications that are concerned with these classes of reactors.

ILLUSTRATIVE EXAMPLE 18.19 Consider the following isothermal, gas-phase reaction

$$A + B \rightleftarrows C + 2D$$

$$-r'_A = (10.0)^5 \exp\left[\frac{-3000(15 + 0.5X)}{1.987(2500 + 70X)}\right] \mathrm{C}^2_{\mathrm{A_0}} \frac{(1 - X)(\theta_\mathrm{B} - X)}{(1 + \varepsilon X)^2}; \quad \frac{\text{gmol A}}{\text{kg cat} \cdot \text{h}}$$

It is desired to produce 10.0 kg per hour of C ($MW = 18$). The feed consists of 20% A, 70% B, and 10% inerts (by mole) and the initial concentration of A is 2.0 gmol/L.

1. Calculate the mass of catalyst required to achieve 90% conversion in a fluidized bed reactor.
2. Calculate the mass of catalyst required to achieve 90% conversion in a fixed bed reactor. Use Simpson's 3-point rule to perform the calculation.
3. Comment on the results.

Solution.

$$F_{A_0} = \frac{10.0}{0.9} = 11.11\,\text{kg/h} = 0.6178\,\text{kg mol/h}$$

$$\delta = 1, \quad \varepsilon = 0.2, \quad \theta_B = 3.5$$

For a fluid bed reactor

$$W_{\text{FLUID}} = \frac{F_{A_0}\Delta X}{-r'_{\text{A}}} = \frac{(11.11)(0.9)}{-r'_{\text{A}}} : r'_{A'} = 8.3296\,\text{gmol A/kg cat} \cdot \text{h}$$

$$W_{\text{FLUID}} = 66.7\,\text{kg cat}$$

For a fixed bed reactor

$$W = F_{A_0} \int_0^X \frac{1}{r'_A} \mathrm{d}X = \int I \mathrm{d}X$$

Apply Simpson's 3-point rule (see Chapter 21)

$$I_0 = 0.00614$$
$$I_1 = 0.0155$$
$$I_2 = 0.12005$$

$$W_{\text{FIXED}} = 0.6178\left[\frac{0.45}{3}\right][0.00614 + 4(0.0155) + 0.12005]$$
$$= 17.45\,\text{kg cat}$$

As expected, the fluid bed unit requires less catalyst.

■

ILLUSTRATIVE EXAMPLE 18.20 Refer to the previous example. *Outline* the calculations to determine the mass of catalyst necessary to achieve half the conversion in a fluidized bed as that in a fixed-bed reactor. If there a solution to this problem and is it unique?

Solution. For this example

$$X_{\text{FLUID}} = X_{FL} = 0.5\,X_{\text{FIXED}} = 0.5X_{FB}$$

Therefore,

$$W_{FL} = \frac{F_{A_0} X_{FL}}{-r'_A|_{X_{FL}}}$$

$$W_{FB} = F_{A_0} \int_0^{X_{FB}} \frac{\mathrm{d}X}{-r'_A}; \quad X_{F_B} = 2X_{FL}$$

with

$$W_{FL} = W_{FB}$$

Therefore,

$$\frac{F_{A_0}X_{FL}}{-r'_A|_{X_{FL}}} = F_{A_0}|_0^{2X_{FL}} \frac{dX}{-r'_A|_{2X_{FL}}}$$

This requires a trial and error calculation for X_{FL}. There may be one solution, many solutions or no solution (if $X_{FL} > 0.5$ or negative). There are no guarantees here. Actually, it is suggested that the interested reader attempt to solve the problem. ■

ILLUSTRATIVE EXAMPLE 18.21 Consider the following gas phase catalytic reaction:

$$A + B \longleftrightarrow 2R$$

The rate equation is:

$$-r'_A = 0.05\frac{\left[1.0 + \left(\frac{1-X}{1+\varepsilon X}\right) + \left(\frac{3.125X}{1+\varepsilon X}\right)\right]}{\left[\left(\frac{1-X}{1+\varepsilon X}\right)\left(\frac{2-X}{1+\varepsilon X}\right) - \left(\frac{X}{0.86(1+\varepsilon X)}\right)\right]}; \text{ gmol A/kg cat}$$

The volumetric flow rate to a fixed bed reactor is 7000 L/min. The volumetric flow rate to a fluidized bed reactor is the same. The temperature is 77°C and the pressure is 7 atm. The feed consists of 25% A, 50% B, and 25% inerts. The desired conversion is 50%. The bulk density of the catalyst in the tubular fixed bed is 2.4 g/cm^3 and the bulk density of the catalyst in the fluidized bed is 0.4 g/cm^3. For the fixed bed reactor, calculate the mass of the catalyst and the volume of the catalyst. Also calculate the mass of the catalyst and the volume of the catalyst for the fluid bed reactor.

Solution. For this gas phase reaction

$$y_{A_0} = 0.25, \delta = 2 - 1 - 1 = 0, \varepsilon = 0$$

$$C_{A_0} = \frac{y_{A_0}P}{RT} = \frac{(0.25)(7.0\text{ atm})}{(0.0845)(350\text{ K})} = 0.0690\text{ gmol A/L}$$

$$F_{A_0} = (C_{A_0})(q) = (0.0609)(7000) = (426.3)(60) = 25{,}578\text{ gmol A/h}$$

For fixed bed:

$$W = F_{A_0}\int_0^X \frac{dX}{-r'_A}; X = 0.50 \qquad (18.8)$$

Use Simpson's 5 point rule.

$$\int_0^{0.5} f(X)\,dX = \frac{h}{3}[f(X_0) + 4f(X_1) + 2f(X_2) + 4f(X_3) + f(X_4)]; \quad h = \frac{0.50}{4} = 0.125$$

The result is (see also Table 18.3)

$$\int_0^{0.5} f(X)\,\mathrm{d}X = \frac{0.125}{3}[20 + (4)(13.158) + (2)(8.065) + (4)(4.149) + 1.101]$$
$$= 4.436\,\text{g cat} \cdot \text{h/gmol A}$$

The mass of catalyst is therefore

$$W = (25{,}578)(4.436) = 113.5\,\text{g cat}$$

The volume is given by

$$V = W/\rho_B \tag{18.8}$$

Substituting,

$$V = \frac{113.5}{2.5}$$
$$= 47.3\,\text{L}$$

For a fluidized bed system

$$W = \frac{F_{A_0} X}{-r'_A} \tag{18.4}$$

Substituting,

$$W = (25{,}578)(0.5)/0.908; \quad X = 0.5$$
$$= 14{,}080\,\text{g cat} = 14.08\,\text{kg cat}$$

and

$$V = W/\rho_B \tag{18.5}$$
$$= 14{,}080/0.4$$
$$= 35{,}200\,\text{cm}^3 = 35.2\,\text{L}$$

TABLE 18.3 Calculation for Illustrative Example 18.21

X	$-r_A$	$f(X) = \left(-\frac{1}{r_A}\right)$
0	0.05	20
0.125	0.076	13.158
0.250	0.124	8.065
0.375	0.241	4.145
0.50	0.908	1.101

A sample calculation is provided below for r_A: If $X = 0.5$

$$-r'_A = 0.05 \frac{\left[1.0 + \left(\frac{1-0.5}{1+0}\right) + \left(\frac{(3.125)(0.5)}{1+0}\right)\right]}{\left[\left(\frac{1-0.5}{1+0}\right)\left(\frac{2-0.5}{1+0}\right) - \left(\frac{0.50}{0.86(1+0)}\right)\right]}$$

$$= 0.908 \text{ gmol A/g cat} \cdot \text{h}$$

■

ILLUSTRATIVE EXAMPLE 18.22 With reference to the previous illustrative example, answer the following four questions

1. Can one safely assume ideal gas behavior? (One sentence)
2. Are the results reasonable? Why? (One sentence)
3. Explain why the mass of catalyst is less for the fluid bed system. (One or two sentences)
4. Comment on what is the controlling "step" for this reaction. (One or two sentences)

Solution.

1. Marginal since the pressure is in excess of 5 atm[8].
2. Not reasonable since $V_{PB} > V_{FB}$.
3. The term $(-\frac{1}{r_A})$ decreases with increasing X (not normally effected).
4. Tough to say not enough information provided. ■

REFERENCES

1. S. Fogler, "*Elements of Chemical Reaction Engineering*," 4th edition, Prentice-Hall, Upper Saddle River, NJ, 2006.
2. J. Reynolds, J. Jeris, and L. Theodore, "*Handbook of Chemical and Environmental Engineering Calculations*," John Wiley & Sons, Hoboken, NJ, 2004.
3. D. Kunii and O. Levenspiel, "*Fluidization Engineering*," John Wiley & Sons, Hoboken, NJ, 1969.
4. L. Theodore, "*Air Pollution Control Equipment Calculations*," John Wiley & Sons, Hoboken, NJ, 2007.
5. L. Theodore, personal notes, East Williston, NY, 1971.
6. L. Theodore, personal notes, East Williston, NY, 1989.
7. J. Happel, personal communication to L. Theodore, 1994.
8. J. Santoleri, J. Reynolds, and L. Theodore, "*Introduction to Hazardous Waste Incineration*," 2nd edition, John Wiley & Sons, Hoboken, NJ, 2000.
9. P. Abulencia, and L. Theodore, "*Fluid Flow for the Practicing Engineer*," John Wiley & Sons, Hoboken, NJ, 2009.
10. D. Green and R. Perry (editors), "*Perry's Chemical Engineers' Handbook*," 8th edition, McGraw-Hill New York City, NY, 2008.
11. I. Farag, "*Fluid Flow*," A Theodore Tutorial, Theodore Tutorial, East Williston, NY, 1996.
12. S. Ergun, "*Fluid Flow Through Packed Columns*," CEP, 48: 49, New York City, NY, 1952.
13. J. Happel, personal communications, 2003.
14. L. Theodore, personal notes, East Williston, NY, 1999.

CHAPTER 19

BIOCHEMICAL REACTORS

INTRODUCTION

Early chapters focused on homogeneous and heterogeneous reactions, since these are the two major systems with which practicing engineers are concerned. However, there are also many important biological/biochemical reactors, and systems. This has become a major growth industry for engineers.

Unfortunately, biochemical and biomedical engineering is a relatively new discipline in the engineering profession and, as one might expect, they have come to mean different things to different people. Terms such as biochemical engineering, biomedical engineering, bioengineering, biotechnology, biological engineering, genetic engineering, etc., have been used interchangeably by many in the technical community. To date, consistent standard definitions have not been created to distinguish these genres. Consequently, the author has lumped them all together for the sake of simplicity. What one may conclude from all of the above is that the general subject of biochemical engineering involves applying the concepts, knowledge, basic fundamentals, and approaches of virtually all engineering disciplines (not only chemical engineering); the opportunities for interaction between engineers, biologists, and healthcare professionals are therefore many and varied.

On a personal note, the author views this topic as the application of engineering, mathematics, and physical sciences to principles in engineering biology and medicine. Interestingly, the terms biophysics and bioengineering either involve the interaction of physics or engineering with either biology or medicine.

Because of the broad nature of this subject, this chapter addresses only introductory matter as it applies to biochemical engineering. The reader is referred to three excellent references in the literature for an extensive and comprehensive treatment of this "new" discipline.[1–3]

In conclusion, this subject area encompasses an important vital interdisciplinary field. The ultimate role of the practicing engineer and the profession is to serve society. The great potential, challenge, and promise in this relatively new endeavor offers both technological and humanitarian benefits. The possibilities appear to be unlimited.

This subject has served as the title for numerous books. Condensing this subject matter into one chapter was a particularly difficult task. In the end, the author decided

Chemical Reactor Analysis and Applications for the Practicing Engineer. By Louis Theodore

to provide a superficial treatment that would introduce the reader to the topic. Subject matter not covered includes reactor systems such as:

Batch fermenter

Chemostat: continuous stirred tank fermenter

Fed–batch fermenter

Fluidized bed bioreactor

Comparison of batch and continuous stirred tank fermenters (CSTF)

Multistage chemostats

Multistage reactors; CSTR and TF in series

Immobilized cell bioreactor

Vasudevan[4] provides an excellent and detailed review of these topics. However the subject of sterilization and environmental applications are included in the presentation.

Following the introductory section, the chapter consists of three sections:

Basic Operations

Design Considerations

Environmental Applications

Specific detail on Michaelis–Menten kinetics, quasi steady-state approximations, competitive and non-competitive inhibitions, substrate inhibition, rate expressions for enzyme catalysis and deactivations, Monod growth kinetics, etc. are not presented in an extensive manner although additional information is available in the work of Vasudevan for the interested reader.[4] Also note that the notation adopted by Vasudevan is employed *throughout this chapter*.

ILLUSTRATIVE EXAMPLE 19.1 List some key activities in the biomedical and biochemical profession.[2,3]

Solution.

1. Development of improved species of plants and animals for food production.
2. Invention of new medical diagnostic tests for diseases.
3. Production of synthetic vaccines from clone cells.
4. Bioenvironmental engineering to protect human, animal, and plant life from toxicants and pollutants.
5. Study of protein-surface interactions.
6. Modeling of flow dynamics.
7. Modeling of mass transfer through membranes.
8. Modeling of the growth of the kinetics of yeast and hybridoma cells.
9. Research in immobilized enzyme technology.
10. Development of therapeutic proteins and monoclonal antibodies.
11. Development of artificial hearts. ■

ILLUSTRATIVE EXAMPLE 19.2 List at least six new applications that have emerged over the last half-century in the bio area.

Solution.

1. Application of engineering system analysis (physiologic modeling, simulation, and control of biological problems).
2. Detection, measurement, and monitoring of physiologic signals (i.e., biosensors and biomedical instrumentation).
3. Diagnostic interpretation via signal-processing techniques of bioelectric data.
4. Therapeutic and rehabilitation procedures and devices (rehabilitation engineering).
5. Devices for replacement or augmentation of bodily functions (artificial organs).
6. Computer analysis of patient-related data and clinical decision-making (i.e., medical information and artificial intelligence).
7. Medical imaging, i.e., the graphical display of anatomic details or physiologic functions.
8. The creation of new biologic products (i.e., biotechnology and tissue engineering). ■

ILLUSTRATIVE EXAMPLE 19.3 List job-related activities of biochemical and biomedical engineers.[2,3]

Solution.

1. Research into new materials for implanted artificial organs.
2. Development of new diagnostic instruments for blood analysis.
3. Writing software for the analysis of medical research data.
4. Analysis of medical device hazards for safety and efficacy.
5. Development of new diagnostic imaging systems.
6. Design of telemetry systems for patient monitoring.
7. Design of biomedical sensors.
8. Development of expert systems for the diagnosis and treatment of diseases.
9. Design of closed-loop control systems for drug administration.
10. Modeling of the physiologic systems of the human body.
11. Design of instrumentation for sports medicine.
12. Development of new dental materials.
13. Design of communication aids for individuals with disabilities.
14. Study of pulmonary fluid dynamics.
15. Study of biomechanics of the human body.

16. Development of material to be used as replacement for human skin.
17. Applications of nanotechnology to many of the above activates. ■

ILLUSTRATIVE EXAMPLE 19.4 Discuss regulatory issues as they apply to this field.

Solution. Regulatory issues are a constant cause for concern and, perhaps, justifiably so. To satisfy safety regulations, most biomedical systems must have documented analyses of risk to show that they were designed, built, tested, delivered, managed, and used according to a planned and approved process. The two key regulatory agencies in the US are the Food and Drug Administration (FDA) and the Consumer Product Safety Commission.[5] ■

BASIC OPERATIONS

Six basic operations are reviewed in this section. Details follow.

Enzyme and Microbial Kinetics

Enzyme and microbial kinetics involve the study of reaction rates and the variables that affect these rates. It is a topic, that is critical for the analysis of enzyme and microbial reacting systems. The rate of a biochemical reaction can be described in many different ways. The most commonly used definition is similar to that employed for traditional reactors. It involves the time change in the amount of one of the components participating in the reaction or of one of the products of the reaction; this rate is also based on some arbitrary factor related to the system size or geometry, such as volume, mass or interfacial area. In the case of immobilized enzyme catalyzed reactions, it is common to express the rate per unit mass or per unit volume of the catalyst.

ILLUSTRATIVE EXAMPLE 19.5 For an enzyme catalyzed reaction, calculate the time required to achieve a given conversion for a reactor operating at ambient temperature; for a microbial reaction, calculate the time required to double the concentration of cells; discuss the Michaelis–Menten equation, and the rate equation for growth of cells. Pertinent information is provided below.
Data:

Temperature = 25°C
Conversion = 90%
Maximum reaction rate = 0.1 moles/L · min
Michaelis constant = 0.035 M
Initial substrate concentration = 0.001 M
Microbial specific growth reaction rate = 0.1 $(h)^{-1}$

Solution. The Michaelis–Menten rate equation is as follows:

$$v = \frac{v_{max}[S]}{K_m + [S]} \tag{19.1}$$

where v = reaction rate = $-\dfrac{d[S]}{dt}$

t = time

v_{max} = constant defined as the maximum reaction rate

K_m = Michaelis constant (dissociation constant)

$[S]$ = substrate or reactant concentration.

Note that when $K_m \gg [S]$, the equation reduces to the following first order rate expression:

$$-\frac{d[S]}{dt} = v = \frac{v_{max}}{K_m}[S]$$

The rate equation for microbial cell growth can be written as:

$$\frac{d[X]}{dt} = \mu[X] \tag{19.2}$$

where $[X]$ = cell concentration or mass

μ = specific growth rate

For an enzyme reaction in which $[S] \ll K_m$,

$$\ln\left(\frac{[S]}{[S]_o}\right) = -\frac{v_{max}}{K_m}t \tag{19.3}$$

For the microbial reaction:

$$\ln\left(\frac{[X]}{[X]_o}\right) = \mu t \tag{19.4}$$

where $[S]_o$ and $[X]_o$ are the initial substrate and cell concentrations at time $t = 0$, respectively.

For a conversion of 90% (see problem statement),

$$[S]/[S]_o = 1.0 - 0.9 = 0.1$$

Since $[S] \ll K_m$, the reaction rate may be assumed to be first order. The time required to achieve the desired conversion of 90% is given by:

$$t = -\frac{K_m}{v_{max}}\ln\left(\frac{[S]}{[S]_o}\right) \tag{19.3}$$

substituting,

$$t = -\frac{0.035}{0.1}\ln(0.1)$$

$$= 0.805 \text{ min}$$

$$= 48.4 \text{ s}$$

For the microbial reaction, calculate the time required to double the concentration of the cells. The doubling time is given by:

$$t = \ln\left(\frac{[X]}{[X]_o}\right)\Big/\mu \tag{19.4}$$

substituting,

$$\begin{aligned} t &= \ln(2)/0.1 \\ &= 6.93\,\text{h} \end{aligned}$$

Unlike chemical reactions, enzyme and microbial reactions are generally complex. The mechanism of enzyme catalyzed reactions is discussed in the next basic operation. ■

Enzyme Reaction Mechanisms

The rate of a chemical or biochemical reaction is similar to that defined earlier, i.e., the time rate of change in the quantity of a particular species participating in a reaction divided by a factor that characterizes the reacting system's geometry. The choice of this factor is also a matter of convenience. For homogeneous media, the factor is almost always the volume of the reacting system. For fluid–solid reaction systems, the factor is often the mass of the solid. For example, in immobilized enzyme reactions, the factor is the mass of the immobilized or supported enzyme catalyst.

ILLUSTRATIVE EXAMPLE 19.6 For a relatively simple enzyme catalyzed reaction, derive an expression for the reaction rate. The reaction scheme is as follows:

$$E + S \underset{k_{-1}}{\overset{k_1}{\rightleftharpoons}} ES \tag{19.5}$$

$$ES \xrightarrow{k} E + P \tag{19.6}$$

where E = free enzyme

S = substrate or reactant

ES = primary enzyme-substrate complex

P = product

Solution. The decomposition of the primary complex ES to the free enzyme E and the product P is assumed to be the rate-determining (slow) step. The expression below is valid for both homogeneous (where the enzyme is used in the native or soluble form) and for immobilized enzyme reactions. The reaction rate v is given by:

$$v = -\frac{d[S]}{dt} = \frac{d[P]}{dt} \tag{19.7}$$

where $[S]$ and $[P]$ are the concentrations of substrate and product, respectively.

There are two approaches in deriving an expression for the reaction rate. In the first approach (Michaelis–Menten approach), the first reaction is assumed to be in equilibrium. The decomposition of the enzyme-substrate complex ES to form E and P is the rate-determining step. In the second approach, it is assumed that after an initial period, the rate of change of the concentration of the enzyme-substrate complex is essentially zero; mathematically, this can be expressed as:

$$\frac{d([ES])}{dt} \approx 0 \tag{19.8}$$

This is known as the quasi-steady state approximation, and is valid for enzyme catalyzed reactions if the total initial enzyme concentration is much less than the initial substrate concentration ($[E]_o \ll [S]_o$). The maximum reaction rate v_{max} is equal to $k[E]_o$ (see equation (19.6)). When $K_m = [S]$, $v = v_{max}/2$—see equation (19.1).

Assume that Michaelis–Menten kinetics is valid and derive and expression for the reaction rate. The total enzyme balance can be written as:

$$[E]_o = [E] + [ES] \tag{19.9}$$

where $[E]_o$ = the total enzyme concentration

$[E]$ = concentration of free enzyme

$[ES]$ = concentration of the enzyme-substrate complex

Since Michaelis–Menten kinetics is valid, the reaction between the free enzyme and substrate to form ES, Equation (19.5) may be assumed to be in equilibrium. Hence:

$$K_m = \frac{k_{-1}}{k_1} = \frac{[E][S]}{[ES]} \tag{19.10}$$

where K_m is the Michaelis–Menten constant. The preceding equation can be combined with the total enzyme balance to provide a relationship between $[ES]$ and the total enzyme concentration, $[E]_o$.

$$[ES] = \frac{[E]_o[S]}{K_m + [S]} \tag{19.11}$$

The reaction rate v then equals:

$$v = k[ES] = \frac{k[E]_o[S]}{K_m + [S]} \tag{19.12}$$

In the quasi steady-state approximation, the constant K_m is known as the dissociation constant. Assuming quasi steady-state, the rate of disappearance of the enzyme-substrate complex, ES is:

$$\frac{d[ES]}{dt} = k_1[E][S] - k_{-1}[ES] - k[ES] = 0 \tag{19.13}$$

Eliminating $[E]$ by combining with the equation for the total enzyme balance, and solving for $[ES]$,

$$[ES] = \frac{[E]_o[S]}{\frac{k + k_{-1}}{k_1} + [S]} \tag{19.14}$$

The reaction rate, v then equals:

$$v = k[ES] = \frac{k[E]_o[S]}{K_m + [S]} \tag{19.15}$$

which is identical to the expression obtained earlier. The only difference lies in the definition of K_m, which is equal to $\frac{k + k_{-1}}{k_1}$, instead of $\frac{k_{-1}}{k_1}$ (equilibrium assumption). ■

Effectiveness Factor

The effectiveness factor is defined as the ratio of the reaction rate in the presence of internal or pore diffusion to the reaction rate in the absence of pore diffusion. The value of the effectiveness factor is a measure of the extent of diffusional limitation. For isothermal reactions (generally true of most biochemical reactions), diffusional limitations are negligible when the effectiveness factor (η) is close to unity. If $\eta < 1$, the reaction is diffusion limited.

The problem of pore diffusion is only limited to immobilized enzyme catalysts, and not enzyme catalyzed reactions in which the enzyme is used in the native or soluble form. Immobilized enzymes are supported catalysts in which the enzyme is supported or immobilized on a suitable inert support such as alumina, kiesulguhr, silica, or microencapsulated in a suitable polymer matrix. The shape of the immobilized enzyme pellet may be spherical, cylindrical, or rectangular (as in a slab). If the reaction follows Michaelis–Menten kinetics discussed previously, then a shell balance around a *spherical* enzyme pellet results in the following second order differential equation:

$$\mathcal{D}_e\left(\frac{d^2[S]}{dr^2} + \frac{2}{r}\frac{d[S]}{dr}\right) = \frac{v_{max}[S]}{K_m + [S]} \tag{19.16}$$

where $\mathcal{D}_e$ is the effective diffusivity (consistent units).
The boundary conditions (*BC*) are:

1. $[S] = [S]_o$ at $r = R$ where R is the radius of the spherical catalyst pellet and $[S]_o$ is the substrate concentration in the bulk liquid, and
2. $d[S]/dr = 0$ at $r = 0$ (due to symmetry of the pellet, the concentration gradient is zero at the center).

The preceding differential equation can be solved numerically or analytically to determine the concentration profile inside the pellet.

From the definition of effectiveness factor, the actual or observed reaction rate, once again v_s (in the presence of pore diffusion), is from (Equation (19.1)) equal to:

$$v_s = \eta v = \eta \frac{v_{max}[S]_o}{K_m + [S]_o} \tag{19.17}$$

When the substrate concentration is low, i.e., when $[S] \ll K_m$, the reaction rate once again becomes first order. In this case, the preceding differential equation can be solved analytically to obtain the concentration profile inside the catalyst pellet. By defining the following dimensionless parameters, the differential equation can be written in a dimensionless form as follows:

$$\bar{S} = [S]/[S]_o,\ \bar{r} = r/R \tag{19.18}$$

so that

$$\left(\frac{d^2\bar{S}}{d\bar{r}^2} + \frac{2}{\bar{r}}\frac{d\bar{S}}{d\bar{r}}\right) = 9\Phi^2\bar{S}$$

where

$$\Phi = \frac{R}{3}\sqrt{\frac{v_{max}/K_m}{\mathcal{D}_e}} \tag{19.19}$$

$\Phi =$ is known as the Thiele modulus.

Once the concentration profile is known, the effectiveness factor can be expressed as a function of the Thiele modulus by the following relationship:

$$\eta = \frac{1}{\Phi}\left[\frac{1}{\tanh 3\Phi} - \frac{1}{3\Phi}\right] \tag{19.20}$$

The observed reaction rate can then be determined from Equation (19.17).

For a rectangular catalyst slab or rectangular membrane in which both sides are exposed to the substrate, the effectiveness factor is related to the Thiele modulus as follows:

$$\eta = \frac{\tanh \Phi}{\Phi} \tag{19.21}$$

where

$$\Phi = \frac{L}{2}\sqrt{\frac{v_{max}/K_m}{\mathcal{D}_e}}$$

and L is the thickness of the membrane or catalyst slab.

ILLUSTRATIVE EXAMPLE 19.7 Obtain expressions for the effectiveness factor for a first order reaction for different shaped catalyst pellets, and calculate the effectiveness factor for an immobilized enzyme catalyst. The following information is provided.

An enzyme reaction follows Michaelis–Menten kinetics.
Radius of spherical immobilized enzyme catalyst = 0.05 cm
$v_{max} = 1.0\ \text{mmol/L} \cdot \text{min}$
$K_m = 0.05$ M
$[S]_o = 0.005$ M
$\mathcal{D}_e = 2 \times 10^{-6}\ \text{cm}^2/\text{s}$

Solution. Even though the reaction follows Michaelis–Menten kinetics, the reaction is assumed to be first order since $[S]_o \ll K_m$. The Michaelis–Menten equation then becomes

$$v = \frac{v_{max}[S]}{K_m}$$

The smaller the value of the Thiele modulus, the greater is the effectiveness factor. The ratio $R/3$ is known as the "characteristic length" of the catalyst pellet. This characteristic length for any shape is the ratio of the volume of the pellet to the surface area exposed to the substrate.

First, calculate the Thiele modulus. For spherical pellets, the Thiele modulus is equal to:

$$\Phi = \frac{R}{3}\sqrt{\frac{v_{max}/K_m}{\mathcal{D}_e}} \tag{19.19}$$

Substituting the values:

$$\Phi = \frac{0.05}{3}\sqrt{\frac{10^{-3}/(0.05 \times 60)}{2 \times 10^{-6}}}$$
$$= 0.2152$$

Since the effectiveness factor is close to unity, pore diffusion is not a problem. From Equation (19.20) the effectiveness factor, η, is equal to

$$\eta = \frac{1}{0.2152}\left[\frac{1}{\tanh 3(0.2152)} - \frac{1}{3(0.2152)}\right]$$
$$= 0.9733$$

Since the effectiveness factor is close to unity, the observed rate is almost equal to the rate in the absence of diffusion. The observed rate is equal to

$$v_s = \eta \frac{v_{max}[S]_o}{K_m + [S]_o} \tag{19.17}$$

Substituting

$$v_s = 0.9733 \frac{1(0.005)}{0.05 + 0.005}$$
$$= 0.0885\ \text{mmol/L} \cdot \text{min}$$

Note once again that the smaller the value of the Thiele modulus, the larger is the value of the effectiveness factor. The effectiveness factor can be increased by decreasing the size of the immobilized enzyme pellet to the extent possible. When the Thiele modulus is large, ($\Phi \geq 10$ for spheres, $\Phi \geq 5$ for rectangular slabs or membranes), $\eta \approx 1/\Phi$. ■

Biological Oxygen Demand (BOD)

One of the most widely used parameters of organic pollutants is the 5-day BOD (BOD_5). This involves the measurement of the dissolved oxygen used by microorganisms in the biochemical oxidation of organic matter. The "carbon strength" of wastewater can be expressed as BOD, chemical oxygen demand (COD) or total organic carbon (TOC)[5]. The BOD_5 is the amount of dissolved oxygen consumed when a wastewater sample is seeded with active bacteria and is incubated at 20°C for 5 days.

BOD tests are used to determine the approximate quantity of oxygen, i.e., needed to biologically stabilize the organic matter present, to measure the efficiency of treatment processes, to determine compliance with wastewater discharge standards, and to determine the size of wastewater treatment plants. When the dilution water is seeded:

$$y = \text{BOD} = \frac{(D_1 - D_2) - (B_1 - B_2)R}{F} \tag{19.22}$$

where D_1 = dissolved oxygen of fresh diluted sample, mg/L

D_2 = dissolved oxygen after 5 d incubation at 20°C, mg/L

B_1 = dissolved oxygen of seed control before incubation, mg/L

B_2 = dissolved oxygen of seed control after incubation, mg/L

R = ratio of seed in sample to seed in control

F = volume fraction of sample used

The kinetics of the BOD reaction may be assumed to be first order and can be expressed as:

$$\frac{dW_t}{dt} = -kW_t \tag{19.23}$$

where W_t = amount of BOD remaining in the water at time t

k = reaction rate constant.

Integrating the preceding equation, one obtains:

$$\frac{W_t}{W_o} = \exp(-kt) \tag{19.24}$$

where W_o is the BOD at time $t = 0$ (total or ultimate BOD present initially). The amount of BOD that has been exerted (spent) at any time t, denoted by y, equals

$$y_t = W_o - W_t = W_o[1 - \exp(-kt)] \tag{19.25}$$

The 5-day BOD equals

$$y_5 = W_o - W_5 = W_o[1 - \exp(-5k)] \tag{19.26}$$

ILLUSTRATIVE EXAMPLE 19.8 The following information is given

Reaction constant $k = 0.2\ (\mathrm{d})^{-1}$.
5 day BOD (BOD_5) = 250 mg/L
Determine the ultimate and 2-day BOD.

Solution. The equation for determining ultimate BOD is:

$$W_t = W_o \exp(-kt) \tag{19.24}$$

The 5-day BOD equals:

$$y_5 = W_o - W_5 = W_o[1 - \exp(-5k)]$$

Since $y_5 = 250$ mg/L and $k = 0.2$:

$$250 = W_o[1 - \exp(-5 \times 0.2)]$$

Solving for W_o, the ultimate BOD is equal to

$$W_o = 395.5\ \mathrm{mg/L}$$

Determine the 2-day BOD. The amount of BOD remaining at any time t once again equals

$$W_t = W_o \exp(-kt) \tag{19.24}$$

The amount of BOD remaining after 2 days equals

$$\begin{aligned} W_2 &= 395.5 \exp(-0.2 \times 2) \\ &= 265.1\ \mathrm{mg/L} \end{aligned}$$

The amount of BOD exerted after 2 days (BOD_2) equals:

$$\begin{aligned} y_2 &= W_o - W_2 \\ &= 395.5 - 265.1 \\ &= 130.4\ \mathrm{mg/L} \end{aligned}$$

Remember that BOD is an indication of the dissolved oxygen consumed by the microorganism. For wastewater, a typical value of k is 0.23 $(\mathrm{d})^{-1}$. This however, varies significantly with the type of waste. The oxygen consumption may with vary with time and with different k values even if the ultimate BOD is the same. ■

Power Number

The power required to drive the impeller is an important consideration in the design of agitated tanks. The power number, N_p is somewhat analogous to the friction factor or drag coefficient.[6] It is proportional to the ratio of the drag force acting on a unit area of the impeller and the inertial stress. To estimate the power requirements for an impeller, empirical correlations relating the power or the power number with other variables of the system such as tank and impeller dimensions, liquid viscosity, liquid density, liquid depth, distance of the impeller from the bottom of the vessel, impeller speed, and baffle dimensions should be available. The number and arrangement of baffles and the number of impeller blades also affect the power consumption. By dimensional analysis, the power number can be related to the Reynolds number, the Froude number and various shape factors.[7] These shape factors are obtained by converting the various linear measurements to dimensionless ratios (usually obtained by dividing these measurements by the impeller diameter). The Froude number provides a measure of the ratio of the inertial stress to the gravitational force per unit area acting on the fluid.

ILLUSTRATIVE EXAMPLE 19.9 The following information is provided for an agitator in a fermenter.

Impeller diameter, $D_I = 1.5$ ft
Liquid viscosity, $\mu = 1$ cp (6.72×10^{-4} lb/ft · s)
Liquid density, $\rho = 62.4$ lb/ft^3
Impeller speed, $N = 90$ rpm $= 1.5$ rps
Power number, $P = 6.0$
$g = 32.2$ ft/s^2
Calculate the Reynolds number and the power consumption.

Solution. By definition of Reynolds number, Re, is

$$Re = \frac{D_I^2 N \rho}{\mu} \tag{19.27}$$

Substituting,

$$Re = \frac{(1.5)^2(1.5)(62.4)}{6.72 \times 10^{-4}}$$
$$= 313{,}393$$

By definition, the Power number, N_p, is

$$N_p = \frac{P g_c}{N^3 D_I^5 \rho} \tag{19.28}$$

where P = Power consumption.

The above equation relating the power number to the power consumption can be rearranged as follows:

$$P = \frac{N_p N^3 D_I^5 \rho}{g_c}$$

Substituting,

$$\begin{aligned} P &= \frac{(6.0)(1.5)^3(1.5)^5(62.4)}{32.2} \\ &= 298\ \text{ft} \cdot \text{lb}_\text{f}/\text{s} \\ &= (298\ \text{ft} \cdot \text{lb}_\text{f}/\text{s})/(550\ \text{ft} \cdot \text{lb}_\text{f}/\text{s} \cdot \text{hp}) \\ &= 0.542\ \text{hp} \end{aligned}$$

The power number is not usually affected by the shape of the tank; the power number is affected if the tank is baffled. However, for Reynolds number below about 300, the power number for baffled and unbaffled tanks are almost identical. Finally, power number calculations are useful in determining the agitation requirements in fermenters. ■

Air Sterilization: Dimensionless Groups

An important problem in air sterilization is particulate carryover. Particulate material (in this case, microbes) are removed by flowing the air through a fibrous filter or equivalent collector. The particles are removed by a combination of three mechanisms: collision or impaction, interception and diffusion.[8] Which of these three mechanisms is predominant depends on the particle size and mass. In the case of impaction, the efficiency of collection is a function of the Reynolds (and Stokes) number. The Reynolds number provides information on flow behavior.

The region of particle Reynolds number between 10^{-4} and 0.5 is known as the Stokes or streamline flow region. When a particle settles by gravity, it reaches a constant settling velocity, known as the terminal velocity. In terms of a force balance, the gravity force on the particle is equal to the sum of the drag force plus the buoyancy force. The terminal velocity for spherical particles in the Stokes' region can be easily derived.[6] The velocity is equal to:

$$v_t = \frac{g d_p^2 \rho_p}{18\mu} \tag{19.29}$$

where d_p = diameter of particle

ρ_p = density of particle

v_t = terminal velocity

g = acceleration due to gravity

μ = viscosity of the medium

The preceding equation is quite accurate for spherical particles with diameters less than 50 μm, and frequently is used with little error for particle sizes up to 100 μm. This range of roughly 1 to 100 μm is an important size range for industrial dusts. For particle sizes less than about 2 μm, the particle diameter starts approaching the mean free path of the gas molecules, and the settling velocity becomes greater than that predicted by Stokes' law. Below this lower limit it is necessary to apply Cunningham's correction to the equation for terminal velocity. The Cunningham correction factor, C_f is given by:

$$C_f = 1 + \frac{2\lambda}{d_p}\left[1.257 + 0.4\exp\left(\frac{-1.10 d_p}{2\lambda}\right)\right]; \lambda = \frac{\mu}{0.499\rho u_m} \tag{19.30}$$

The term λ is the mean free path of the molecules in the gas phase, u_m = the mean molecular speed, μ = the gas viscosity, and ρ = the gas density. From the kinetic theory of gases u_m is given by:

$$u_m = \left[\frac{8RT}{P(\mathrm{MW})}\right]^{1/2} \tag{19.31}$$

where MW = molecular weight of the gas

P = pressure

T = temperature in absolute units

The aforementioned Stokes number is given by:

$$St = \frac{C_f \rho_p d_p^2 u}{18\mu D} \tag{19.32}$$

where (in addition to the symbols already defined),

D = diameter of the collection device

u = fluid velocity *within* the filter void space.

$$u = \frac{u_s}{1 - \varepsilon} \tag{19.33}$$

where u_s = upstream fluid velocity

ε = volume of fiber per unit volume of filter bed.

The Peclet number is an important dimensionless parameter in convective transfer, and is the product of the Reynolds number and Schmidt number. The Peclet number, $Pe = 0$ corresponds to complete back mixing, while $Pe = \infty$ corresponds to plug flow. Large values of the Peclet number result when mixing is therefore not extensive.

The Schmidt number is the ratio of the kinematic viscosity to the molecular diffusivity. In this case, the diffusivity is due to the Brownian motion of submicron particles. The Schmidt number is therefore given by:

$$Sc = \frac{\mu}{\rho D'} \tag{19.34}$$

The term D' is the diffusivity due to Brownian motion and is given by:

$$\mathcal{D} = \frac{C_f kT}{3\pi\mu d_p} \tag{19.35}$$

where k is the Boltzmann constant (1.38054×10^{-23} J/K).
The Peclet number is also the product of the Reynolds number ($Du\rho/\mu$) and the Schmidt number ($\mu/\rho D'$), and is equal to uD/D'.

ILLUSTRATIVE EXAMPLE 19.10 Air flow to a fermenter (25°C) containing 10,000 microbial particles per cubic meter is sterilized by passing it through a filter bed. The diameter of the particles are about 1 μm and the density of the particles is 1 g/cm³. The velocity of the air before it enters the filter bed is 10 cm/s. Data is provided below

$$\varepsilon = 0.02$$

Diameter of the collection device, $D = 20$ μm
The physical properties of air at 20°C and atmospheric pressure are $\rho = 1.2 \times 10^{-3}$ g/cm³ and $\mu = 1.8 \times 10^{-4}$ g/cm · s.

Solution. The velocity of air within the void space of the filter is

$$u = \frac{u_s}{1 - \varepsilon} \tag{19.33}$$

$$u = \frac{10}{1 - 0.02} = 10.204 \text{ cm/s}$$

This velocity should be used in the calculation of the Reynolds number and the Stokes number.

The Cunningham's correction factor has to be applied since the particle size is less than 5 μm.[6] The mean free path, λ, equals

$$\lambda = \frac{1.8 \times 10^{-4}}{0.499(1.2 \times 10^{-3})} \sqrt{\frac{29P}{8 \times (8.314 \times 10^{7})(298)}}$$
$$= 6.44 \times 10^{-6} \text{cm}$$

The Cunningham correction factor, C_f, equals[6]

$$C_f = 1 + \frac{2(6.44 \times 10^{-6})}{1.0 \times 10^{-4}}\left[1.257 + 0.4 \exp\left(\frac{-1.10(1.0 \times 10^{-4})}{2(6.44 \times 10^{-6})}\right)\right] = 1.162$$

The Stokes number, St, equals

$$St = \frac{1.162(1)(1.0 \times 10^{-4})^2(10.204)}{18(1.8 \times 10^{-4})(2 \times 10^{-3})}$$
$$= 0.0183$$

The diffusivity can be estimated from Equation (19.35):

$$\mathcal{D} = \frac{C_f kT}{3\pi\mu d_p}$$

Substituting

$$\mathcal{D} = \frac{1.162(1.38 \times 10^{-16})298}{3\pi(1.8 \times 10^{-4})(1.0 \times 10^{-4})}$$
$$= 2.82 \times 10^{-7}$$

The Peclet number can also be calculated from as the product of the Reynolds number and the Schmidt number. A large value of the Peclet number suggests that there is very little back mixing. The Peclet number is

$$Pe = \frac{Du}{\mathcal{D}} \tag{19.36}$$

Substituting,

$$Pe = \frac{20 \times 10^{-4}(10.204)}{2.82 \times 10^{-7}}$$
$$= 72{,}370$$

The Schmidt number, Sc, equals

$$Sc = \frac{1.8 \times 10^{-4}}{(1.2 \times 10^{-3})(2.82 \times 10^{-7})}$$
$$= 531{,}900$$

These dimensionless groups are very useful in solving air sterilization problems. ■

Batch Sterilization: Time-Temperature Profiles

In batch sterilization, the medium is heated to the sterilization temperature, held at that temperature for a definite time, and then cooled to the fermentation temperature. The time-temperature profile depends upon the type of heating and cooling. The numerous equations for different types of heating and cooling are available in the literature.(9)

Depending upon the type of heating, the appropriate time-temperature equation can be selected for determining the time needed to reach the sterilization temperature when the medium is heated from ambient temperature. In case steam sparging is used, it must be remembered that this steam will condense. While calculating the time taken to cool the medium from the sterilization temperature to the fermentation temperature, this amount must be added to the time to cool the mass of the medium. Obviously, the time taken to cool the medium will be larger in this case.

ILLUSTRATIVE EXAMPLE 19.11 Calculate the time for a batch sterilization unit contents to cool from 121°C to 35°C. Data is provided below.

Cooling: Water at 20°C; circulation rate $= 4000\ \text{kg/min}$
Heat transfer area of the fermenter, $A = 240\ \text{m}^2$
Overall heat transfer coefficient[9], $U = 600\ \text{kcal/m}^2 \cdot \text{h} \cdot {}^\circ\text{C}$
Medium quantity $= 60{,}000\ \text{kg}$

The heat capacity and specific gravity of the medium are $1.0\ \text{kcal/kg} \cdot {}^\circ\text{C}$ and 1.0, respectively.

Solution. Since the medium is being cooled, the time (t)-temperature (T) profile is exponential.[9]

The equation is

$$T = T_{co}[1 + b\ \exp(-\alpha t)] \tag{19.37}$$

with

$$\alpha = \frac{wc}{M_\rho C_p}[1 - \exp(-UA/wc)] \tag{19.38}$$

and

$$b = \frac{T_o - T_{co}}{T_{co}} \tag{19.39}$$

where w = covalent flow rate

c = covalent heat capacity

M = initial sterlizer contents

ρ = specific gravity of catalysts

T_o = initial content temperature

T_{co} = initial covalent temperature

First calculate the values of α and b.

$$\alpha = \frac{(4000)(1.0)}{(60000)(1.0)}\left[1 - \exp\left(-\frac{(600)(240)}{(4000)(60)(1.0)}\right)\right]$$

$$\alpha = 0.03(\text{min})^{-1}$$

$$b = \left(\frac{394 - 293}{293}\right) = 0.345$$

Substituting in Equation (19.37) provides the time to reach the desired temperature

$$308 = 293[1 + 0.345\exp(-0.03t)]$$

Solving for t provides the time taken to reach the desired temperature

$$t = 63.6\,\text{min}$$

■

DESIGN CONSIDERATIONS

There are a host of design topics associated with biochemical reactors. Discussing these in any detail is beyond the scope of this book. Topics of interest to the practicing engineer are primarily concerned with sterilization. Four specific areas include:

Design of a batch sterilization unit
Design of a continuous sterilization unit
Design of an air sterilizer
Scale-up of a fermentation unit

Each of these briefly receives qualitative treatment below with Vasudevan[4] providing additional quantitative details and analyses.

Design of a Batch Sterilization Unit

Sterilization is the process of inactivation or removal of viable organisms. Sterilization by steaming of equipment, of medium, and of additives is an important operation in the fermentation industry. The main objective of media sterilization is to kill all living organisms present before inoculation, to eliminate any possible competition or interference with the growth and metabolism of the desired microorganism. This objective must be accomplished with minimal damage to the media ingredients. The thermal inactivation procedure is designed on a probability basis. A typical design value might be a 1 in 1000 chance that at least 1 organism will survive the sterilization process. One of the drawbacks of batch sterilization is that the Del factors[4] for heating and cooling are scale dependent. Large fermenters require longer heating and cooling times. The consequences of longer times can be severe if the medium is thermolabile. This destruction arises since the heat is dependent on the value of the thermal rate constant. The best alternative for heat sensitive material is to use continuous sterilization.

Design of a Continuous Sterilization Unit

The consequences of long heat-up and cool-down times in batch sterilization can be severe if the medium components are heat-sensitive. This arises because the destruction due to heat is dependent on the value of the thermal death constant. It can be shown that the spore *B. stearothermophilus* is only significantly inactivated above 110°C due to its high activation energy of 67.7 kcal/mol. On the other hand, many organic nutrients which also follow Arrhenius relationship for thermal degradation, have a much lower activation energy of around 25–30 kcal/mol. This implies that longer exposure to lower sterilization temperatures, due to slow heat-up or cool-down can cause more damage to nutrients. The best alternative for heat sensitive materials is to use continuous sterilization. In continuous sterilization, the raw medium is mixed with water, and then continuously pumped through the sterilizer to a sterilized fermenter. In the sterilizer, the media is instantaneously heated by either direct or indirect contact with steam, and held at a very high temperature (about 140°C) for a short time. The residence time or holding time of the medium is fixed by adjusting the flow rate and length of the insulated holding pipe. The hot stream from the sterilizer is rapidly cooled by a heat exchanger (with or without heat recovery), and/or by flash-cooling before it enters the fermenter. The design of continuous sterilizers must allow some flexibility in operating conditions to adapt the system to different medium. The system must also incorporate automatic recycle to recirculate the medium if the temperature falls below the design value. The design objectives should include:

1. the ability to fill the fermenter within 2–3 hours,
2. the recovery of 60–70% of the heat,
3. plug flow in the holding section, and
4. the option of either direct or indirect heating.

The continuous sterilization process has several advantages, including

1. the temperature profile of the medium is almost one of instant heating and cooling allowing an easy estimate of the Del factor required,
2. scale-up is very simple since the medium is exposed to high temperatures for very short times, thereby minimizing nutrient degradation, and
3. the energy requirements of the sterilization process can be dramatically reduced by using the incoming raw medium to cool the hot sterile medium.

The difficulties with continuous sterilizers are typically due to exchanger fouling or control instability. In general, medium containing starches require special attention.

Design of an Air Sterilizer

In aerobic fermentations, it is necessary to sterilize air since the volume of air required in aerobic fermentations is usually large, conventional techniques of heat sterilization are uneconomical. Effective and viable alternatives include the use of membrane or fibrous filters. An important consideration of the filter medium is that it should not be wetted, since this can lead to contamination. Materials such as glass fibers can be used to avoid this problem. As discussed in a previous subsection, the mechanisms by which particles suspended in a flowing stream of air are removed include impaction, diffusion and interception[6]. Impaction occurs when particles in the air collide with the fibrous filter due to their higher momentum as compared to air. Smaller particles, on the other hand, travel towards the fiber as a result of diffusion caused by Brownian motion. Particles less than 1 μm are collected by this mechanism. Particles that are not small or heavy, but are large in size, are intercepted by the fiber. The efficiency of air filtration is therefore a combination of the three mechanisms. The mathematical equations for designing air sterilizers may be developed by considering the effect of each of the above mechanisms separately, and then developing a combined expression. In the case of collision or impaction, the efficiency of the process is a function of the Reynolds (*Re*) and Stokes (*St*) numbers with the Reynolds number based on the diameter of the filter or collection device. The reader should note that the effect of the air velocity on efficiency is considerable. Removal of particles by collision and interception is enhanced as the air velocity is increased, whereas the efficiency of particulate removal by diffusion is lowered. This is of course also dependent on the particle size. In general, it should be remembered that the equations for efficiency are empirical.(6)

Scale-Up of a Fermentation Unit

Scale-up is a fundamental problem in the fermentation industry because of the need to perform microbial operations in different size equipment. The scale-up problem arises from the difference in transport phenomena when the scale and geometry of the equipment are changed. Describing transport and kinetic phenomena in fermenters requires knowledge of kinetics and flow patterns. Since the balances are very complex and the flow patterns largely unknown, the use of fundamental principles for design is limited. Instead, various essentially empirical procedures are employed.

The first step in scaling up a fermentation process is to use the production requirements to determine the size and number of fermenters that will be required. In sizing a new unit, or evaluating an existing fermenter for a biological process, it is necessary to establish the desired product rate R on an annual (or daily) basis. Then the productivity—defined as the mass of product that the fermenter can produce per unit volume per unit time—of the individual fermenter may be determined. In addition to the fermentation phase, the overall productivity must account for the time spent in activities such as cleaning, filling and sterilizing. This is important if the process is to be operated batchwise. The total installed fermenter capacity V_t required can then be calculated by dividing the desired product rate by the productivity. The number of fermenters can then be easily calculated. The next step is to size the number of inoculum stages. Fermenter precultures must be prepared in order to have sufficient inoculum for a large fermenter. The production fermenter volume and the optimal inoculum levels will determine how many preculture stages are necessary. Typical inoculum concentrations are:

Bacteria (0.1–3%)

Actinomycetes (5–10%)

Fungi (5–10%)

In practice, the first stage in inoculum production, usually a shaker flask, and the final production fermenter, are fixed. The number of intermediate levels is usually a compromise between the optimal inoculum levels and available equipment. Due to the requirement for many preculture stages in scaling up to a large volume, the culture may go through a large number of generations. It is therefore critical to establish that the culture is genetically stable for many generations. Assuming exponential growth, the number of generations required to achieve a certain biomass concentration can be easily calculated. The number of generations is proportional to the logarithm of the fermenter volume. It is recommended that the calculation should be verified experimentally.

BIO ENVIRONMENTAL APPLICATIONS

This last section addresses the general subject of bio environmental engineering, particularly as it applies to environmental reactions and reactors. Nine illustrative examples follow.

ILLUSTRATIVE EXAMPLE 19.12 Briefly describe the anaerobic process and explain its difference from an aerobic process.

Solution. The term *anaerobic digestion*, as opposed to *aerobic digestion*, is applied to a process in which organic material is decomposed biologically in an environment devoid of oxygen. Decomposition results from the activities of two major groups of bacteria. One group, called the acid formers, consists of facultative bacteria, which are also found in many aerobic environments and which, in a multistep anaerobic environment, convert carbohydrates, fats, and proteins to organic acids. The other group, the methane bacteria, converts the organic acids produced by the acid formers to methane and carbon dioxide. Some organic materials such as lignin are quite resistant to the activity of both groups and, hence, pass through the process relatively unaltered. Although cellulose is a relatively difficult material to degrade, it is treated due to the long detention time provided. The role of predator populations in anaerobic digestion is considered to be minor compared to that in aerobic processes. A detailed discussion of the biochemistry of anaerobic digestion is available in the literature.[4] ■

ILLUSTRATIVE EXAMPLE 19.13 Provide a qualitative overview of the design of aerobic biological wastewater treatment processes.

Solution. Through many years of designing and operating publicly owned treatment works (POTWs), typical values that reflect good wastewater treatment have been identified for most of the common biological treatment design parameters. Average values and typical ranges are often listed in wastewater engineering textbooks. However, "textbook" values for any given parameter are only applicable if the character of the wastewater being treated is also considered typical. A municipal wastewater might be considered typical if the common wastewater characterization parameters such as biochemical oxygen demand (BOD) and total suspend solids (TSS) fall within certain ranges (BOD_5, 110–300 mg/L; TSS, 100–350 mg/L). There must also be insignificant impacts from industrial wastewater that are inhibitory to biological treatability. If a wastewater can be characterized as typical, it may be possible to design a full-scale wastewater treatment system with a minimum of laboratory and pilot-scale treatability studies.

The cumulative wastewater engineering experience has evolved to the extent where the design of plants for aerobic suspended growth biological treatment systems can be based on previous design records and textbook values. Treatment processes typically begin with some form of bulk solids removal such as screening, grit removal, and/or primary sedimentation. Aerobic, suspended biological growth treatment disperses mechanically pressurized air supply within the treatment tank. The air is generally dispersed within the treatment tank "mixed liquor" (wastewater with a high concentration of process microorganisms) by means of either coarse bubble or fine bubble diffusers. The air serves two purposes. First, oxygen from air injected into the mixed liquor is dissolved and, therefore, is available to the microorganisms for the biological oxidation of substrate. Second, the injected air causes agitation of the liquid in the wastewater treatment tank and helps suspend and mix the microorganisms for efficient absorption and substrate oxidation. ■

ILLUSTRATIVE EXAMPLE 19.14 Microbial regrowth can be defined as an increase in viable microorganism concentrations in drinking water downstream of the point of disinfection after treatment. These microorganisms may be coliform bacteria, bacteria enumerated by the heterotrophic plate count (HPC bacteria), other bacteria, fungi, or yeasts. Regrowth of bacteria drinking water can lead to numerous associated problems including multiplication of pathogenic bacteria such as *legionella pneumophila*, deterioration of taste, odor, and color of treated water, and intensified degradation of the water mains, particularly cast iron, by creating anaerobic conditions and reducing pH in a limited area. To obtain stable drinking water (i.e., to control regrowth), one needs to understand the sources of biological instability.

Explore the possible factors that affect regrowth in a water distribution system.

Solution. The following items can affect microbial regrowth in a drinking water distribution system:

Water quality (e.g., concentration of organic carbon and nutrients, temperature, disinfectant residual, etc.). Water quality certainly influences the potential for bacterial regrowth in the distribution system. The proliferation of microorganisms depends upon the presence of an adequate food supply and a favorable environment for their growth. Though the majority of organic compounds in water can be removed through treatment processes, low concentrations may remain that sufficiently stimulate bacterial growth. Some groups of bacteria (generally known as *oligotrophic bacteria*) are capable of surviving under low carbon and nutrient conditions. Examples of water quality parameters affecting microbial regrowth are temperature, pH, and dissolved oxygen. In general, disinfectant residual is used to suppress growth during storage and transportation of potable water.

Physical condition. The physical condition of distribution system pipes influences their tendency to foster biological regrowth. Distribution pipes that have tubercules or other surface irregularities commonly harbor microbial encrustations. Certain pipe materials and conditions can lead to a heavy accumulation of bacteria on their walls, a so-called biofilm. By attaching to the surface, microorganisms can be protected from washout and can exploit larger nutrient resources either accumulated at the surface or in the passing water. Moreover, attached bacteria appear to be less affected by disinfectants than those suspended in the disinfected liquid.

Flow conditions in the distribution system (e.g., detention time, flow velocity, wall shear stress, flow reversals, etc.). Long detention times and low flow velocities tend to encourage bacterial growth in the distribution system.

Water treatment processes. The efficiency of organic compound and nutrient removal depends upon the type of treatment processes utilized at a given plant. Generally, a residual disinfectant in the water is required to prevent regrowth in the distribution system. Ozonation is known to increase bacterial nutrients by converting some of the nonbiodegradable organic matter into oxidized, degradable compounds, making it more available for uptake by microorganisms colonizing the distribution system.

Physical state of bacteria (type, concentration, and physiological state). It is evident that bacteria or other organisms must be present in a distribution system for biological regrowth to occur. The possible sources of bacteria in the water are:

1. recovery of injured or dormant bacteria from disinfection,
2. cross-connections that allow back siphonage of contaminated water,
3. improperly protected distribution system storage, and
4. line breaks and subsequent repair operations that fail to redisinfect the repaired lines effectively before placing them back into service. ■

ILLUSTRATIVE EXAMPLE 19.15 Bacteria known to be present in a source of liquid with a mean number of 3 per cubic centimeter. Ten 1-cm^3 test tubes are filled with the liquid. Calculate the probability that all ten test tubes will show growth (i.e., contain at least one bacterium each). Assume a Poisson distribution applies.[10]

Solution. Let X denote the number of bacteria in 1 cm^3. Substitute $\mu = 3$ in the Poisson pdf gives:

$$f(X) = \frac{e^{-3}(3)^X}{X!}; \quad X = 0, 1, 2, \ldots \tag{19.40}$$

The probability that a test tube will show growth is

$$\begin{aligned} P(X \geq 1) &= 1 - P(X = 0) \\ &= 1 - e^{-3} \\ &= 1 - 0.0498 \\ &= 0.9502 \end{aligned}$$

The probability that all 10 test tubes will show growth is

$$\begin{aligned} P(X = 10) &= (0.9502)^{10} \\ &= 0.60 = 60\% \end{aligned}$$

■

ILLUSTRATIVE EXAMPLE 19.16 Biochemical engineers have recently simulated a 1.0 gmol/s flow of blood through arteries as flow through pipes. The diffusion of oxygen into the blood stream may be expressed as a chemical reaction of the form:

$$B_{\text{rich } O_2} \longrightarrow B_{\text{difficient } O_2}$$

or

$$B_r \longrightarrow B_d$$

The reaction velocity constant is 0.09 L/mol $B_r \cdot$ s and the order of the reaction with respect to B_r is 2nd and elementary. What is the conversion if the volume of the arteries is 21 L?

Assume the flow through an artery is analogous to a tubular flow reactor and the initial concentration of $B_{\text{rich } O_2}$ is 0.25 mol/L

Solution. The describing equation for a tubular flow reactor is

$$V = F_{BO} = \int_0^X \frac{dX}{-r_B}$$

Substituting,

$$21 = 1.0 \int_0^X \frac{dX}{-k_B C_{Br_0}^2 (1 - X)^2}$$

$$21 = \frac{1.0}{(0.09)(0.25)^2}\left(\frac{X}{1 - X}\right) = 177.8\left(\frac{X}{1 - X}\right)$$

Solving for X,

$$X = 0.106 = 10.6\%$$

■

ILLUSTRATIVE EXAMPLE 19.17 The lung surface(s) may be considered as a catalyst for the reaction between oxygen and hemoglobin that is given by:

$$\text{Hb}_{(l)} + \text{O}_{2(g)} \rightarrow \text{HbO}_{2(l)}$$

The lung volume capacity is approximately 2 liters and the reaction may be analyzed assuming a batch reaction equation applies. If a breath is held for 10 seconds the rate expression is:

$$-r = 1.57 \times 10^{-3} \times \frac{(1 - X)^2}{(1 - 0.21X)} = \frac{\text{gmol O}_2}{\text{ft}^2 \text{ lung surface} \cdot \text{s}}$$

1. What is the lung surface area required for a conversion of 20%?
2. How long must a person hold his/her breath to absorb 50% of the oxygen?

Solution. 1. For air,

$$V_{O_2} = (0.21)(2) = 0.42\ \text{L}$$

Applying the ideal gas law (100°F, 1 atm)

$$N_{O_2} = \frac{(1)(0.42)}{(0.0821)(311)}; \quad T = 100°F = 311\ \text{K}$$
$$= 0.0165\ \text{gmol}\ O_2$$

For a total reactor, with S (the surface) replacing V (the volume)

$$t = \frac{N_{A_0}}{S}\int_0^{0.2}\frac{dX}{-r_A}$$
$$= \frac{N_{A_0}}{(S)(1.57\times 10^{-3})}\int_0^{0.2}\frac{(1-0.21X)}{(1-X)^2}dX; \quad S = 10$$

Integrating and substituting,

$$S = \frac{0.0165}{(10)(1.57\times 10^{-3})}\left[\frac{(1-0.21)(0.2)}{(1-0.2)} + 0.2\ln\frac{1}{(1-0.2)}\right]$$
$$= 1.051[0.1975 + 0.00446]$$
$$= 0.254\ \text{ft}^2$$

2. The process a reversed to calculate the time for 50% conversion.

$$t = \frac{(0.0165)}{(0.254)(1.57\times 10^{-3})}\left[\frac{(1-0.21)(0.5)}{(1-0.5)} + 0.21\ln\left(\frac{1}{1-0.5}\right)\right]$$
$$= 38.7\ \text{s}$$

■

ILLUSTRATIVE EXAMPLE 19.18 The following reaction takes place in the small intestine of a human:

$$\text{Protein} + \text{Cofactor} \xrightarrow{\text{ENZYME}} \text{Digested Products}$$

Assuming that the protein (A) is 90% digested, that there is three times the amount of cofactor (B) than protein, $k = 0.905\ \text{L/g}\cdot\text{min}$ (enzyme), and the concentration of protein (50 mg/L) moves at a rate of 100 mL/min through the small intestine, determine the length of intestine needed for an average intestinal diameter of 2 cm. Also assume that the reaction is elementary with respect to the equation given above and model the small intestine as a tubular reactor.[10]

Solution. For this reactor, $\theta_B = 3$ so that

$$-r_A = k_A C_A C_B = k_A C_{A_0}(1 - X) C_{A_0}(\theta_B - X)$$
$$= k_A C_{A_0}^2 (1 - X)(\theta_B - X)$$

For a tubular reactor

$$V = F_{A_0} \int_0^X \frac{dX}{-r_A}$$
$$V = q_0 C_{A_0} \int \frac{dX}{k_A C_{A_0}^2 (1 - X)(\theta_B - X)} = \frac{Q_0}{r_A C_{A_0}} \int_0^X \frac{dX}{(1 - X)(\theta_B - X)}$$
$$V = \frac{q_0}{K C_{A_0}} \left[\frac{1}{(\theta_B - 1)} \ln\left(\frac{\theta_B - X}{\theta_B(1 - X)} \right) \right]$$
$$V = \frac{0.1}{(0.905\,(\text{g/l})^{-1}\,(\text{min})^{-1})(0.050\text{g/l})}[0.973] = \frac{0.1}{(0.905)(0.05)}[0.973]$$
$$V = 2.15\ \text{L} = 0.00215\ \text{m}^3$$

Since

$$V = \left(\frac{\pi d^2}{4} \right) L$$
$$L = \frac{0.00215(4)}{\pi(2 \cdot 10^{-2})^2} = 6.84\ \text{m}$$

■

ILLUSTRATIVE EXAMPLE 19.19 At 98.6°F (T_1), an important neural activator is found to have a specific reaction rate of 0.25 $(\text{min})^{-1}$ with $E_a = 20{,}000$ Btu/lbmol. It has been determined that after the reaction rate reaches 0.32 $(\text{min})^{-1}$, the activator irreversibly denatures, causing coma and eventually death. How high a fever (T_2) can a human withstand before this condition occurs?

Solution. Apply Arrhenius equation between two temperatures

$$k(T_2) = k(T_1) e^{(-E_a/R)[(1/T_2) - (1/T_1)]}$$

or

$$\ln\left[\frac{k(T_2)}{k(T_1)} \right] = (-E_a/R)[(1/T_2) - (1/T_1)]$$

Solve for T_2

$$\frac{1}{T_2} = \left[\ln\frac{k(T_2)}{k(T_1)}\left(-\frac{R}{E_a}\right)\right] + \frac{1}{T_1}$$

$$T_2 = \frac{1}{\left[\ln\frac{k(T_2)}{k(T_1)}\left(-\frac{R}{E_a}\right)\right] + \frac{1}{T_1}}$$

Substituting,

$$\begin{aligned} k(T_1) &= 0.25 \\ k(T_2) &= 0.32 \\ T_1 &= 98.6^\circ\text{F } (558.6^\circ\text{R}) \\ E_\text{a} &= 20{,}000\ \text{Btu/lb} \cdot \text{mol} \\ R &= 1.987\ \text{Btu/lb} \cdot \text{mol} \cdot {}^\circ\text{R} \end{aligned}$$

leads to

$$T_2 = \frac{1}{\left[\ln\left(\frac{0.32}{0.25}\right)\left(\frac{-1.987}{20{,}000}\right)\right] + \frac{1}{558.6}}$$

$$\begin{aligned} T_2 &= 556.4^\circ\text{R} \\ &= 106.4^\circ\text{F} \end{aligned}$$

■

ILLUSTRATIVE EXAMPLE 19.20 A certain antibiotic is administered intravenously to attack a virulent bacteria in the bloodstream. The flow of antibiotic into the system is 0.5 ml/min and the antibiotic concentration is 1,000,000 molecules/L. The drug that binds to the bacteria destroys it, and the rest is excreted. The rate of excretion of antibiotic is 30 mL/h at a concentration of 250,000 molecules/L.

Assuming a CSTR model applies, and a total blood volume of 4.7 L, what is the rate of destruction of the bacteria in molecules/L · min?

Solution. The antibiotic entering (F_{IN}) is

$$F_{\text{IN}} = \left(0.5 \times 10^{-3}\,\frac{\text{L}}{\text{min}}\right)\left(\frac{1 \times 10^6\ \text{molecules}}{\text{L}}\right) = \frac{500\ \text{molecules}}{\text{min}}$$

The antibiotic excreted (F_{OUT}) is

$$F_{\text{OUT}} = \left(30 \times 10^{-3}\,\frac{\text{L}}{\text{h}}\right)\left(\frac{1\ \text{h}}{60\ \text{min}}\right)\left(\frac{250{,}000\ \text{molecules}}{\text{L}}\right) = \frac{125\ \text{molecules}}{\text{min}}$$

Assuming steady state conditions and noting that "r" is the rate of utilization of antibiotic and therefore the destruction rate of the bacteria,

$$r = \frac{F_{OUT} - F_{IN}}{V} = \frac{[125 - 500] \text{ molecules/min}}{4.7 \text{ L}}$$

$$r = -80 \frac{\text{molecules}}{\text{min} \cdot \text{L}}$$

The negative sign indicates a loss or destruction of bacteria to the blood streams. ■

ILLUSTRATIVE EXAMPLE 19.21 A 100,000-gal per day protein-contaminated wastewater with a COD of 4000 mg/L is generated by a meat-processing plant. This waste is to be treated by an anaerobic contact process using a loading of 0.15 lbCOD/ft^3 · day and a sludge detention time of 20 days. Assuming 80% efficiency, determine/provide the following.

1. A flow diagram of the process.
2. Hydraulic detention time, hours.
3. Daily solids accumulation, pounds (assume a 6% synthesis of solids from BOD).
4. Nitrogen and phosphorous requirements, pounds (10 and 1.5%, respectively, in solids).
5. Mixed liquor suspended solids concentration.

Solution. A simple flow diagram of the process is provided in Figure 19.1. The daily COD removed is

$$\begin{aligned} \text{COD} &= (4000 \text{ mg/L})(0.1 \times 10^6 \text{ gal/day})(10^{-6} \text{ L/mg})(8.34 \text{ lb/gal}) \\ &= 3336 \text{ lb COD/day} \end{aligned}$$

The volume of the anaerobic contact tank is

$$\begin{aligned} V &= (3336 \text{ lb COD/day})(7.48 \text{ gal/ft}^3)(\text{ft}^3 \cdot \text{day}/0.15 \text{ lb}) \\ &= 166{,}000 \text{ gal} \end{aligned}$$

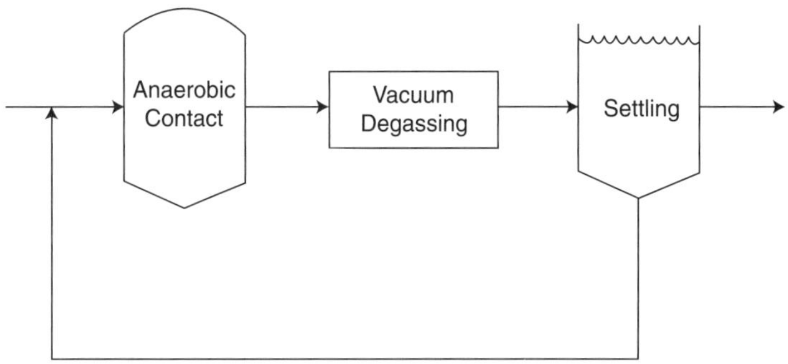

Figure 19.1 Flow diagram for Illustrative Example 19.21.

The hydraulic detention time, t_H, is then

$$t_H = (166{,}000\,\text{gal})\left(\frac{\text{day}}{0.1 \times 10^6\,\text{gal}}\right)$$
$$= 1.66\,\text{days}$$
$$= 39.8\,\text{h}$$

The solids growth (*SG*) may be calculated using the 6% synthesis of solids and 80% efficiency:

$$SG = (3336)(0.8)(0.06)$$
$$= 160\,\text{lb/day}$$

The *SG* represents the mass of suspended solids and is the active cell population. This concentration may be converted to mg/L as follows:

$$(1.0\,\text{lb/gal}) = (454\,\text{g/lb})(1000\,\text{mg/g})(264\,\text{gal/m}^3)(\text{m}^3/1000\,\text{L})$$
$$= 120{,}000\,\text{mg/L}$$

This number is equivalent to the conversion constant

$$\left(\frac{10^6\,\text{mg/L}}{8.34\,\text{lb/gal}}\right) \approx 120{,}000$$

The latter conversion constant (120,000) is often used in industry. Thus,

$$\text{MLVSS} = \frac{(160)(20)}{166{,}000}(120{,}000)$$
$$= 2300\,\text{mg/L}$$

Since 10% of the solids are nitrogen,

$$N = (160)(0.1)$$
$$= 16\,\text{lb N/day}$$

Since 1.5% of solids is phosphorous,

$$P = (160)(0.015)$$
$$= 2.4\,\text{lb P/day}$$

Since the wastewater is proteinaceous (rich in nitrogen and phosphorous) *N* and *P* will not be required. ■

REFERENCES

1. J. Enderle, S. Blanchard, and J. Bronzino, "*Introduction to Biomedical Engineering*," 2nd edition, Elsevier/Academic Press, New York City, NY, 2000.

2. J. Bronzino (editor), "*Biomedical Engineering Fundamentals*," 3rd edition, CRC/Taylor & Francis, Boca Raton, FL, 2000.
3. S. Vogel, "*Life in Moving Fluids*," 2nd edition, Princeton University Press, Princeton, NJ, 1994.
4. P. Vasudevan, "*Biochemical Engineering*," A Theodore Tutorial, East Williston, NY, 1994.
5. L. Stander and L. Theodore, "*Environmental Regulatory Calculations Handbook*," John Wiley & Sons, Hoboken, NJ, 2008.
6. L. Theodore, "*Air Pollution Control Equipment Calculations*," John Wiley & Sons, Hoboken, NJ, 2008.
7. P. Abulencia and L. Theodore, "*Fluid Flow for the Practicing Engineer*," John Wiley & Sons, Hoboken, NJ, 2008.
8. L. Theodore and F. Ricci, "*Mass Transfer Operations for the Practicing Engineer*," John Wiley & Sons, Hoboken, NJ, 2010.
9. L. Theodore, "*Heat Transfer for the Practicing Engineer*," John Wiley & Sons, Hoboken, NJ, 2011.
10. S. Shaefer and L. Theodore, "*Probability and Statistics Applications for Environmental Science*," CRC Press/Taylor & Francis, Boca Raton, FL 2007.
11. P. Atonecchia, problem assignment submitted to L. Theodore, Manhattan College, Bronx, NY, 12/17/1985.

CHAPTER 20

OPEN-ENDED PROBLEMS

INTRODUCTION

The educational literature provides frequent references to individuals, particularly practicing engineers and scientists, that have different learning styles, and in order to successfully draw on these different styles, a variety of approaches can be employed. One such approach involves the use of open-ended problems.

The term "open-ended problem" has come to mean different things to different people. It basically describes an approach to the solution of a problem and/or situation for which there is usually not a unique solution. Three literature sources[1–3] provide sample problems that can be used when this educational tool is employed.

The author of this book has applied this somewhat unique approach and included numerous open-ended problems in several earlier course offerings at Manhattan College. Student comments for the graduate course "Accident and Emergency Management" were tabulated for one such course. Student (unedited) responses to the question "What aspects of this course were most beneficial to you?" are listed below.

1. "The open-ended questions gave engineers a creative license. We don't come across many of these opportunities."
2. "Open-ended questions allowed for candid discussions and viewpoints that the class may not have been otherwise exposed to."
3. "The open-ended questions gave us an opportunity to apply what we were learning in class with subjects we have already learned and gave us a better understanding of the course."
4. "Much of the knowledge that was learned in this course is applicable to everyday situations and our professional lives."
5. "Open-ended problems made me sit down and research the problem to come up with ways to solve them."
6. "I thought the open-ended problems were inventive and made me think about problems in a better way."
7. "I felt that the open-ended problems were challenging. I, like most engineers, am more comfortable with quantitative problems than qualitative."

Engineers bring mathematics and science to bear on practical problems, molding materials and harnessing technology for human benefit. Creativity is often a key

Chemical Reactor Analysis and Applications for the Practicing Engineer. By Louis Theodore

component in this synthesis; it is the spark motivating efforts to devise solutions to novel problems, design new products, and improve existing practices. In the competitive marketplace, it is a crucial asset in the bid to win the race to build better machines to decrease product delivery times, and anticipate the needs of future generations;[4] one of the keys to the success of a practicing engineer or a scientist is to generate fresh approaches, processes and products. In effect, they need to be creative.

This approach requires individuals to ask questions, to not always accept things at face value, and to select a methodology that provides the most effective and efficient solution. Those who conquer this topic have taken the first step toward someday residing in an executive suite.

The remainder of this chapter addresses a host of illustrative examples that most could label as open-ended problems.

APPLICATIONS

ILLUSTRATIVE EXAMPLE 20.1 Using terms which a liberal arts major could understand, briefly explain the concepts of:

1. unit conversion(s) or conversion factors,
2. dimensional analysis,
3. dimensionless numbers,
4. kinetic energy,
5. chemical kinetics,
6. reaction rate.

Solution. Qualitative questions like these are often difficult to answer. Perhaps (1) is the easiest of the six. For this case, one might draw an analogy between 5 pennies and 1 nickel, or 2 nickels and 1 dime, or 12 inches and 1 foot, etc. ■

ILLUSTRATIVE EXAMPLE 20.2 Discuss some key design specifications for combustion devices.

Solution. A key design specification for any combustion device, including boilers and incinerators, is the operating temperature. Materials of construction must withstand the operating temperature without experiencing any damage for safe and efficient plant operation. In addition, the volumetric flow rate is a strong function of temperature and plays an important role in properly sizing combustion/reactor equipment. ■

ILLUSTRATIVE EXAMPLE 20.3 It has been proposed to use the sensible heat from a reactor discharge stream to reduce the energy needs of an overall process. A young engineer has proposed to recover the heat in a heat exchanger and use the recovered heat to preheat the air required for combustion in another part of the process. Comment on the advantages and disadvantages of the young engineer's proposal.

Solution. The total energy content of the gas should be taken into consideration since the mass involved is (relatively speaking) small (i.e., the density of gas is low). ■

ILLUSTRATIVE EXAMPLE 20.4 The Alaska Department of Environmental Protection (DEP) has requested your consulting firm to provide "the best estimate" of the oil requirements to sustain a proposed remote outpost in the middle of a pristine region. If both the heating and electrical requirements for the site are known, *outline* how to calculate the average daily oil needs. Assume the energy value of the oil is given. Indicate if any additional information is required. Note that no calculations are required.(5)

Solution. When calculating the average daily oil needs for the outpost, several factors come into play. The most important is knowing the heating and electrical requirements for the site and the heating value of the oil. It is also important to take into account how the oil will be transported to the outpost. It is one thing to purchase the oil and another entirely different thing to transport and store the oil at the outpost. As this nation knows, oil prices are governed by international and national events as well as the demand for the oil. Thus, the fluctuations in the oil prices also need to be accounted for. The thermal conversion efficiencies are also required to be known. The following four-step approach is suggested.

Step 1: Break down the given heating and electrical requirements into a per month basis and from this the average daily oil needs can be obtained. (Take the total monthly requirements and divide it by the number of days in a month to get the average daily requirements.) One reason for this is that winter months tend to use more electricity, heating and oil than summer months. Also, one must know when the "peak hours are." This is what is referred to as "demand."

Step 2: Use conversion factors to convert the heating and electrical units into oil units. Oil is usually sold by the barrel. Once all the units are converted into Btu units, the heating value of oil can be used (such as one barrel = 6.287 million Btu) as a way to see how many barrels of oil are required.

Step 3: Find out the cost of oil after calculating how much the daily and monthly consumption will be.

Step 4: Decide how the oil will be transported to the outpost and how/where it will be stored.

Include all these factors in the analysis. ■

ILLUSTRATIVE EXAMPLE 20.5 A vertical fixed bed catalytic reactor (see Figure 20.1) is designed to operate with a conversion of 80%. Once the unit is installed and running, the unit operates with a conversion of 60%. Rather than purchase a new reactor, what options are available to bring the unit into compliance with the specified design conversion?

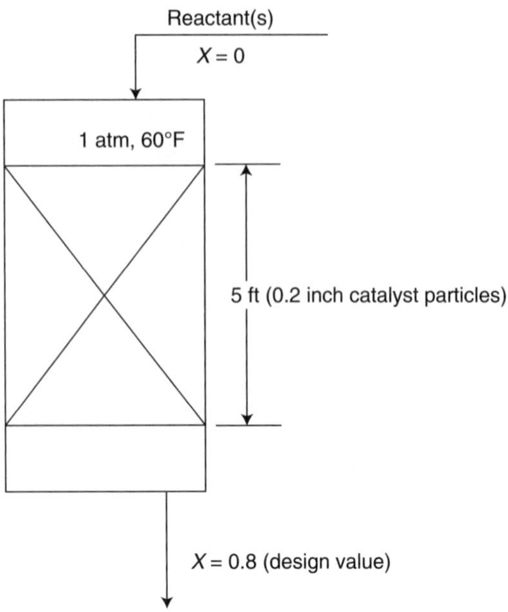

Figure 20.1 Vertical fixed bed reactor.

Solution

1. Use a smaller catalyst size. A different size may produce a higher conversion.
2. Use a different type of catalyst. A different type may produce a higher conversion.
3. Use a different physical shaped catalyst.
4. Make sure there is no channeling inside the reactor. The catalyst should be randomly distributed and there should be no open spots.
5. Consider increasing or decreasing the fluid flow rate entering the reactor. If the flow is increased, there will be a greater mass transfer between the fluid and catalyst, and the conversion of the reactor may increase. If the flow rate is decreased, the increased residence time could increase conversion.
6. Use a different fluid as the reactant (if possible).
7. Check to see if the pressure drop is excessive.
8. Lower the initial concentration of the reactant (if possible).
9. Increase the height of the reactor bed. This will provide more catalyst mass or surface area for reaction and a longer residence time.
10. Additional catalyst could be added at the top of the column.
11. Increase the temperature. This can increase the reaction rate.
12. Modify the process that may be producing the reactant(s). Since the details of this process are not included in the problem statement, no specific recommendations can be made.

13. Finally, before considering changes to the system, one should undertake a thorough inspection of the reactor and surrounding components. The emission monitoring system should be recalibrated. All valves, fittings, and pipes should be checked for plugging or leaks. Any gas or liquid distributor should be checked to make sure it is functioning properly. The distribution of the catalyst should be inspected to make sure it is as uniform as possible. Any problems encountered during this inspection should be corrected immediately. Following the maintenance check, the performance of the reactor should be re-evaluated.

The reader should note that some of the recommendations above could lead to higher pressure drops and potential problems with the flow. An example of this problem would be the implementation of suggestion (1). The reader is left the exercise of determining what other steps could lead to flow/pressure drop problems. *Hint*: There are at least six suggestions that fall into this category, including a consideration to fire the engineer who designed the reactor.[6] ■

ILLUSTRATIVE EXAMPLE 20.6 Refer to Illustrative Example 20.5. A similar situation exists with a fluidized bed catalytic reactor. Rather than purchase a new reactor, what options are available to bring the unit into compliance with the specified design conversion?

Solution Some of the available options are similar to those provided in the previous example.

1. Check/increase the quantity of the catalyst. Make up any catalyst lost to carry-over (as necessary).
2. Increasing the depth of the bed will increase the conversion of the column.
3. Change the catalyst size.
4. Reducing the flow will increase the residence time and may allow for more reaction.
5. Increase the inlet temperature; this will favor the reaction process.
6. Regenerate the catalyst for a longer period.
7. Replace the catalyst.
8. Consider changing the flow rate.

Again, pressure drop considerations should be included in the analysis. ■

ILLUSTRATIVE EXAMPLE 20.7 A 24-year-old batch reactor is no longer delivering the degree of conversion required for the process. Rather than replace the unit, you have been asked to recommend what other possible steps can be taken to the existing reactor to get it back "on line". Unfortunately, it is not currently possible to simply replace the reactor. Some other solution(s) need to be found.

Solution. The issue with the current operation is its inability to achieve the conversion at its designed level.

The first thing that needs to be investigated is what is the current state of the reactor? How was it originally designed? Were safety margins included in the design that allow for some "tweaking" of the actual operation of the reactor? How much change in reactant quantity can be tolerated? What will the materials stand up to?

It also might be possible to increase the reaction time. It is possible after 24 years of service that the reactor is not working at optimum condition. An examination should be made of the various contributions to the failure to achieve the deserved conversion.

The composition and conditions of the feed stream also need to be examined. Is it the same as what the reactor was originally designed to operate? Has the process been altered upstream in such a way that the conversion is no longer feasible? Perhaps it has changed through the years in such a manner that the reactor is no longer designed to deal with the stream. This feed could be altered to adhere to the original conditions to make the reactor operate correctly again.

Finally, it is possible that the reactor needs cleaning and that could increase the reactor's conversion efficiency. ■

ILLUSTRATIVE EXAMPLE 20.8 Refer to Illustrative Example 20.5. A similar situation exists with a tubular flow reactor. Rather than purchase a new reactor, what options are available to bring the reactor into "compliance" with the specified design conversion?

Solution. The solution to this problem, in many respects is similar to that presented in Illustrative Examples 20.5 and 20.6. ■

ILLUSTRATIVE EXAMPLE 20.9 Consider the liquid phase irreversible reaction

$$\text{a}A + \text{b}B \rightarrow \text{c}C + \text{d}D$$

If this is a new reaction, outline what steps you would take as a practicing engineer to determine the rate of expression?

Solution. First, assume the rate equation is given by

$$-r_A = k_A C_A^{\alpha} C_B^{\beta}$$

Employ a batch reactor for experimentation. Saturate the system with reactant B, so that the concentration change is negligible for differential runs. The rate expression now takes the form

$$-r_A = k_A^* C_A^{\alpha}$$

Conduct several runs, altering the initial concentration of A for short run times. Follow the procedure detailed in Chapter 13 to obtain a value for α. Repeat the experiment but now saturate the system with A. This will generate a value for β.

Return to the original rate expression:

$$-r_A = kC_A^{\alpha} C_B^{\beta}$$

The terms α and β are now known. Use any suitable method to obtain k_A, noting that

$$\ln(-r_A) = \ln k_A + \alpha \ln C_A + \beta \ln C_B$$

This is an equation of the form

$$Y = A_0 + A_1 X_1 + A_2 X_2$$

What guarantees are there that the above model is correct? The reader is referred to the literature for the answer. ■

ILLUSTRATIVE EXAMPLE 20.10 Dr. Mark Corapi, the author's primary physician, has determined that the author can tolerate a level of 10 mg of a toxic agent per kilogram of body weight, and that the metabolism follows a first order reaction whose half-life for the agent in a human is three hours. If it is assumed that all the agent that is inhaled (worst case scenario) is absorbed by the body, what continuous concentration in the atmosphere could the author, with a body mass of 77 kg (170 lb) tolerate? Assume that his average breathing rate is 40 L/min.

Solution. It is assumed that the agent will be absorbed in its entirety by the body. The objective is to find the concentration in the atmospheric air that a 77 kg individual can withstand, knowing that the average breathing rate is 40 L/min. Since the metabolic function follows 1st order kinetics, the following equation applies

$$\frac{dC}{dt} = kC \tag{1}$$

The kinetic constant, k, for the model above, can be obtained from the half life where the half life equation is

$$k = \frac{\ln(2)}{t_{1/2}} = \frac{\ln(2)}{3} = 0.231\ (\mathrm{h})^{-1} \tag{2}$$

The tolerable concentration of the agent in the body can be expressed as a function of unit body volume

$$C = \left(10\frac{\mathrm{mg_{agent}}}{\mathrm{kg}}\right)\left(\frac{77\ \mathrm{kg}}{V_{\mathrm{individual}}}\right) = 770\frac{\mathrm{mg_{agent}}}{V_{\mathrm{individual}}} \tag{3}$$

where: $V_{\mathrm{individual}}$ = body volume for the individual
Equations (2) and (3) can now be directly substituted in Equation (1), yielding the time rate of change of concentration of the reagent in the body.

$$\frac{dC}{dt} = (0.231\ \mathrm{h}^{-1})\left(770\frac{\mathrm{mg_{agent}}}{V_{\mathrm{individual}}}\right) = 177.9\frac{\mathrm{mg_{agent}}}{V_{\mathrm{individual}} \cdot h}$$

Hence, the maximum intake rate of the toxic agent for the author's body volume is 177.9 mg/h. For air inhalation of 40 L/min, the maximum concentration of an agent in the air that the author can tolerate is

$$\frac{177.9\,\dfrac{\text{mg}_{\text{agent}}}{\text{h}}}{40\,\dfrac{\text{L}}{\text{min}}\left(\dfrac{60\text{ min}}{\text{h}}\right)} = 0.0741\,\frac{\text{mg}}{\text{L}} = 74.1\text{ mg/m}^3$$

The reader is left the exercise of solving the problem for a different reaction order. ■

ILLUSTRATIVE EXAMPLE 20.11 Refer to the Illustrative Example in the CSTR, Chapter 9. Assign an enthalpy of reaction (assume it to be exothermic) for the reaction under consideration. *Outline* how to redesign the two-stage CSTR system taking enthalpy of reaction effects into account. Assume the enthalpy of reaction is constant (and independent of temperature), the operation is adiabatic, and the Arrehenius equation is known.

Solution. Refer to Chapter 12. The "thermal" equation for each reactor are

$$X_1 = \frac{\Sigma\theta_i C_{pi}(T_1 - T_0)}{\Delta H_r^0} \tag{1}$$

$$X_2 = \frac{\Sigma\theta_i C_{pi}(T_2 - T_0)}{\Delta H_r^0}; \quad X_2 = 0.875 \tag{2}$$

Refer Chapter 9. The CSTR "design" equations are

$$V_1 = \frac{F_{\text{A}_0}X_1}{-rA_1} = \frac{\text{F}_{\text{A}_0}X_1}{k_1 C_{\text{A}_0}^2(1 - X_1)^2}; \; k_{A1} = k_1 \tag{3}$$

$$V_2 = \frac{F_{\text{A}_0}(X_2 - X_1)}{k_2 C_{\text{A}_0}^2(X_2 - X_1)^2}; \; k_{A2} = k_2 \tag{4}$$

The system constraint is:

$$V_1 = V_2 \tag{5}$$

In addition,

$$k_1 = A_e^{-E/RT_1}$$

$$k_2 = A_e^{-E/RT_2}$$

There are five equations (1)–(5) and five unknowns—X_1, T_1, T_2, V_1 and V_2. A solution is possible. One possible approach to a solution is as follows:

1. Guess T_1; Calculate k_1
2. Calculate X_1 from (1)
3. Calculate V_1 from (3)
4. Calculate T_2 from (2); calculate k_2

5. Calculate V_2 from (4)
6. Compare $V_1 = V_2$
7. Exit or upgrade T_1 and repeat until $V_1 = V_2$. ■

ILLUSTRATIVE EXAMPLE 20.12 You have been hired to design the optimal chemical reactor for a homogeneous chemical reaction of your own choice, e.g.,

$$A + 2B \rightleftharpoons 5C$$

Indicate the reaction phase, whether the reaction is reversible, all pertinent physical/chemical properties for the system, etc. Also indicate the ancillary (peripheral) equipment required for the entire reactor system.

Solution. Look up and obtain all physical and chemical property data. Use engineering judgement in assigning an operating temperature and pressure. Select a reactor, i.e., batch, CSTR, or TF. Include decisions on pump(s), cooling coils, insulation, rotameters, safety equipment, instrumentation, etc. Perform a safety study and give thought to environmental considerations. Complete the initial analysis with an economic study. The calculation can then be repeated for different operating conditions and different reactors. ■

ILLUSTRATIVE EXAMPLE 20.13 Refer to Illustrative Example 9.9 in the CSTR set in Chapter 9. You have been hired as a hazardous waste management consultant to consider methods of treating the hazardous waste in the product mixture since reactant B is both hazardous and toxic. Outline design considerations, operational procedures and treatment methods to help reduce and/or eliminate the environmental problem.

Solution. Here are 18 suggestions:

1. Obtain an MSDS sheet (or the equivalent) on B.
2. Make reactant B the "limiting" reactant.
3. The system should be leak-proof.
4. Operate below atmospheric pressure if the reactor system is "closed".
5. Provide appropriate training for operators, engineers, administrators, maintenance/design personnel, etc.
6. Post warning signs near and in the vicinity of the reactor.
7. Consider operating at a higher conversion to reduce the presence of B.
8. Minimize the inventory of reactant B.
9. Try to replace B with another compound.
10. Treat unreacted B in the product stream.
11. The system should be capable of handling accidental overflows.
12. Increase the temperature in order to increase the conversion of B.

13. Consider moving the process to an isolated area.
14. Maintain the area in the immediate vicinity of the reactor clear and free of other chemicals.
15. Ensure that the waste is disposed of in a safe and timely manner.
16. Consider recycling the product stream containing *B*.
17. Consider adding another reactor to increase the conversion.
18. The reactor material (stainless steel?) should be able to handle the participating species.

Any other ideas? Can the reader provide other suggestions? ■

ILLUSTRATIVE EXAMPLE 20.14 You have been hired as a accident and emergency management consultant to consider the exothermic Diels–Alder reaction treated earlier. Based on laboratory data, you have been informed that the reactor can explode if the reaction mixture temperature exceeds 140°F. Outline what additional design factors need to be considered to help prevent and/or reduce the possibility of an accident.[6]

Solution. Here are a baker's dozen suggestions.

1. Develop an accident emergency response plan.[6]
2. Perform routine maintainance, safety and hazards checks.
3. Perform routine equipment and personnel checks.
4. Practice emergency response drills.
5. Provide routine training for personnel.
6. How exothermic is the process?
   a. temperature control required
   b. saftey and relief equipment required
   c. cooling coils
   d. jacketed tank
7. Are there other side reactions?
   a. temperatures, pressures
   b. identify by-products
   c. nature of by-products
   d. treatment of by-products
   e. catalyst
8. Can the reaction be run at temperatures >140°F?
9. Materials of construction.
   a. explosion proof process units
   b. is the process corrosive, toxic, hazardous?
   c. process units for excessive temperatures and pressures

10. Plant geographical location (isolated area).
   a. flooding
   b. tornadoes
   c. hurricanes
   d. earthquakes
   e. tsunamis

11. Physical design considerations.
   a. reinforced concrete
   b. raised platforms, adequate drainage, etc.

12. Plant climate.
   a. humidity
   b. temperature
   c. pressure
   d. tropics
   e. subtropics

   (b–e) } equipment operation
   operating temperature and pressure.

13. Plant layout.
   a. adequate space for routine maintainance, emergency response, and materials transport
   b. storage tanks

Regarding (1) here are some safety equipment/considerations:

1. Provide adequate drainage, relief/spill collection systems, waste disposal systems, available emergency services, utilities, access and the alarms, automatic shutdown compability, feedback controller(s), etc.
2. Equipment, automate where possible, pressure relief valves, rupture disks, temperature control devices (valves, cooling coils, jacketed tanks, etc.), flares, warning signals and alarms, etc. ■

ILLUSTRATIVE EXAMPLE 20.15 You are an engineer at a specialty chemical plant. The feedstream to the reactor in the process is water previously purified using reverse osmosis (RO). One of the RO units at the reactor site is no longer operating to the required "separation" efficiency. What steps should be taken to avoid purchasing a new purification unit?[7]

Solution

1. Check the membrane to see if it needs to be replaced.
2. If the membrane is satisfactory, check the concentration of the feed stream to the reactor. The feed stream may be more concentrated due to a problem before the RO unit.

3. Check that the appropriate pressure is being applied in order to overcome the osmotic pressure (i.e., check the pumps).
4. Check that none of the other components of the system have been damaged by the membrane. ■

ILLUSTRATIVE EXAMPLE 20.16[8] Consider a chemical reactor that uses two liquid feeds of different densities as provided in Table 20.1. It produces four different liquid chemical products of varying density following chemical reaction and separation, see Table 20.2. The plant storage requirements call for maintaining 4–5 weeks supply of each feed, 4–6 weeks supply of products A, B, and C, and 1–2 weeks supply of product D. The plant operates year round but each tank must be emptied once a year for a week for maintenance. Tanks are normally dedicated to one feed or product and one or two could be used as "swing" tanks; however, one day of cleaning is required between uses with different liquids.

Specify a set of tanks from the "standard" sizes given in Table 20.3 to minimize this plant's needs.

Solution. There is no single, simple method for determining the optimum mix of stage tanks for the chemical reactor in question. Most often, estimates are made of the minimum and maximum amounts of feeds, intermediates, and products that must be kept on hand. Then, some additional allowance is made to permit periodic cleaning and maintenance of the feed tanks to the reactor. The minimum number of tanks may not always be optimum if the tanks are extremely large. Several smaller

TABLE 20.1 Feed Data

Feed 1	110,000 lb/day	$\rho = 49$ lb/ft^3
Feed 2	50,000 lb/day	$\rho = 68$ lb/ft^3

TABLE 20.2 Product Data

Product A	40,000 lb/day	$\rho = 52$ lb/ft^3
Product B	25,000 lb/day	$\rho = 62$ lb/ft^3
Product C	10,000 lb/day	$\rho = 52$ lb/ft^3
Product D	95,000 lb/day	$\rho = 47$ lb/ft^3

TABLE 20.3 Tank Data

Standard Tank Sizes (gallons):			
2800	11,200	56,100	561,000
5600	16,800	140,000	1,123,000
8400	28,100	281,000	

TABLE 20.4 Storage Requirements

Material	lb/day	Gal/ day	Days	Gals storage required
Feed 1	110,000	16,800	28–35	470,200–587,800
Feed 2	50,000	5500	28–35	154,000–192,500
Product A	40,000	5750	28–42	161,100–241,700
Product B	25,000	3020	28–42	84,500–126,700
Product C	10,000	1440	28–42	40,300–60,400
Product D	95,000	15,120	7–14	105,800–211,700

tanks may cost somewhat more initially but they offer more flexibility in use. One solution and some calculational details follow.

Determine the maximum and minimum amounts of each material to be stored (see Table 20.4). The conversion to gallons requires dividing the rate in lb/day by the density in lb/ft^3, then multiplying by 7.481 gal/ft^3.

For each material, select a set of tanks:

Feed 1:	Use 2	281,000 gallon tanks
Feed 2:	Use 4	56,100 gallon tanks
Product A:	Use 2	140,000 gallon tanks
Product B:	Use 2	56,100 gallon tanks
Product C:	Use 1	56,100 gallon tanks
Product D:	Use 2	140,000 gallon tanks

This set is 5% short on the maximum of Feed 1, 11% short on Product B, and 7% short on Product C. For most situations, this would be acceptable.

Select spare and/or "swing" tanks to provide for maintenance:

Feed 1:	Use an additional 281,000 gallon "swing" tank
Feed 2 and all products:	Use one 56,100 gallon "swing" tank

This combination provides adequate auxiliary storage for maintenance periods.

Total tanks required: 3	281,000 gallons
4	140,000 gallons
8	56,100 gallons

The number of tanks required will be quite large in this application. If market forces such as fluctuating demand require this much storage, they may all be necessary. More modern commercial operations such as "just-in-time" manufacturing, call for reducing in-plant inventory to the absolute minimum possible (often for safety concerns). ■

REFERENCES

1. A. M. Flynn and L. Theodore, "*An Air Pollution Control Equipment Design Course for Chemical and Environmental Engineering Students Using an Open-Ended Problem Approach*," ASEE Meeting, Rowan University, NJ, 2001.
2. A. M. Flynn, J. Reynolds, and L. Theodore, "*Courses for Chemical and Environmental Engineering Students Using an Open-Ended Problem Approach*." AWMA Meeting, San Diego, CA, 2003.
3. L. Thedore, personal notes, 2002–2004.
4. K. Gibney, "*Awakening Creativity*," ASEE Promo, Washington, DC, March, 1988.
5. A. Baez, adapted from homework assignment submitted to L. Theodore, Manhattan College, Bronx, NY, 2007.
6. L. Theodore and R. DuPont, " Environmental Health and Hazard Risk Assessment Calculations," CRC Press/Taylor & Francis, Boca Raton, FL, 2012.
7. L. Theodore and F. Ricci, "Mass Transfer Operations for the Practicing Engineer," John Wiley & Sons, Hoboken, NJ, 2010.
8. D. Kauffman, "Process Synthesis and Design," A Theodore Tutorial, Theodore Tutorials, East Williston, NY, 1992.

CHAPTER 21

ABET-RELATED TOPICS

INTRODUCTION

This last chapter of the book is concerned with five topics that the Accreditation Board for Engineering and Technology (ABET) has recently indicated should be included in any engineering curriculum. The section contents of these topics are briefly described below.

The first section is concerned with environmental management; this section contains a broad discussion of environmental issues facing today's practicing engineers and presents some of the more recent technology to deal with the issues at hand. The next section is concerned with accident and emergency management; it deals with ways to ensure both employee and public safety, the determination of the severity of accidents, and determining the causes and potential causes of accidents. The third section is concerned with ethics; the case study approach is employed to make the reader think about the ethical questions involved, to reflect on their past decisions, and to project forward to their future decisions with a higher degree of thought and insight when faced with an ethical dilemma. The penultimate section is concerned with numerical methods: a brief overview of numerical methods is given to provide practicing engineer with some insight into what many of the currently used software packages (Mathcad, Mathematica, MatLab, etc.) are actually doing. The last section is concerned with economics and finance; this subject area provides material that can ultimately dictate the decisions made by the practicing engineer and his/her company.

Finally, it should be noted that the previous chapter introduced the reader to open ended questions; although engineers at their very essence are problem solvers, most problems in real life do not come fully defined with a prescribed methodology to arrive at a solution. These open-ended questions are exercises in using brain power—and like any muscle, you use it or lose it. The author believes that those who conquer or become adept with this topic will have taken the first step toward someday residing in an executive suite (or the equivalent); and, it is for this reason that a separate chapter was assigned to this important topic.

ENVIRONMENTAL MANAGEMENT

The first section of this chapter is concerned with environmental management. This section contains a broad discussion of the environmental issues facing today's

Chemical Reactor Analysis and Applications for the Practicing Engineer. By Louis Theodore

practicing engineers and presents some of the more recent technology to deal with issues at hand.

In the past four decades, there has been an increased awareness of a wide range of environmental issues covering all resources: air, land, and water. More and more people are becoming aware of these environmental concerns, and it is important that professional people, many of whom do not possess an understanding of environmental problems, have the proper information available when involved with environmental issues. All professionals should have a basic understanding of the technical and scientific terms related to these issues. In addition to serving the needs of the professional, this chapter examines how one can increase his or her awareness of and help solve the environmental problems facing both industry and society.

This section is titled Environmental Management; it provides a broad discussion of environmental issues facing today's engineers. Some of the topics covered in this chapter include air pollution, water pollution, solid waste, etc., and six Illustrative Examples dealing with these topics are presented.

This chapter is not intended to be all-encompassing. Rather, it is to be used as a starting point. Little is presented on environmental regulations because of the enormity of the subject matter; in a very real sense, it is a moving target that is beyond the scope of this text. Further, this material primarily keys on traditional environmental topics. Although much of the material is qualitative in nature, some quantitative material and calculations are presented in the six Illustrative Examples that are presented in material that follows.

ILLUSTRATIVE EXAMPLE 21.1 In order to meet recently updated environmental regulations for discharging hydrocarbons to the atmosphere, a gas stream must be reduced by 99.5% of its present hydrocarbon concentration. Due to economic considerations, it is proposed to meet the above requirement by combusting the hydrocarbons in an incinerator operating at 1500°F. The gas and methane (fuel) are to be fed to the incinerator at 80°F and 1 atm. Design the proposed incinerator using kinetic principles.

Flue gas flow rate (from fuel combustion) = 2500 scfm (80°F, 1 atm)

Process gas flow rate = 7200 scfm

Hydrocarbon: essentially toluene

Reaction rate constant, $k = 7.80(\text{s})^{-1}$ at 1500°F

Solution. Calculate the volumetric flow rate into the reactor at 1500°F, including the volume increase associated with both the combustion of fuel (2500 scfm) and process gas stream. See Chapter 3.

$$\begin{aligned} q_a &= (2500 + 7200)(460 + 1500)/(460 + 80) \\ &= 35{,}200 \text{ acfm} \end{aligned}$$

Calculate the cross-sectional area, S, of the reactor assuming an average velocity of 20 ft/s. Employ equations presented earlier in the text and the notation q to represent the volumetric flow rate.

$$S = q_a/v$$
$$= 35200\,\text{acfm}/(20)(60)\,\text{fpm}$$
$$= 29.3\,\text{ft}^2$$

For the reactor diameter,

$$D = (4S/\pi)^{1/2}$$
$$= 6.11\,\text{ft}$$

Finally, calculate the length of the reactor (along coordinate z), L, assuming a plug flow reactor

$$L = -(v/k)\ln(C_f/C_o)$$
$$= -(20/7.80)\ln(0.5/100)$$
$$= 13.6\,\text{ft}$$

■

ILLUSTRATIVE EXAMPLE 21.2 Provide a comparative analysis—in terms of advantages and disadvantages—between thermal and catalytic combustion devices that are employed for environmental control.

Solution. Some of the advantages of catalytic combustion reactors over thermal reactors are:

1. lower fuel requirements,
2. lower operating temperatures,
3. little or no insulation requirements,
4. reduced fire hazards,
5. reduced flashback problems.

The disadvantages include:

1. higher initial cost,
2. catalyst poisoning,
3. (large) particles must first be removed,
4. some liquid droplets must first be removed,
5. catalyst regeneration problems.

■

ILLUSTRATIVE EXAMPLE 21.3 To apply risk assessments to large groups of individuals, certain assumptions are usually made about an "average" person's attributes. List the average or standard values used for :

1. body weight,
2. daily drinking water intake,
3. amount of air breathed per day,

4. expected life span,
5. dermal contact area.

Solution

1. The average body weight is 70 kg for an adult and 10 kg for a child.
2. The average daily drinking water intake is 2 liters for an adult and 1 liter for a child.
3. The average amount of air breathed per day is 20 m^3 for an adult and 10 m^3 for a child.
4. The average expected life span is 70 years.
5. The average dermal contact area is 1000 cm^2 for an adult and 300 cm^2 for a child. ■

ILLUSTRATIVE EXAMPLE 21.4 Incineration is one of the major types of treatment for environmental control. An important parameter in incineration is the operating temperature. The operating temperature in an incinerator is a function of many variables. For most hazardous waste incinerators, the operating temperature is calculated by determining the flame temperature under adiabatic or near-adiabatic conditions. From a calculational point of view, the flame temperature has a strong dependence on the excess air requirement and the heating value of the combined waste–fuel mixture. The Theodore–Reynolds equation shown below can be used to estimate the average temperature in the incinerator in lieu of using a rigorous model that may require extensive experimental data and physical/chemical properties.[1]

$$T = 60 + \frac{\text{NHV}}{(0.325)[1 + (1 + \text{EA})(7.5 \times 10^{-4})(\text{NHV})]} \tag{21.1}$$

where T = temperature, °F

NHV = net heating value of the inlet mixture, Btu/lb

EA = excess air on a fractional basis

The value of EA may be estimated by

$$\text{EA} = \frac{0.95Y}{(21 - Y)} \tag{21.2}$$

where Y is the dry mol% O_2 in the combustion (incinerated) gas.
Additional details on these equations (including their derivations) are available in the literature.[1]

Estimate the theoretical flame temperature of a hazardous waste mixture containing 25% cellulose, 35% motor oil, 15% water (vapor), and 25% inerts, by mass. Assume 5% radiant heat losses. The flue gas contains 11.8% CO_2, 13 ppm CO, and 10.4% O_2 (dry basis), by volume.

$$\begin{aligned}\text{NHV of cellulose} &= 14{,}000\,\text{Btu/lb}\\ \text{NHV of motor oil} &= 25{,}000\,\text{Btu/lb}\\ \text{NHV of water} &= 0\,\text{Btu/lb}\\ \text{NHV of inerts (effective)} &= -1000\,\text{Btu/lb}\end{aligned}$$

Solution. The NHV for the mixture is obtained by multiplying the component mass fractions by their respective NHVs and taking the sum of the products. Thus,

$$\text{NHV} = \sum_{i=1}^{n} (\text{NHV})_i w_i \tag{21.3}$$

Substituting,

$$\begin{aligned}\text{NHV} &= 0.25(14{,}000\ \text{Btu/lb}) + 0.35(25{,}000\ \text{Btu/lb})\\ &\quad + 0.15(0.0\ \text{Btu/lb}) + 0.25(-1000\ \text{Btu/lb})\\ &= 12{,}000\ \text{Btu/lb}\end{aligned}$$

The excess air employed is obtained from the equation provided in the problem statement.

$$\text{EA} = \frac{0.95Y}{21 - Y} \tag{21.2}$$

Substituting,

$$\begin{aligned}\text{EA} &= \frac{0.95(10.4)}{21 - 10.4}\\ &= 0.932\end{aligned}$$

The flame temperature is estimated using the Theodore–Reynolds equation provided in the problem statement:

$$T = 60 + \frac{\text{NHV}}{(0.325)[1 + (1 + \text{EA})(7.5\times10^{-4})(\text{NHV})]} \tag{21.1}$$

Substituting,

$$\begin{aligned}T &= 60 + \frac{12{,}000}{(0.325)[1 + (1 + 0.932)(7.5\times10^{-4})(12{,}000)]}\\ &= 2068^\circ\text{F}\end{aligned}$$

The reader should also note that this equation is sensitive to the value assigned to the coefficient which represents the average heat capacity of the flue gas, in this case, 0.325. This is a function of both the temperature (T) and excess air fraction (EA) and also depends on the flue products since the heat capacities of air and CO_2 are about half that of H_2O. In addition, the 7.5×10^{-4} term may vary slightly with the composition of the waste–fuel mixture incinerated. The overall relationship between operating temperature and composition is therefore rather complex, and its prediction not necessarily as straightforward as shown here. ■

ILLUSTRATIVE EXAMPLE 21.5 A chemical reactor produces 15,000 acfm of gas at 700°F and 1 atm. HCL and SO_2 concentrations are 10,000 ppm and 250 ppm, respectively. HCl must be controlled to 99% collection efficiency or 4 lb/h. SO_2 emissions are to be controlled at 70% collection efficiency. A spray dryer is used to control the HCl and SO_2 emissions. $Ca(OH)_2$ is the sorbent that will react with HCl and SO_2 to form $CaCl_2$ and $CaSO_4$, respectively. Assume that it is necessary to provide 10% and 30% excess lime feed for the required HCl and SO_2 removal, respectively . What is the required feed rate of $Ca(OH_2)$? What is the total mass production rate of solids from the spray dryer? Assume that the excess solids in the spray dryer are $Ca(OH)_2$.

Solution. Determine the mass flow rate of HCL and SO_2 in the fuel gas.

$$PV = nRT$$

$$Pq = \dot{n}RT$$

$$\dot{m} = \dot{n}(MW) = Pq(MW)/RT$$

$$\begin{aligned}\dot{m}_{HC_1} &= (1\text{ atm})(15{,}000\text{ acfm})(0.01\text{ mol fraction HCl})(36.45\text{ lb/lbmol HCl})\\ &\quad /(0.7302\text{ atm}\cdot\text{ft}^3/\text{lbmol}\cdot{}^\circ\text{R})(460+700^\circ\text{R})(1\text{h}/60\text{ min})\\ &= 387.3\text{ lb HCl/h}\end{aligned}$$

$$\begin{aligned}\dot{m}_{SO_2} &= (1\text{ atm})(15{,}000\text{ acfm})(2.5\times10^{-4}\text{ mol fraction SO}_2)(64\text{ lb SO}_2/\text{lbmol})\\ &\quad /(0.7302\text{ atm}\cdot\text{ft}^3/\text{lbmol}\cdot{}^\circ\text{R})(460+700^\circ\text{R})(1\text{ h}/60\text{ min})\\ &= 17.0\text{ lb SO}_2/\text{h}\end{aligned}$$

Determine which appropriate regulation applies for HCl control.

$$(0.01)(387.3\text{ lb/h}) = 3.87\text{ lb/h};\ \text{this is less than 4 lb/h}$$

Therefore, the 4 lb/h rule applies.

Write the two balanced chemical reaction equations (one for HCl and one for SO_2), including the excess feed.

$$1.3\,Ca(OH)_2 + SO_2 + \frac{1}{2}O_2 \longrightarrow CaSO_4 + H_2O + 0.3\,Ca(OH)_2$$

$$1.1\,Ca(OH)_2 + 2HCl \longrightarrow CaCl_2 + 2H_2O + 0.1\,Ca(OH)_2$$

Determine the molar amount of $Ca(OH)_2$ needed for neutralization. SO_2 removal requires 1.3 mol lime/mol SO_2, while HCl removal requires 1.1/2 or 0.55 mol lime/mol HCl. Also determine the feed rate of $Ca(OH)_2$ required in lbmol/h.

$$\begin{aligned}\text{Lime feed rate} &= [(17\text{ lb SO}_2/\text{h})/(64\text{ lb SO}_2/\text{lbmol SO}_2)](1.3\text{ lbmol lime/lbmol SO}_2)\\ &\quad + [(387.3-4.0\text{ lb HCl/h})/(36.45\text{ lb HCl/lbmol})]\\ &\quad (0.55\text{ lbmol lime/lbmol HCl})\\ &= 6.12\text{ lbmol lime/h}\end{aligned}$$

The $Ca(OH)_2$ feed rate in lb/h is

$$\text{Lime feed rate} = (6.12)(74) = 453\ \text{lb/h}$$

There is one mol of $CaSO_4$ produced per mol of SO_2 reacted. Since 70% of the SO_2 reacts,

$$\begin{aligned}\text{Production rate of } CaSO_4 &= (17)(0.7)/64 = 0.186\ \text{lbmol/h}\\ &= (0.186)(136)\\ &= 25.3\ \text{lb/h}\end{aligned}$$

One mol of $CaCl_2$ is produced per two mol of HCl reacted. 4 lb of HCl are not reacted. Therefore, the $CaCl_2$ produced is

$$\begin{aligned}\text{Production rate} &= (387.3 - 4.0)(0.5)/36.45 = 5.258\ \text{lbmol/h}\\ &= (5.258)(110.9)\\ &= 583.1\ \text{lb/h}\end{aligned}$$

One mol of $Ca(OH)_2$ is required to react in the production of either one mol of $CaSO_4$ or one mol of $CaCl_2$. Using the above results,

$$\begin{aligned}Ca(OH)_2 \text{ unreacted} &= 6.12 - 5.258 - 0.186 = 0.676\ \text{lbmol/h}\\ &= (0.676)(74) = 50.0\ \text{lb/h}\end{aligned}$$

Finally calculate the total solids produced in lb/h.

$$\begin{aligned}\text{Total solids} &= CaCl_2 + CaSO_4 + Ca(OH)_2(\text{unreacted})\\ &= 583.1 + 25.3 + 50.0\\ &= 658.4\ \text{lb/h}\end{aligned}$$

■

ILLUSTRATIVE EXAMPLE 21.6 Consider the water system pictured in Figure 21.1. The upstream flowrate at point 1 and downstream flowrate at point 2 are 20 cfs and 28 cfs, respectively. If the decay (reaction) rate of the chemical present in the system can be described by a first-order reaction with a reaction velocity constant of 0.2 $(\text{day})^{-1}$, determine the concentration profile of the chemical if both the upstream and infiltrating flows do not contain the chemical.

Assume the cross-sectional area available for flow is constant at 40 ft^2 and that the flow variation with distance can be described by a linear relationship. Perform the calculation for the following cases.

1. Use the inlet velocity and assume it to be constant. Neglect concentration variations arising due to the infiltration.
2. Use the average velocity (inlet, outlet) and assume it to be constant. Neglect concentration variations arising due to the infiltration.
3. Account for both velocity and concentration variation due to infiltration.

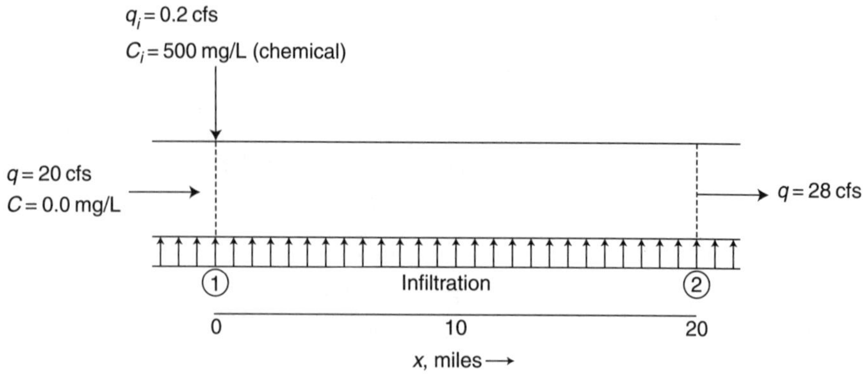

Figure 21.1 Water system for Illustrative Example 21.6.

Solution.

1. The inlet water velocity is

$$v = \frac{20 + 0.2}{40} = \frac{20.2}{40}$$
$$= 0.505\ \text{ft/s}$$
$$= 8.28\ \text{mpd (miles per day)}$$

The velocity at point 2 is

$$v = \frac{28}{40}$$
$$= 0.7\ \text{ft/s}$$
$$= 11.5\ \text{mpd}$$

Since the cross-sectional area for the flow is constant, one may employ the "plug" flow model described in Chapter 10 with a constant inlet velocity. For this case,

$$\frac{dC}{dx} = -\frac{kC}{v}$$

with,

$$C_0 = \frac{(0.2)(500)}{20.2}$$
$$= 4.95\ \text{mg/L}$$

Integrating the above,

$$C = C_0 e^{-k(x/v)}$$
$$= 4.95 e^{-0.2(x/8.28)}$$
$$= 4.95 e^{-0.0241x}$$

2. Using the average velocity of 9.89 mpd (rather than 8.28 mpd) results in the following concentration profile equation:

$$C = 4.95e^{-0.0202x}$$

3. Theodore[2] has provided an approach to account for these two variations. Assume the increase in velocity (and volumetric flowrate) occurs linearly down the system. For this condition

$$v = 8.28(1 + 0.161x)$$

Since the volumetric flowrate correspondingly increases, a concentration reduction will occur due to its "dilution" effect. This accounting is obtained by dividing the concentration by the factor $(1 + 0.161x)$. The describing equation now becomes

$$\frac{dC}{dx} = -k\left(\frac{C}{1 + 0.161x}\right)\left(\frac{1}{1 + 0.161x}\right)$$

Separating the variables leads to

$$\frac{dC}{C} = \frac{-k\,dx}{(1 + 0.161x)^2}$$

Integrating gives

$$\ln\frac{C}{C_0} = k\left[-\frac{1}{(0.161)(1 + 0.161x)}\right]_0^x$$

$$C = C_0 \exp\left\{-k\left[\frac{1}{(0.161)(1 + 0.161x)} - \frac{1}{0.161}\right]\right\}$$

$$= (4.95)\exp\left(1.242 - \frac{1.242}{1 + 0.161x}\right)$$

Thomann and Mueller,[3] provide other approaches in solving this type of problem. ■

ACCIDENT AND EMERGENCY MANAGEMENT

Accidents are a fact of life, whether they are a careless mishap at home, an unavoidable collision on the freeway, or a miscalculation at a chemical plant. Even in prehistoric times, long before the advent of technology, a club-wielding caveman might have swung at his prey and inadvertently toppled his friend in what can only be classified as an "accident."

As man progressed, so did the severity of his misfortunes. The "Modern Era" has brought about assembly lines, chemical manufacturers, nuclear power plants, etc. all carrying the capability of disaster. To keep pace with the changing times, safety precautions must constantly be upgraded. It is no longer sufficient, as with the caveman, to shout the warning, "Watch out with that thing!" Today's problems require more elaborate systems of warnings and controls to minimize the chances of serious accidents.

Industrial accidents occur in many ways—a chemical spill, an explosion, a nuclear power plant melt down, etc. There are often problems in transport, with trucks overturning, trains derailing, or ships capsizing. There are "acts of God," such as earthquakes and storms. The one common thread through all of these situations is that they are rarely expected and frequently mismanaged.

Most industrial process plants are safe to be around. Plant management, aided by reliable operators, who are in turn backed up by still-more-reliable automatic controls, does its best to keep operations moving along within the limits usually considered reasonably safe to man and machine in order to have a "safe operating envelope." Occasionally, however, there is a whoosh or a bang that is invariably to the detriment of the operation, endangering investment and human life, and rudely upsetting the plant's loss expectancy.[4]

Accidents have occurred since the birth of civilization. Anyone who crosses a street, rides in a car, or swims in a pool runs the risk of injury through carelessness, poor judgment, ignorance, or other circumstances. This has not changed throughout history. In the following pages, a number of "accidents and disasters" illustrative examples will be examined.

ILLUSTRATIVE EXAMPLE 21.7 Discuss the major differences between rupture disks and relief valves.

Solution. Overpressure protection is one of the major requirements to prevent accidents from occuring in modern chemical operations. The two common devices used for overpressure protection are rupture disks and relief valves. Relief valves are usually used for process protection and rupture disks for reactor and vessel protection.

The relief system is designed to relieve the pressure in equipment if the process/reactor gets out of control. To avoid an explosion, the relief system should be designed to relieve the pressure at a rate determined by the heat transfer to the reactor or vessel during a fire.

Table 21.1 provides a comparison of relief valves and rupture disks. ■

ILLUSTRATIVE EXAMPLE 21.8 A reactor is located in a relatively large laboratory at 22°C and 1 atm with a volume of 1100 m^3. The reactor can emit as much as 0.75 gmol of hydrocarbon (HC) into the room if a safety valve ruptures. A

TABLE 21.1 Relief/Rupture Disk Information

Relief Valve	Rupture Disk
May leak	Zero leakage
Reusable	Need to replace
Requires maintenance	Low maintenance
Responds to pressure slowly	Responds to pressure quickly
Set pressure (adjustable)	Burst pressure (nonadjustable)
More expensive	Less expensive

hydrocarbon mole fraction in the air greater than 425 parts per billion (ppb) constitutes a health and safety hazard.

Suppose the reactor valve ruptures and the maximum amount of HC is released instantaneously. Assume the air flow in the room is sufficient to cause the room to behave as a batch stirred tank reactor, i.e., the air composition is spatially uniform. Calculate the ppb of hydrocarbon in the room. Is there a health risk? From a treatment point of view, what can be done to decrease the environmental hazard or to improve the safety of the reactor?

Solution. Calculate the total number of gmols of air in the room, n_{air}. Assuming that air is an ideal gas, 1 gmol of air occupies 22.4 liters (0.0224 m^3) at standard temperature and pressure (273 K, 1 atm). Since the room temperature is not 273 K,

$$n_{\text{air}} = (1100\,\text{m}^3)\left(\frac{1\,\text{gmol}}{0.224\,\text{STP}\,\text{m}^3}\right)\left(\frac{273\,\text{K}}{295\,\text{K}}\right)$$

$$= 45{,}445\ \text{gmols}$$

Note: STP m^3 indicates the volume (in m^3) the gas would have at standard temperature and pressure.

The mole fraction of hydrocarbon in the room, x_{HC}, is

$$x_{\text{HC}} = \frac{0.75\,\text{gmol HC}}{45{,}445\,\text{gmol air} + 0.75\,\text{gmol HC}} = 16.5\,\text{ppm} = 16{,}500\,\text{ppb}$$

Since 16,500 ppb $\gg$ 850 ppb, the hazard presents a health risk.

To implement safety measures, the potential rupture area should be vented directly into a hood or a duct to capture any leakage in the event of a rupture. Another alternative is input substitution, a source reduction measure. Input substitution is the replacement of the material in the reactor with material with a lower vapor pressure and/or is less toxic.(5) ■

ILLUSTRATIVE EXAMPLE 21.9 Many systems consisting of several components can be classified as series, parallel, or a combination of both. A series system is one in which the entire system fails to operate if any one of its components fail to operate. If such a system consists of n components which function independently, then the reliability of the system is the product of the reliabilities of the individual components. If R_s denotes the reliability of a series system and R_i denotes the reliability of the ith component; $i = 1, 2, \ldots, n$; then

$$R_s = R_1 R_2 \ldots R_n \tag{21.4}$$

A parallel system is one which fails to operate only if all of its components fail to operate. If R_i is the reliability of the ith component, then $(1-R_i)$ is the probability that the ith component fails. Assuming all n components function independently, the probability that all n components fail is $(1-R_1)\ (1-R_2)\ \ldots\ (1-R_n)$.

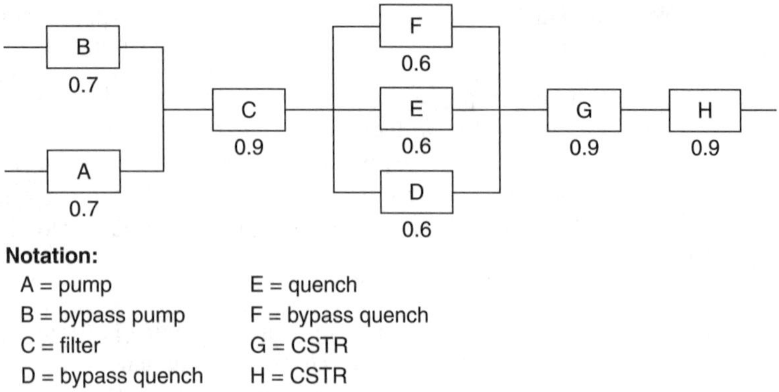

Figure 21.2 Series and parallel systems.

Subtracting this product from 1 yields the following formula for R_p, the reliability of a parallel system:

$$R_p = 1 - (1 - R_1)(1 - R_2)\dots(1 - R_n) \tag{21.5}$$

The reliability formulas for series and parallel systems can be used to obtain the reliability of a system which combines features of a series and a parallel system.

Two CSTRs in series are required to produce a chemical critical for a process. The chemical reactor system is diagrammed above in Figure 21.2, with annual reliabilities indicated under the various components. Determine the reliability of the system.

Solution. The reliability of the parallel subsystem consisting of components A and B in the problem statement is obtained by applying Equation (21.5) for a parallel system:

$$\begin{aligned} R_p &= 1 - (1 - 0.7)(1 - 0.7) \\ &= 0.91 \end{aligned}$$

The reliability of the parallel subsystem consisting of components D, E, and F is

$$\begin{aligned} R_p &= 1 - (1 - 0.6)(1 - 0.6)(1 - 0.6) \\ &= 0.936 \end{aligned}$$

The reliability of the entire system is obtained by applying Equation (21.4) for a series system:

$$\begin{aligned} R_p &= (0.91)(0.9)(0.936)(0.9)(0.9) \\ &= 0.621 \end{aligned}$$

Note that the reliability of a component will frequently depend on the length of time it has been in service. The reader is referred to the literature for more details on reliability calculations and for information on hazard risk management.[5] ■

ILLUSTRATIVE EXAMPLE 21.10 A coolant sprinkler system in a reactor has 20 independent spray components, each of which fails with probability of 0.1. The coolant system is considered to "fail" only if 4 or more of the sprays fail. What is the probability that the sprinkler system will fail?

Solution. Several probability distributions figure prominently in not only residence time distribution calculations (see Chapter 15) but also in reliability calculations. The binomial distribution is one of them. Consider n independent performances of a random experiment with mutually exclusive outcomes that can be classified "success" and "failure". These outcomes do not necessarily have the ordinary connotation of success or failure. Assume that P, the probability of success on any performance of the random experiment, is constant. Let $q = 1 - P$ be the probability of failure. The probability distribution of X, the number of successes in n performances of the random experiment, is a binomial distribution with probability distribution function (pdf) specified by

$$f(x) = \frac{P^x q^{n-x} n!}{x!(n-x)!} \qquad x = 0, 1, 2, \ldots, n \tag{21.6}$$

where $f(x)$ = probability of x successes in n performances

n = number of independent performances of a random experiment

The binomial distribution can therefore be used to calculate the reliability of a redundant system. A redundant system consisting of n identical components is a system that fails only if more than r components fail. Typical examples include single-usage equipment such as missile engines, short-life batteries, and flash bulbs that are required to operate for one time period and not to be reused.[6]

Assume that the n components of the spray system are independent with respect to failure, and that the reliability of each is $1 - P$. One may associate "success" with the failure of a component. Then X, the number of failures, has a binomial pdf and the reliability of the random system is

$$P(X \leq r) = \sum_{x=0}^{r} \frac{P^x q^{n-x} n!}{x!(n-x)!} \tag{21.7}$$

For this problem, X has a binomial distribution with $n = 20$ and $P = 0.10$; the probability that the system fails is given by

$$P(X \geq 4) = \sum_{x=4}^{20} \frac{(0.10^x)(0.90^{20-x})20!}{x!(20-x)!}$$

$$= 1 - P(X \leq 3)$$

This simplifies the problem:

$$P(X \geq 4) = 1 - \sum_{x=0}^{3} \frac{(0.10^x)(0.90^{20-x})20!}{x!(20-x)!} = 0.13$$

■

ILLUSTRATIVE EXAMPLE 21.11 Thomann and Mueller[3] have provided simple, easy-to-use equations that can be employed to describe the concentrations of species in different bodies of water for (a large number of) various conditions. For the case of a steady continuous discharge of a pollutant species into a lake or reservoir undergoing an irreversible first-order reaction, they have shown that the concentration of the pollutant can be described by the following equation[3]

$$C = \frac{(\dot{m})\left\{1 - \exp\left[-\left(\frac{1}{t_d} + k\right)t\right]\right\}}{q(1 + kt_d)} \tag{21.8}$$

where C = concentration of the pollutant at time t

$\dot{m}$ = mass flowrate of pollutant discharge

q = net volume flowrate through lake or reservoir

t_d = lake retention time= V/q

V = lake or reservoir volume

k = reaction velocity constant, $(\text{time})^{-1}$

The concentration at a time θ following termination of the discharge may be calculated using the following equation:[3]

$$C = C_0\left\{1 - \exp\left[-\left(\frac{1}{t_d} + k\right)\theta\right]\right\} \tag{21.9}$$

where C_0 is the concentration that would be achieved if the pollution discharge continued indefinitely.

A near core meltdown at a nuclear power plant brought about the implementation of an emergency response procedure. Part of the response plan resulted in the step discharge of a radioactive effluent into a nearby reservoir. The mass flowrate of discharge was 120,000 lb/h with a radioactive waste concentration of 10^6 picocuries/L (pCi/L) over an 11-h period. (One gram of radium undergoes 3.7×10^6 nuclear disintegrations in 1 s. This number of disintegrations is known as a curie, Ci, which is a unit used to measure nuclear activity.) The reservoir volume and the net throughput volumetric flowrate are approximately (annual average) 3.6×10^8 ft^3 and 200 ft^3/s. respectively, If the nuclear waste decays in a first order manner with a decay constant of 0.23 (h)$^{-1}$, determine the following.

1. The equilibrium concentration associated with the steady waste discharge.
2. The maximum concentration.
3. The time after termination of the waste discharge for the concentration to reach an acceptable level of 10 pCi/L.

Solution. The describing equation for the equilibrium concentration is obtained using the equation provided above.

$$C = \frac{(\dot{m})\left\{1 - \exp\left[-\left(\frac{1}{t_d} + k\right)t\right]\right\}}{q(1 + kt_d)} \tag{21.8}$$

Setting $t = \infty$ leads to

$$C_{eq} = \frac{\dot{m}}{q(1 + kt_d)}$$

Values may now be substituted in this equation. Assume that the effluent has the properties of water. Also, note that $t_d = V/q$.

$$\begin{aligned}
\dot{m} &= \left(120{,}000\,\frac{\text{lb}}{\text{h}}\right)\left(\frac{1}{62.4\,\frac{\text{lb}}{\text{ft}^3}}\right)\left(\frac{\text{L}}{0.0353\,\text{ft}^3}\right)\left(\frac{10^6\,\text{pCi}}{\text{L}}\right)\\
&= 5.45 \times 10^{10}\,\text{pCi/h}\\
q &= (200\,\text{ft}^3/\text{s})(3600\,\text{s/h})\\
&= 7.2 \times 10^5\,\text{ft}^3/\text{h}\\
t_d &= (3.6 \times 10^8\,\text{ft}^3)/(7.2 \times 10^5\,\text{ft}^3/\text{h})\\
&= 500\,\text{h}\\
k &= 0.23\,\text{h}^{-1}
\end{aligned}$$

Substituting the values of $\dot{m}$, q, t_d, and k into the equation

$$C_{eq} = \frac{\dot{m}}{q(1 + kt_d)}$$

yields

$$\begin{aligned}
C_{eq} &= \frac{5.45 \times 10^{10}\,\frac{\text{pCi}}{\text{h}}}{\left(7.2 \times 10^5\,\frac{\text{ft}^3}{\text{h}}\right)\left[1 + \left(0.23\,\frac{1}{\text{h}}\right)(500\,\text{h})\right]}\\
&= 652\,\text{pCi/ft}^3\\
&= 652\,\text{pCi/ft}^3(0.0353\,\text{ft}^3/\text{L})\\
&= 23.02\,\text{pCi/L}
\end{aligned}$$

The maximum concentration C_{max}, is achieved when the discharge is stopped, i,e, when $t = 11$ h. This is obtained from the following equation:

$$C_{max} = \frac{(\dot{m})\left\{1 - \exp\left[-\left(\frac{1}{t_d} + k\right)t\right]\right\}}{q(1 + kt_d)} \qquad (21.8)$$

Substituting yields

$$\begin{aligned} C_{max} &= C_{eq}\left\{1 - \exp\left[-\left(\frac{1}{t_d + k}\right)t\right]\right\} \\ &= 23.02\frac{\text{pCi}}{\text{L}}\left\{1 - \exp\left[-\left(\frac{1}{500\,\text{h}} + 0.23\frac{1}{\text{h}}\right)(11\,\text{h})\right]\right\} \\ &= 21.23\,\text{pCi/L} \end{aligned}$$

The time, θ, after the termination of the waste discharge, that the concentration will reach an acceptable level is given by

$$C = C_{max}\left\{1 - \exp\left[-\left(\frac{1}{t_d} + k\right)t\right]\right\}$$

Set $t = \theta$.

$$\begin{aligned} 10\frac{\text{pCi}}{\text{L}} &= 21.23\frac{\text{pCi}}{\text{L}}\left\{1 - \exp\left[-\left(\frac{1}{500\,\text{h}} + 0.23\frac{1}{\text{h}}\right)\theta\right]\right\} \\ \frac{10}{21.23} &= 1 - \exp(-0.232\theta) \\ 0.529 &= \exp(-0.232\theta) \\ \ln(0.529) &= -0.232\theta \\ \theta &= -0.638/-0.232 \\ &= 2.75\,\text{h} \end{aligned}$$

Since the waste is discharged for 11 h, the time from the start of the discharge that the concentration will reach the acceptable level is

$$\begin{aligned} t &= \theta + 11 \\ &= 2.75 + 11 \\ &= 13.75\,\text{h} \end{aligned}$$

In reality, a concentration profile will exist in the body of water. An average or mean concentration may not be realistic over this short period of time since complete mixing will probably not occur.

Thomann and Mueller's work[3] is an exceptionally informative and well written text/reference book that is presented in a clear, concise manner. A significant amount of information is provided for describing the fate of pollutant species in oceans, rivers, lakes, estuaries etc. ■

ILLUSTRATIVE EXAMPLE 21.12 A reactor explosion precipitates a fire in the building housing the reactor. If a building fire occurs, a smoke alarm sounds with probability 0.9. The sprinkler system functions with probability 0.7 whether or not the smoke alarm sounds. The consequences are minor fire damage (alarm sounds, sprinkler works), moderate fire damage with few injuries (alarm sounds, sprinkler fails), moderate fire damage with many injuries (alarm fails, sprinkler works), and major fire damage with many injuries (alarm fails, sprinkler fails). Construct an event tree and indicate the probabilities for each of the four consequences.

Solution. An event tree provides a diagrammatic representation of event sequences that begin with a so-called *initiating* event and terminate in one or more undesirable consequences. In contrast to a fault tree (considered in the next example), which works backward from an undesirable consequence to possible causes, an event tree works forward from the initiating event to possible undesirable consequences. The initiating event may be equipment failure, human error, power failure, or some other event that has the potential for adversely affecting the environment or an ongoing process and/or equipment.

Note that for each branch in an event tree, the sum of probabilities must equal 1.0. Note again that an event tree:

1. works forward from the initial event, or an event that has the potential for adversely affecting an ongoing process, and ends at one or more undesirable consequences;
2. is used to represent the possible steps leading to a failure or accident;
3. uses a series of branches that relate the proper operation and/or failure of a system with the ultimate consequences;
4. provides a quick identification of the various hazards that could result from a single initial event;
5. is beneficial in examining the possibilities and consequences of a failure;
6. usually does not quantify (although it can) the potential of the events occurring; and,
7. can be incomplete if all the initial events are not identified.[5]

Thus, the use of event trees is sometimes limiting for hazard analysis because it usually does not quantify the potential of the event occurring. It may also be incomplete if all the initial occurrences are not identified. Its use is beneficial in examining, rather than evaluating, the possibilities and consequences of a failure. For this reason, a fault tree analysis should supplement this model to establish the probabilities of the event tree branches. This topic is introduced in the next example.

Regarding the problem statement, the first consequence(s) of the building fire and the probabilities of the first consequence(s) are shown in Figure 21.3. The second consequence(s) of the building fire and the probabilities of the consequence(s)

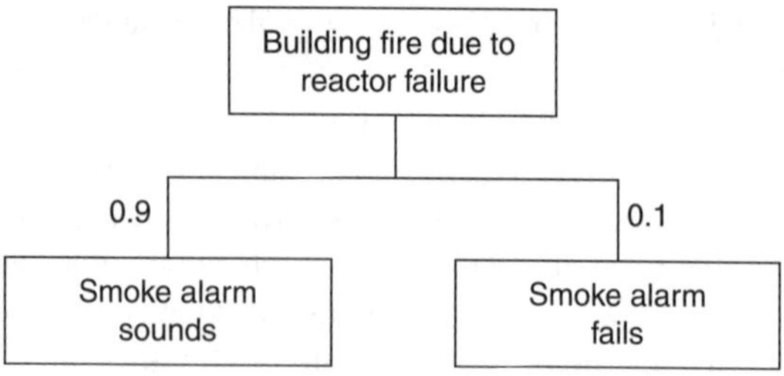

Figure 21.3 Event tree with first set of consequences.

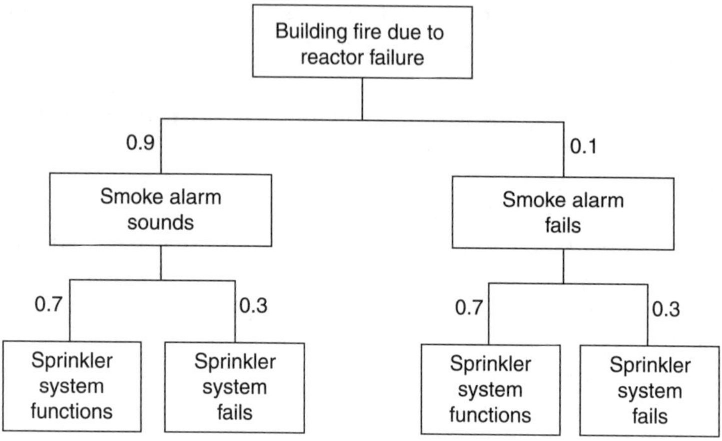

Figure 21.4 Event tree with second set of consequences.

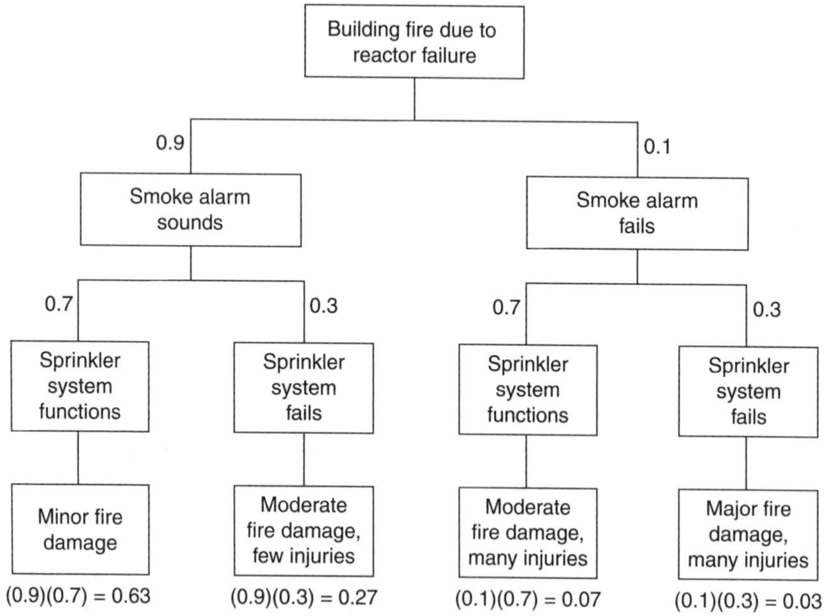

Figure 21.5 Event tree with final set of consequences.

are shown in Figure 21.4. The final consequence(s) and the probabilities of minor fire damage, moderate fire damage with few injuries, moderate fire damage with many injuries, and major fire damage with many injuries are shown in Figure 21.5. ■

ILLUSTRATIVE EXAMPLE 21.13 A runaway chemical reaction can occur if coolers fail (A) or there is a bad chemical batch (B). Cooler failure occurs only if both cooler 1 fails (C) and cooler 2 fails (D). A bad chemical batch occurs if there is a wrong mix (E) or there is a process upset (F). A wrong mix occurs only if there is an operator error (G) and instrument failure (H).

1. Construct a fault tree.
2. If the following annual probabilities are provided by the plant engineer, calculate the probability of a runaway chemical reaction occurring in a year's time given the following probabilities:

   $P(\text{C}) = 0.05$

   $P(\text{D}) = 0.08$

   $P(\text{F}) = 0.06$

   $P(\text{G}) = 0.03$

   $P(\text{H}) = 0.01$

Solution.

1. As noted earlier, fault tree analysis seeks to relate the occurrence of an undesired event to one or more antecedent events. The undesired event is called the *top event* and the antecedent events are called *basic events*. The top event may be, and usually is, related to the basic events via certain intermediate events. The fault tree diagram exhibits the casual chain linking of the basic events to the intermediate events and the latter to the top event. In this chain, the logical connection between events is illustrated by so-called *logic gates*. The principal logic gates are the *AND* gate, symbolized on the fault tree by ⌓ and the *OR* gate, symbolized by ⋒.

   The reader should note that a fault tree:

   1. works backward from an undesirable event or ultimate consequence to the possible causes and failures;
   2. relates the occurrence of an undesired event to one or more preceding events;
   3. "chain links" basic events to intermediate events that are in turn connected to the top event;
   4. is used in the calculation of the probability of the top event;
   5. is based on the, most likely or credible events that lead to a particular failure or accident, and,
   6. analysis includes human error as well as equipment failure.(5)

Begin with the top event for this problem, as shown in Figure 21.6. Generate the first branch of the fault tree, applying the logic gates (see Figure 21.7). Generate the

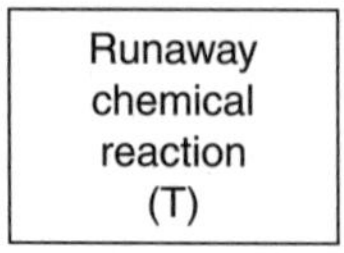

Figure 21.6 Top event of fault tree.

second branch of the fault tree, applying the logic gates as shown in Figure 21.8. Generate the last branch of the fault tree, applying the logic gates as shown in Figure 21.9.

2. Finally, calculate the probability that the runaway reaction will occur:

$$P = (0.5)(0.08) + (0.01)(0.03) + 0.06$$
$$= 0.064$$

Note that the process upset, F, is the major contributor to the probability to a potential runaway reaction. ■

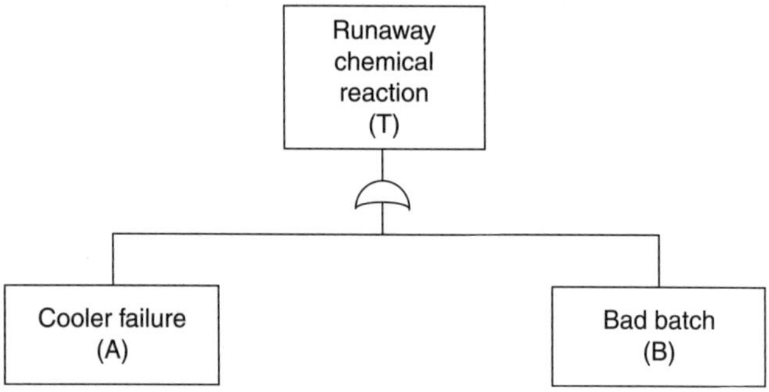

Figure 21.7 Fault tree with first branch.

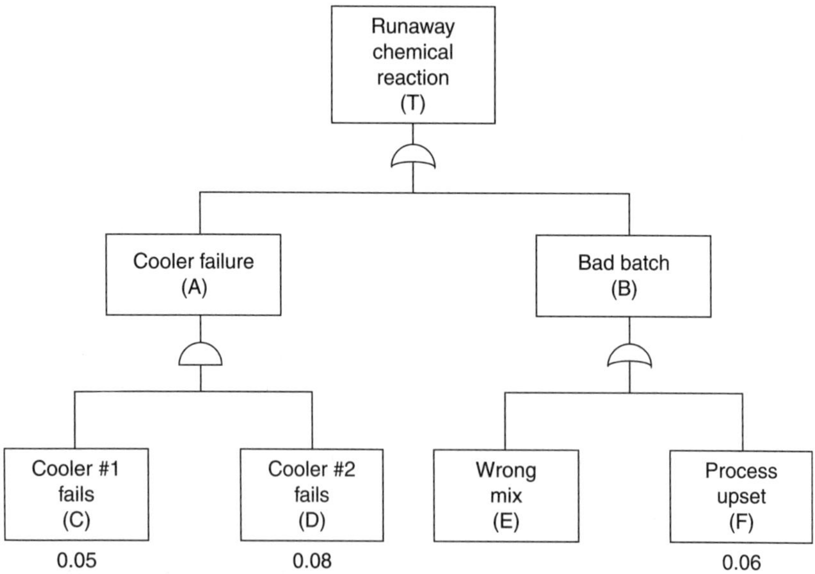

Figure 21.8 Fault tree with second branch.

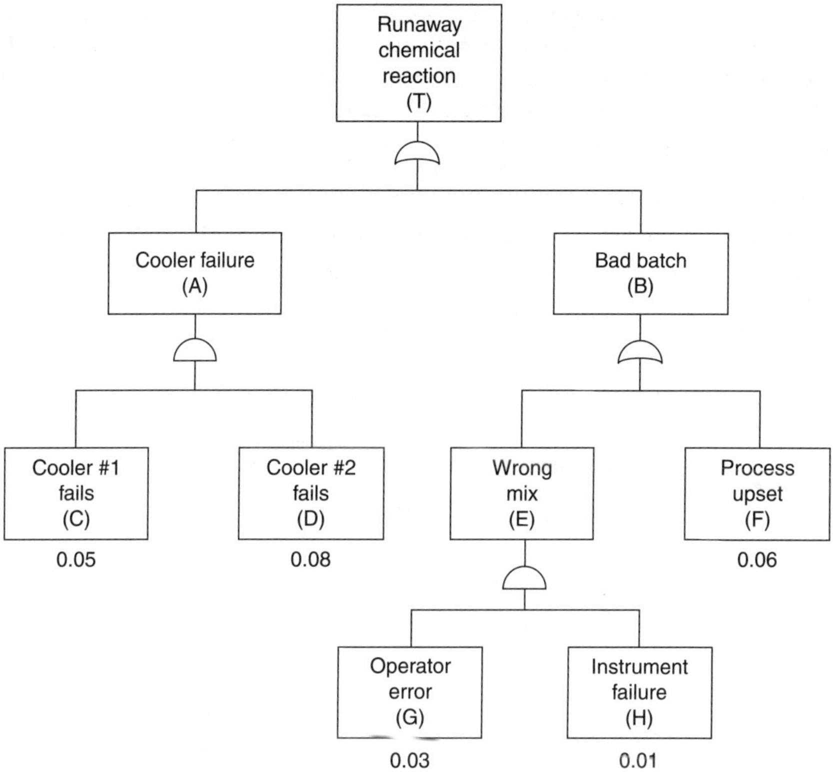

Figure 21.9 Fault tree with third branch.

ETHICS

This section is concerned with ethics. For engineers, the question of ethics typically boils down to the struggle between the responsibility to ensure public health and safety and the responsibility to the employers, clients, and shareholders of the company. The case study approach is employed to make the reader think about ethical questions, to reflect on their past decisions, and to project forward to their future decisions a higher degree of thought and insight in determining when one is faced with an ethical dilemma.

The primary responsibility of an engineering professional is to protect public health and safety. However, engineering professionals also have a responsibility to their employers or clients, to their families, to themselves, and to the environment. Meeting these responsibilities will challenge the practicing engineer to draw upon a system of ethical values.

Well, what about ethics? Ethics means "doing the right thing" as opposed to "what you have the right to do." But, doing the right thing is not always obvious or easy. In fact, ethical decisions are often difficult and may involve a certain amount of self-sacrifice. Doing the right thing for a practicing engineer can be especially challenging. Furthermore, the corporate and government world has confused this concept by developing ethics programs that emphasize only what you have the right to do. An organization, for example, may have a list—often called a Code of Ethics or Code of Conduct—of what an employee can and cannot get away with. Employees are required to sign an acknowledgment that they have read and understood the list. The company unfortunately calls this "ethics training."[7]

One difficulty in some situations is recognizing when a question of ethics is involved. Frequently, in the area of environmental management, a breach of ethics involves a practice that endangers public health and safety or covers up a violation of a rule or regulation. Occasionally, however, a breach may involve a case of the exact opposite. This might seem an unlikely scenario. How can someone be too honest, too caring, or too professional?

Regarding the above, one example is lying to save a life. Suppose you are standing on a street and a woman runs past you chased by two men. She screams, "They are trying to attack me!" as she dashes into the entry of a building around a corner. The men ask you, "Which way did she go?" What do you tell them? Clearly, the right thing is to lie. In this case, the value of caring overrides the value of honesty. This situation is exaggerated to illustrate that sometimes it is appropriate to violate certain values to protect public health and safety. In doing the right thing, ideally one should not have to make snap decisions and should take the time to investigate all of the facts, e.g., whether or not the woman was a thief and the men were police.

Sometimes one must decide how much to sacrifice to ensure public health and safety. In establishing environmental regulations, the regulating agency must decide how safe and how stringent to make the regulations. For example, in the case of air toxic regulations, one standard may result in 10 cancer cases per one million people. But why isn't it for none ? Who should decide?[8]

This section, in line with its title, addresses ethics and ethical issues relevant to the practicing engineer, and contains three Illustrative Examples directly or indirectly related to chemical reactors. These case studies have been primarily drawn from the work of Wilcox and Theodore.[9]

ILLUSTRATIVE EXAMPLE 21.14

Fact Pattern

Tom works in an engineering company and is responsible for managing a batch reactor that produces a specialty petroleum product. His boss, Ken, informs him of a need to increase output of that product because customers have been placing more orders than usual.

Tom continues to socialize with his coworkers as usual. He thinks that speeding production is just a matter of cutting some procedures short; Tom decides to eliminate the extra cleaning step in the process to obtain a quicker turnaround time.

The company later receives customer complaints regarding the quality of the petroleum product; it was contaminated with cleaning chemicals that were supposed to be removed prior to shipment. Ken meets with Tom to discuss the issue.

"Tom, we've been getting customer complaints regarding our product, "says Ken. "They have found the petroleum contaminated with other chemicals. What's going on?"

Tom is left confused on how to get out of this predicament. What can he say to Ken without risking his job and losing his production team? Should he blame his team? Or should Tom tell Ken the truth?

Questions for Discussion

1. What are the facts in this case?
2. How should Tom respond to Ken's question?
3. How could Tom have avoided his problem? ■

ILLUSTRATIVE EXAMPLE 21.15

Fact Pattern

Jane works in the quality control group in the engineering section of her company. She is responsible for testing reactor samples in her group's backlog. Much of the backlog contains similar samples that all need to have the same testing. This testing procedure takes hours to conduct for a single sample.

Jane has worked for three weeks straight on the same batch of samples; all of the tested samples have met quality control standards. With a holiday weekend approaching, she decides to end her week early by entering similar results for untested samples.

Glen also works in the quality control group with Jane. He sees her forging numbers in the computer and pulls her to the side.

"Why are you putting in results when you haven't finished testing the remaining samples?" inquires Glen.

"I've got a big weekend planned, and the samples in this batch seem to be routinely passing the tests," Jane replies. "The rest of the batch is probably going to pass testing anyway. I'm just saving myself some time to prepare for the weekend."

"What if the rest of the samples really don't pass the testing?" asks Glen.

"I test these types of samples all the time," responds Jane. "They always pass quality control testing."

Glen is troubled with the issue and wonders whether he should just leave it up to Jane.

Questions for Discussion

1. What are the facts in this case?
2. What should Glen do?
3. Are there any other options for Jane? For Glen? ■

ILLUSTRATIVE EXAMPLE 21.16. Andrew is an engineer working in a pollution control company who has been asked by his boss to do an evaluation of a certain landfill they covered a few years ago. His main task is to run a check for methane gas at the landfill and on some of the blocks in the surrounding neighborhoods.

When Andrew goes to the landfill, he notices that most of the vents are emitting high concentrations of methane, and these concentrations are accumulating in pockets at some places along the fence-line of the landfill.

Andrew comes back from the landfill with a 50-page set of data and hands it to Monica, a temporary employee, to present as a proposal to the city. Before leaving Monica's desk, Andrew comments in a brusk manner that even though the numbers seem high in some areas, he could not detect much of an odor. He also mentions under his breath they would have to pay a high fine if the city does not like some of their numbers.

As Monica is reducing the data into readable tables, she notices that not only some but *most* of the numbers are extremely high around the adjacent neighborhoods.

Monica asks her coworker Judy, "Can't high levels of methane gas be harmful to people?"

Judy tells Monica, "Of course they can. Methane not only is harmful to health but also can cause explosions in the presence of fire or sparks."

Knowing that she should talk to Andrew about this first before saying anything else to Judy, Monica approaches Andrew.

She tells him, "These numbers for methane are very high, even in the neighborhood areas. I am going to have to report it that way."

Andrew tells Monica, "If you report it that way, we could lose a lot of money/time redoing the landfill cover, and even possibly our jobs. Look, no one is complaining, and there is no odor, so how is anyone going to find out, right?"

"Yeah, I guess so," replies Monica.

"It *is* okay, believe me. It doesn't matter if the numbers are a little off," retorts Andrew.

Questions for Discussion

1. What are the facts in this case?
2. What effect will Judy's comments have on Monica's decision?
3. What might happen if Monica does not report high numbers?
4. What might happen if Monica *does* report high numbers?
5. What final action do you think Monica will take? ■

NUMERICAL METHODS

This fourth section is concerned with numerical methods. This subject was taught in the past as a means of providing engineers and scientists with ways to solve complicated mathematical expressions that they could not otherwise solve. However, with the advent of computers, these solutions are now readily obtained. A brief overview of numerical methods is given to provide the practicing engineer with some insight into what many of the currently used software packages (MathCad, Mathematica, MatLab, etc.) are actually doing.

Early in one's career, the engineer/scientist learns how to use equations and mathematical methods to obtain exact answers to a large range of relatively simple problems. Unfortunately, these techniques are often not adequate for solving real-world problems, although the reader should note that one rarely needs exact answers in technical practice. Most real-world applications are usually inexact because they have been generated from data or parameters that are measured, and hence represent only approximations. What one is likely to require in a realistic situation is not an exact answer but rather one having reasonable accuracy from an engineering point of view.

The solution to an engineering or scientific problem usually requires an answer to an equation or equations, and the answer(s) may be approximate or exact. Obviously an exact answer is preferred, but because of the complexity of some equations, exact solutions may not be attainable. Furthermore, to engineers, an answer that is precise may not be necessary and can waste time. For this condition, one may resort to another method that has come to be defined as a numerical method. Unlike the exact solution, which is continuous and in closed form, numerical methods provide an inexact (but often reasonably accurate) solution. The numerical method leads to discrete answers that are almost always acceptable.

Numerical methods provide a step-by-step procedure that ultimately leads to an answer and a solution to a particular problem. The method usually requires a large number of calculations and is therefore ideally suited for digital computation.

High speed computing equipment has had a tremendous impact on engineering design, scientific computation, and data processing. The ability of computers to handle large quantities of data and to perform mathematical operations described above at tremendous speeds permits the examination of many more cases and more engineering variables than could possibly be handled on the slide rule—once employed by the author and the trademark of engineers of yesteryear. Scientific calculations previously estimated in lifetimes of computation time are currently generated in seconds and, in many instances, microseconds.[10]

As noted above, this section is concerned with numerical methods. A brief overview of three numerical methods is given to provide the practicing engineer with some insight into what many of the currently used software packages are actually doing. The author has not attempted to cover all the topics of numerical methods. Topics that traditionally fall in the domain of this subject include:

Regression analyses (see Chapter 13)

Differentiation (Illustrative Example 21.17)

Integration (Illustrative Example 21.18)

Simultaneous linear algebraic equations (Illustrative Example 21.19)

Nonlinear algebraic equations

Ordinary differential equation(s) (Illustrative Example 21.20)

Partial differential equation(s)

Optimization

Since detailed treatment of each of the above topics is beyond the scope of this text, the reader is referred to the literature[11–13] for a more extensive analysis and additional information. The remainder of this section consists of illustrative examples that examine the four topics listed on previous page. Also note that the illustrative examples primarily address chemical reactor topics.

ILLUSTRATIVE EXAMPLE 21.17 Derivatives of a function, $f(x)$, may be evaluated numerically by any one of several different methods. Three simple formulas that may be employed to evaluate the derivative, $d[f(x)]/dx$ or $f'(x)$, are given below.

$$f'(x) = (1/h)[f(x+h) - f(x)] \quad \text{forward} \tag{21.10}$$

$$f'(x) = (1/2h)[f(x+h) - f(x-h)] \quad \text{central} \tag{21.11}$$

$$f'(x) = (1/h)[f(x) - f(x-h)] \quad \text{backward} \tag{21.12}$$

The term h once again represents the incremental change in x. The reader is referred to any numerical analysis text for a detailed review of numerical differentiation methods.

Refer to Table 21.2. Generate expressions and the corresponding calculated values for the benzene concentration gradient, dC/dt, at $t = 4.0$ s. Use several differentiation methods.[14]

Solution

1. The first method consists of choosing any three data points and calculating the slope of the two extreme points. This slope is approximately equal to the slope at the point lying in the middle. The value obtained is the equivalent of the derivative at that point 4.

   Using data points from the time interval 3.0 to 5.0,

$$\text{Slope} = \frac{C_5 - C_3}{t_5 - t_3}$$

$$= \frac{1.63 - 2.70}{5.0 - 3.0} = -0.535$$

TABLE 21.2 Illustrative Example 21.17

Time (s)	Concentration of benzene (mg/L)
0.0	7.46
1.0	5.41
2.0	3.80
3.0	2.70
4.0	2.01
5.0	1.63
6.0	1.34
7.0	1.17

2. The second method involves taking the average of two slopes. Using the same points chosen above plus point 4, two slopes are calculated—one for points 3 and 4 and the other for points 4 and 5. Adding the two results and dividing them by two provides an approximation of the derivative at point 4. For the points used in this method, the results are:

$$\text{Slope}_1 = \frac{C_4 - C_3}{t_4 - t_3}$$

$$= \frac{2.01 - 2.70}{4.0 - 3.0} = -0.69$$

$$\text{Slope}_2 = \frac{C_5 - C_4}{t_5 - t_4}$$

$$= \frac{1.63 - 2.01}{5.0 - 3.0} = -0.38$$

$$\text{Slope}_{\text{avg}} = \frac{-0.69 + (-0.38)}{2} = -0.535$$

3. Method three consists of using any three data points (in this case the same points chosen before) and fitting a curve to it. The equation for the curve is obtained by employing a second-order equation and solving it with the three data points. The result is:

$$C = 0.155t^2 - 1.775t + 6.63$$

The derivative of the equation is then calculated and evaluated at any point (here, point 4 is used):

$$\frac{dC}{dt} = 0.31t - 1.775$$

Evaluate the dervative at $t = 4.0$ s:

$$\text{Slope} = \frac{dC}{dt} = 0.31(4) - 1.775 = -0.535$$

4. The following method uses the method of least squares. In this case, all data points are used to generate a second-order polynomial equation. This equation is then differentiated and evaluated at the point where the value of the derivative is required. For example, Microsoft Excel can be employed to generate the regression equation. Once all the coefficients are known, the equation has only to be analytically differentiated:

$$C = 0.1626t^2 - 1.9905t + 7.3108$$

$$\frac{dC}{dt} = 0.3252t - 1.9905$$

At $t = 4.0$ s:

$$\frac{dC}{dt} = 0.3252(4.0) - 1.9905 = -0.6897$$

5. The last two methods are very similar to each other. They are based on five data points used to generate coefficients. For this development, represent C and t by f and x (as it appeared in the literature[15]) respectively.

The fifth method uses five data points to generate a five coefficient (fourth order) model using an equation of the form $f = A + Bx + Cx^2 + Dx^3 + Ex^4$ equation. This method is known as interpolating. A set of equations is used to evaluate numerical derivatives from the interpolating polynomial. The equations are listed below:

$$f'(x_0) = \frac{(-25f_0 + 48f_1 - 36f_2 + 16f_3 - 3f_4)}{12h} \tag{21.13}$$

$$f'(x_1) = \frac{(-3f_0 - 10f_1 + 18f_2 - 6f_3 + f_4)}{12h} \tag{21.14}$$

$$f'(x_i) = \frac{(f_{i-2} - 8f_{i-1} + 8f_{i+1} - f_{i+2})}{12h} \tag{21.15}$$

$$f'(x_{n-1}) = \frac{(-f_{n-4} + 6f_{n-3} - 18f_{n-2} + 10f_{n-1} + 3f_n)}{12h} \tag{21.16}$$

$$f'(x_n) = \frac{(3f_{n-4} - 16f_{n-3} + 36f_{n-2} - 48f_{n-1} + 25f_n)}{12h} \tag{21.17}$$

where $h = x_{i+1} - x_i$

f_i = function evaluated at i

For example, the equation obtained for "the five data set" from 1.0 to 5.0 s, i.e., $t = 1.0$, 2.0, 3.0, 4.0, and 5.0 s, using the author's approach

$$f(x) = -0.0012x^4 + 0.002x^3 + 0.2616x^2 - 2.34x + 7.467$$

All these equations can be evaluated for each value of x and $f(x)$. The value for the derivative for point 4.0 is -0.5448.

6. The last method also uses five data points but only three coefficients are generated for a second-order polynomial equation of the form $f = A + Bx + Cx^2$. Another set of equations are used to evaluate the derivative at each point using this method. The equations are provided below:

$$f'(x_0) = \frac{(-54f_0 + 13f_1 + 40f_2 + 27f_3 - 26f_4)}{70h} \tag{21.18}$$

$$f'(x_1) = \frac{(-34f_0 + 3f_1 + 20f_2 + 17f_3 - 6f_4)}{70h} \tag{21.19}$$

$$f'(x_i) = \frac{(-2f_{i-2} - f_{i-1} + f_{i+1} + 2f_{i+2})}{10h} \tag{21.20}$$

$$f'(x_{n-1}) = \frac{(6f_{n-4} - 17f_{n-3} - 20f_{n-2} - 3f_{n-1} + 34f_n)}{70h} \tag{21.21}$$

$$f'(x_n) = \frac{(26f_{n-4} - 27f_{n-3} - 40f_{n-2} - 13f_{n-1} + 54f_n)}{70h} \tag{21.22}$$

At point 4.0, the solution for the derivative using this method is -0.6897.

Comparing all the values obtained for the derivative at $t = 4.0$ s, it can be observed that the answers are very close to each other. It is important to remember that these are approximate values and that they vary depending on the approach and the number of data points used to generate the equations.

Some useful *analytical* derivatives in reactor calculations are provided below.

1. $d(a)/dx = 0; \quad a = \text{constant}$
2. $d(x)/dx = 1$
3. $d(ax)/dx = a$
4. $d(ax^n)/dx = nax^{n-1}$
5. $d(au)/dx = a(du/dx); \quad u = f(x)$
6. $d(uv)/dx = u(dv/dx) + v(du/dx); \quad v = f(x)$
7. $d(u/v)/dx = (1/v)(du/dx) - (u/v^2)(dv/dx)$
8. $d(u^n)/dx = nu^{n-1}(du/dx)$
9. $d(u^{1/2})/dx = [1/2u^{1/2}](du/dx)$
10. $d(1/u)/dx = -(1/u^2)(du/dx)$
11. $d(1/u^n)/dx = -(n/u^{n+1})(du/dx)$
12. $d(\ln u)/dx = (1/u)(du/dx)$
13. $d(a^u)dx = (a^u)(\ln a)(du/dx)$
14. $d(e^u)/dx = e^u(du/dx)$
15. $d(\sin u)/dx = (\cos u)(du/dx)$
16. $d(\cos u)/dx = -(\sin u)(du/dx)$ ■

ILLUSTRATIVE EXAMPLE 21.18 Many chemical reactor applications require solving differential equations. Some of these equations can be solved numerically or analytically by integrating the describing differential equation. Two simple numerical integration methods that are commonly employed are the trapezoidal rule and Simpson's rule.

In order to use the *trapezoidal* rule to evaluate the integral, I, given by

$$I = \int_a^b f(x)\,dx \tag{21.23}$$

use the equation

$$I = (h/2)[y_0 + 2y_1 + 2y_2 + \cdots + 2y_{n-1} + y_n] \tag{21.24}$$

where h = incremental change in x, i.e., Δx

y_1 = value of $f(x)$ at x_1, i.e., $f(x_1)$

Thus,

$$\begin{aligned} y_0 &= f(x_0) = f(x = a) \\ y_n &= f(x_n) = f(x = b) \\ h &= (a - b)/n \end{aligned} \tag{21.25}$$

For *Simpson's* 3-point (or one-third) rule, use the equation

$$I = (h/3)(y_a + 4y_{(a-b)/2} + y_b) \tag{21.26}$$

For the general form of Simpson's rule (n is an even integer), use the equation

$$I = (h/3)[y_0 + 4y_1 + 2y_2 + 4y_3 + 2y_4 + \cdots + 2y_{n-2} + 4y_{n-1} + y_n] \tag{21.27}$$

The volume requirment of a reactor undergoing conversion X for the reactant (the principal component from the bottom of a distillation column) is described by the following integral:

$$V = 6.0 \times 10^{-3} \int_0^{0.45} \frac{(1 - 0.125X)^3 dX}{10^{-4}(1 - X)(1 - 0.5X)^2}; \quad \text{liters}$$

Calculate the volume using the trapezoidal rule method of integration. Discuss the effect of varying the increment in ΔX (e.g., if $\Delta X = 0.45$, 0.09, 0.05, 0.01, 0.005, 0.001).[13]

Solution. Regarding integration, an algorithm for applying the trapezoid rule is given in Figure 21.10. For an increment size of $\Delta X = 0.45$, the step size for the reactor volume is:

$$h = X_1 - X_0 = 0.45 - 0.00 = 0.45$$

Evaluate the function at X_0 and X_1:

$$V_0 = f(X_0) = 6.0 \times 10^{-3} \left[\frac{(1 - 0.125X)^3 dX}{10^{-4}(1 - X)(1 - 0.5X)^2} \right]_{X=0.0}$$

$$= 6.0 \times 10^{-3} \frac{(1)^3}{10^{-4}(1)(1)^2} = 60$$

$$V_1 = f(X_1) = 6.0 \times 10^{-3} \left[\frac{(1 - 0.125X)^3 dX}{10^{-4}(1 - X)(1 - 0.5X)^2} \right]_{X=0.45}$$

$$= 6.0 \times 10^{-3} \frac{[1 - 0.125(0.45)]^3}{10^{-4}(1 - 0.45)[1 - 0.5(0.45)]^2} = 152.67$$

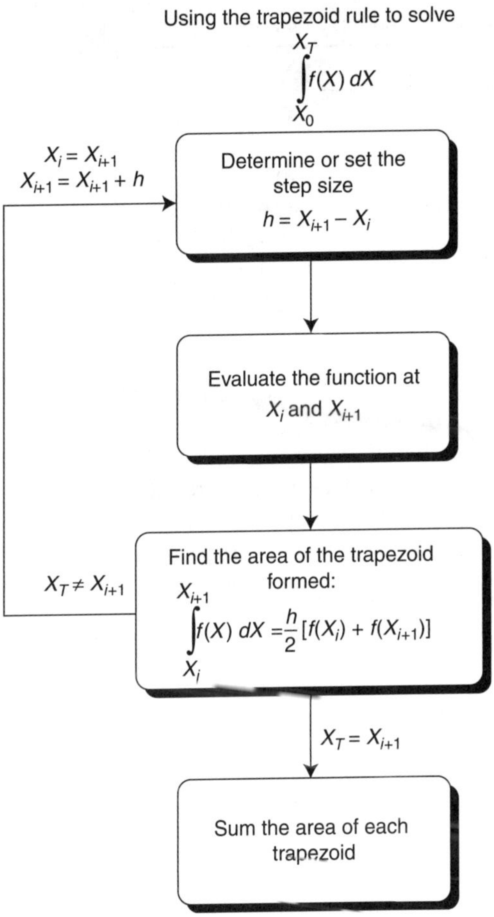

Figure 21.10 Algorithm for trapezoid rule.

The two-point trapezoid rule is given by

$$\int_{X_0}^{X_1} f(X)\,dX = \frac{h}{2}[f(X_0 + f(X_1)] \tag{21.24}$$

Therefore,

$$\begin{aligned} V &= \frac{h}{2}[f(X_0 + f(X_1)] \\ &= \frac{h}{2}[V_0 + V_1] \\ &= \frac{0.45}{2}[60 + 152.67] \\ &= 47.85\,\text{L} \end{aligned}$$

TABLE 21.3 Trapezoid Rule for Various Step Sizes

Step size (ΔX)	Volume
0.45	47.85094
0.09	43.13532
0.05	42.98842
0.01	42.92521
0.005	42.92324
0.001	42.92260

The trapezoid rule is often the quickest but least accurate way to perform a numerical integration by hand. However, if the step size is decreased, the answer should converge to the analytical solution. Note that for smaller step sizes, the results of each numerical integration must be added together to obtain the final answer. The results (to seven significant figures) for various step sizes are listed in Table 21.3.

Some useful *analytical* integrals in reactor calculations are provided below; note that some equations are indefinite integrals, i.e., the upper and lower limits are not specified.[17]

1. $\int_0^X a\,dx = ax$
2. $\int_0^X [1/(1-x)]\,dx = \ln[1/(1-x)]$
3. $\int [1/(1-x)^2]\,dx = 1/(1-x)$
4. $\int_0^X [1/(1-x)^2]\,dx = x/(1-x)$
5. $\int [1/x]\,dx = \ln x$
6. $\int e^x\,dx = e^x$
7. $\int e^{ax}\,dx = e^{ax}/a$
8. $\int \sin x\,dx = -\cos x$
9. $\int \cos x\,dx = \sin x$
10. $\int_0^X [1/(1+\varepsilon x)]\,dx = (1/\varepsilon)\ln(1+\varepsilon x)$
11. $\int_0^X [(1+\varepsilon x)/(1-x)]\,dx = (1+\varepsilon)\ln[1/(1-x)] - \varepsilon x$
12. $\int_0^X [(1+\varepsilon x)/(1-x)^2]\,dx = [(1+\varepsilon)(x)/(1-x)] - (\varepsilon)\ln[1/(1-x)]$
13. $\int_0^X [(1+\varepsilon x)^2/(1-x)^2]\,dx = 2\varepsilon(1+\varepsilon)\ln(1-x) + \varepsilon^2 x$

    $+ (1+\varepsilon)^2(x)/(1-x)$
14. $\int_0^X \{1/[(1-x)(\theta_B - x)]\}\,dx = [1/(\theta_B - 1)]\ln[(\theta_B - x)/(\theta_B)(1-x)]$

    for $\theta_B \neq 1$

15. $\int_0^X [1/(ax^2 + bx + c)]\,dx = -[2/(2ax + b)] + (2/b)$; for $b^2 = 4ac$

16. $\int_0^X [1/(ax^2 + bx + c)]\,dx = [1/(a)(p - q)\ln[(q)(x - p)/(p)(x - q)]$;

for $b^2 > 4\text{ac}$

where p and q are roots of the equation

$ax^2 + bx + c = 0$ that is, $\text{p, q} = [-b \pm (b^2 - 4\text{ac})^{1/2}]/2\text{a}$

17. $\int_0^X [(a + bx)/(c + gx)]\,dx = (bx/g) + [(ag - bc)/g^2]\ln|c + gx|$

18. $\int_0^X [x/(ax + b)]\,dx = (x/a) + (b/a^2)\ln(ax + b)$

19. $\int_0^X [x^2/(ax + b)]\,dx = [(ax + b)^2/2a^3] - [(2b/a^2)(ax + b)]$

$+ (b^2/\text{a}^3)\ln(ax + b)$

20. $\int_0^X \{1/[x(ax + b)]\}\,dx = (1/b)\ln[x/(ax + b)]$

Note, once again, that there are no limits on some of the integrals. ■

ILLUSTRATIVE EXAMPLE 21.19 The concentration, C, variation with length, z, in a 3 ft continuous flow tubular reactor (TF) is described by the equation:

$$\frac{dC}{C} = -\frac{k}{v}dz$$

where k is 50 $(\text{h})^{-1}$, v is 50 ft/h, C is in lbmol/ft^3, z is in feet, and $C = 1.0\ \text{lbmol/ft}^3$ at $z = 0$.

It has been proposed to represent the TF with 10 CSTR staged units 0.3 ft in length. Develop solutions to this problem using a finite difference method of solving an ordinary differential equation and a lumped parameter model employing a method of solution of simultaneous linear algebraic equations.

Solution

Finite Difference Method

A finite difference procedure is first applied to the equation. The first derivative of C with respect to z is equivalent to the finite difference in the z-direction, i.e.,

$$\frac{dC}{dz} = \frac{\Delta C}{\Delta z} = \frac{C_{n+1} - C_n}{\Delta z}$$

Inserting this into the describing equation and noting that the average concentration over the increment should be used leads to

$$\frac{C_{n+1} - C_n}{\Delta z} = -\left(\frac{k}{v_z}\right)C_{\text{av}}$$

Knowing the initial condition, each successive concentration may be found by rearranging the above equation:

$$C_{n+1} = -\left(\frac{k\Delta z}{v_z}\right)C_{av} + C_n$$

The application of this equation requires a trial-and-error procedure where the average concentration over the increment is approximated and then checked when, C_{n+1} is calculated. One should note that the approximation part of this calculation can be removed by replacing C_{av} by

$$\frac{C_{n+1} + C_n}{2}$$

and solving for C_{n+1} directly in terms of C_n.

Lumped-parameter Method

In the lumped-parameter method, the reactor is divided into equal stages and each segment is considered to be a perfectly mixed vessel. This is shown schematically in Figure 21.11. The equation is now written as

$$\frac{C_{out} - C_{in}}{\Delta z} = -\frac{k}{v_z}C_{out}$$

Rearranging the above equation gives

$$C_{out} = \frac{C_{in}}{1 + \dfrac{k\Delta z}{v_z}}$$

If the initial concentration is C_0 and the term

$$1 + \frac{k\Delta z}{v_z} = R$$

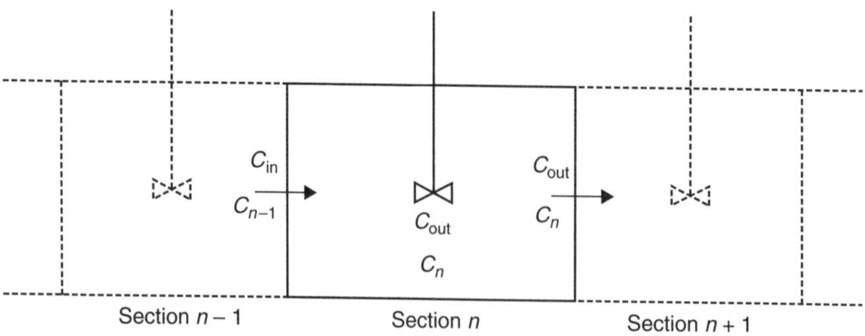

Figure 21.11 Lumped-parameter method.

then the following set of equations are generated for each stage:

$$C_1 - RC_0 = 0; \quad \text{first stage}$$
$$C_2 - RC_1 = 0; \quad \text{second stage}$$
$$\vdots$$
$$C_{\text{final}} - RC_{\text{final}-1} = 0; \quad \text{final stage}$$

The above represents a series of (simultaneous) linear algebraic equations. These equations can be solved by a direct step-by-step hand calculation. Alternatively, a Gauss–Jordan or Gauss–Seidel method can also be used.[16]

The results of the analyses by the two methods are combined and presented in tabular form for comparison along with the analytical solution (see Table 21.4). The reader is left the exercise of calculating the values at $z = 3.0$ ft. ■

ILLUSTRATIVE EXAMPLE 21.20[16] The Runge–Kutta method is one of the most widely used techniques for solving first order ordinary differential equations (ODE).

An ODE is a relation between more than one variable and the derivatives with respect to *only one* of the variables. A partial differential equation (PDE) contains derivatives with respect to *more than one* variable. The *order* of an ODE is the order of the highest derivative present.

The equation describing the concentration of a reactant in a tubular flow reactor is given by

$$\frac{dC}{dt} = -0.580C + 6 \times 10^{-5}$$

with $C_0 = 5.0 \times 10^{-5}$ gmol/cm^3 at $t = 0$ min. Estimate C in gmol/cm^3 at 1 and 5 minutes. Use the Runge–Kutta method analysis.[17]

TABLE 21.4 Concentration Profile via Three Methods

z, ft	Analytical	Finite-difference	Lumped-parameter
0.0	1.0000	1.0000	1.0000
0.3	0.8354	0.8358	0.8474
0.6	0.6980	0.7001	0.7181
0.9	0.5830	0.5863	0.6086
1.2	0.4870	0.4910	0.5157
1.5	0.4070	0.4112	0.4371
1.8	0.3400	0.3444	0.3704
2.1	0.2840	0.2884	0.3139
2.4	0.2370	0.2415	0.2660
2.7	0.1980	0.2022	0.2254

The Runge–Kutta (R–K) method is one of the most widely used techniques for solving first-order differential equations. For the equation

$$\frac{dy}{dx} = f(x, y) \tag{21.28}$$

the solution takes the form

$$y_{n+1} = y_n + \frac{h}{6}(D_1 + 2D_2 + 2D_3 + D_4) \tag{21.29}$$

$$\begin{aligned} \text{where} \quad D_1 &= hf(x, y) \\ D_2 &= hf\left(x_n + \frac{h}{2}, y_n + \frac{D_1}{2}\right) \\ D_3 &= hf\left(x_n + \frac{h}{2}, y_n + \frac{D_2}{2}\right) \\ D_4 &= hf(x_n + h, y_n + D_3) \end{aligned} \tag{21.30}$$

The term h once again represents the increment in x. The term y_n is the solution to the equation at x_n while y_{n+1} is the solution to the equation at x_{n+1} where $x_{n+1} = x_n + h$. Thus, the R–K method provides a straightforward means for developing expressions for Δy in terms of the function $f(x, y)$ at various "locations" along the interval in question.

For a simple equation of the form

$$\frac{dC}{dt} = a + bC \tag{21.31}$$

where at $t = 0$, $C = C_0$, the R–K algorithm given above becomes (for $t = h$)

$$C_1 = C_0 + \frac{h}{6}(D_1 + 2D_2 + 2D_3 + D_4) \tag{21.32}$$

$$\begin{aligned} \text{where} \quad D_1 &= hf(x, y) = h(a + bC_0) \\ D_2 &= hf\left(x_n + \frac{h}{2}, y_n + \frac{D_1}{2}\right) = h[a + b(C_0 + D_1/2)] \\ D_3 &= hf\left(x_n + \frac{h}{2}, y_n + \frac{D_2}{2}\right) = h[a + b(C_0 + D_2/2)] \\ D_4 &= hf(x_n + h, y_n + D_3) = h[a + b(C_0 + D_3)] \end{aligned} \tag{21.33}$$

The same procedure is repeated to obtain values for C_2 at $t = 2h$, C_3 at $t = 3h$, etc.

Solution. Based on the data provided, evaluate the R–K coefficients for $t = h$:

$$D_1 = 1.0(6.0 - 2.9) = 3.1$$
$$D_2 = 1.0[6.0 - 0.58(5.0 + 3.1/2)] = 2.2$$
$$D_3 = 1.0[6.0 - 0.58(5.0 + 2.2/2)] = 2.46$$
$$D_4 = 1.0[6.0 - 0.58(5.0 + 2.46)] = 1.67$$

Calculate C_1:

$$C_1 = C_0 + \frac{h}{6}(D_1 + 2D_2 + 2D_3 + D_4)$$
$$= 5.0 + \frac{1}{6}[3.1 + 2(2.2) + 2(2.46) + 1.67]$$
$$= 7.35 \text{ gmol/cm}^3$$

To calculate C_2:

$$D_1 = 1.0[6.0 - 0.58(7.35)] = 1.74$$
$$D_2 = 1.24$$
$$D_3 = 1.38$$
$$D_4 = 0.9$$

and

$$C_2 = C_1 + \frac{h}{6}(D_1 + 2D_2 + 2D_3 + D_4)$$
$$= 7.34 + \frac{1}{6}[1.74 + 2(1.24) + 2(1.38) + 0.94]$$
$$= 8.66 \text{ gmol/cm}^3$$

Also calculate C_3, C_4, and C_5:

$$C_3 = 9.40 \quad \text{at } t = 3 \text{ min}$$
$$C_4 = 9.81 \quad \text{at } t = 4 \text{ min}$$
$$C_5 = 10.04 \quad \text{at } t = 5 \text{ min}$$

The reader is left the exercise of comparing the numerical solution above (including that at $t = \infty$) with that provided by the analytical solution.

The R–K method can be used if the function also contains the independent variable. Consider the following equation:

$$\frac{dC}{dt} = f_1(C, t) \tag{21.34}$$

For this situation,

$$C_1 = C_0 + \frac{1}{6}(D_1 + 2D_2 + 2D_3 + D_4) \tag{21.35}$$

with

$$\begin{aligned} D_1 &= hf(C, t) \\ D_2 &= hf\left(C_0 + \frac{D_1}{2}, t_0 + \frac{h}{2}\right) \\ D_3 &= hf\left(C_0 + \frac{D_2}{2}, t_0 + \frac{h}{2}\right) \\ D_4 &= hf(C_0 + D_3, t_0 + h) \end{aligned} \tag{21.36}$$

If, for example,

$$\frac{dC}{dt} = 10C - e^{-Ct} \tag{21.37}$$

then,

$$D_2 = h\left[10\left(C_0 + \frac{D_1}{2}\right) - e^{-[C_0+(D_1/2)][t_0+(h/2)]}\right] \tag{21.38}$$

etc.

Situations may arise when there is a need to simultaneously solve more than one ordinary differential equation (ODE). In a more general case, one could have n dependent variables $y_1\ y_2, \ldots, y_n$ with each related to a single independent variable x by the following system of n simultaneous first-order ODEs:

$$\begin{aligned} \frac{dy_1}{dx} &= f_1(x, y_1, y_2, \ldots, y_n), \\ \frac{dy_2}{dx} &= f_2(x, y_1, y_2, \ldots, y_n), \\ &\vdots \\ \frac{dy_n}{dx} &= f_n(x, y_1, y_2, \ldots, y_n), \end{aligned} \tag{21.39}$$

Note that the equations in Equation (21.39) are interrelated, i.e., they are dependent on each other. This is illustrated in the following development.

Consider the following two equations:

$$\frac{dC}{dt} = -Ae^{-ET/R}C = f(C, t) \tag{21.40}$$

$$\frac{dT}{dt} = -kC\frac{\Delta H}{\rho C_P} = g(C, t) \tag{21.41}$$

or, in a more general sense,

$$\frac{dy}{dx} = f(x, y, z); \quad \text{(e.g., } xyz) \tag{21.42}$$

$$\frac{dz}{dt} = g(x, y, z); \quad \text{(e.g., } x^2y^2e^{-z}) \tag{21.43}$$

The R–K algorithm for Equations (21.42) and (21.43) is

$$y_1 = y_0 + \frac{1}{6}(RY_1 + 2RY_2 + 2RY_3 + RY_4) \tag{21.44}$$

$$z_1 = z_0 + \frac{1}{6}(RZ_1 + 2RZ_2 + 2RZ_3 + RZ_4) \tag{21.45}$$

where $y_1 - y_0 = \Delta y$, $z_1 - z_0 = \Delta z$, $h = \Delta x$ and

$$\begin{aligned}
RY_1 &= h \times f(x_0, y_0, z_0) \\
RZ_1 &= h \times g(x_0, y_0, z_0) \\
RY_2 &= h \times f(x_0 + h/2, y_0 + RY_1/2, z_0 + RZ_1/2) \\
RZ_2 &= h \times g(x_0 + h/2, y_0 + RY_1/2, z_0 + RZ_1/2) \\
RY_3 &= h \times f(x_0 + h/2, y_0 + RY_2/2, z_0 + RZ_2/2) \\
RZ_3 &= h \times g(x_0 + h/2, y_0 + RY_2/2, z_0 + RZ_2/2) \\
RY_4 &= h \times f(x_0 + h, y_0 + RY_3, z_0 + RZ_3) \\
RZ_4 &= h \times g(x_0 + h, y_0 + RY_3, z_0 + RZ_3)
\end{aligned} \tag{21.46}$$

Although the R–K approach (and other companion methods) have traditionally been employed to solve first-order ODEs, it can also treat higher ODEs. The procedure requires reducing an nth order ODE to n first-order ODEs. For example, if the equation is of the form[17]

$$\frac{d^2y}{dx_2} = f(y, x) \tag{21.47}$$

set

$$z = \frac{dy}{dx} \tag{21.48}$$

so that

$$\frac{dz}{dx} = \frac{d^2y}{dx^2} \tag{21.49}$$

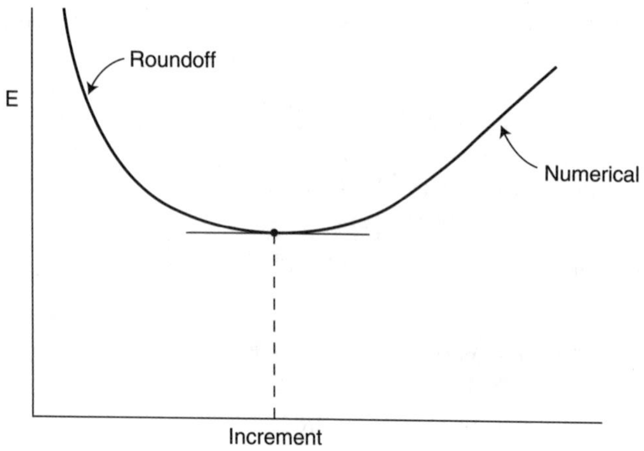

Figure 21.12 Error analysis.

The second-order equation in Equation (21.49) has now been reduced to the two first-order ODEs in Equation (21.50):

$$\frac{d^2y}{dx^2} = \frac{dz}{dx} = f(y, x)$$
$$\frac{dy}{dx} = z \qquad (21.50)$$

The procedure set forth in Equations (21.44) and (21.45) can be applied to generate a solution to Equation (21.47). Note, however, that the first derivative (i.e., dy/dx or its estimate) is required at the start of the integration. Extending the procedure to higher-order equations is left as an exercise for the reader.

The selection of increment size remains a variable to the practicing engineer. Most numerical analysis methods provided in the literature are not concerned with error analysis. In general, roundoff errors appear as demonstrated in Figure 21.12. In the limit, when the increment $\rightarrow$ 0, one obtains a solution approaching that analytically. Roundoff error then increases exponentially as the increment $\rightarrow$ 0 and the number of computations correspondingly increases. However, selecing the increment size that will minimize the error is rarely a problem; in addition, computing it is also rarely a problem. ■

ECONOMICS AND FINANCE

This last section is concerned with economics and finance. Economics and finance ultimately dictate many of the decisions made by practicing engineers and their companies. For example, a company may decide that because of the rising price of the feedstock to a chemical reactor, they will explore the possibility of producing the feed material from a cheaper raw material. A decision will then be based on whether it makes sense economically in the short- and long-term. Furthermore, economic evaluations are a major part of process and plant design.

This section provides introductory material to this vast field within engineering; it is devoted to a discussion on the need for economic analyses and definitions. This is followed with Illustrative Examples. Both the qualitative and quantitative viewpoints are emphasized although it is realized that the broad subject of engineering economics cannot be fit into any rigid set of formulas.

A company or individual hoping to increase profitability must carefully assess a range of investment opportunities and select the most profitable options from those available. Increasing competitiveness also requires that efforts need to be made to reduce the costs of existing processes. In order to accomplish this, engineers should be fully aware of not only technical factors but also economic factors, particularly those that have the largest effect on profitability.

In earlier years, engineers concentrated on the technical side of projects and left the financial studies to the economist. In effect, engineers involved in making estimates of the capital and operating costs had often left the overall economic analysis and investment decision-making to others. This approach is no longer acceptable.

Some engineers are not equipped to perform a financial and/or economic analysis. Furthermore, many engineers already working for industry have never taken courses in this area. This short-sighted attitude is surprising in a group of people who normally go to great lengths to obtain all the available technical data and information before making an assessment of a project or study. The attitude is even more surprising when one notes that data are readily available to enable an engineer to assess the prospects on both his/her own company and on those of his/her particular industry.[18]

As noted above, the purpose of this chapter is to provide a working tool to assist the student or practicing engineer in not only understanding economics and finance but also in applying technical information to the economic design and operation of not only reactors but also processes and plants. The material to follow will often focus on industrial and/or plant applications. Hopefully, this approach will provide the reader with a better understanding of some of the fundamentals and principles.

Bridging the gap between theory and practice is often a matter of experience acquired over a number of years. Even then, methods developed from experience must all too often be re-evaluated in the light of changing economic conditions if optimum designs are to result. The approach presented here therefore represents an attempt to provide a consistent and reasonably concise method for the solution of these problems involving economic alternatives.[18]

The term *economic analysis* in engineering problems generally refers to calculations made to determine the conditions for realizing maximum financial return for a design or operation. The same general principles apply whether one is interested in the choice of alternatives for competing projects or in the design of a particular type of chemical reactor. General considerations that form the framework on which sound decisions must be made are often simple. Sometimes their application to the problems encountered in the development of a commercial reactor involves too many intangibles to allow exact analysis; in that case, judgment must be intuitive. Often, however, such calculations may be made with a considerable degree of exactness. This section will attempt to develop a relatively concise method for applying these principles.

Concern with maximum financial return implies that the criterion for judging projects involved is profit. While this is usually true, there are many important objectives which, though aimed at ultimate profit increase, cannot be immediately evaluated in quantitative terms. Perhaps the most significant of these is the recent increased concern with environmental degradation and sustainability. Thus, there has been some tendency in recent years to regard management of commercial organizations as a profession with social obligations and responsibilities; considerations other than the profit motive may govern business decisions. However, these additional social objectives are for the most part, often not inconsistent with the economic goal of satisfying human wants with the minimum effort. In fact, even in the operation of primarily non-profit organizations, it is still important to determine the effect of various policies on profit.[18]

Other indices for construction, labor, buildings, engineering, and so on are also available in the literature. Generally it is not wise to use past cost data older than 5 to 10 years, even with the use of the cost indices. Within that time span, the technologies used in the processes may have changed drastically. The use of the indices could cause the estimates to be much greater that the actual costs. Such an error might lead to the choice of an alternative reactor/equipment other than the least costly.

ILLUSTRATIVE EXAMPLE 21.21 A simple procedure is available to estimate equipment cost from past cost data. The method consists of adjusting the earlier cost data to present values using factors that correct for inflation. A number of such indices are available; one of the most commonly used is the *Chemical Engineering Fabricated Equipment Cost Index* (FECI), outdated past values of which are listed in Table 21.5.

Plans for the construction of a new process were initiated in 1975. The reactor element in the process was to be a CSTR. The cost for this unit at that time was determined to be $136,000. In 1993, the author was asked to estimate the cost of the CSTR in terms of 1984 dollars. Also estimate the cost in terms of 1991 dollars.

TABLE 21.5 Fabricated Equipment Cost Index

Year	Index
1991	393.0
1990	385.0
1989	355.4
1988	342.5
1987	323.8
1986	318.4
1985	325.3
1984	322.7
1980	289.3
1975	192.2
1970	122.7
1960	101.2

Solution. Express the 1984 cost in terms of the 1975 cost and the FECIs for 1975 and 1984. Refer to Table 21.5.

$$\text{Cost} = (\text{cost in 1975})\,(1984\ \text{FECI}/1975\ \text{FECI})$$

Calculate the cost in 1984 dollars.

$$\begin{aligned}\text{Cost (1984)} &= (136{,}000)\,(322.7/192.2)\\ &= \$228{,}300\end{aligned}$$

Estimate the cost of the CSTR in 1991 dollars.

$$\begin{aligned}\text{Cost (1991)} &= (136{,}000)\,(393.0/192.2)\\ &= \$278{,}100\end{aligned}$$

■

ILLUSTRATIVE EXAMPLE 21.22 A fluidized bed reactor discharge of 50,000 acfm of gas contains a catalyst at a loading of 2.0 gr/ft^3. A cyclone is employed for catalyst capture and the catalyst recovered from the unit is worth \$0.03/lb. Experimental data have shown that the collection efficiency, E, is related to the cyclone pressure drop, ΔP, by the formula:

$$E = \frac{\Delta P}{\Delta P + 15.0}$$

where E = fractional collection efficiency

ΔP = pressure drop, lb_f/ft^2

If the overall fan is 55% efficient (overall) and electric power costs \$0.18/kW · h, at what collection efficiency is the cost of power equal to the value of the recovered catalyst? What is the pressure drop in inches of water at this condition?

Solution. The value of the recovered catalyst (RC) may be expressed in terms of the collection efficiency E, the volumetric flowrate q, the inlet dust loading C, and the value of the dust (DV):

$$\text{RC} = (q)(C)(\text{DV})(E)$$

Substituting yields

$$\text{RC} = \left(\frac{50{,}000\ \text{ft}^3}{\text{min}}\right)\left(\frac{2.0\ \text{gr}}{\text{ft}^3}\right)\left(\frac{1\,\text{lb}}{7000\ \text{gr}}\right)\left(\frac{0.03\$}{\text{lb}}\right)(E) = 0.429E\ E/\text{min}$$

The recovered value can be expressed in terms of pressure drop, i.e., replace E by ΔP:

$$\text{RC} = \frac{(0.429)(\Delta P)}{\Delta P + 15.0};\ \$/\text{min}$$

The cost of power (CP) in terms of q, ΔP, the cost of electricity (CE), and the fan efficiency, (E_f), is

$$\text{CP} = (q)(\Delta P)(\text{CE})/(E_f)$$

Substitution yields

$$\text{CP} = \left(\frac{50{,}000\ \text{ft}^3}{\text{min}}\right)\left(\frac{\Delta P\ \text{lb}_f}{\text{ft}^2}\right)\left(\frac{0.18\$}{\text{kWh}}\right)\left(\frac{1\ \text{min}\cdot\text{kW}}{44{,}200\ \text{ft}\cdot\text{lb}_f}\right)\left(\frac{1}{0.55}\right)\left(\frac{1\text{h}}{60\ \text{min}}\right)$$
$$= 0.006\,\Delta P;\ \$/\text{min}$$

The pressure drop at which the cost of power is equal to the value of the recovered catalyst is obtained by equating RV with CP:

$$\text{RV} = \text{CP}$$

Solving for ΔP,

$$\Delta P = 66.5\ \text{lb}_f/\text{ft}^2$$
$$= 12.8\ \text{in}\ H_2O$$

The collection efficiency corresponding to the above calculated ΔP is

$$E = \frac{\Delta P}{\Delta P + 15.0}$$
$$= \frac{66.5}{66.5 + 15.0}$$
$$= 0.82$$
$$= 82.0\%$$

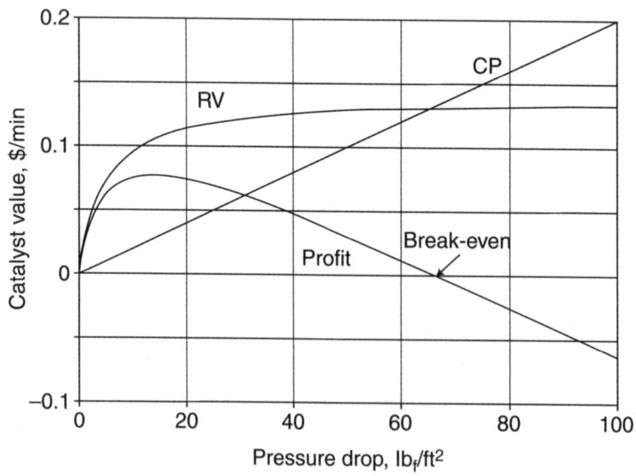

Figure 21.13 Profit as a function of pressure drop.

The reader should note that operating below this efficiency (or the corresponding pressure drop) will produce a profit; operating above this value leads to a loss.

Figure 21.13 shows the variation of RC, CP, and profit with pressure drop. The operating condition for maximum profit can be estimated from the graph. Calculating this value is left as an exercise for the reader. [*Hint*: Set the first derivative of the profit (i.e., RC – CP) with respect to ΔP equal to zero. The answer is 13.9 lb_f/ft^2.] ■

ILLUSTRATIVE EXAMPLE 21.24 Plans are underway to purchase and operate some type of chemical reactor. The company is still undecided as to whether to install a CSTR or a TF unit. The CSTR unit is less expensive to purchase and operate than a comparable tubular system, primarily because of energy costs. However, projected income from the TF unit is higher since it will handle a larger quantity of process liquid and provide a purer product. Based on the economic and financial data given in Table 21.6, select the reactor unit that will yield the higher annual profit. Calculations should be based on an interest rate of 12% and a process lifetime of 12 years for both units.(19)

Solution. Calculate the capital recovery factor, CRF:(18)

$$\begin{aligned}\text{CRF} &= (0.12)(1+0.12)^{12}/[(1+0.12)^{12}-1]\\ &= 0.1614\end{aligned}$$

Determine the annual capital and installation costs for the CSTR (CR) unit:

$$\begin{aligned}\text{COST(CR)} &= (2{,}625{,}000 + 1{,}575{,}000)(0.1614)\\ &= \$677{,}880/\text{yr}\end{aligned}$$

Determine the annual capital and installation costs for the TF unit:

$$\begin{aligned}\text{COST(TF)} &= (2{,}975{,}000 + 1{,}700{,}000)(0.1614)\\ &= \$754{,}545/\text{yr}\end{aligned}$$

TABLE 21.6 Economic Data for Illustrative Example 21.24

Costs/credits	CSTR	TF
Reactor unit	\$750,000	\$800,000
Peripherals	\$1,875,000	\$2,175,000
Total capital	\$2,625,000	\$2,975,000
Installation	\$1,575,000	\$1,700,000
Operation	\$400,000/yr	\$550,000/yr
Maintenance	\$650,000/yr	\$775,000/yr
Income	\$2,000,000/yr	\$2,500,000/yr

TABLE 21.7 Cost Analysis

	CSTR	TF
Total installed ($/yr)	678,000	755,000
Operation ($/yr)	400,000	550,000
Maintenance ($/yr)	650,000	775,000
Total annual cost ($/yr)	1,728,000	2,080,000
Income credit ($/yr)	2,000,000	2,500,000

See Table 21.7 for a comparison of costs and credits for both reactors.

Finally, calculate the profit for each reactor on an annualized basis:

$$\text{PROFIT(CR)} = 2{,}000{,}000 - 1{,}728{,}000 = +272{,}000\,\$/\text{yr}$$
$$\text{PROFIT(TF)} = 2{,}500{,}000 - 2{,}080{,}000 = +420{,}000\,\$/\text{yr}$$

A TF unit should be selected based on the above economic analysis.

Detailed cost estimates are beyond the scope of this text. Such procedures are capable of producing accuracies in the neighborhood of $\pm 5\%$. However, such estimates generally require significant time of engineering work. This type of analysis is designed to give the reader a basis for a preliminary cost analysis only.

Although this is a relatively simple comparative analysis of two reactor, it does provide some quantitative information. The reader should be aware that there are other factors that can, and often are, included in this type of calculation. Any standard economics text can provide additional details. ■

REFERENCES

1. J. Santoleri, J. Jeris, and L. Theodore, "*Introduction to Hazardous Waste Incineration*", 2nd edition, John Wiley & Sons, Hoboken, NJ, 2000.
2. L. Theodore, personal notes, East Williston, NY, 1972.
3. R. Thomann and R. Müeller, "*Principles of Surface Water Quality Modeling and Control,*" Harper Collins, New York City, NY, 1987.
4. H. de Heer, "*Calculating How Much Safety is Enough,*" Chemical Engineering, New York City, NY, 1973.
5. L. Theodore and R. Dupont, "*Environmental Health Risk and Hazard Risk Assessement Calculations,*" CRC Press/Taylor & Francis, Boca Raton, FL, 2012.
6. S. Shaefer and L. Theodore, "*Probability and Statistics Calculations for Engineering Science,*" CRC Press/Taylor & Francis, Boca Raton, FL, 2007.
7. H. Taback, "*Ethics Corner: Doing the Right Thing,*" Environmental Management (EM), Pittsburgh, PA, January, 2002.
8. H. Taback, "*Ethics Corner: When Ego Gets in the Way,*" Environmental Management, Pittsburgh, PA, January, 2002.
9. J. Wilcox and L. Theodore, "*Engineering and Environmental Ethics: A Case Study Approach,*" John Wiley & Sons, Hoboken, NJ, 1998.
10. M. Moyle, "*Introduction to Computers for Engineers,*" John Wiley & Sons, Hoboken, NJ, 1967.
11. B. Carnahan and J. Wilkes, "*Digital Computing and Numerical Methods,*" John Wiley & Sons, Hoboken, NJ, 1973.

12. R. Ketter and S. Prawler, "*Modern Methods of Engineering Computations*," McGraw-Hill, New York City, NY, 1969.
13. J. Reynolds, J. Jeris, and L. Theodore, "*Handbook of Chemical and Environmental Engineering Calculations*," John Wiley & Sons, Hoboken, NJ, 2004.
14. L. Perez, *Homework Assignment*, submitted to L. Theodore, Manhattan College, Bronx, NY, 2003.
15. F. Lavery, "*The Perils of Differentiating Engineering Data Numerically*," Chem. Eng., New York City, NY, Jan. 15, 1979.
16. L. Theodore, class notes, Manhattan College, Bronx, NY, 1991.
17. L. Theodore, class notes, Manhattan College, Bronx, NY, 1971.
18. L. Theodore and K. Neuser, "*Engineering Economics and Finance*," A Theodore Tutorial, Theodore Tutorials, East Williston, NY, 1996.
19. L. Theodore, F. Ricci, and T. Van Vliet, (adapted from) "*Thermodynamics for the Practicing Engineer*," John Wiley & Sons, Hoboken, NJ, 2009.

APPENDIX

SI UNITS

1. The Metric System
2. The SI System
3. Seven Basic Units
4. Two Supplementary Units
5. SI Multiples and Prefixes
6. Conversion Constants
7. Selected Common Abbreviations

THE METRIC SYSTEM

The need for a single worldwide coordinated measurement system was recognized over 300 years ago. Gabriel Mouton, Vicar of St. Paul in Lyons, proposed in 1670 a comprehensive decimal measurement system based on the length of one minute of arc of a great circle of the earth. In 1671 Jean Picard, a French astronomer, proposed the length of a pendulum beating seconds as the unit of length. (Such a pendulum would have been fairly easily reproducible, thus facilitating the wide-spread distribution of uniform standards.) Other proposals were made, but over a century elapsed before any action was taken.

In 1790, in the midst of the French Revolution, the National Assembly of France requested the French Academy of Sciences to "deduce an invariable standard for all the measures and weights." The Commission appointed by the Academy created a system that was, at once, simple and scientific. The unit of length was to be a portion of the earth's circumference. Measures for capacity (volume) and mass (weight) were to be derived from the unit of length, thus relating the basic units of the system to each other and to nature. Furthermore, the larger and smaller versions of each unit were to be created by multiplying or dividing the basic units by 10 and its multiples. This feature provided a great convenience to users of the system by eliminating the need for multiplying and dividing by 16 (to convert ounces to pounds) or by 12 (to convert inches to feet), respectively. Similar calculations in the metric system could be performed simply by shifting the decimal point. Thus, the metric system is a *base-10* or *decimal* system.

Chemical Reactor Analysis and Applications for the Practicing Engineer. By Louis Theodore

The Commission assigned the name *metre* (which is now spelled *meter*) to the unit of length. This name was derived from the Greek word *metron* meaning "a measure." The physical standard representing the meter was to be constructed so that it would equal one ten-millionth of the distance from the north pole to the equator along the meridian of the earth running near Dunkirk in France and Barcelona in Spain.

The metric unit of mass, called the *gram*, was defined as the mass of one cubic centimeter (a cube that is 1/100 of a meter on each side) of water at its temperature of maximum density. The cubic decimeter (a cube 1/10 of a meter on each side) was chosen as the unit of fluid capacity. This measure was given the name *liter*.

Although the metric system was not accepted with enthusiasm at first, adoption by other nations occurred steadily after France made its use compulsory in 1840. The standardized character and decimal features of the metric system made it well suited to scientific and engineering work. Consequently, it is not surprising that the rapid spread of the system coincided with an age of rapid technological development. In the United States, by Act of Congress in 1866, it was made "lawful throughout the United States of America to employ the weights and measures of the metric system in all contracts, dealings, or court proceedings."

By the late 1860s, even better metric standards were needed to keep pace with scientific advances. In 1875, the "Treaty of the Meter" international treaty set up well-defined metric standards for length and mass, and established permanent machinery to recommend and adopt further refinements in the metric system. This treaty, known as the *Metric Convention*, was signed by 17 countries, including the United States.

As a result of the Treaty, metric standards were constructed and distributed to each nation that ratified the Convention. Since 1893, the internationally agreed to metric standards have served as the fundamental weights and measures standards of the United States.

By 1900 a total of 35 nations—including the major nations of continental Europe and most of South America—had officially accepted the metric system. Today, with the exception of the United States and a few small countries, the entire world is using predominantly the metric system or is committed to such use. In 1971 the Secretary of Commerce, in transmitting to Congress the results of a 3-year study authorized by the Metric Study Act of 1968, recommended that the U.S. change to predominant use of the metric system through a coordinated national program.

The International Bureau of Weights and Measures located at Sevres, France, serves as a permanent secretariat for the Metric Convention, coordinating the exchange of information about the use and refinement of the metric system. As measurement science develops more precise and easily reproducible ways of defining the measurement units, the General Conference of Weights and Measures—the diplomatic organization made up of adherents to the Convention—meets periodically to ratify improvements in the system and the standards.

THE SI SYSTEM

In 1960, the General Conference adopted an extensive revision and simplification of the system. The name *Le Systeme International d'Unites* (International System of

Units), with the international abbreviation *SI*, was adopted for this modernized metric system. Further improvements in and additions to SI were made by the General Conference in 1964, 1968, and 1971.

The basic units in the SI system are the *kilogram* (mass), *meter* (length), *second* (time), *Kelvin* (temperature), *ampere* (electric current), *candela* (the unit of luminous intensity), and *radian* (angular measure). All are commonly used by the engineer. The Celsius scale of temperature (0°C−273.15 K) is commonly used with the absolute Kelvin scale. The important derived units are the *newton* (SI unit of force), the *joule* (SI unit of energy), the *watt* (SI unit of power), the *pascal* (SI unit of pressure), and the *hertz* (unit of frequency). There are a number of electrical units: *coulomb* (charge), *farad* (capacitance), *henry* (inductance), *volt* (potential), and *weber* (magnetic flux). One of the major advantages of the metric system is that larger and smaller units are given in powers of ten. In the SI system a further simplification is introduced by recommending only those units with multipliers of 10^3. Thus for lengths in engineering, the *micrometer* (previously *micron), millimeter*, and *kilometer* are recommended, and the *centimeter* is generally avoided. A further simplification is that the decimal point may be substituted by a comma (as in France, Germany, and South Africa), while the other number, before and after the comma, will be separated by spaces between groups of three, i.e., one million dollars will be $1 000 000.00. More details are provided below.

SEVEN BASIC UNITS

a. *Length—meter (m)* The meter (common international spelling, *metre*) is defined as 1 650 763.00 wavelengths in vacuum of the orange–red line of the spectrum of krypton-86. The SI unit of area is the *square meter* (m^2) . The SI unit of volume is the *cubic meter* (m^3) . The *liter* (0.001 cubic meter, L), although not an SI unit, is commonly used to measure fluid volume.

b. *Mass—kilogram (kg)* The standard for the unit of mass, the *kilogram*, is a cylinder of platinum–iridium alloy kept by the International Bureau of Weights and Measures at Paris. A duplicate in the custody of the National Bureau of Standards serves as the mass standard for the United States. This is the only base unit still defined by an artifact. The SI unit of force is the *Newton* (N). One Newton is the force which, when applied to a 1 kilogram mass, will give the kilogram mass an acceleration of 1 (meter per second per second). $1\ N = 1\ kg \cdot m/s^2$. The SI unit for pressure is the *Pascal* (Pa). $1\ Pa = 1\ N/m^2$. The SI unit for work and energy of any kind is the *joule* (J). $1\ J = 1\ N \cdot m$. The SI unit for power of any kind is the *watt* (W). $1\ W = 1\ J/s$.

c. *Time—second (s)* The *second* is defined as the duration of 9 192 632 770 cycles of the radiation associated with a specified transition of the cesium-133 atom. It is realized by tuning an oscillator to the resonance frequency of cesium-133 atoms as they pass through a system of magnets and a resonant cavity into a detector. The number of periods or cycles per second is call *frequency*. The SI unit for frequency is the *hertz* (Hz). One hertz equals one cycle per second. The SI unit for speed is the *meter per second* (m/s). The SI unit for acceleration is the *meter per second per second* (m/s^2).

d. *Electric current—ampere (A)* The *ampere* is defined as that current which, if maintained in each of two long parallel wires separated by one meter in free space, would produce a force between the two wires (due to their magnetic fields) of 2×10^{-7} Newton for each meter of length. The SI unit of voltage is the *volt* (V), 1 V = 1 W/A. The SI unit of electrical resistance is the *ohm* (Ω), $1\ \Omega = 1$ V/A.

e. *Temperature—Kelvin (K)* The *Kelvin* is defined as the fraction 1/273.16 of the thermodynamic temperature of the triple point of water. The temperature 0 K is called *absolute zero.* On the commonly used Celsius temperature scale, water at atmospheric pressure freezes at about 0°C and boils at about 100°C. The °C is defined as an interval of 1 K, and the Celsius temperature 0°C is defined as defined as 273.15 K. 1.8 Fahrenheit scale degrees are equal to 1.0°C or 1.0 K; the Fahrenheit scale uses 32°F as a temperature corresponding to 0°C.

f. *Amount of substance—mole (mol)* The *mole* is the amount of substance of a system that contains as many elementary entities as there are atoms is 0.012 kilogram of carbon-12. When the mole is used, the elementary entities must be specified and may be atoms, molecules, ions, electrons, other particles, or specified groups of such particles. The SI unit of concentration (of amount of substance) is the *mole per cubic meter* (mol/m^3).

g. *Luminous intensity—candela (cd)* The *candela* is defined as the luminous intensity of 1/600 000 of a square meter of a blackbody at the temperature of freezing platinum (2045 K). The SI unit of light flux is the *lumen* (lm). A source having an intensity of 1 candela in all directions radiates a light flux of 4π lumens.

TWO SUPPLEMENTARY UNITS

a. *Phase angle—radian (rad)* The *radian* is the plane angle with its vertex at the center of a circle that is subtended by an arc equal in length to the radius.

b. *Solid angle—steradian (sr)* The *steradian* is the solid angle with its vertex at the center of a sphere that is subtended by an area of the spherical surface equal to that of a square with sides equal in length to the radius.

SI MULTIPLES AND PREFIXES

Multiples and submultiples		Prefixes	Symbols
100 000 000 000	10^{12}	tera (ter'a)	T
100 000 000	10^{9}	giga (ji'ga)	G
100 000	10^{6}	mega (meg'a)	M
1 000	10^{3}	kilo (kil'o)	k
100	10^{2}	hecto (hek'to)	h
10	10^{1}	deka (dek'a)	da
Base unit 1	10^{0}		
0.1	10^{-1}	deci (des'i)	d
0.01	10^{-2}	centi (sen'ti)	c
0.001	10^{-3}	milli (mil'i)	m
0.000 001	10^{-6}	micro (mi'kro)	μ
0. 000 000 001	10^{-9}	nano (nan'o)	n
0.000 000 000 001	10^{-12}	pico (pe'ko)	p
0.000 000 000 000 001	10^{-15}	femto (fem'to)	f
0.000 000 000 000 000 001	10^{-18}	atto (at'to)	a

CONVERSION CONSTANTS (SI)

To convert from	To	Multiply by
Length		
m	cm	100
m	mm	1000
m	microns (μm)	10^{6}
m	angstroms (Å)	10^{10}
m	in.	39.37
m	ft	3.281
m	mi	6.214×10^{-4}
ft	in	12
ft	m	0.3048
ft	cm	30.48
ft	mi	1.894×10^{-4}
Mass		
kg	g	1000
kg	lb	2.205
kg	oz	35.24
kg	ton	2.268×10^{-4}

(Continued)

Conversion Constants Table (*Continued*)

To convert from	To	Multiply by
kg	grains	1.543×10^4
lb	oz	16
lb	ton	5×10^{-4}
lb	g	453.6
lb	kg	0.4536
lb	grains	7000
Time		
s	min	0.01667
s	h	2.78×10^{-4}
s	day	1.157×10^{-7}
s	week	1.653×10^{-6}
s	yr	3.171×10^{-8}
Force		
N	$kg \cdot m/s^2$	1
N	dynes	10^5
N	$g \cdot cm/s^2$	10^5
N	lb_f	0.2248
N	$lb \cdot ft/s^2$	7.233
lb_f	N	4.448
lb_f	dynes	4.448×10^5
lb_f	$g \cdot cm/s^2$	4.448×10^5
lb_f	$lb \cdot ft/s^2$	32.17
Pressure		
atm	N/m^2 (Pa)	1.013×10^5
atm	kPa	101.3
atm	bars	1.013
atm	$dynes/cm^2$	1.013×10^6
atm	lb_f/in^2 (psi)	14.696
atm	mm Hg @ 0°C (torr)	760
atm	in Hg @ 0°C	29.92
atm	ft H_2O @ 4°C	33.9
atm	in H_2O @ 4°C	406.8
psi	atm	6.80×10^{-2}
psi	mm Hg @ 0°C (torr)	51.71
psi	in H_2O @ 4°C	27.70
in H_2O @ 4°C	atm	2.458×10^{-3}
in H_2O @ 4°C	psi	0.0361
in H_2O @ 4°C	mm Hg @ 0°C (torr)	1.868
Volume		
m^3	L	1000

(*Continued*)

Conversion Constants Table (*Continued*)

To convert from	To	Multiply by
m^3	cm^3 (cc, mL)	10^6
m^3	ft^3	35.31
m^3	gal (U.S.)	264.2
m^3	qt	1057
ft^3	in^3	1728
ft^3	gal (U.S.)	7.48
ft^3	m^3	0.02832
ft^3	L	28.32
Energy		
J	N-m	1
J	erg	10^7
J	dyne · cm	10^7
J	kW · h	2.778×10^{-7}
J	cal	0.2390
J	$ft \cdot lb_f$	0.7376
J	Btu	9.486×10^{-4}
cal	J	4.186
cal	Btu	3.974×10^{-3}
cal	$ft \cdot lb_f$	3.088
Btu	$ft \cdot lb_f$	778
Btu	hp · h	3.929×10^{-4}
Btu	cal	252
Btu	kW · h	2.93×10^{-4}
$ft \cdot lb_f$	cal	0.3239
$ft \cdot lb_f$	J	1.356
$ft \cdot lb_f$	Btu	1.285×10^{-3}
Power		
W	J/s	1
W	cal/s	0.2390
W	$ft \cdot lb_f/s$	0.7376
W	kW	10^{-3}
kW	Btu/s	0.949
kW	hp	1.341
hp	$ft \cdot lb_f/s$	550
hp	kW	0.7457
hp	cal/s	178.2
hp	Btu/s	0.707
Concentration		
$\mu g/m^3$	lb/ft^3	6.243×10^{-11}
$\mu g/m^3$	lb/gal	8.346×10^{-12}
$\mu g/m^3$	gr/ft^3	4.370×10^{-7}

(*Continued*)

Conversion Constants Table (*Continued*)

To convert from	To	Multiply by
gr/ft^3	$\mu g/m^3$	2.288×10^6
gr/ft^3	g/m^3	2.288
lb/ft^3	$\mu g/m^3$	1.602×10^{10}
lb/ft^3	$\mu g/L$	1.602×10^7
lb/ft^3	lb/gal	7.48
Viscosity		
P (poise)	g/cm-s	1
P	cP (centipoise)	100
P	kg/m · h	360
P	lb/ft · s	6.72×10^{-2}
P	lb/ft · h	241.9
P	lb/m · s	5.6×10^{-3}
lb/ft · s	P	14.88
lb/ft · s	g/cm · s	14.88
lb/ft · s	kg/m · h	5.357×10^3
lb/ft · s	lb/ft · h	3600
Heat Capacity		
cal/g · °C	Btu/lb · °F	1
cal/g · °C	kcal/kg · °C	1
cal/g · °C	cal/gmol · °C	Molecular weight
cal/gmol · °C	Btu/lbmol · °F	1
J/g · °C	Btu/lb · °F	0.2389
Btu/lb · °F	cal/g · °C	1
Btu/lb · °F	J/g · °C	4.186
Btu/lb · °F	Btu/lbmol · °F	Molecular weight

SELECTED COMMON ABBREVIATIONS

Å, A	angstrom unit of length
abs	absolute
amb	ambient
app. MW, M	apparent molecular weight
atm	atmospheric
at. wt.	atomic weight
b.p.	boiling point
bbl	barrel
Btu	British thermal unit
cal	calorie
cg	centigram

(*Continued*)

Selected Common Abbreviations Table (*Continued*)

cm	centimeter
cgs system	centimeter-gram-second system
conc.	concentrated, concentration
cc, cm^3	cubic centimeter
cu ft, ft^3	cubic feet
cfh	cubic feet per hour
cfm	cubic feet per minute
cfs	cubic feet per second
m^3, M^3	cubic meter
°	degree
°C	degree Celsius, degree Centigrade
°F	degree Fahrenheit
K	degree Kelvin
°R	degree Reamur, degree Rankine
ft	foot
ft-lb	foot pound
fpm	feet per minute
fps	feet per second
fps system	foot · pound · second system
f.p.	freezing point
gr	grain
g, gm	gram
h	hour
in	inch
kcal	kilocalorie
kg	kilogram
km	kilometer
liq	liquid
L	liter
log	logarithm (common)
ln	logarithm (natural)
m.p.	melting point
m, M	meter
m	pound
μm	micrometer (micron)
mks system	meter · kilogram · second system
mph	miles per hour
mg	milligram
ml	milliliter
mm	millimeter
mμ	millimicron
min	minute
mol wt, MW, M	molecular

(*Continued*)

Selected Common Abbreviations Table (*Continued*)

oz	ounce
ppb	parts per billion
pphm	parts per hundred million
ppm	parts per million
psi	pounds per square inch
psia	pounds per square inch absolute
psig	pounds per square inch gage
q, Q	cubic feet per second
Q	Btu
rpm	revolutions per minute
s	second
sp gr	specific gravity
sp ht	specific heat
sp wt	specific weight
sq	square
scf	standard cubic foot
STP	standard temperature and pressure
t	time, temperature
T, temp.	temperature
wt	weight

INDEX